QUANTUM MECHANICS, STATISTICAL MECHANICS

SOLID STATE PHYSICS

(AN INTRODUCTION)

For B.Sc. Physics (Honours)
Also for B.E./B.Tech. Engineering Students

Dr. D. CHATTOPADHYAY
Ph.D., D.Sc., P.R.S.
Professor

Dr. P.C. RAKSHIT
Ph.D.
Professor

Institute of Radio Physics and Electronics
Calcutta University
Kolkata

NINTH REVISED EDITION

S. CHAND
PUBLISHING

S Chand And Company Limited
(ISO 9001 Certified Company)

S Chand And Company Limited
(ISO 9001 Certified Company)

Head Office: Block B-1, House No. D-1, Ground Floor, Mohan Co-operative Industrial Estate, New Delhi – 110 044 | Phone: 011-66672000

Registered Office: A-27, 2nd Floor, Mohan Co-operative Industrial Estate, New Delhi – 110 044
Phone: 011-49731800

www.schandpublishing.com; e-mail: info@schandpublishing.com

Branches

Chennai	:	Ph: 23632120; chennai@schandpublishing.com
Guwahati	:	Ph: 2738811, 2735640; guwahati@schandpublishing.com
Hyderabad	:	Ph: 40186018; hyderabad@schandpublishing.com
Jalandhar	:	Ph: 4645630; jalandhar@schandpublishing.com
Kolkata	:	Ph: 23357458, 23353914; kolkata@schandpublishing.com
Lucknow	:	Ph: 4003633; lucknow@schandpublishing.com
Mumbai	:	Ph: 25000297; mumbai@schandpublishing.com
Patna	:	Ph: 2260011; patna@schandpublishing.com

© S Chand And Company Limited, 1989

All rights reserved. No part of this publication may be reproduced or copied in any material form (including photocopying or storing it in any medium in form of graphics, electronic or mechanical means and whether or not transient or incidental to some other use of this publication) without written permission of the copyright owner. Any breach of this will entail legal action and prosecution without further notice.

Jurisdiction: All disputes with respect to this publication shall be subject to the jurisdiction of the Courts, Tribunals and Forums of New Delhi, India only.

First Edition 1989
Subsequent Editions and Reprints 1993, 96, 98, 2001, 2003, 2006, 2008
Ninth Revised Edition 2009
Reprints 2010, 2013 (Twice), 2016, 2017, 2018, 2020

Reprint 2021

ISBN : 978-81-219-0931-0   **Product Code:** H6QMS68PHYS10ENAI09O

PRINTED IN INDIA

By Vikas Publishing House Private Limited, Plot 20/4, Site-IV, Industrial Area Sahibabad, Ghaziabad – 201 010 and Published by S Chand And Company Limited, A-27, 2nd Floor, Mohan Co-operative Industrial Estate, New Delhi – 110 044.

PREFACE TO THE NINTH EDITION

In this edition a few changes have been made at places to improve the presentation. Some more questions and problems have also been included in the light of recent examination papers.

The users of the book are sincerely thanked for their unstinting support.

AUTHORS

PREFACE TO THE FIRST EDITION

Courses on the related fields of Quantum Mechanics, Statistical Mechanics, and Solid State Physics have been recently introduced in the B.Sc. Physics (Honours) syllabus of many universities. This treatise is intended to be a textbook covering these courses in a single volume. It is written in such a way as to be useful also to a wider readership.

We have attempted to deal with the subject-matter in a clear, concise, and easy-to-read manner. Emphasis has been given to the clarification of the physical principles. Mathematical details have also been worked out to the extent suitable for the undergraduate students. Many solved problems have been incorporated to make the students familiar with the applications of the principles and formulae they have come across in the text. A variety of questions and problems have also been given for the practice of the students. Many of these have been taken from recent examination papers.

We shall deem our labour amply rewarded if the book proves useful to those for whom it is intended. Suggestions for the improvement of the book are welcome.

AUTHORS

CONTENTS

QUANTUM MECHANICS

Chapter 1 : Introductory Ideas ... 3–26
1.1. The Origin of Quantum Mechanics ... 3
1.2. The Hypothesis of de Broglie ... 4
1.3. The Compton Effect ... 5
1.4. Electron Diffraction Experiments ... 8
1.5. Wave Packets : Phase Velocity and Group Velocity ... 12
1.6. Heisenberg's Uncertainty Principle ... 13
1.7. Gamma-Ray Microscope ... 14
1.8. Some Applications of the Uncertainty Principle ... 15
1.9. Worked-out Problems ... 16
Questions ... 24

Chapter 2 : The Schrödinger Wave Equation ... 27–51
2.1. Development of Schrödinger's Wave Equation ... 27
2.2. Physical Meaning of the Wave Function ... 29
2.3. Normalization of the Wave Function ... 29
2.4. Expectation Values ... 30
2.5. The Ehrenfest Theorem ... 31
2.6. Time-independent and Time-dependent Schrödinger's Equation ... 33
2.7. Quantum Mechanical Operators ... 34
2.8. Orthogonality of Eigenfunctions ... 39
2.9. Properties of Eigenfunctions and Boundary Conditions ... 40
2.10. Probability Current Density ... 41
2.11. Postulates of Quantum Mechanics ... 42
2.12. Worked-out Problems ... 43
Questions ... 49

Chapter 3 : Free Particles and Wave Packets ... 52 - 64
3.1. The Free Particle ... 52
3.2. Particle Confined in an Enclosure ... 53
3.3. Wave Packets ... 56
3.4. Worked-out Problems ... 58
Questions ... 62

Chapter 4 : Application of Quantum Mechanics to Some Potential Problems ... 65 - 78
4.1. Particle at a Rectangular Potential Barrier ... 65
4.2. Particle in a One-dimensional Square Well ... 70
4.3. Particle in an Infinitely Deep Potential Well ... 73
4.4. Emission of α-Particles from a Radioactive Element ... 74
4.5. Worked-out Problems ... 75
Questions ... 77

Chapter 5 : The Harmonic Oscillator ... 79 - 97
5.1. Harmonic Oscillator and Energy Eigenfunctions ... 79
5.2. Hermite Polynomials and the Eigenfunctions ... 81
5.3. Correspondence with Classical Oscillator ... 82
5.4. Uncertainty Principle and the Zero-point Energy ... 84

(v)

5.5	Raising and Lowering Operators	...	86
5.6.	Worked-out Problems	...	88
	Questions	...	95

Chapter 6 : Spherically Symmetric Potential and the Hydrogen Atom ... **98 - 113**

6.1.	Schrödinger's Equation and its Separation	...	98
6.2.	Solutions of θ and φ Equations	...	99
6.3.	Radial Equation and the Angular Momentum	...	101
6.4.	The Hydrogen Atom	...	103
6.5.	Worked-out Problems	...	107
	Questions	...	112

STATISTICAL MECHANICS

Chapter 1 : Basic Concepts ... **117 - 143**

1.1.	Microscopic and Macroscopic Systems	...	117
1.2.	Calculation of Probabilities	...	117
1.3.	Phase Trajectory	...	119
1.4.	Statistics of an Assembly of Particles	...	119
1.5.	Entropy	...	123
1.6.	Perfect Gas Law	...	124
1.7.	Maxwell-Boltzmann Statistics of a System of Particles	...	124
1.8.	Thermodynamic Quantities from Partition Function	...	128
1.9.	Approach to Equilibrium	...	129
1.10.	Equipartition Theorem and its Applications	...	130
1.11.	Gibbs Paradox	...	133
1.12.	Limitations of MB Statistics	...	134
1.13.	Worked-out Problems	...	135
	Questions	...	141

Chapter 2 : Fermi-Dirac and Bose-Einstein Statistics ... **144 – 164**

2.1.	Fermi-Dirac Statistics	...	144
2.2.	Specific Heat of Conduction Electrons in Metals	...	148
2.3.	Bose-Einstein Statistics	...	151
2.4.	Symmetry Properties	...	154
2.5.	Planck's Law of Radiation	...	155
2.6.	Deductions from Planck's Law of Radiation	...	157
2.7.	Worked-out Problems	...	158
	Questions	...	162

Chapter 3 : Third Law of Thermodynamics ... **165-167**

3.1.	Introduction	...	165
3.2.	Consequences of the third law	...	165
	Questions	...	167

SOLID STATE PHYSICS

Chapter 1 : Crystals and their Properties ... **171 - 193**

1.1.	Structure of Solids	...	171
1.2.	Classification of Crystals	...	171
1.3.	Some Crystallographic Terms	...	175

1.4.	Bravais Lattices	...	176
1.5.	Some Crystal Structures	...	176
1.6.	Crystal Planes and Miller Indices	...	181
1.7.	Spacing Between Adjacent Planes in the Lattice	...	183
1.8	Anisotropy of the Physical Properties of Single Crystals	...	185
1.9.	Worked-out Problems	...	185
	Questions	...	191

Chapter 2 : X-Ray Crystal Analysis ... 194 - 213

2.1.	Introduction	...	194
2.2.	Reasons for using X-Ray	...	194
2.3.	The Bragg Diffraction Law	...	195
2.4.	Reciprocal Lattice	...	196
2.5.	Laue Condition of X-Ray Diffraction	...	198
2.6.	Experimental Methods of X-Ray Diffraction	...	201
2.7.	Amplitude of the Scattered Wave	...	203
2.8.	Steps for Analysing a Crystal Structure	...	206
2.9.	Determination of Crystal Structures of KCl and NaCl from Bragg's Law	...	207
2.10.	Worked-out Problems	...	209
	Questions	...	212

Chapter 3 : Band Theory of Solids ... 214 - 236

3.1.	Introduction	...	214
3.2.	A Simple Discussion on the Formation of Energy Bands	...	214
3.3.	Periodic Potential in a Crystalline Solid	...	215
3.4.	The Kronig Penney Model	...	215
3.5.	Effective Mass	...	219
3.6.	Number of Electrons in a Band	...	221
3.7.	Electrons and Holes	...	222
3.8.	Metals, Insulators and Semiconductors	...	223
3.9.	Intrinsic Semiconductors	...	224
3.10.	Extrinsic Semiconductors	...	225
3.11.	Carrier Concentrations and Fermi Levels in Semiconductors	...	228
3.12.	Worked-out Problems	...	232
	Questions	...	235

Chapter 4 : Transport Phenomena in Metals and Semiconductors ... 237 - 255

4.1.	The Drude Theory	...	237
4.2.	Electrical Conductivity	...	237
4.3.	Thermal Conductivity of a Metal	...	240
4.4.	Quantum Mechanical Free Electron Theory	...	241
4.5.	Electrical Conductivity of Semiconductors	...	243
4.6.	Hall Effect	...	244
4.7.	The Boltzmann Transport Equation	...	248
4.8.	Worked-out Problems	...	250
	Questions	...	253

Chapter 5 : Specific Heat of Solids and Lattice Vibrations ... 256 - 271

5.1.	Classical Calculation of Lattice Specific Heat	...	256
5.2.	Einstein's Theory of Specific Heat	...	257
5.3.	Debye's Theory of Specific Heat	...	259
5.4.	Vibrations of a One-dimensional Lattice	...	264
5.5.	Vibrational Mode as a Linear Harmonic Oscillator	...	266
5.6.	Vibrational Modes of a Crystal	...	267

5.7.	Worked-out Problems	...	268
	Questions	...	269

Chapter 6 : Dielectric Properties of Solids — 272 - 283

6.1.	The Static Dielectric Constant	...	272
6.2.	Dipole Moment and Polarisation	...	273
6.3.	Types of Polarisation	...	274
6.4.	Electronic Polarisation	...	274
6.5.	Ionic Polarisation	...	276
6.6.	Orientational Polarisation	...	277
6.7.	Static Dielectric Constant of Gases	...	279
6.8.	Internal Field in Solids	...	279
6.9.	Static Dielectric Constant of Solids	...	280
6.10.	Ferroelectricity	...	281
6.11.	Worked-out Problems	...	282
	Questions	...	283

Chapter 7 : Magnetic Properties of Solids — 284 - 316

7.1.	Force on a Current-Carrying Conductor in a Magnetic Field	...	284
7.2.	Magnetic Dipole Moment of a Current Loop and Magnetisation	...	285
7.3.	Potential Energy of a Dipole in a Magnetic Field	...	286
7.4.	Magnetic Field Intensity due to a Current-carrying Conductor	...	287
7.5.	Magnetic Susceptibility and the Relation between $\vec{B}$, $\vec{H}$ and $\vec{M}$	...	287
7.6.	Elementary Magnet	...	288
7.7.	Origin of Permanent Magnetic Dipoles in Materials	...	290
7.8.	Diamagnetism	...	294
7.9.	Paramagnetism	...	296
7.10.	Ferromagnetism	...	301
7.11.	Antiferromagnetism and Ferrimagnetism	...	308
7.12.	Brillouin Theory (or the Quantum Theory) of Paramagnetism	...	309
7.13.	The Quantum Theory of Ferromagnetism	...	312
7.14.	Comparison of Diamagnetic, Paramagnetic, and Ferromagnetic Materials	...	313
7.15.	Worked-out Problems	...	314
	Questions	...	315

Chapter 8 : Superconductivity — 317 - 330

8.1.	Superconducting State	...	317
8.2.	Destruction of Superconductivity by Magnetic Field	...	317
8.3.	Meisner Effect	...	318
8.4.	Specific Heat	...	319
8.5.	London Equations	...	320
8.6.	Penetration Depth	...	321
8.7.	Fluxoid	...	322
8.8.	Type I and Type II Superconductors	...	323
8.9.	Microscopic Theory	...	323
8.10.	Josephson Junction	...	325
8.11.	Applications	...	327
8.12.	Worked-out Problems	...	329
	Questions	...	330
	Miscellaneous Problems	...	**331**
	Objective-type Questions	...	**335**
	Index	...	**339 - 344**

QUANTUM MECHANICS

Chapter 1

Introductory Ideas

1.1. THE ORIGIN OF QUANTUM MECHANICS

Quantum mechanics is a systematic theory of the behaviour of matter and light and, in particular, of atomic and subatomic phenomena. It is founded on a set of self-consistent mathematical rules aided by suitable physical interpretation. It is different in many respects from Newtonian mechanics; however, in the limit when the masses and the energies of the particles in question are relatively large, the results of quantum mechanics reduce to those of Newtonian mechanics. This is known as the *correspondence principle*.

It was Newton who first thought that light was made up of particles. In the year 1803 Young discovered that light behaves like a wave. In the beginning of the 20th century experimental observations on the spectral distribution of thermal radiation from a black body, Compton effect and photoelectric effect established that electromagnetic radiation indeed sometimes behaves like a particle. Historically, the electron was considered to be a particle. Its wave-like character was later established through the diffraction experiments of Davisson and Germer (1927) and of G.P. Thomson (1928).

The particle nature of electromagnetic radiation was described by Planck in 1900 by assuming that emission and absorption of electromagnetic radiation take place in discrete quanta (singular: quantum), and that each quantum has an amount of energy E, where

$$E = h\nu \qquad \ldots(1.1)$$

Here h is a universal constant, called *Planck's constant*, and ν is the frequency of the electromagnetic radiation.

The wave-like character of particles of matter was considered by de Broglie (1924). He propounded that a material particle of momentum p has associated with it a wave of wavelength λ, where

$$\lambda = \frac{h}{p} \qquad \ldots(1.2)$$

The particle and the wave nature are *complementary*, but they are mutually exclusive. That is, the same experiment cannot reveal the particle and the wave characters simultaneously. This mutual exclusiveness is referred to as the *principle of complementarity*.

Besides the experimental evidence on the *dual character* of matter and light, it was demonstrated that the measurable parameters of atomic systems might possess discrete values. For example, we may cite the Einstein-Debye theory on the specific heat of solids, the Franck-and-Hertz experiment on the discrete energy losses of electrons during collision with atoms, and the Stern-and-Gerlach experiment showing the existence of discrete values for the component of the magnetic moment of an atom along an external magnetic field.

Classical theories failed to explain the above observations. The confusion was finally resolved by Schrödinger, Heisenberg, Dirac, and others during 1925-1930. They developed the subject of quantum mechanics as the fundamental theory of phenomena occurring in atomic and subatomic scales. Since human experience is confined to large objects, and atomic and subatomic phenomena fall beyond the range of direct human perception, the ideas of quantum mechanics appear strange and peculiar when we encounter them.

1.2. THE HYPOTHESIS OF DE BROGLIE

In 1924 de Broglie made a hypothesis that like an electromagnetic radiation, a material particle, such as, electron, proton, atom or molecule has a dual character, particle-like and wave-like. According to him, the momentum p of a particle and the wavelength λ of the associated wave, also called the *de Broglie wave*, are related through Eq. (1.2), which may be rewritten as

$$p = \frac{h}{\lambda} = \frac{h}{2\pi} \cdot \frac{2\pi}{\lambda} = \hbar k, \qquad \ldots(1.3)$$

where $\hbar = h/2\pi$ and $k = 2\pi/\lambda$. The quantity $\hbar$ is termed *reduced Planck's constant* or *Dirac's constant* and k, the wave vector. The wavelength λ given by Eq. (1.2), is called the de Broglie wavelength. If the particle has a mass m and a velocity v, then

$$p = mv = \frac{h}{\lambda},$$

or,

$$\lambda = \frac{h}{p} = \frac{h}{mv} \qquad \ldots(1.4)$$

Suppose that the particle is an electron which has acquired a velocity v on falling through a potential difference of V volt under nonrelativistic conditions.

Then

$$\frac{1}{2} m_0 v^2 = eV, \qquad \ldots(1.5)$$

where e is the charge and m_0 is the mass of the electron.

Therefore,

$$v = \sqrt{\frac{2eV}{m_0}},$$

and

$$\lambda = \frac{h}{m_0 \sqrt{2eV/m_0}} = \frac{h}{\sqrt{2eVm_0}} \qquad \ldots(1.6)$$

In SI units, $e = 1.6 \times 10^{-19}$ coulomb, $m_0 = 9.11 \times 10^{-31}$ kg and $h = 6.62 \times 10^{-34}$ J.s. Substituting these values in Eq. (1.6) we obtain

$$\lambda = 12.26/\sqrt{V} \text{ Å}, \qquad \ldots(1.7)$$

where V is expressed in volt.

The velocity v is one-tenth the velocity of light in free space when $V = 2.56$ kilo volt. Thus, unless V is some tens of kilo volt or larger, relativistic corrections are not important and Eq. (1.7) can be used. The de Broglie wavelengths under relativistic conditions will be considered later in this chapter.

We shall now see that Eq. (1.2) also holds for a light particle, called *photon*. The relation between the energy E of the photon and the frequency ν of the electromagnetic radiation is given by Eq. (1.1). If the mass of the associated particle is m, then according to the Special Theory of Relativity, we have

$$E = mc^2 \qquad \ldots(1.8)$$

where c is the velocity of the electromagnetic radiation in free space.

Therefore from Eqs. (1.1) and (1.8) we get

$$E = h\nu = mc^2 \qquad \ldots(1.9)$$

The momentum p of the photon is given by

$$p = mc \qquad \ldots(1.10)$$

Hence from Eqs. (1.9) and (1.10) we find

$$p = mc = \frac{h\nu}{c} = \frac{h}{\lambda}, \qquad \ldots(1.11)$$

Introductory Ideas

where $\lambda = c/\nu$ is the wavelength of the radiation. In general, for matter and also for radiation the total energy E of a particle is related to the frequency ν of the wave associated with its motion by Eq. (1.1) and the momentum p of the particle is related to the wavelength λ of the corresponding wave by Eq. (1.2).

1.3. The Compton Effect

In 1924, the discovery of the Compton effect gave an evidence in support of the photon character of electromagnetic radiation. Compton observed that monochromatic X-rays scattered by carbon in a certain direction contained a component with the frequency of the incident radiation, and a second component with a lower frequency or longer wavelength. The lower-frequency component originates from the inelastic scattering of the incident radiation. Such scattering is referred to as the *Compton scattering* and the effect as the *Compton effect*. Apart from X-rays, the Compton effect was also observed with γ-rays.

While the classical electromagnetic theory can explain the component with unchanged frequency in the scattered radiation, it fails to account for the lower-frequency Compton component. The particle or photon character of the radiation must be invoked to explain the Compton scattering.

To obtain a theoretical expression for the change in wavelength in Compton scattering, consider the collision of a photon with an electron of rest mass m_0, initially at rest (Fig. 1.1A). After the collision, the photon is scattered in a direction at an angle ϕ with the incident direction, and the electron moves with a velocity v in a direction θ with the direction of the incident photon.

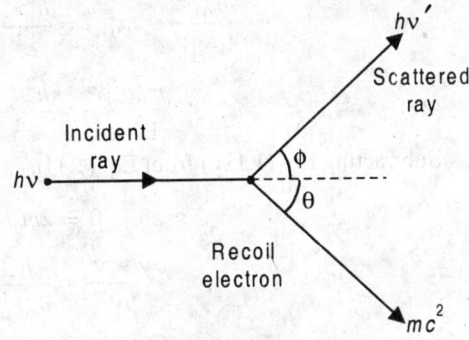

Fig. 1.1 A. Compton scattering

Let ν be the frequency of the incident photon and ν' be that of the scattered photon. Then the energy and the momentum of the incident photon are $h\nu$ and $h\nu/c$, respectively; the corresponding quantities for the scattered photon are $h\nu'$ and $h\nu'/c$, respectively. If m is the mass of the electron moving with the velcoity v, its kinetic energy is $E_k = mc^2 - m_0c^2$, and its momentum is $mv = m\beta c$, where $\beta = v/c$.

The *law of conservation of energy* gives

$$h\nu = h\nu' + E_k \qquad \ldots(1.11a)$$

which shows that the energy of the incident photon is the sum of the energy of the scattered photon and the kinetic energy gained by the electron. As the incident photon loses some energy to the electron, $\nu' < \nu$. Since $m = m_0/\sqrt{1-\beta^2}$, Eq. (1.11a) yields

$$h\nu = h\nu' + \frac{m_0c^2}{\sqrt{1-\beta^2}} - m_0c^2 \qquad \ldots(1.11b)$$

Applying the *law of conservation of momentum* along and perpendicular to the direction of the incident photon, we get

$$\frac{h\nu}{c} = \frac{h\nu'}{c}\cos\phi + \frac{m_0\beta c}{\sqrt{1-\beta^2}}\cos\theta \qquad \ldots(1.11c)$$

and

$$0 = \frac{h\nu'}{c}\sin\phi - \frac{m_0\beta c}{\sqrt{1-\beta^2}}\sin\theta \qquad \ldots(1.11d)$$

or,

$$\frac{m_0\beta c}{\sqrt{1-\beta^2}}\cos\theta = \frac{h\nu}{c} - \frac{h\nu'}{c}\cos\phi \qquad (1.11e)$$

and
$$\frac{m_0 \beta c}{\sqrt{1-\beta^2}} \sin\theta = \frac{h\nu'}{c} \sin\phi \qquad \ldots(1.11f)$$

Squaring and adding Eqs. (1.11e) and (1.11f), we find
$$\frac{m_0^2 \beta^2 c^2}{1-\beta^2} = \frac{h^2\nu^2}{c^2} + \frac{h^2\nu'^2}{c^2} - \frac{2h^2\nu\nu'}{c^2} \cos\phi \qquad \ldots(1.11g)$$

From Eq. (1.11b), we get after transposing and squaring
$$\left(\frac{m_0 c^2}{\sqrt{1-\beta^2}}\right)^2 = (h\nu - h\nu' + m_0 c^2)^2$$
$$= h^2\nu^2 + h^2\nu'^2 + m_0^2 c^4 - 2h^2\nu\nu' + 2m_0 c^2 h(\nu - \nu')$$

Dividing by c^2 and rearranging, we obtain
$$\frac{m_0^2 c^2}{1-\beta^2} - m_0^2 c^2 = \frac{h^2\nu^2}{c^2} + \frac{h^2\nu'^2}{c^2} - \frac{2h^2\nu\nu'}{c^2} + 2m_0 h(\nu - \nu')$$

or,
$$\frac{m_0^2 c^2 \beta^2}{1-\beta^2} = \frac{h^2\nu^2}{c^2} + \frac{h^2\nu'^2}{c^2} - \frac{2h^2\nu\nu'}{c^2} + 2m_0 h(\nu - \nu') \qquad \ldots(1.11h)$$

Subtracting Eq. (1.11g) from Eq. (1.11h) gives
$$0 = 2m_0 h(\nu - \nu') - \frac{2h^2\nu\nu'}{c^2}(1 - \cos\phi)$$

or,
$$\frac{\nu - \nu'}{\nu\nu'} = \frac{h}{m_0 c^2}(1 - \cos\phi)$$

or,
$$\frac{c}{\nu'} - \frac{c}{\nu} = \frac{h}{m_0 c}(1 - \cos\phi) \qquad \ldots(1.11i)$$

We have $c/\nu' = \lambda'$ and $c/\nu = \lambda$, where λ and λ' are the incident wavelength and the scattered wavelength, respectively. The *Compton wavelength* of the electron is defined as $\lambda_c = h/(m_0 c)$. Hence Eq. (1.11i) gives for the change (increase) in wavelength in Compton scattering at an angle ϕ

$$\delta\lambda = \lambda' - \lambda = \lambda_c(1 - \cos\phi) = 2\lambda_c \sin^2\frac{\phi}{2} \qquad \ldots(1.11j)$$

Obviously, $\delta\lambda$ depends on the scattering angle ϕ and is independent of the wavelength of the incident radiation and of the nature of the scatterer containing the free electrons. Furthermore, λ_c is a *universal constant*, being determined by the fundamental constants h, m_0, and c. Substituting the values $h = 6.62 \times 10^{-34}$ J.s, $m_0 = 9.11 \times 10^{-31}$ kg, and $c = 3 \times 10^8$ m/s, we have $\lambda_c = 0.024 \times 10^{-10}$ m = 0.024 Å.

If $\phi = 180°$, $\delta\lambda$ attains its maximum value of $2\lambda_c \simeq 0.05$ Å. For $\phi = 90°$, $\delta\lambda = \lambda_c$. So, for a detectable wavelength change, the incident wavelength λ must not exceed a few Å. If $\lambda = 1$ Å, $\delta\lambda/\lambda \simeq 0.05 = 5\%$. For light in the visible range, λ lies between 4000 to 8000 Å, and hence the wavelength shift is not perceptible. In fact, for light in the visible range, the photon energy is not larger than the binding energy of even the loosely bound electrons in the scatterer. Therefore, it is hard to detect a Compton shift.

From Eq. (1.11i) one can express the scattered photon energy in terms of the incident photon energy as follows :

$$h\nu' = \frac{h\nu}{1 + \alpha(1 - \cos\phi)} \qquad \ldots(1.11k)$$

where
$$\alpha = h\nu/(m_0 c^2).$$

If the free electron in the scatterer is not at rest, but in motion, the Compton radiation is *broadened*.

Introductory Ideas

The unmodified line, called the *Rayleigh line*, in the scattered radiation appears due to scattering from a bound electron of the scatterer, whereas the Compton line occurs due to a free electron. For scattering from a bound electron, the recoil momentum is taken up by the whole atom which is much heavier than the electron. Hence the wavelength shift is negligible, accounting for the presence of the unmodified line.

A relationship between θ and ϕ, and an expression for the kinetic energy of the recoil electron can be obtained as follows.

Dividing Eq. (1.11 *f*) by Eq. (1.11*e*) gives

$$\tan \theta = \frac{v' \sin \phi}{v - v' \cos \phi} = \frac{\sin \phi}{\dfrac{v}{v'} - \cos \phi} \qquad \ldots(1.11l)$$

Also Eq. (1.11*i*) yields

$$\frac{1}{v'} = \frac{1}{v} + \frac{h}{m_0 c^2}(1 - \cos \phi) = \frac{1}{v} + \frac{2h}{m_0 c^2} \sin^2 \frac{\phi}{2}$$

$$= \frac{1}{v}\left(1 + 2\alpha \sin^2 \frac{\phi}{2}\right),$$

Hence

$$v' = \frac{v}{1 + 2\alpha \sin^2(\phi/2)} \qquad \ldots(1.11m)$$

So, from Eq. (1.11 *l*) we have

$$\tan \theta = \frac{2 \sin(\phi/2) \cos(\phi/2)}{(1 - \cos \phi) + 2\alpha \sin^2(\phi/2)}$$

or,

$$\cot \theta = (1 + \alpha) \tan \frac{\phi}{2} \qquad \ldots(1.11n)$$

Equation (1.11 *n*) is the desired relationship between the angle of scattering (ϕ) of the photon and the angle of emission (θ) of the recoil electron.

The scattering angle ϕ lies between 0° and 180°. Equation (1.11 *n*) shows that for such values of ϕ, θ lies between 90° and 0°. So, the electron cannot scatter at an angle greater than 90°.

The kinetic energy of the recoil electron is $E_k = hv - hv'$. Substituting for v' from Eq. (1.11 *m*) we obtain

$$E_k = hv\left(1 - \frac{1}{1 + 2\alpha \sin^2(\phi/2)}\right) = hv \frac{2\alpha \sin^2(\phi/2)}{1 + 2\alpha \sin^2(\phi/2)} \qquad \ldots(1.11p)$$

$$= hv \frac{\alpha(1 - \cos \phi)}{1 + \alpha(1 - \cos \phi)} \qquad \ldots(1.11q)$$

Expressing ϕ in terms of θ with the help of Eq. (1.11 *n*), Eq. (1.11 *p*) can be rewritten as

$$E_k = hv \frac{2\alpha \cos^2 \theta}{(1 + \alpha)^2 - \alpha^2 \cos^2 \theta} \qquad \ldots(1.11r)$$

The kinetic energy E_k of the recoil electron is a maximum when $\phi = 180°$ or $\theta = 0°$. The maximum value of E_k is $E_{km} = 2\alpha\, hv/(1 + 2\alpha)$.

Experimental study :

Figure 1.1B shows the experimental arrangement to study the Compton effect. Monochromatic K_α radiation from a molybdenum target T in an X-ray tube is scattered by a carbon scatterer S through a known angle ϕ. The scattered radiation on passing through a number of slits is diffracted by the crystal C in a Bragg spectrometer. The diffracted X-rays enter the ionization chamber I, connected to an electrometer to measure the intensity of the diffracted beam. The wavelength of the X-rays scattered by S at a given angle can be found by measuring the angle of diffraction at which the intensity maximum occurs by using Bragg's equation.

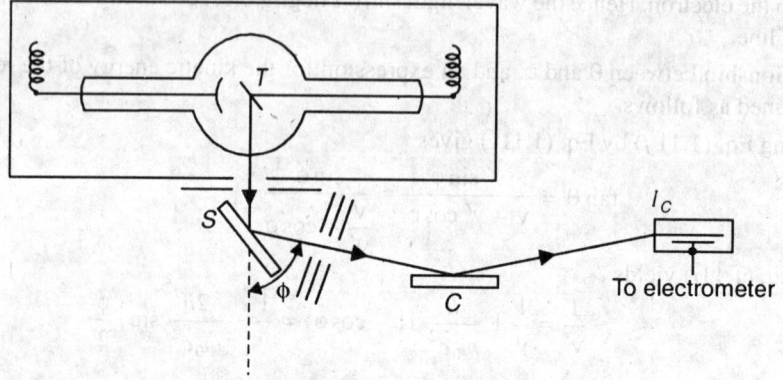

Fig. 1.1 B. Experimental set-up for Compton scattering

Figure 1.1C displays the variation of the intensity I against the scattered wavelength λ in Compton's experiment for a scattering angle $\phi = 90°$. Two intensity peaks are observed : one at a wavelength of about 0.708Å which is the wavelength of molybdenum K_α X-rays, and the other at a wavelength of about 0.730 Å. The wavelength difference of 0.022Å between the two peaks agrees very well with the Compton shift $\delta\lambda = \lambda_c = 0.024$Å for $\phi = 90°$.

Employing monochromatic X-rays of different wavelengths, the wavelength difference of the two intensity peaks has been determined. In all the cases, the observed wavelength difference is found to agree with the theoretical value of $\delta\lambda$.

The measured values at other angles of scattering have also confirmed the theoretical predictions of Compton.

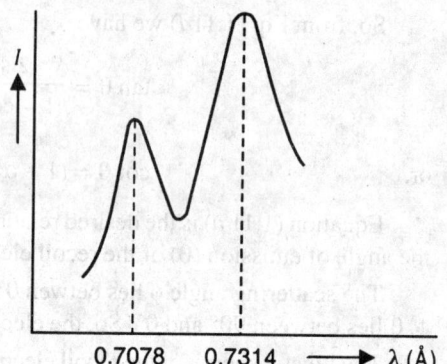

Fig. 1.1 C. Observed intensity maxima

1.4. ELECTRON DIFFRACTION EXPERIMENTS

An experimental verification of the famous de Broglie hypothesis, given mathematically by Eq. (1.7), for the wavelengths of the waves associated with electrons moving in vacuum, was provided by Davission and Germer in 1927. In the following year, further experiments confirming the existence of matter waves were performed by G.P. Thomson. He obtained diffraction patterns by shooting electrons through thin metal foils. Subsequently several other experimenters observed diffraction of neutral particles, such as, neutrons, atoms, and molecules. We shall describe below the experiments of Davission and Germer, and of G.P. Thomson.

(i) **Davission and Germer Experiment :** The apparatus consists of a tungsten filament F mounted in an electron gun G. The electron gun contains a set of metallic diaphragms and produces a collimated beam of electrons when voltages are applied between the diaphragms and the filament. A target T made of a single crystal of nickel is placed on the path of the electron beam (Fig. 1.2). The electrons are scattered from the target and are then collected by a type of Faraday bucket C connected to a sensitive galvanometer. The bucket consists of a double-walled metallic cylinder with an aperture at the front. A retarding potential is applied between the inner and the outer walls of this cylinder, which are insulated from each other. As a result, only the fastest

Introductory Ideas

electrons can reach the inner cylinder. The bucket C is mounted on an arm A which passes through the narrow slit of the cylindrical envelope S. The hanging weight W maintains the target in its original position. The whole assembly is in high vacuum. The angle θ between the incident beam and the scattered beam is varied by moving the collector.

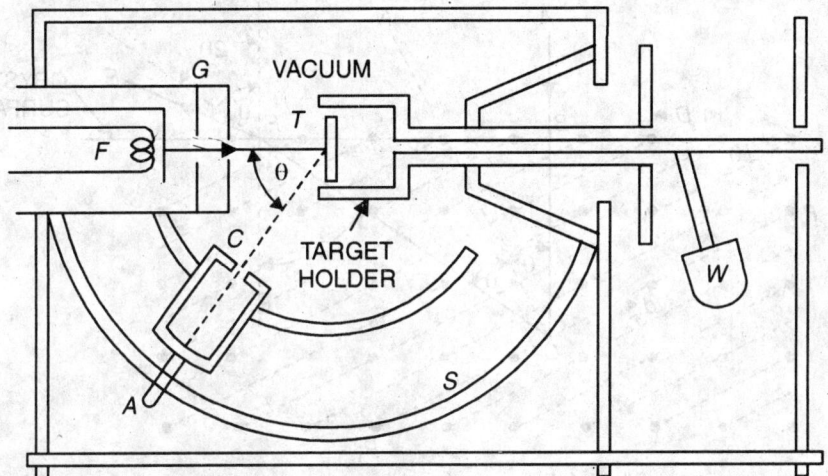

Fig. 1.2. Experimental arrangement for the study of diffraction of electrons.

Davisson and Germer studied the distribution of scattered electrons as a function of θ for various values of the energy of the incident beam. In one set of experiments they used (111) face of the nickel crystal as target, applied 40V initially between the diaphragms and the filament, and measured the collector current for various values of the angle θ. By increasing the voltage between the diaphragms and the filament in some convenient steps they measured the collector current against θ at each step. Their results are shown in Fig. 1.3. In each graph the length of the radius vector r corresponding to an angle θ is directly proportional to the collector current at that angle. It is evident from the plots that the intensity of the scattered beam reaches a pronounced maximum at 54 V and at θ = 50°. Above 54V the spur reduces with increasing voltage and becomes insignificant at about 68V. The occurrence of the pronounced maximum at 54V, 50° confirms the existence of the diffraction of the electrons by the crystal. The occurrence of this maximum is explained below.

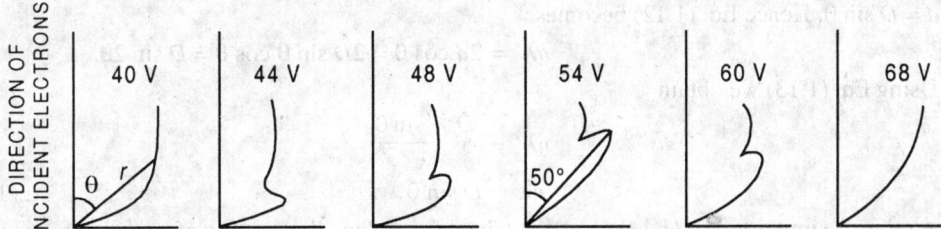

Fig. 1.3. Polar graphs showing the intensity of the scattered beam against θ.

Figure 1.4 shows the diffraction of a beam of electrons by a crystal. The lattice spacing D for the (111) planes of the nickel crystal as determined from X-rays is 2.15Å. The beam of electrons incident normally on the surface of the crystal is shown to traverse several layers of the surface atoms of the crystal. The parallel planes of atoms within the crystal are separated from each other by the distance d. Suppose the incident beam AB reaches the point C of the plane PCP_1 inside the crystal. Let the angle of incidence i.e., ∠ACN be θ, where NC is the normal to PCP_1 at C. Because of the presence of the electrostatic fields of force between the atoms (ions), the space within the crystal is at an average potential which is different from that of the surrounding vacuum. As a result, the electrons on entering the crystal will experience an acceleration and will be associated with waves of wavelength λ' different from those of λ associated with the incident beam (AB) in vacuum. If the refractive index of the crystal for electrons is n_c, then $n_c = v/v' = \lambda/\lambda'$, where v and v' represent the electron velocities

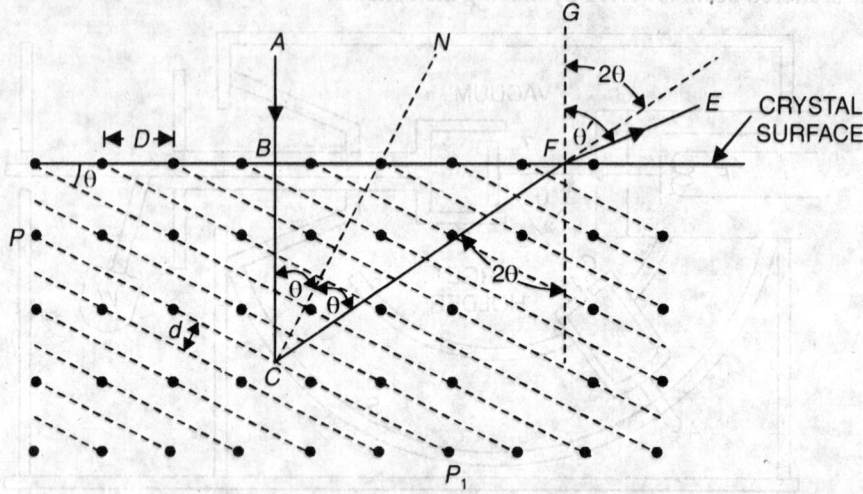

Fig. 1.4. Diffraction of a beam of electrons by crystal.

outside and inside the crystal, respectively. Now according to Huyghen's construction considerable intensity in the scattered electrons will occur when the angle of incidence equals the angle of reflection, i.e., $\angle ACN = \angle FCN$. Further, according to Bragg's law, successive parallel planes of atoms, i.e. the planes parallel to PCP_1 yield constructive interference when

$$n\lambda' = 2d \cos \theta \qquad \ldots(1.12)$$

where d is the distance between two reflecting planes as shown in Fig. 1.4.

The reflected beam CF corresponding to maximum intensity will be refracted at F while leaving the crystal and will trace the path FE. Thus

$$n_c = \frac{\lambda}{\lambda'} = \frac{\sin \theta'}{\sin 2\theta}, \qquad \ldots(1.13)$$

where θ' is the colatitude angle and was measured by Davisson and Germer. From Fig. 1.4, we find that $d = D \sin \theta$. Hence Eq. (1.12) becomes

$$n\lambda' = 2d \cos \theta = 2D \sin \theta \cos \theta = D \sin 2\theta.$$

Using Eq. (1.13) we obtain

$$n\lambda' = \frac{D \lambda' \sin \theta'}{\lambda}$$

or,
$$n\lambda = D \sin \theta' \qquad \ldots(1.14)$$

A relation similar to Eq. (1.14) can be obtained for a plane line grating of spacing D.

Substituting the experimental values of Davisson and Germer in Eq. (1.14) for the first order in the diffraction pattern ($n = 1$), i.e., $D = 2.15$Å, $V = 54$V, and $\theta' = 50°$, we get

$$\lambda = 2.15 \sin 50° = 2.15 \times 0.766$$
$$= 1.65 \text{Å}.$$

Using $V = 54$V in the de Broglie equation (Eq. 1.7) we obtain

$$\lambda = \frac{12.26}{\sqrt{54}} = 1.67 \text{Å}.$$

The excellent agreement between these two results for the wavelength confirms the de Broglie hypothesis in the case of electrons.

Introductory Ideas

(ii) G.P. Thomson's Experiment : The apparatus used by Thomson consists of a long glass tube T which is evacuated to a pressure of about 10^{-5} mm of mercury by connecting a mercury diffusion rotary-pump combination at the two points A and B (Fig. 1.5). A fine-bore metallic tube C separates the evacuated tube T into two regions D and E. The tube C serves to canalise the cathode rays and also provides a high impedance to the flow of the gas from the region D to the region E of the tube T. By admitting a gas through the inlet tube G sufficient pressure is produced in D. This permits cathode rays to be generated in a cold discharge. This however does not break the vacuum condition of the region E. The cathode rays are generated by an induction coil across the cathode K with the tube C acting as the anode. The cathode rays after traversing the tube C hit the target F which is a very thin film of metal. A photographic plate P is placed at a distance of 32.5 cm behind F. The arrangement also permits to take two exposures by lowering the photographic plate in two stages. A fluorescent screen of willemite at H enables direct observations to be made when the photographic plate is raised. A spark gap between two aluminium spheres enables measurement of the potential difference across the discharge tube. The electron beam beyond the anode C is very nearly monoenergetic.

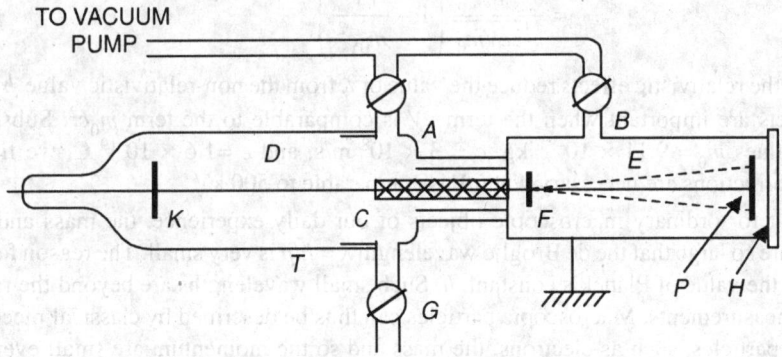

Fig. 1.5. Apparatus used by G.P. Thomson for studying diffraction.

In this experiment G.P. Thomson observed rings due to electron diffraction. He also observed a central spot formed by those electrons that penetrated the small holes in the very thin film. Thomson determined the crystal lattice spacing d by measuring the diameters of these rings and using the de Broglie equation for the electrons of known energy. The result obtained by him for the crystal lattice spacing d was found to agree within 1 per cent with that determined by using X-rays of known wavelength. In this way Thomson verified the de Broglie hypothesis for electrons.

In Thomson's experiment the electrons are accelerated through a potential difference of a few tens of kilo volt. As a result, the velocity of the electrons becomes comparable to that of light. The relativistic increase of mass of the electrons should, therefore, be considered. From Eq. (1.5) we obtain

$$\lambda = \frac{h}{mv} = \frac{h(1 - v^2/c^2)^{1/2}}{m_0 v}, \qquad \ldots(1.15)$$

where c is the velocity of light, m_0 is the rest mass of the electron, and m is the electron mass at the velocity v.

Using the binomial theorem we get

$$\lambda = \frac{h}{m_0 v}\left[1 - \frac{v^2}{2c^2} + \frac{(1/2)(-1/2)}{2!}\left(\frac{v^2}{c^2}\right)^2 - \ldots + \ldots\right]$$

$$\cong \frac{h}{m_0 v}\left(1 - \frac{v^2}{2c^2}\right), \qquad \ldots(1.16)$$

where the term (v^4/c^4) and terms with higher powers have been neglected.

Again, equating the potential energy eV of the electron to its kinetic energy $(m - m_0)c^2$ we get after incorporating the same order of approximation as before

$$eV = mc^2 - m_0c^2 = m_0c^2\left[\left(1 - \frac{v^2}{c^2}\right)^{-1/2} - 1\right]$$

$$\cong m_0c^2\left(1 + \frac{1}{2}\frac{v^2}{c^2} - 1\right) = \frac{1}{2}m_0v^2 \qquad \ldots(1.17)$$

Note that the right-hand side of Eq. (1.17) has the usual non-relativistic form of kinetic energy at this order of approximation.

Substituting the expression for v from Eq. (1.17) into Eq. (1.16) we obtain

$$\lambda = \frac{h}{m_0\sqrt{2eV/m_0}}\left(1 - \frac{eV}{m_0c^2}\right)$$

$$= \frac{h}{\sqrt{2eVm_0}}\left(1 - \frac{eV}{m_0c^2}\right) \qquad \ldots(1.18)$$

Clearly, the relativistic effects reduce the value of λ from the non-relativistic value $h/\sqrt{2eVm_0}$, and the effects are important when the term eV is comparable to the term m_0c^2. Substituting the numerical values $m_0 = 9.11 \times 10^{-31}$ kg, $c = 3 \times 10^8$ m/s, and $e = 1.6 \times 10^{-19}$ C, we find that the relativistic corrections are necessary when V is comparable to 500 kV.

Note that for ordinary macroscopic objects of our daily experience, the mass and hence the momentum are so large that the de Broglie wavelength $\lambda = h/p$ is very small. The reason for this is the smallness of the value of Planck's constant, h. Such small wavelengths are beyond the range of experimental measurements. Macroscopic particles can thus be described by classical mechanics. For microscopic particles, such as electrons, the mass and so the momentum are small even when the velocity is large. Hence their de Broglie wavelength can be measured experimentally.

1.5. WAVE PACKETS: PHASE VELOCITY AND GROUP VELOCITY

Groups of waves or wave packets are formed by combining waves of different wavelengths which are close to each other. Wave packets are confined to a small region of space and can, therefore, represent a particle in motion at any instant of time.

The *phase velocity* v_p of the de Broglie wave associated with a moving particle is given by

$$v_p = \frac{\omega}{k}, \qquad \ldots(1.19)$$

where ω is the angular frequency and k is the wave vector. The energy and the momentum of the particle are given by

$$E = \hbar\omega \text{ and } p = \hbar k,$$

where $\hbar$ is the reduced Planck's constant. Therefore

$$v_p = \frac{E}{p} \qquad \ldots(1.20)$$

We now specialise to consider the cases of a nonrelativistic and a relativistic particle. For a *nonrelativistic* particle of mass m, its velocity v is much less than the velocity of light, and

$$E = \frac{p^2}{2m} \qquad \ldots(1.21)$$

Therefore, $$v_p = \frac{E}{p} = \frac{p}{2m} = \frac{mv}{2m} = \frac{v}{2} \qquad \ldots(1.22)$$

That is, the phase velocity is half the particle velocity.

Introductory Ideas

For a *relativistic* particle, the particle velocity v is comparable to the velocity of light. In this case

$$v_p = \frac{E}{p} = \frac{mc^2}{mv} = \frac{c^2}{v} \qquad \ldots(1.23)$$

As v is always less than c, the velocity of light in free space, the phase velocity is greater than c. As the phase velocity of the de Broglie wave is greater than c, a particle in motion cannot be represented by such a wave. The particle has to be represented by a superposition of such waves, slightly differing in wavelengths, *i.e.* by a wave packet.

We shall now establish a relation between the phase velocity and the de Broglie wavelength under relativistic condition. We have

$$E = \sqrt{p^2c^2 + m_0^2 c^4} = pc\sqrt{1 + m_0^2 c^2/p^2} \qquad \ldots(1.24)$$

Hence

$$v_p = \frac{E}{p} = c\sqrt{1 + m_0^2 c^2/p^2} = c\sqrt{1 + (m_0^2 c^2/h^2)\lambda^2}, \qquad \ldots(1.25)$$

where the relation $p = h/\lambda$ has been used. Equation (1.25) shows that $v_p > c$ and v_p increases with λ.

The *group velocity*, *i.e.*, the velocity of the wave packet is given by

$$v_g = \frac{d\omega}{dk} = \frac{dE}{dp}, \qquad \ldots(1.26)$$

since $E = \hbar\omega$ and $p = \hbar k$. For a *nonrelativistic* particle moving with velocity v, we have

$$E = \frac{1}{2} mv^2 = \frac{p^2}{2m},$$

so that

$$\frac{dE}{dp} = \frac{p}{m} = v \qquad \ldots(1..27)$$

From Eqs. (1.26) and (1.27) we get

$$v_g = v, \qquad \ldots(1.28)$$

i.e., the particle velocity is equal to the group velocity. For a *relativistic* particle we use Eq. (1.24) and get

$$E^2 = p^2 c^2 + m_0^2 c^4$$

or

$$2E \frac{dE}{dp} = 2pc^2$$

Hence

$$v_g = \frac{dE}{dp} = \frac{pc^2}{E} = \frac{mvc^2}{mc^2} = v \qquad \ldots(1.29)$$

Again, the group velocity is found to be equal to the particle velocity.

1.6. Heisenberg's Uncertainty Principle

This principle states that *it is impossible to mention accurately and simultaneously the values of both the members of particular pairs of physical quantities that dictate the behaviour of an atomic system.* Quantities, canonically conjugate to each other in the Hamiltonian sense, constitute the members of the pairs of the quantities involved in the uncertainty principle. Some examples of such conjugate quantities are given below :

(i) The position x of a particle in a rectangular coordinate and the corresponding momentum component p_x, (*ii*) The angular position ϕ of a particle in the perpendicular *y-z* plane and the corresponding angular momentum component J_x.

(*iii*) The energy E of a particle and the time t of its measurement.

Quantitatively, the uncertainty principle states that *the order of magnitude of the product of the uncertainties in the values of the two conjugate quantities must be at least equal to the reduced Planck's constant* $\hbar$. Thus, if α and β are two canonically conjugate quantities, then

$$\Delta\alpha \cdot \Delta\beta \gtrsim \hbar, \qquad ...(1.30)$$

where $\Delta\alpha$ and $\Delta\beta$ are the uncertainties in α and β.

For the examples given above, the *uncertainty relations are* :

$$\Delta x \cdot \Delta p_x \gtrsim \hbar, \qquad ...(1.31)$$
$$\Delta\phi \cdot \Delta J_x \gtrsim \hbar, \qquad ...(1.32)$$
$$\Delta t \cdot \Delta E \gtrsim \hbar. \qquad ...(1.33)$$

Equation (1.30) shows that if $\Delta\alpha$ is small, $\Delta\beta$ is large, and *vice versa*. That is, an accurate determination of one of the quantities leads to an enhanced uncertainty in the value of the other. Since the value of $\hbar$ is small, the *uncertainty principle assumes importance in the systems of atomic size*.

The uncertainties in the simultaneous measurement of the canonically conjugate physical quantities are not due to defects in the method of measurement, but they appear from an inherent limitation of nature. If a method of measurement tries to determine one quantity precisely, it cannot simultaneously determine the other quantity so precisely. This inherent limitation for atomic and subatomic particles is referred to by Bohr as the *complementarity principle*.

The uncertainty relation between position and momentum for a wave packet representing a particle is derived in Sec. 3.3. There it is shown that the product $\Delta p \, \Delta x$ can have the minimum value of $\hbar/2$. This minimum value is also obtained for a harmonic oscillator in the ground state (vide Sec. 5.4.). The uncertainty principle is beautifully illustrated with the help of the gamma-ray microscope which we consider below.

1.7. Gamma-Ray Microscope

A usual practice of measuring the position of a particle is to observe it with a microscope. The absolute limit to the accuracy Δx with which a position can be determined by a microscope is given by the resolving power of the microscope :

$$\Delta x = \frac{\lambda}{2\sin\alpha} \qquad ...(1.34)$$

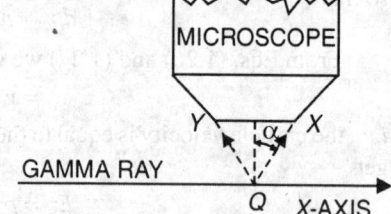

Fig. 1.6. Position determination by gamma-ray microscope.

where λ is the wavelength of the light used for illuminating the object and α is the half-angle subtended by the objective lens of the microscope at the position of the object being observed (Fig. 1.6).

From the above expression it follows that the position can be determined as accurately as desired by using light of small enough wavelength since the value of $\sin\alpha$ cannot be larger than unity. As gamma rays have extremely short wavelengths, the name *gamm-ray microscope* has been used to this conceptual method of measuring the position very accurately.

To see the particle in the microscope, the incident γ-rays used for illuminating the particle must be scattered into the objective of the microscope. During such scattering the particle will recoil. For simplicity, let us assume that minimum amount of γ-ray energy, *i.e.*, one photon, is used. This also gives minimum recoiling of the particle.

We shall now find out the effect of this kind of measurement of position of the particle on the accuracy of its momentum determination by using the theory of Compton effect. Due to the impact of the photon on the particle, a scattered photon of wavelength, say λ, will enter the microscope

Introductory Ideas 15

objective. This photon of momentum h/λ can enter the objective along QX or QY or between QX and QY. If the photon enters along QX then its momentum component along x-axis will be $(h/\lambda)\sin\alpha$. Thererfore the momentum transferred in the x-direction to the particle is

$$\frac{h}{\lambda'} - \frac{h}{\lambda}\sin\alpha \qquad ...(1.35)$$

where λ' is the wavelength of the incident photon. On the other hand, when the scattered photon enters the objective along QY, the momentum imparted to the particle in the x-direction is given by

$$\frac{h}{\lambda'} + \frac{h}{\lambda}\sin\alpha \qquad ...(1.36)$$

Evidently, the momentum transferred to the particle can have any value between those given by (1.35) and (1.36). Therefore, the uncertainty in the momentum of the particle is

$$\Delta p_x = \frac{2h}{\lambda}\sin\alpha \qquad ...(1.37)$$

Hence from Eqs. (1.34) and (1.37) we obtain

$$\Delta x \cdot \Delta p_x = h \qquad ...(1.38)$$

That is, the measurement of position of the particle and its momentum with a gamma-ray microscope takes place in accordance with Heisenberg's uncertainty principle.

> **Note.** We see from Eqs. (1.34) and (1.37) that if optical wavelengths are used instead of γ-rays, the uncertainty in position Δx will increase while that in momentum Δp_x will decrease. However, the product $\Delta x\, \Delta p_x$, being independent of the wavelengh λ, will remain unchanged.

1.8. SOME APPLICATIONS OF THE UNCERTAINTY PRINCIPLE

(a) Nonexistence of free electrons in an atomic nucleus

It is known that the rest mass of an electron is $m_0 = 9.11 \times 10^{-31}$ kg, the diameter of an atomic nucleus is about 2×10^{-14} m, and the maximum possible kinetic energy of an electron emitted from a radioactive nucleus is 4 MeV.

If it is possible for an electron to exist in the nucleus, it would lie anywhere within the diameter d of the nucleus. Thus the maximum uncertainty in the position of the electron would be

$$\Delta x = d = 2 \times 10^{-14} \text{ m}$$

According to Heisenberg's uncertainty principle, we have

$$\Delta x\, \Delta p \geq \hbar,$$

where Δp is the uncertainty in the momentum of the electron. The minimum uncertainty in the momentum will be

$$\Delta p = \frac{\hbar}{\Delta x} = \frac{1.054 \times 10^{-34} \text{ J.s}}{2 \times 10^{-14} \text{ m}} = 0.527 \times 10^{-20} \text{ kg.m.s}^{-1}$$

Therefore, the minimum momentum of the electron, if it exists in the nucleus, would be

$$p_{min} = \Delta p = 0.527 \times 10^{-20} \text{ kg.m.s}^{-1}.$$

The total relativistic energy E of the electron is expressed by

$$E^2 = p^2 c^2 + m_0^2 c^4,$$

where c is the velocity of light in free space.

For the electron with the minimum momentum p_{min}, the minimum energy E_{min} would hence be given by

$$E_{min}^2 = p_{min}^2 c^2 + m_0^2 c^4$$

Substituting $c = 3 \times 10^8$ m.s^{-1} and the values of m_0 and p_{min} given above, we get

$$E_{min} = 1.58 \times 10^{-12} \text{ J} = \frac{1.58 \times 10^{-12}}{1.6 \times 10^{-19}} \text{eV} = 9.88 \text{ MeV}$$

That is, a free electron must have an energy of at least 9.88 MeV in order to exist in the nucleus. As the maximum kinetic energy of an electron emitted from a radioactive nucleus is only 4 MeV, it follows that free electrons cannot exist in atomic nuclei.

(b) First Bohr radius of the hydrogen atom

Let r denote the orbital radius of the electron revolving round the nucleus in the hydrogen atom. The maximum uncertainty in the position of the electron with respect to the nucleus is $\Delta x = r$. Using the uncertainty principle, the minimum uncertainty in the momentum of the electron is found to be

$$\Delta p = \frac{\hbar}{\Delta x} = \frac{\hbar}{r}$$

The minimum possible value of the momentum would be

$$p_{min} = \Delta p = \frac{\hbar}{r}$$

The kinetic energy of the electron is

$$K.E. = \frac{p_{min}^2}{2m} = \frac{\hbar^2}{2mr^2},$$

and the potential energy of the electron is (in SI units)

$$P.E. = -\frac{e^2}{4\pi\varepsilon_0 r},$$

where ε_0 is the permittivity of free space. The total energy of the electron is

$$E = K.E. + P.E. = \frac{\hbar^2}{2mr^2} - \frac{e^2}{4\pi\varepsilon_0 r}$$

In the ground state, the energy E would attain a minimum value E_{min} and the radius r would attain the value a_0 where a_0 is the first Bohr radius. For the energy E to be a minimum, we have

$$\left(\frac{dE}{dr}\right)_{r=a_0} = 0,$$

which yields

$$a_0 = \frac{4\pi\varepsilon_0 \hbar^2}{me^2} = \frac{\varepsilon_0 h^2}{\pi me^2}$$

This is the familiar expression for the first Bohr radius of the hydrogen atom. Substituting $\varepsilon_0 = 8.854 \times 10^{-12}$ F.m^{-1}, $h = 6.62 \times 10^{-34}$ J.s, $m = 9.11 \times 10^{-31}$ kg, and $e = 1.6 \times 10^{-19}$ C, we obtain $a_0 = 5.3 \times 10^{-11}$ m.

If we try to squeeze an atom, the electrons must be confined to a smaller region. So, by the uncertainty principle, their momenta and hence energy increases. The resistance to atomic compression is thus a quantum mechanical phenomenon.

1.9. Worked-Out Problems

1. An electron falls through a potential difference of 100 volt. Calculate the momentum of the electron and the length of the wave associated with the electron in motion. How could these waves be detected?

Ans. The velocity v of the electron is given by

$$v = \sqrt{2eV/m_0}$$

where $e = 1.6 \times 10^{-19}$ C, $m_0 = 9.11 \times 10^{-31}$ kg, and $V = 100$V.

Therefore the momentum of the moving electron is

$$p = m_0 v = \sqrt{2eVm_0}$$

$$= \sqrt{2 \times 1.6 \times 10^{-19} \times 100 \times 9.11 \times 10^{-31}}$$
$$= 10^{-24} \sqrt{3.2 \times 9.11} = 5.4 \times 10^{-24} \text{ kg.m.s.}^{-1}$$

The wavelength associated with the electron is given by

$$\lambda = \frac{h}{p} = \frac{6.62 \times 10^{-34}}{5.4 \times 10^{-24}} \cong 1.226 \times 10^{-10} \text{ m}$$
$$= 1.226 \text{Å}$$

The value of this wavelengh is about the same as that of a typical X-ray. This suggests the possibility of detecting the matter waves by defraction by a crystal.

2. Find the angle of incidence of the electrons of energy 100 eV on the lattice planes of a metal crystal so that a strong Bragg reflection of the first order occurs. Given: $d = 2.15$Å, mass of the electron = 9.11×10^{-31} kg, $e = 1.6 \times 10^{-19}$ coulomb, and $h = 6.62 \times 10^{-34}$ J.s.

Ans. The Bragg equation is

$$n\lambda = 2d \sin \theta,$$

where θ is the angle of incidence of the electrons on the lattice planes.

The wavelength λ associated with the electron is given by

$$\lambda = \frac{h}{\sqrt{2eVm_0}}$$

Putting the values we get

$$\lambda = \frac{6.62 \times 10^{-34}}{\sqrt{2 \times 1.6 \times 10^{-19} \times 100 \times 9.11 \times 10^{-31}}}$$
$$= \frac{6.62 \times 10^{-10}}{\sqrt{3.2 \times 9.11}} = 1.23 \times 10^{-10} \text{ m}$$

Hence from Bragg equation we get

$$\sin \theta = \frac{\lambda}{2d} \quad \text{(since } n = 1 \text{ for first order reflection)}$$

or,
$$\sin \theta = \frac{1.23 \times 10^{-10}}{2 \times 2.15 \times 10^{-10}} = 0.286$$

or,
$$\theta = 16.6°.$$

3. The velocity of an electron and that of a rifle bullet of mass 30 gm are measured each with an uncertainty of $\Delta v_x = 10^{-3}$ m/s. Determine the minimum uncertainties in their positions using Heisenberg's uncertainty relation.

Ans. Using the relation $\Delta p_x = m \Delta v_x$, we get for the minimum position uncertainty

$$\Delta x = \frac{\hbar}{m \Delta v_x}$$

(i) For the electron, $m = 9.11 \times 10^{-31}$ kg.

Therefore $\Delta x = \dfrac{1.054 \times 10^{-34}}{9.11 \times 10^{-31} \times 10^{-3}} = 0.116 \text{ m}$

(ii) For the bullet, $m = 30 \times 10^{-3}$ kg.

Therefore $\Delta x = \dfrac{1.054 \times 10^{-34}}{30 \times 10^{-3} \times 10^{-3}} = 3.5 \times 10^{-30} \text{ m}$

Note that for the bullet, uncertainty principle does not impose any effective limit on experiments.

This is because the errors in position measurements are always much larger than 10^{-30} m. For the electron, however, the principle does impose a limit. For example, the spacing between the atoms in a solid is about 10^{-9} m. Therefore, a position measurement with $\Delta x = 0.1$ m means that the electron may be anywhere among billions of atoms. The uncertainty principle is thus more important for lighter particles than for heavier particles.

4. 4 keV electrons from a source fall straight on to a target 50 cm away. Find the radius of the electron spot (order only) due to Heisenberg's uncertainty principle assuming the focussing to be perfect. **[CU 1984]**

Ans. Let r be the required radius. If D be the distance between the source and the target (see Fig. 1.7), then $\sin\theta = r/D$. Also, if Δp_y is the uncertainty in momentum in the y-direction, then $\sin\theta = \Delta p_y / p$, where p is the momentum in the x-direction. The uncertainty in the position in the y-direction is $\Delta y = r$. From the uncertainty principle, we have $\Delta p_y \Delta y \sim \hbar$, i.e., $\Delta p_y \cdot r \sim \hbar$.

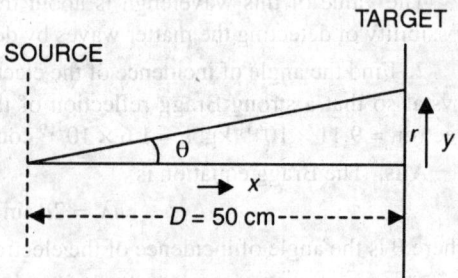

Fig. 1.7.

Thus $\dfrac{r}{D} = \dfrac{\Delta p_y}{p} \sim \dfrac{\hbar}{pr}$, or $r^2 \sim \dfrac{\hbar D}{p}$

Now, $p = \sqrt{2mE} = \sqrt{2 \times 9.11 \times 10^{-31} \times 4 \times 10^3 \times 1.6 \times 10^{-19}}$

$= 3.414 \times 10^{-23}$ kg.m.s.$^{-1}$.

Hence $r^2 \sim \dfrac{1.054 \times 10^{-34} \times 0.5}{3.414 \times 10^{-23}} = 1.54 \times 10^{-12}$ m^2,

or, $r \sim 10^{-6}$ m.

5. (*a*) Show that the de Broglie wavelength of a particle of rest mass m_0 and kinetic energy E_k is given by

$$\lambda = \dfrac{h}{\sqrt{2m_0 E_k}} \left(1 + \dfrac{E_k}{2m_0 c^2}\right)^{-1/2}$$

(N. Beng. U. 2001)

(*b*) Hence find the de Broglie wavelength of an electron having an energy of 1 MeV.

Ans. (*a*) The total relativistic energy of the particle is given by

$$E = \sqrt{p^2 c^2 + m_0^2 c^4},$$

where p is the momentum and m_0 is the rest mass.

The kinetic energy is

$$E_k = E - m_0 c^2 = \sqrt{p^2 c^2 + m_0^2 c^4} - m_0 c^2$$

or $\sqrt{p^2 c^2 + m_0^2 c^4} = E_k + m_0 c^2$.

Squaring both sides, we get

$$p^2 c^2 + m_0^2 c^4 = E_k^2 + 2 E_k m_0 c^2 + m_0^2 c^4$$

or, $p = \dfrac{\sqrt{E_k(E_k + 2m_0 c^2)}}{c}$

Therefore, the de Broglie wavelength is

$$\lambda = \dfrac{h}{p} = \dfrac{hc}{\sqrt{E_k(E_k + 2m_0 c^2)}}$$

$$= \frac{h}{\sqrt{2m_0 E_k}} \left(1 + \frac{E_k}{2m_0 c^2}\right)^{-1/2}$$

[If the particle is an electron accelerated through a potential difference V, then $E_k = eV$.]

(b) For an electron, we have $m_0 c^2 = 9.11 \times 10^{-31} \times (3 \times 10^8)^2 = 0.8199 \times 10^{-13}$ J. If $E_k = 1$ MeV $= 10^6$ eV $= 1.6 \times 10^{-13}$ J, we find that E_k is comparable with $m_0 c^2$ so that relativistic corrections are important. Substituting the values of E_k and $m_0 c^2$ in the above expression for λ, we have

$$\lambda = \frac{6.62 \times 10^{-34} \times 3 \times 10^8}{\sqrt{1.6 \times 10^{-13} \, (1.6 \times 10^{-13} + 2 \times 0.8199 \times 10^{-13})}}$$

$$= 0.0087 \times 10^{-10} \text{ m} = 0.0087 \text{ Å}.$$

6. Calculate the de Broglie wavelength corresponding to the most probable velocity of thermal neutrons at 300K. Can you detect it? Given, mass of a neutron $m = 1.676 \times 10^{-27}$ kg. **(C.U. 1992)**

Ans. The most probable velocity* is given by

$$v = \sqrt{\frac{2k_B T}{m}},$$

where k_B is Boltzmann's constant, T is the absolute temperature, and m is the mass of a neutron. Given, T = 300 K and $m = 1.676 \times 10^{-27}$ kg.

Hence, $$v = \sqrt{\frac{2 \times 1.38 \times 10^{-23} \times 300}{1.676 \times 10^{-27}}} = 2.22 \times 10^3 \text{ m/s}$$

The de Broglie wavelength is

$$\lambda = \frac{h}{mv} = \frac{6.62 \times 10^{-34}}{1.676 \times 10^{-27} \times 2.22 \times 10^3}$$

$$= 1.779 \times 10^{-10} \text{ m}$$

$$= 1.779 \text{ Å}$$

This wavelength falls in the X-ray range, and can therefore be detected by diffraction of neutrons through a crystal.

7. Show that electron diffraction through a narrow slit takes place in accordance with Heisenberg's uncertainty principle.

Ans. The diffraction of electrons through a slit S takes place due to the deflection of the electrons at the slit (Fig. 1.8). The electrons which produce the diffraction pattern on the photographic plate P undoubtedly pass through the slit, but the exact point in the slit through which a particular electron passes is not precisely known. The uncertainty in the position of the electron is given by Δx, the width of the slit. As the electron is deflected at the slit, it gains a component of momentum perpendicular to the original direction of its movement. If p is the momentum of the electron, the momentum component perpendicular to the original direction is $p_x = p \sin \theta$, where θ denotes the mean angle of deflection. Considering the central band one can say that the electron may be anywhere within this band; in other words, the uncertainty in the momentum component p_x is

$$\Delta p \cong p \sin \theta$$

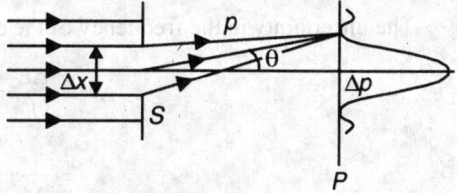

Fig. 1.8.

---

* See Eq. (1.60) of 'Statistical Mechanics'

From optical analysis of diffraction, we get
$$\Delta x \sin \theta = \lambda$$
where λ is the wavelength

or
$$\Delta x = \frac{\lambda}{\sin \theta}$$

Therefore,
$$\Delta x \, \Delta p = \frac{\lambda}{\sin \theta} p \sin \theta$$
$$= \lambda p = h,$$

since $\lambda = h/p$, from de Broglie's hypothesis. Thus Heisenberg's uncertainty principle is obeyed.

8. Obtain the uncertainty relation between energy and time from that between momentum and position of a particle for a wave packet.

Ans. A moving particle can be represented by a wave packet; the particle velocity v being equal to the group velocity v_g of the wave packet. The uncertainty in the x-coordinate of the particle is given by Δ_x, the width of the wave packet. The uncertainty in time is $\Delta t = \Delta x/v_g$. If Δ_E is the uncertainty in energy, we have

$$\Delta E = \frac{\partial E}{\partial p} \Delta p = v_g \, \Delta p,$$

where p is momentum and Δp is the uncertainty in p.

Hence $\Delta E \, \Delta t = v_g \, \Delta p \, \dfrac{\Delta x}{v_g} = \Delta p \, \Delta x$. But $\Delta p \, \Delta x \geq \hbar$ according to the uncertainty relation between momentum and position. Therefore $\Delta E \, \Delta t \geq \hbar$ which is the uncertainty relation between energy and time.

9. What is the limit, imposed by Heisenberg's uncertainty principle, on the accuracy with which the frequency of the radiation emitted by an atom can be determined ?

Ans. An "excited" atom gets rid of its excess energy by emitting one or more photons of appropriate frequency. The "mean life time" of the excited state, *i.e.*, the average time that elapses between the excitation of an atom and the emission of radiation, is about 10^{-8} sec. Hence the uncertainty in the photon energy is

$$\Delta E = \frac{\hbar}{\Delta t} = \frac{1.054 \times 10^{-34} \text{ J.s}}{10^{-8} \text{ s}}$$
$$= 1.054 \times 10^{-26} \text{J}.$$

The uncertainty in the frequency of the emitted radiation is

$$\Delta \nu = \frac{\Delta E}{h} = \left(\frac{\hbar}{h}\right) \frac{1}{\Delta t} = \frac{1}{2\pi \Delta t}$$

$$= \frac{1}{2 \times 3.1416 \times 10^{-8}} = 1.59 \times 10^7 \text{ Hz}$$

This is the minimum limit to the accuracy with which the frequency of the emitted radiation can be determined.

10. From the uncertainty relation between momentum and position of a particle moving in a circle, determine the uncertainty relation between its angular momentum and angular position. What is the uncertainty in the angular position of an electron moving in a Bohr orbit of a hydrogen atom ?

Ans. As the particle describes a circle, the position momentum uncertainty is $\Delta p \, \Delta x \geq \hbar$, where p is the momentum along a tangent and x is measured along the circumference of the circle. The

Introductory Ideas

angular momentum is given by $L = pr$, where r is the radius of the circle. Hence $\Delta p = \Delta L/r$. The angular position is given by $\theta = x/r$, so that $\Delta x = r\Delta\theta$. Therefore,

$$\Delta p \cdot \Delta x = \frac{\Delta L}{r} \cdot r\Delta\theta = \Delta L \cdot \Delta\theta \geq \hbar$$

which is the desired uncertainty relation.

For an electron in a Bohr orbit of a hydrogen atom, the angular momentum is fixed (see Chap. 6). Therefore, $\Delta L = 0$. Hence $\Delta\theta$ is infinite, implying that the position of the electron in the orbit cannot be determined.

11. A spectral line of wavelength 6000 Å has a width of 5×10^{-5} Å. Determine the minimum time spent by the atomic system in the associated energy state.

Ans. The relationship between the photon energy E and the wavelength λ is $E = hc/\lambda$, where c is the velocity of light in free space. If ΔE and $\Delta\lambda$ are the uncertainties in E and λ, respectively, we have $|\Delta E| = hc\,\Delta\lambda/\lambda^2$. Also, $\Delta E \cdot \Delta t \geq \hbar$, where Δt is the time spent by the atomic system in the corresponding energy state. We have $\Delta t \geq \hbar/\Delta E = \lambda^2/(2\pi c\,\Delta\lambda)$

$$= \frac{(6 \times 10^{-7}\text{ m})^2}{2\pi\,(3 \times 10^8\text{ m.s}^{-1})\,(5 \times 10^{-15}\text{ m})} = 3.82 \times 10^{-8}\text{ s}$$

Hence the minimum value of Δt is 3.82×10^{-8} s.

12. Find the minimum energy in eV of the photon required to observe an object of size 4 Å.

Ans. For scattering to take place, the wavelength of the waves should be of the same order of magnitude as the size of the object. Therefore, the maximum wavelength that can be used here is fixed at $\lambda_{max} = 4$ Å. The required minimum energy is

$$E_{min} = \frac{hc}{\lambda_{max}} = \frac{6.62 \times 10^{-34} \times 3 \times 10^8}{4 \times 10^{-10}}\text{ J}$$

$$= \frac{6.62 \times 10^{-34} \times 3 \times 10^8}{4 \times 10^{-10} \times 1.6 \times 10^{-19}}\text{ eV} = 3.1 \times 10^3\text{ eV}$$

13. Photons of wavelength 662 nm are incident normally on a surface and are totally reflected from it. If the force exerted by the photons on the surface is 0.5 N, calculate the photon flux.

Ans. The momentum of the incident photon is $\hbar k = 2\pi\hbar/\lambda$. The change in momentum upon reflection is $\Delta p = 2\hbar k$. If Δt is the time of interaction with the reflecting surface and F is the force exerted on it, we have $F = \Delta p/\Delta t$. If N is the number of photons incident in time Δt, the total force exerted on the surface is

$$F_0 = NF = \frac{N\,\Delta p}{\Delta t} = \frac{2\hbar k\,N}{\Delta t} = 2\hbar k\,N_0,$$

where $N_0\,(= N/\Delta t)$ is the *photon flux*, i.e., the number of photons incident per second. Hence

$$N_0 = \frac{F_0}{2\hbar k} = \frac{F_0\lambda}{4\pi\hbar} = \frac{F_0\lambda}{2h} = \frac{0.5 \times 662 \times 10^{-9}}{2 \times 6.62 \times 10^{-34}}$$

$$= 2.5 \times 10^{26}\text{ photons per second.}$$

14. Show how one can arrive at Bohr's quantization condition on the basis of de Broglie's hypothesis of matter waves. **(C.U. 1998)**

Ans. Let an electron of mass m rotate about the nucleus of a hydrogen atom in an orbit of radius r. Since the electron comes back to the same point or state in going once round the orbit, the circumference $2\pi r$ of the orbit contains n number of de Broglie wavelengths λ, where n is an integer. So,

$$2\pi r = n\lambda = \frac{nh}{mv},$$

where v is the electron velocity. Hence the angular momentum of the electron is

$$mvr = \frac{nh}{2\pi} = n\hbar,$$

which is Bohr's quantization condition.

15. Find the number of photons emitted per second by a He - Ne laser of 1 mW power, if the radiated wavelength is 633 nm.

Ans. The energy of a photon is $E = h\nu = hc/\lambda$, where $\lambda = 6.33 \times 10^{-7}$ m. The laser power is $P = 1$ mW $= 10^{-3}$ W. So, the number of photons emitted per second is

$$n = \frac{P}{E} = \frac{\lambda P}{hc} = \frac{6.33 \times 10^{-7} \times 10^{-3}}{6.626 \times 10^{-34} \times 3 \times 10^8} = 3.19 \times 10^{15}$$

16. Monochromatic γ-rays of energy 2.62 MeV are incident on a beryllium target. If the γ-rays counter for the scattered photon is kept at an angle of 30°, find the angle at which the electron counter should be positioned. **(C.U. 1991)**

Ans. The energy of the incident γ-ray photon is $h\nu = 2.62$ MeV $= 2.62 \times 10^6 \times 1.6 \times 10^{-19}$ J. From the theory of Compton scattering, if θ is the angle of emission of the recoil electron, we have $\cot \theta = (1 + \alpha) \tan \frac{\phi}{2}$. Here $\phi = 30°$ and

$$\alpha = \frac{h\nu}{m_0 c^2} = \frac{2.62 \times 1.6 \times 10^{-13}}{9.11 \times 10^{-31} \times 9 \times 10^{16}} = 5.11$$

So, $\cot \theta = (1 + 5.11) \tan 15° = 1.64$

or, $\theta = 31.4°$.

Therefore, the electron counter should be positioned at an angle of 31.4°.

17. The molybdenum K_α line ($\lambda = 0.0712$ nm) suffers a Compton scattering in carbon. Find the wavelength shift of the line scattered at 90° if the scattering particle is (a) an electron, (b) the whole carbon atom. Determine the scattered wavelength in each case.

Ans. (a) For $\phi = 90°$, the wavelength shift for an electron scatterer is $\delta\lambda = \lambda_c = 0.024$ Å $= 0.0024$ nm.

(b) Here m_0 in the expression $\lambda_c = h/(m_0 c)$ is the mass of the carbon atom. Since the atomic weight of carbon is 12, by Avogadro's hypothesis, 12 gm of carbon contains N_A carbon atoms where N_A = Avogadro's number = 6.02×10^{23}. Hence the mass of a carbon atom is

$$m_0 = \frac{12}{6.02 \times 10^{23}} \simeq 2 \times 10^{-23} \text{ gm} = 2 \times 10^{-26} \text{ kg}.$$

The wavelength shift is

$$\delta\lambda = \frac{h}{m_0 c} = \frac{6.62 \times 10^{-34}}{2 \times 10^{-26} \times 3 \times 10^8} = 1.1 \times 10^{-16} \text{ m} = 1.1 \times 10^{-7} \text{ nm}.$$

The scattered wavelength is $\lambda' = \lambda + \delta\lambda$.

For the electron scatterer $\lambda' = 0.0712 + 0.0024 = 0.0736$ nm.

For the carbon atom scatterer, $\lambda' = 0.0712 + 1.1 \times 10^{-7} \simeq 0.0712$ nm.

18. A γ-ray of energy 0.511 MeV strikes an electron head–on and is scattered straight backward. Find the energy of the scattered γ-ray, the recoil energy of the electron, and the ratio v/c for the latter. **(C.U. 1994)**

Ans. Here $\phi = 180°$ and so the wavelength shift in Compton scattering is $\delta\lambda = 2\lambda_c = 0.048$ Å. The energy of the scattered γ-ray photon is $h\nu' = hc/\lambda' = hc/(\lambda + \delta\lambda)$, where λ is the wavelength of the incident photon. We have

$$\lambda = \frac{c}{\nu} = \frac{hc}{h\nu} = \frac{6.62 \times 10^{-34} \times 3 \times 10^8}{0.511 \times 10^6 \times 1.6 \times 10^{-19}} = 2.429 \times 10^{-12} \text{ m}$$

So,
$$hv' = \frac{hc}{\lambda'} = \frac{6.62 \times 10^{-34} \times 3 \times 10^8}{(2.429 + 4.8) \times 10^{-12}} = 2.747 \times 10^{-14} \text{ J}$$
$$= \frac{2.747 \times 10^{-14}}{1.6 \times 10^{-19}} = 1.717 \times 10^5 \text{ eV} \approx 0.172 \text{ MeV}$$

The recoil energy of the electron is
$$E_k = hv - hv' = 0.511 - 0.172 = 0.339 \text{ MeV}$$

Also,
$$E_k = mc^2 - m_0 c^2 = m_0 c^2 \left(\frac{1}{\sqrt{1-\beta^2}} - 1 \right)$$

or,
$$\frac{E_k}{m_0 c^2} + 1 = \frac{1}{\sqrt{1-\beta^2}} \quad \text{or,} \quad 1 - \beta^2 = \frac{1}{\left(\frac{E_k}{m_0 c^2} + 1 \right)^2}$$

or,
$$\beta^2 = 1 - \frac{1}{\left(\frac{0.339 \times 10^6 \times 1.6 \times 10^{-19}}{9.11 \times 10^{-31} \times 9 \times 10^{16}} + 1 \right)^2} = 1 - \frac{1}{(1 + 0.662)^2} = 0.638$$

or,
$$\beta = \frac{v}{c} = 0.799.$$

19. A 0.75 MeV photon is scattered at an angle of 60°. Find the energy of the recoil electron. Take $m_0 = 0.5$ MeV. **(C.U. 1989)**

Ans. We have for the energy of the recoil electron
$$E_k = hv \frac{2\alpha \sin^2(\phi/2)}{1 + 2\alpha \sin^2(\phi/2)}$$

where,
$$\alpha = \frac{hv}{m_0 c^2} = \frac{0.75 \text{ MeV}}{0.5 \text{ MeV}} = 1.5, \text{ and } \phi = 60°$$

So,
$$E_k = 0.75 \frac{2 \times 1.5 \times \sin^2 30°}{1 + 2 \times 1.5 \times \sin^2 30°} = 0.321 \text{ MeV}$$

20. An electron has a speed of 500 m/s with an accuracy of 0.005%. Calculate the uncertainty in the position of the electron. Given $h = 6.6 \times 10^{-34}$ J.s. **(Garhwal 1996)**

Ans. The uncertainty in the velocity v is
$$\Delta v = \frac{0.005}{100} v = \frac{0.005}{100} \times 500 = 0.025 \text{ m/s}.$$

If Δp is the uncertainty in momentum and Δx is that in position, we have by Heisenberg's uncertainty principle
$$\Delta p \, \Delta x = \hbar = \frac{h}{2\pi}$$

As $\Delta p = m \Delta v$, where m is the electron mass, we have
$$\Delta x = \frac{h}{2\pi \, m \Delta v} = \frac{6.6 \times 10^{-34}}{2 \times 3.1416 \times 9.11 \times 10^{-31} \times 0.025}$$
$$= 0.00461 \text{ m} = 4.61 \text{ mm}.$$

21. Show that if a component of the angular momentum of the electron in a hydrogen atom is $2\hbar$ within an error of 5%, its angular orbital position in the plane perpendicular to that component cannot be specified at all. **(Bundelkhand 1997)**

Ans. If ΔJ is the uncertainty in the angular momentum and $\Delta\theta$ that in the corresponding angular position, we have by Heisenberg's uncertainty principle

$$\Delta J\, \Delta\theta = \hbar$$

Here
$$\Delta J = \frac{5}{100} \times 2\hbar = \frac{\hbar}{10}.$$

Hence
$$\Delta\theta = \frac{\hbar}{\Delta J} = 10 \text{ rad.}$$

As the angle in the plane perpendicular to the angular momentum component cannot be greater than 2π rad, clearly the orbital angular position of the electron cannot be specified at all.

QUESTIONS

1. (a) What is the hypothesis of de Broglie? **(C.U. 2008)**
   (b) Why is the wave nature of matter not apparent for macroscopic particles?
   (c) Derive the de Broglie relation for a photon from the principle of mass-energy equivalence.
   (d) Show that the group velocity of the de Broglie wave is equal to the velocity of the particle.
   (C.U. 1995, 2005)

2. (a) Show that for de Broglie waves representing a relativistic particle the product of the phase velocity and the group velocity is equal to c^2.
   (b) Show that the phase velocity of the de Broglie waves for a particle is a function of the wavelength. **(C.U. 2001)**

   Describe the Davisson–and–Germer experiment to demonstrate the wavelike behaviour of moving electrons. Discuss how the dual character of electrons can be reconciled on the basis of de Broglie's theory. **(cf. C.U. 1999, 2008. K.U. 2004)**

4. Give an account of G.P. Thomson's experiment on electron diffraction and derive an expression for the wavelength associated with the moving electrons taking into account the relativistic increase of mass of the electron with its speed.

5. (a) State and explain Heisenberg's uncertainty principle. **(C.U. 1985, 94, 2004)**
   (b) Explain why free electrons cannot exist in atomic nuclei. **(C.U. 1999)**
   (c) What is the correspondence principle? **(C.U. 1985, Burd. U. 1999)**

6. (a) Describe the thought experiment with gamma ray microscope. Could the experiment be done with optical microscope? **(C.U. 1987)**
   (b) Can such an experiment be actually performed? Explain. **(C.U. 2006)**

7. (a) What is the wavelength of the wave associated with an electron having kinetic energy of 100 eV?
   (Nagpur Univ., 1980) [Ans. 1.226 Å]
   (b) Calculate de Broglie's wavelength of electrons accelerated through a potential difference of 150 volt. Given: $h = 6.62 \times 10^{-27}$ erg sec, $m = 9.11 \times 10^{-28}$ gm, $e = 4.8 \times 10^{-10}$ esu.
   (C.U. 1986) [Ans. 1.001 Å]

8. The mass of an oxygen molecule is 5.4×10^{-26} kg. If this molecule moves with a speed of 500 m/s, calculate the de Broglie wavelength of the wave associated with the molecule. [Ans. 0.245 Å]

9. The lengths of the matter waves associated with a moving electron and a proton are each equal to 1Å. Calculate their momenta and kinetic energies. Assume that the mass of a proton is 1835 times that of an electron. **(cf.C.U. 1989)**
   [Ans. 6.62×10^{-24} kg.m/s, 2.37×10^{-17} joule and 1.29×10^{-20} joule]

10. An electron with a kinetic energy of 2000 eV is confined to a region of atomic dimensions 10^{-10} m. Find the uncertainty in its linear momentum and compare this value with the momentum of the electron. [Ans. $\Delta p = 1.05 \times 10^{-24}$ kg. m/s, about 4% of p]

11. In an experiment, an elementary particle is observed at intervals as short as 10^{-7}s. Find the minimum uncertainty in energy of the particle, expressed in electron volt. [Ans. 0.66×10^{-8} eV]

Introductory Ideas 25

12. The uncertainty in the energy of a particle in an atomic nucleus is about 100 MeV. Find the minimum uncertainty of its location in time. If the particle is assumed to move with essentially the speed c, what is the uncertainty in its position ? **[Ans. 0.66×10^{-23} s; 2×10^{-15} m]**

13. The average period that elapses between the excitation of an atom and the time it emits radiation is 10^{-8} sec. Find the uncertainty in the energy emitted and the uncertainty in the frequency of light emitted. (Given $\hbar = 1.054 \times 10^{-34}$ joule sec.) **(Nagpur Univ. 1981)**
[Ans. 1.054×10^{-26} joule; 1.59×10^{7} Hz]

14. What is the de Broglie wavelength of an automobile of mass 1000 kg moving with a velocity of 80 km/hour? Can you detect it? Explain your answer. **(C.U. 1984) [Ans. 2.98×10^{-38} m]**

15. What is the minimum uncertainty in the position of an electron moving with a speed of 3×10^9 cm/s ?
[Ans. 0.038Å]

[**Hint :** Take the greatest uncertainty in momentum Δp to be equal to $m_0 v$ where v is the given speed, *i.e.* 3×10^9 cm/s. Apply the uncertainty relation $\Delta p \, \Delta x = \hbar$, to obtain Δx, the minimum uncertainty in position. The relativistic correction is not very important here.]

16. Find the de Broglie wavelength of an electron travelling with the velocity $0.6c$.

$\left[\textbf{Hint: } \text{Apply the relation } \lambda = \dfrac{h}{m_0 v} \sqrt{1 - v^2/c^2} \right]$ **[Ans. 0.032 Å]**

17. Show that the ratio of the de Broglie wavelength to the Compton wavelength of a particle is $\sqrt{(c/v)^2 - 1}$.
(C.U. 1984)

[**Hint :** The de Broglie wavelength is $\dfrac{h}{m_0 v}\sqrt{1 - v^2/c^2}$ and the Compton wavelength is $h/(m_0 c)$]

18. The mean life time of the excited state of a certain nucleus emitting γ-rays is 5×10^{-9} s. Estimate the width of the distribution in the energies (in eV) of the emitted photons. **(C.U. 1985)**
[Ans. 1.32×10^{-7} eV]

19. An electron is confined to a box of length 10^{-8} cm. Calculate the minimum uncertainty in the measurement of its velocity. **(C.U. 1992) [Ans. 1.16×10^8 cm/s]**

20. A meson has a lifetime of 10^{-23} s. To what accuracy can its mass be known? **(C.U. 1994)**
[Ans. 1.17×10^{-28} kg]

21. Calculate the number of photons emerging per second from a 50 W source, if the wavelength is 500 nm. **[Ans. 1.26×10^{20} s^{-1}]**

22. A neutron takes 1.0 ms to traverse a distance of 20 m. Obtain its velocity, kinetic energy, and de Broglie wavelength. (Rest mass of the neutron = 940 MeV/c^2). **(C.U. 1995)**
[Ans. 2×10^4 m/s, 2.09 eV, 0.0198 nm]

23. An electron has de Broglie wavelength of 0.15 Å. Compute the phase and the group velocities of the de Broglie waves. Find the kinetic energy of the electron. (Rest mass of the electron = 0.511 MeV/c^2).
(C.U. 1995) [Ans. $v_g = 4.86 \times 10^7$ m/s, $v_p = v_g/2$ (since $v_g \ll c$), 6700 eV]

24. Calculate the de Broglie wavelength of electrons of energy 10^4 eV and compare it with the wavelength of electromagnetic radiation for which the photon has the same energy. **(C.U. 1997)**
[Ans. 0.0123 nm, 0.1242 nm]

25. An electron of energy 200eV is passed through a circular hole of radius 10^{-4} cm. What is the uncertainty introduced in the angle of emergence? What would be the corresponding uncertainty for a 0.1 gm lead ball thrown with a velocity of 10^3 cm/s through a hole 1 cm in radius? **(C.U. 1997, 2004)**
[Ans. 1.38×10^{-5} rad, 1.054×10^{-29} rad]

26. Find the de Broglie wavelength of an electron in the ground state of the hydrogen atom, given that the ionization potential of hydrogen is 13.6 eV. (Take the kinetic energy as half the magnitude of the potential energy.) **(C.U. 2002) [Ans. 0.33 nm]**

27. What is the Compton effect? Show that

$$h\nu' = \dfrac{h\nu}{1 + \varepsilon(1 - \cos\theta)}$$

where $h\nu'$ and $h\nu$ are the energies of the scattered and the incident photons, respectively, and $\varepsilon = h\nu/(m_0 c^2)$, m_0 being the rest mass of the electron, θ is the scattering angle of the photon.

If the energy of the incident photon is 1.22 MeV and that of the scattered one is 0.511 MeV, what is the photon scattering angle? **(C.U. 2000) [Ans. 65.2°]**

28. In connection with the Compton effect, answer the following:
    (a) What would you observe, if the electrons are bound in atoms? **(C.U. 2002)**
    (b) What is the Compton wavelength?
    (c) If the free electrons are not initially at rest, but in motion, what would happen?
    (d) What is the maximum angle at which an electron can scatter?
    (e) Will there be any Compton shift for light in the visible range?
    (f) When is the kinetic energy of the recoil electron a maximum? What is this maximum energy?
    (g) Explain the presence of the unmodified line in Compton scattering. **(C.U. 2005)**
    (h) Explain why Compton effect is dominant in scattering by lighter elements. **(C.U. 2007)**
    [Ans. In lighter elements, there are lesser number of bound electrons. So, the possibility of scattering from a free electron, producing Compton effect, is greater.]

29. (a) What is Compton scattering and what is its importance?
    (b) Show that the wavelength shift in the Compton effect is $\delta\lambda = 2\lambda_c \sin^2(\phi/2)$, the symbols being standard.
    (c) Derive the relationship between the angles of scattering of the photon and the electron in the Compton effect. Also obtain an expression for the kinetic energy of the recoil electron.

    (C.U. 2005)

    (d) How would you study experimentally the Compton effect?

30. (a) If the radiation scattered in the Compton effect is viewed at 90° to the incident radiation, what is the Compton wavelength shift? **(C.U. 1996) [Ans. 0.024Å]**
    (b) A photon of energy 110 keV suffers a wavelength change of 10% on being scattered by a stationary free electron. Find the kinetic energy of the recoil electron. **(C.U. 1985) [Ans. 10 keV]**
    (c) A 0.3 MeV X-ray photon makes a head-on collision with an electron initially at rest and is scattered backward. What is the recoil velocity of the electron? **[Ans. 0.65 c]**

31. Calculate the de Braglie wavelength of a neutron of energy 28.8 eV. Given, $h = 6.62 \times 10^{-34}$ J.s and mass of a neutron $= 1.67 \times 10^{-27}$ kg. **(Agra 1998) [Ans. 4.2Å]**

32. The interatomic distance D of a nickel crystal is 2.15Å. When electrons of kinetic energy 54 eV are scattered by this crystal, the principal maximum occurs at a colatitude angle of 50°. Show that the data are in conformity with the de Broglie relation. **(cf. Garhwal, 1996)**
    [Hint: Refer to Davission and Germer experiment.]

33. Find the wavelength of the softest X-radiation which produces recoil electrons of energy 5 KeV in a Compton scattering experiment. **(C.U. 2006)**
    [Hint. Here $\phi = 180°$] **[Ans. 0.032 nm]**

Chapter 2

The Schrödinger Wave Equation

In this chapter we shall consider in a quantitative manner the properties of the one-dimensional wave function $\psi(x, t)$ by first developing the Schrödinger wave equation. Some useful quantum mechanical operators and their properties are also briefly discussed in this chapter. Note that the branch of quantum mechanics in which the behaviour of microscopic systems are studied using Schrödinger's equation is known as *wave mechanics*.

2.1. Development of Schrödinger's Wave Equation

A wave function $\psi(x, t)$ that describes a particle of entirely unknown position moving in the positive x-direction with precisely known momentum p and kinetic energy E may assume any one of the following forms:

$$\sin(kx - \omega t), \cos(kx - \omega t), e^{i(kx - \omega t)}, e^{-i(kx - \omega t)} \qquad ...(2.1)$$

or some linear combination of them. All of these forms represent continuous harmonic waves.

We shall consider an equation that will give both harmonic and more complicated waves as solutions. Such an equation must have two fundamental properties: (*a*) it must be linear so that its solutions on superposition will produce the effects of interference and will permit the construction of wave packets, (*b*) the equation will contain constant coefficients such as $\hbar$ and mass and charge of the particle, and will not contain the parameters like momentum, energy, propagation number, and frequency as coefficients; otherwise, the equation will describe a particular nature of motion of the particle. The latter requirement, when fulfilled, provides the possibility of superposing solutions that result for different values of these parameters.

Consider the most familiar one-dimensional equation for wave motion

$$\frac{\partial^2 \psi}{\partial t^2} = \gamma \frac{\partial^2 \psi}{\partial x^2}, \qquad ...(2.2)$$

where γ is the square of the wave velocity. Any of the forms (2.1), when put in Eq. (2.2), shows that each of the four harmonic waves and hence any linear combination of them will be the solution of the differential equation [Eq. (2.2)] provided

$$\gamma = \frac{\omega^2}{k^2} = \frac{4\pi^2 \nu^2}{k^2}, \qquad ...(2.3)$$

since $\omega = 2\pi\nu$, ν being the frequency.

Using Eqs. (1.3) and (1.1) we obtain from Eq. (2.3)

$$\gamma = \frac{E^2}{p^2} = \frac{p^2}{4m^2}, \qquad \text{[since } \hbar = h/(2\pi)\text{]} \qquad ...(2.4)$$

where m is the mass of the particle whose motion is described by Eq. (2.2). However, since the coefficient γ involves E and p, i.e., the parameters of the motion, we are forced to discard this differential equation.

From the forms (2.1) of the function we note that differentiation with respect to x performed on each of them has the general effect of multiplication of the function by k and sometimes interchanging cosine and sine, and differentiation with respect to t produces the general effect of multiplication by ω. Using the relation $E = \dfrac{p^2}{2m}$ and Eq. (1.3) we find that $\omega = \dfrac{\hbar k^2}{2m}$. This suggests at once that the desired differential equation must contain a second derivative with respect to x and a first derivative with respect to t. That is,

$$\frac{\partial \psi}{\partial t} = \gamma \frac{\partial^2 \psi}{\partial x^2} \qquad \ldots(2.5)$$

Substitution of the first two of the forms (2.1) of the wave function ψ indicates that they cannot be the solutions of Eq. (2.5). On the other hand, either of the last two (but not both simultaneously) may be considered as the solution for a proper choice of the constant γ. Thus, when

$$\gamma = \frac{i\omega}{k^2} = \frac{i\hbar E}{p^2} = \frac{i\hbar}{2m} \qquad \ldots(2.6)$$

is chosen we find that the third form of the wave function (2.1) satisfies Eq. (2.5). Also, the constant γ is obtained in terms of the constants $\hbar$ and m. Therefore, the *one-dimensional form of the Schrödinger wave equation for a free particle of mass m* is obtained by substituting the relation (2.6) in Eq. (2.5). That is,

$$i\hbar \frac{\partial \psi}{\partial t} = -\frac{\hbar^2}{2m} \frac{\partial^2 \psi}{\partial x^2} \qquad \ldots(2.7)$$

Evidently, the harmonic solution of (2.7) is the third form of the wave function (2.1), *i.e.* $e^{i(kx-\omega t)}$ which is complex. However, while expressing the predicted results of possible physical observations we have to use the real numbers.

Three-Dimensional Schrödinger's Equation:

Equation (1.3) in three dimensions may be written

$$\vec{p} = \hbar \vec{k}; \ k = \frac{2\pi}{\lambda}, \qquad \ldots(2.8)$$

where $\vec{k}$ is termed the *propagation vector*. Similarly, the function $e^{i(kx-\omega t)}$ in three dimensions can be written as

$$\exp[i(\vec{k} \cdot \vec{r} - \omega t)], \qquad \ldots(2.9)$$

where $\vec{r}$ represents the position vector of the particle. Now extending the reasoning required to derive Eq. (2.7) to the three dimensional case we obtain the three-dimensional Schrödinger wave equation for a free particle of mass m, *i.e.*

$$i\hbar \frac{\partial \psi}{\partial t} = -\frac{\hbar^2}{2m} \nabla^2 \psi \qquad \ldots(2.10)$$

where $\psi(\vec{r}, t)$ represents the wave function of the particle and ∇^2 is the Laplacian operator.

Substituting the form (2.9) for ψ in the left-hand side of Eq. (2.10) we get

$$i\hbar \frac{\partial \psi}{\partial t} = i\hbar(-i\omega)\psi = \hbar\omega\psi = E\psi \qquad \ldots(2.11\ a)$$

Similarly, the right-hand side of Eq. (2.10) gives

$$-\frac{\hbar^2}{2m}\nabla^2\psi = -\frac{\hbar^2}{2m}(ik)^2\psi = \frac{\hbar^2 k^2}{2m}\psi = \frac{p^2}{2m}\psi \qquad \ldots(2.11\ b)$$

The above results suggest that the energy and momentum of a free particle can be represented by differential operators that operate on the wave function ψ of the particle:

$$E \to i\hbar \frac{\partial}{\partial t} \quad \text{and} \quad \vec{p} \to -i\hbar \ \text{grad} \qquad \ldots(2.12)$$

The above energy and momentum operators are also valid for a particle which is not free.

The Schrödinger Wave Equation

Wave Equation in a Force Field: Consider that the particle moves under the influence of a force $\vec{F}$ that is derivable from a potential energy V. That is,

$$\vec{F}(\vec{r}, t) = - \text{grad } V(\vec{r}, t) \qquad ...(2.13)$$

Now the total energy H of the particle is the sum of the kinetic energy $(p^2/2m)$ and the potential energy $V(\vec{r}, t)$:

$$H = \frac{p^2}{2m} + V(\vec{r}, t) \qquad ...(2.14)$$

Since V is independent of $\vec{p}$ or H, the relations (2.12) and Eq. (2.14) put Eq. (2.10) into the following generalized form

$$i\hbar \frac{\partial \psi}{\partial t} = -\frac{\hbar^2}{2m} \nabla^2 \psi + V(\vec{r}, t) \psi. \qquad ...(2.15)$$

This is the *Schrödinger wave equation* describing the motion of a particle of mass m under the influence of a force $\vec{F}$ given by Eq. (2.13). Although this equation cannot claim a high degree of plausibility as Eq. (2.10), the agreement of solutions of Eq. (2.15) with experimental results in certain cases demonstrates its validity and usefulness.

2.2. PHYSICAL MEANING OF THE WAVE FUNCTION

The wave function $\psi(\vec{r}, t)$ is the solution of the Schrödinger wave equation and gives a complete quantum-mechanical description of the behaviour of a moving particle of mass m. The particle in question may be free or under the influence of a force. The function ψ cannot be measured directly by any physical experiment. However, the usual dynamical variables, such as position, momentum, potential energy, kinetic energy, etc. of the particle are obtained by performing suitable mathematical operations on the wave function ψ.

The most important property of ψ is that it gives a measure of the probability of finding a particle at a particular position. In general, the wave function ψ is a complex quantity whereas the probability must be real and positive. Therefore, a term called the *position probability density*, denoted by $P(\vec{r}, t)$ is defined. This is given by the product of the wave function ψ and its complex conjugate ψ^*. This is,

$$P(\vec{r}, t) = \psi^* \psi = |\psi(\vec{r}, t)|^2 \qquad ...(2.16)$$

$P(\vec{r}, t)\, d\tau$ is the probability of finding a particle in the elemental volume $d\tau$ about the point $\vec{r}$ at time t. Thus *the position probability density $P(r, t)$ is the probability of a particle lying in a unit volume about the point $\vec{r}$ at time t.*

2.3. NORMALIZATION OF THE WAVE FUNCTION

The probability that the particle will be found somewhere in the region is unity. This implies that the wavefunction be normalized so that

$$\int |\psi(\vec{r}, t)|^2 \, d\tau = 1, \qquad ...(2.17)$$

the integral being taken over the entire region. Equation (2.17) gives the condition of normalization, and the wavefunction which satisfies this condition is said to be normalized. The normalization is accomplished by adjusting the numerical coefficient of ψ so that the integral is unity. This however does not alter the fact that ψ is a solution of Eq. (2.10), which is homogeneous in ψ. However, wave functions similar to Eq. (2.9), when used in Eq. (2.17), do not give convergence, when integrated over an infinite volume. But, if we think that the region of space in which such a wave function is defined is arbitrarily large and finite, then Eq. (2.17) converges when integrated over the finite volume of this region. This suggests that normalization is always possible.

The coefficient of ψ that makes normalization possible must be time invariant so that ψ may satisfy the wave equation. Therefore, if Eq. (2.17) is satisfied at one instant of time, the position probability density given by $|\psi|^2$ requires that the normalization integral be constant in time.

Salient Features of Schrödinger's Equation and its Comparison with the Classical Wave Equation :

From the above discussion, we summarise below the salient features of Schrödinger's equation and compare it with the classical wave equation.

(*i*) Schrödinger's wave equation can describe the motion of a particle using the concept of wave-particle duality; the classical wave equation connot. In the wave-particle duality, the momentum $\vec{p}$ and the energy E are related to the wave vector $\vec{k}$ and the angular frequency ω by $\vec{p} = \hbar \vec{k}$ and $E = \hbar \omega$, where $\hbar$ is Planck's constant divided by 2π.

(*ii*) In one dimension, the classical wave equation for a physical quantity ϕ is

$$\frac{\partial^2 \phi}{\partial t^2} = v^2 \frac{\partial^2 \phi}{\partial x^2},$$

where v is the wave velocity. The Schrödinger equation for a free particle of mass m moving in a force-free region is

$$i\hbar \frac{\partial \psi}{\partial t} = -\frac{\hbar^2}{2m} \frac{\partial^2 \psi}{\partial x^2}$$

where ψ is the wave function.

(*iii*) The classical wave equation admits solutions like $\sin(kx - \omega t)$, $\cos(kx - \omega t)$, $e^{i(kx-\omega t)}$ and $e^{-i(kx-\omega t)}$, where $v = \omega/k$. The first two forms do not satisfy Schrödinger's equation for a free particle in a force-free region, but either of the last two forms can satisfy the same with $E = p^2/(2m) = \hbar^2 k^2/(2m)$.

(*iv*) ϕ is a physical quantity and is therefore real. But ψ can be a complex quantity. However, $\psi^*\psi$ is real and is the position probability density of the particle. Also, E and $\vec{P}$ can be represented by operators :

$$E \to i\hbar \frac{\partial}{\partial t}, \quad \vec{p} \to -i\hbar \vec{\nabla}$$

(*v*) When the particle moves in a force field, Schrödinger's equation in three dimensions is

$$i\hbar \frac{\partial \psi}{\partial t} = -\frac{\hbar^2}{2m} \nabla^2 \psi + V(\vec{r}, t)\psi,$$

where $V(\vec{r}, t)$ is the potential energy of the particle. This equation asserts that the total energy is the sum of the kinetic and potential energies.

2.4. Expectation Values

Dynamical quantities such as position, momentum, kinetic enegy, potential energy, etc. which form the objectives of measurement in experimental physics, are called *observables*. In quantum mechanics, each observable is represented by an *operator* which acts on the wave function ψ to produce a new function.

The *average* or the *expectation value* $\langle \alpha \rangle$ of any observable α is defined by

$$\langle \alpha \rangle = \int \psi^*(\vec{r}, t) \hat{\alpha} \psi(\vec{r}, t) \, d\tau, \qquad \ldots(2.18)$$

where $\hat{\alpha}$ is the operator associated with the observable α, and the integral is taken over the entire volume.

The expectation value of the position vector $\vec{r}$ of a particle is given by

$$\langle \vec{r} \rangle = \int \psi^*(\vec{r}, t) \vec{r} \psi(\vec{r}, t) \, d\tau \qquad \ldots(2.19)$$

The Schrödinger Wave Equation

The expectation value of the position vector is a vector and is a function of time only because the space coordinates disappear after the integration is carried out. Equation (2.19) is again equivalent to three component equations

$$\langle x \rangle = \int \psi^* x \psi d\tau, \quad \langle y \rangle = \int \psi^* y \psi d\tau, \quad \langle z \rangle = \int \psi^* z \psi d\tau \quad \ldots(2.20)$$

Similarly, the expectation value of the potential energy V is given by

$$\langle V \rangle = \int \psi^* (\vec{r}, t) \, V(\vec{r}, t) \, \psi(\vec{r}, t) \, d\tau \quad \ldots(2.21)$$

The expectation values of the energy and the momentum of a particle are given respectively by

$$\langle E \rangle = \int \psi^* \, i\hbar \frac{\partial \psi}{\partial t} \, d\tau \quad \text{and} \quad \langle \vec{p} \rangle = \int \psi^* (-i\hbar) \, \text{grad } \psi \, d\tau, \quad \ldots(2.22)$$

since the differential operators of the energy and momentum of the particle are $i\hbar \dfrac{\partial}{\partial t}$ and $-i\hbar$ grad, respectively [see Eq. (2.12)]. The second of Eq. (2.22) is equivalent to the three component equations:

$$\langle p_x \rangle = -i\hbar \int \psi^* \frac{\partial \psi}{\partial x} d\tau,$$

$$\langle p_y \rangle = -i\hbar \int \psi^* \frac{\partial \psi}{\partial y} d\tau, \quad \ldots(2.23)$$

and

$$\langle p_z \rangle = -i\hbar \int \psi^* \frac{\partial \psi}{\partial z} d\tau$$

For one-dimensional cases restricted to x-, y-, and z- axes only, $d\tau$ in Eqs. (2.20) and (2.23) is replaced by dx, dy, and dz, respectively.

2.5. THE EHRENFEST THEOREM

Ehrenfest's theorem states that *in quantum mechanics the expectation or average values of observables behave in the same manner as the observables themselves do in classical mechanics*. This theorem furnishes an example of the correspondence principle.

Consider, for simplicity, a one-dimensional case where the dynamical system is confined to the x-axis only. The veocity is the rate of change of the expectation value of the position coordinate x with respect to time t. Therefore,

$$\frac{d\langle x \rangle}{dt} = \frac{d}{dt} \int_{-\infty}^{\infty} \psi^*(x,t) \, x \psi(x,t) \, dx$$

$$= \int_{-\infty}^{\infty} \psi^* x \frac{\partial \psi}{\partial t} dx + \int_{-\infty}^{\infty} \frac{\partial \psi^*}{\partial t} x \psi \, dx \quad \ldots(2.24)$$

The value of $\dfrac{\partial \psi}{\partial t}$ is obtained from the one-dimensional form of Eq. (2.15). The quantity $\dfrac{\partial \psi^*}{\partial t}$ is also obtained from the Schrödinger wave equation as follows:

The wave function can be written as the sum of its real and imaginary parts;

$$\psi(x,t) = u(x,t) + iv(x,t) \quad \ldots(2.25)$$

Substituting this in Eq. (2.15) and equating the real and the imaginary parts we get

$$-\hbar \frac{\partial v}{\partial t} = -\frac{\hbar^2}{2m} \nabla^2 u + Vu \quad \ldots(2.26)$$

and

$$\hbar \frac{\partial u}{\partial t} = -\frac{\hbar^2}{2m} \nabla^2 v + Vv \quad \ldots(2.27)$$

Multiplying Eq. (2.27) by $-i$ and adding to Eq. (2.26) we get a wave equation for $\psi^*(= u - iv)$:

$$-i\hbar \frac{\partial \psi^*}{\partial t} = -\frac{\hbar^2}{2m} \nabla^2 \psi^* + V\psi^* \qquad \ldots(2.28)$$

In one dimension, we have

$$-i\hbar \frac{\partial \psi^*}{\partial t} = -\frac{\hbar^2}{2m} \frac{\partial^2 \psi^*}{\partial x^2} + V\psi^* \qquad \ldots(2.28a)$$

Putting the values of $\frac{\partial \psi}{\partial t}$ and $\frac{\partial \psi^*}{\partial t}$ in Eq. (2.24) we get

$$\frac{d}{dt}\langle x \rangle = -\frac{i}{\hbar} \left[\int_{-\infty}^{\infty} \psi^* x \left(-\frac{\hbar^2}{2m} \frac{\partial^2 \psi}{\partial x^2} + V\psi \right) dx - \int_{-\infty}^{\infty} \left(-\frac{\hbar^2}{2m} \frac{\partial^2 \psi^*}{\partial x^2} + V\psi^* \right) x\psi \, dx \right]$$

$$= \frac{i\hbar}{2m} \int_{-\infty}^{\infty} \left[\psi^* x \frac{\partial^2 \psi}{\partial x^2} - \frac{\partial^2 \psi^*}{\partial x^2} x\psi \right] dx \qquad \ldots(2.29)$$

Integrating by parts, the integral on the right-hand side of Eq. (2.29) reduces to

$$\left[\psi^* x \frac{\partial \psi}{\partial x} \right]_{-\infty}^{\infty} - \int_{-\infty}^{\infty} \left[\frac{\partial \psi^*}{\partial x} x \frac{\partial \psi}{\partial x} + \psi^* \frac{\partial \psi}{\partial x} \right] dx - \left[\frac{\partial \psi^*}{\partial x} x\psi \right]_{-\infty}^{\infty}$$

$$+ \int_{-\infty}^{\infty} \left[\frac{\partial \psi^*}{\partial x} x \frac{\partial \psi}{\partial x} + \frac{\partial \psi^*}{\partial x} \psi \right] dx$$

Since both ψ and $\frac{\partial \psi}{\partial x}$ vanish as $x \to \infty$ or $x \to -\infty$, the above integral reduces to

$$\int_{-\infty}^{\infty} \left[\frac{\partial \psi^*}{\partial x} \psi - \psi^* \frac{\partial \psi}{\partial x} \right] dx$$

Again, $\int_{-\infty}^{\infty} \frac{\partial \psi^*}{\partial x} \psi \, dx = [\psi^* \psi]_{-\infty}^{\infty} - \int_{-\infty}^{\infty} \psi^* \frac{\partial \psi}{\partial x} dx = -\int_{-\infty}^{\infty} \psi^* \frac{\partial \psi}{\partial x} dx$

Therefore we obtain from Eq. (2.29)

$$\frac{d}{dt}\langle x \rangle = \frac{i\hbar}{2m} \left[-2 \int_{-\infty}^{\infty} \psi^* \frac{\partial \psi}{\partial x} dx \right] = -\frac{i\hbar}{m} \int_{-\infty}^{\infty} \psi^* \frac{\partial \psi}{\partial x} dx = \frac{\langle p_x \rangle}{m} \qquad \ldots(2.30)$$

That is, the velocity is equal to the expectation value of the momentum divided by the mass. This corresponds to the classical equation $dx/dt = p_x/m$.

In a similar way, we can calculate the time rate of change of x component of the momentum as follows:

$$\frac{d}{dt}\langle p_x \rangle = -i\hbar \frac{d}{dt} \int_{-\infty}^{\infty} \psi^*(x, t) \frac{\partial}{\partial x} \psi(x, t) \, dx$$

$$= -i\hbar \left(\int_{-\infty}^{\infty} \psi^* \frac{\partial}{\partial x} \frac{\partial \psi}{\partial t} dx + \int_{-\infty}^{\infty} \frac{\partial \psi^*}{\partial t} \frac{\partial \psi}{\partial x} dx \right) \qquad \ldots(2.31)$$

Putting the values of $\frac{\partial \psi}{\partial t}$ and $\frac{\partial \psi^*}{\partial t}$ we get

$$\frac{d}{dt}\langle p_x\rangle = -\frac{\hbar^2}{2m}\int_{-\infty}^{\infty}\left[\frac{\partial^2\psi^*}{\partial x^2}\frac{\partial\psi}{\partial x} - \psi^*\frac{\partial^3\psi}{\partial x^3}\right]dx$$

$$+ \int_{-\infty}^{\infty}\left[V\psi^*\frac{\partial\psi}{\partial x} - \psi^*\frac{\partial}{\partial x}(V\psi)\right]dx$$

$$= -\frac{\hbar^2}{2m}\int_{-\infty}^{\infty}\left[\frac{\partial}{\partial x}\left(\frac{\partial\psi}{\partial x}\frac{\partial\psi^*}{\partial x} - \psi^*\frac{\partial^2\psi}{\partial x^2}\right)\right]dx$$

$$- \int_{-\infty}^{\infty}\psi^*\frac{\partial V}{\partial x}\psi\,dx$$

$$= -\frac{\hbar^2}{2m}\left[\frac{\partial\psi^*}{\partial x}\frac{\partial\psi}{\partial x} - \psi^*\frac{\partial^2\psi}{\partial x^2}\right]_{-\infty}^{\infty} - \int_{-\infty}^{\infty}\psi^*\frac{\partial V}{\partial x}\psi\,dx$$

$$= -\int_{-\infty}^{\infty}\psi^*\frac{\partial V}{\partial x}\psi\,dx = \left\langle -\frac{\partial V}{\partial x}\right\rangle = \langle F_x\rangle, \qquad \ldots(2.32)$$

which is Newton's second law of motion : $\frac{dp_x}{dt} = F_x$. Thus we find that the classical mechanics agrees with the quantum mechanics so far as the expectation values are concerned. However, the measured values of the observables fluctuate over their expectation values. This is the characteristic feature of the quantum theory.

2.6. Time-Independent and Time-Dependent Schrödinger's Equation

The Schrödinger wave equation [Eq. 2.15)] can be solved easily when the potential energy $V(\vec{r})$ is assumed to be time invariant. The solutions may be obtained by separating the variables. Let us assume a particular solution of Eq. (2.15) as

$$\psi(\vec{r}, t) = u(\vec{r})f(t), \qquad \ldots(2.33)$$

where u depends on the position vector $\vec{r}$ only, and f on time t only. The general solution is then equal to the sum of such separated solutions. Differentiating (2.33) and substituting back into Eq. (2.15) we get

$$i\hbar u\frac{df}{dt} = f\left[-\frac{\hbar^2}{2m}\nabla^2 u + V(\vec{r})u\right] \qquad \ldots(2.34)$$

Dividing Eq. (2.34) by $u(r)f(t)$ we obtain

$$i\hbar\left(\frac{1}{f}\frac{df}{dt}\right) = \frac{1}{u}\left[-\frac{\hbar^2}{2m}\nabla^2 u + V(\vec{r})u\right] \qquad \ldots(2.35)$$

Evidently, the left-hand side does not vary with $\vec{r}$ while the right-hand side does not vary with t. The equation can only be satisfied for all $\vec{r}$ and t if each side of the equation is equal to a constant, called a separation constant, and denoted by E. Therefore the equations for f and u become

$$i\hbar\frac{df}{dt} = Ef \qquad \ldots(2.36)$$

and

$$\left[-\frac{\hbar^2}{2m}\nabla^2 + V(\vec{r})\right]u(\vec{r}) = Eu(\vec{r}) \qquad \ldots(2.37)$$

The integration of Eq. (2.36) yields

$$f(t) = Ce^{-iEt/\hbar}$$

where C is an arbitrary constant. It is seen that Eq. (2.37) is homogeneous in u; therefore, the value of the constant C can be chosen to normalize u. The particular solution of the wave equation is then given by

$$\psi(\vec{r}, t) = u(\vec{r}) e^{-iEt/\hbar} \qquad ...(2.38)$$

Equation (2.37) is termed the *time-independent Schrödinger equation* and the solutions $u(\vec{r})$ of this equation are called the *time-independent wave functions*.

When the time-derivative operator in Eq. (2.12) representing the total energy acts on ψ of Eq. (2.38) we obtain

$$i\hbar \frac{\partial \psi}{\partial t} = E\psi \qquad ...(2.39)$$

Equation (2.39) is the *time-dependent Schrödinger equation*. Clearly, the separation constant E is identified as the total energy.

2.7. QUANTUM MECHANICAL OPERATORS

We have defined in Sec. 2.4 the terms observable and operator. Usually, an operator is represented by placing a $\wedge$ ('hat') on the top of the letter or symbol that is used to denote the corresponding observable. Thus, if an observable is denoted by the letter a the corresponding operator will be represented by $\hat{a}$.

The operators in quantum mechanics are *linear*. The definition of a linear operator, say $\hat{l}$, is that it must obey the rules;

$$\hat{l}(\psi_1 + \psi_2) = \hat{l}\psi_1 + \hat{l}\psi_2 \qquad ...(2.40)$$

$$\hat{l}(c\psi_1) = c(\hat{l}\psi_1),$$

where ψ_1 and ψ_2 are two given functions, and c is complex constant.

Here $\hat{l}\psi$ means that $\hat{l}$ is applied to ψ.

The equality of two operators $\hat{l}$ and $\hat{m}$, i.e., $\hat{l} = \hat{m}$ means that they generate identical result when they are applied to an arbitrary wave function ψ. The sum $\hat{l} + \hat{m}$ or the difference $\hat{l} - \hat{m}$ of the two operators satisfies the equation

$$(\hat{l} \pm \hat{m})\psi = \hat{l}\psi \pm \hat{m}\psi \qquad ...(2.41)$$

Clearly, $\hat{l} + \hat{m} = \hat{m} + \hat{l}$ and $\hat{l} - \hat{m} = -\hat{m} + \hat{l}$, i.e., the *commutative law* holds for the addition and subtraction of operators.

The rule of applying the product $\hat{l}\hat{m}$ of two operators on a function is that $\hat{m}$ will operate on the given function first, and then $\hat{l}$ on the result. That is,

$$(\hat{l}\hat{m})\psi = \hat{l}(\hat{m}\psi) \qquad ...(2.42)$$

Similarly, $\hat{l}(\hat{m}\hat{n})$ means that $\hat{n}$ will act first, then $\hat{m}$, and finally $\hat{l}$. Equation (2.42) indicates that multiplication of linear operators is *associative*.

We may also write

$$\hat{l}(\hat{m} + \hat{n}) = (\hat{l}\hat{m}) + (\hat{l}\hat{n}) \qquad ...(2.43)$$

This equation shows that the *distributive law* is valid for the multiplication of linear operators. Here both sides operate on ψ to give $\hat{l}(\hat{m}\psi) + \hat{l}(\hat{n}\psi)$.

The Schrödinger Wave Equation   35

The operators $\hat{l}$ and $\hat{m}$, and the corresponding observables are said to *commute* if

$$\hat{m}\hat{l} = \hat{l}\hat{m} \qquad ...(2.44)$$

Eigenfunction and Eigenvalue: An operator, say $\hat{l}$, applied to a particular function, say u_s, may generate a new function which will differ from u_s only by a constant multiplicative factor l_s. That is

$$\hat{l} u_s = l_s u_s \qquad ...(2.45)$$

Obviously, a measurement of the observable quantity l certainly gives the numerical value l_s. The particle is said to be in an *eigenstate* (eigen ≡ own, characteristic) of l with the eigenvalue l_s. The function u_s is referred to as the *eigenfunction* of the operator $\hat{l}$. The quantity l_s is called the *eigenvalue* of the operator $\hat{l}$.

We have the time-independent Schrödinger equation [Eq. (2.37)]:

$$\left[-\frac{\hbar^2}{2m}\nabla^2 + V(\vec{r})\right] u(\vec{r}) = E u(\vec{r}).$$

This is an eigenvalue equation. Here the operator $\left[-\frac{\hbar^2}{2m}\nabla^2 + V(\vec{r})\right]$ representing the total energy operates on the function $u(\vec{r})$ to generate a constant E multiplied by $u(\vec{r})$. The various solutions of $u(\vec{r})$ are the energy eigenfunctions and the corresponding E are energy eigenvalues.

An energy eigenfunction similar to ψ in Eq. (2.38), is said to represent a *stationary state* of the particle because the position probability density $\psi^*\psi = u^*u$ is constant in time.

Equation (2.37) does not put any restriction on the value of E, since for each value of E there exists a value of u that satisfies this equation. The boundary conditions imposed on $u(\vec{r})$ however dictate the energy eigenvalues E which are of physical interest.

Hermitian Operator: If the operator $\hat{l}$ is such that

$$\int \psi_1^* (\hat{l}\psi_2) \, d\tau = \int (\hat{l}\psi_1)^* \psi_2 \, d\tau \qquad ...(2.46)$$

where ψ_1 and ψ_2 are two arbitrary wave functions, and the integrals are taken over the entire region of space available to the particle, the operator $\hat{l}$ is termed *Hermitian*. The *eigenvalues of Hermitian operators are real*, for if ψ_1 and ψ_2 are both replaced by u_s, an eigenfunction of $\hat{l}$, we obtain

$$\int u_s^* (\hat{l} u_s) \, d\tau = \int u_s^* (l_s u_s) \, d\tau = l_s \int u_s^* u_s \, d\tau = l_s$$
$$\int (\hat{l} u_s)^* u_s \, d\tau = \int (l_s u_s)^* u_s \, d\tau = l_s^* \int u_s^* u_s \, d\tau = l_s^* \qquad ...(2.47)$$

Thus $l_s = l_s^*$ and this necessitates l_s to be real.

Hamiltonian Operator: The classical Hamiltonian expression for the total energy is

$$H = \frac{p^2}{2m} + V \qquad ...(2.48)$$

where $\frac{p^2}{2m}$ represents the kinetic energy of the particle of mass m and momentum p, and V is the potential energy of the particle.

For a free particle the potential energy may be taken to be zero everywhere. So the Hamiltonian for a free particle consists only of the kinetic energy term $p^2/(2m)$.

The operator for the momentum p is $-i\hbar$ grad [see Eq. (2.12)]. Therefore the operator for the kinetic energy is

$$\hat{T} = -\frac{\hbar^2}{2m}\nabla^2$$

Similarly, the potential energy V can be represented by an operator $\hat{V}$ which indicates that a wavefunction is to be multiplied by V.

The total energy of the particle is $T + V$. Thus the operator representing the total energy is

$$\hat{H} = \hat{T} + \hat{V} = -\frac{\hbar^2}{2m}\nabla^2 + \hat{V} \qquad \ldots(2.49)$$

The Schrödinger equation is then

$$\hat{H}\psi = i\hbar\frac{\partial\psi}{\partial t} \qquad \ldots(2.50)$$

$\hat{H}$ is called the *Hamiltonian operator* in quantum mechanics.

For the case of a free particle the Hamiltonian operator is the kinetic energy operator $\hat{T}$:

$$\hat{T} = -\left(\frac{\hbar^2}{2m}\right)\nabla^2$$

Mathematical Properties of Operators: The position operator $\hat{\vec{r}}$ and the potential energy operator $\hat{V}$ are Hermitian, since both are multiplicative and real. This is evident from the following example:

$$\int \psi_1^*(\hat{\vec{r}}\psi_2)\,d\tau = \int \psi_1^*\,\vec{r}\,\psi_2\,d\tau$$

$$= \int (\vec{r}\,\psi_1)^*\psi_2\,d\tau = \int (\hat{\vec{r}}\,\psi_1)^*\psi_2\,d\tau \qquad \ldots(2.51)$$

The operators $\hat{p}$ and $\hat{T}$ can also be shown to be Hermitian. In one dimension we have $\hat{p} = -i\hbar\frac{\partial}{\partial x}$. Therefore

$$\int_{x_1}^{x_2} \psi_1^*(\hat{p}\psi_2)\,dx = \int_{x_1}^{x_2} \psi_1^*\left(-i\hbar\frac{\partial\psi_2}{\partial x}\right)dx$$

$$= \int_{x_1}^{x_2}\left(-i\hbar\frac{\partial\psi_1}{\partial x}\right)^*\psi_2\,dx - i\hbar\left[\psi_1^*\psi_2\right]_{x_1}^{x_2} \qquad \ldots(2.52)$$

In a similar way, $\hat{T} = \left(-\frac{\hbar^2}{2m}\right)\frac{\partial^2}{\partial x^2}$ in one dimension. Therefore, after two integrations we get

$$\int_{x_1}^{x_2} \psi_1^*(\hat{T}\psi_2)\,dx = \int_{x_1}^{x_2}\psi_1^*\left(-\frac{\hbar^2}{2m}\frac{\partial^2\psi_2}{\partial x^2}\right)dx$$

$$= \int_{x_1}^{x_2}\left(-\frac{\hbar^2}{2m}\frac{\partial^2\psi_1}{\partial x^2}\right)^*\psi_2\,dx - \frac{\hbar^2}{2m}\left[\psi_1^*\frac{\partial\psi_2}{\partial x} - \frac{\partial\psi_1^*}{\partial x}\psi_2\right]_{x_1}^{x_2} \qquad \ldots(2.53)$$

In the above it has been assumed that the particle lies between x_1 and x_2. Evidently, $\hat{p}$ and $\hat{T}$ are Hermitian if the second term of each of Eqs. (2.52) and (2.53) can be discarded. These terms do vanish for localized particles so that the wave functions vanish as $x_1 \to \infty$ and $x_2 \to \infty$, or when periodic boundary conditions are used: $\psi(x_1) = \psi(x_2)$ and $\left(\frac{\partial\psi}{\partial x}\right)_{x_1} = \left(\frac{\partial\psi}{\partial x}\right)_{x_2}$. Since the operators $\hat{T}$ and $\hat{V}$ are Hermitian, the Hamiltonian operator $\hat{H}$ is also so.

The one-dimensional position and momentum operators of a particle do not commute as can be

The Schrödinger Wave Equation

seen below.

$$(\hat{p}\hat{x})\psi = -i\hbar \frac{\partial}{\partial x}(x\psi) = -i\hbar x \frac{\partial \psi}{\partial x} - i\hbar \psi = (\hat{x}\hat{p} - i\hbar)\psi$$

or
$$\hat{x}\hat{p} - \hat{p}\hat{x} = i\hbar \qquad ...(2.54)$$

The quantity $\hat{l}\hat{m} - \hat{m}\hat{l}$ defines the commutation of two operators $\hat{l}$ and $\hat{m}$ and is written as $[\hat{l}, \hat{m}]$. Therefore, we can write

$$[\hat{x}, \hat{p}] = i\hbar \qquad ....(2.55)$$

Clearly, $\quad [\hat{l}, \hat{m}] = \hat{l}\hat{m} - \hat{m}\hat{l} = -(\hat{m}\hat{l} - \hat{l}\hat{m}) = -[\hat{m}, \hat{l}]$

Compatible Observables and Commutation

When the determination of an observable introduces an uncertainty in another observable, the two observables are said to be *incompatible*. The position and the momentum of a particle are thus incompatible. The observables that can be simultaneously measured precisely without influencing each other are termed *compatible*.

Let $\hat{l}$ and $\hat{m}$ be two operators representing the observables l and m, respectively. If l_s and m_s are the eigenvalues of $\hat{l}$ and $\hat{m}$, respectively, and u_s is the corresponding eigenfunction, measurements of l and m certainly give the values l_s and m_s, respectively, with the system in the state u_s. Thus l and m can be simultaneously measured precisely, and are compatible. We have

$$\hat{l} u_s = l_s u_s, \quad \hat{m} u_s = m_s u_s$$

So,
$$\hat{l}\hat{m} u_s = m_s \hat{l} u_s = m_s l_s u_s \qquad ...(2.55a)$$

and
$$\hat{m}\hat{l} u_s = l_s \hat{m} u_s = l_s m_s u_s \qquad ...(2.55b)$$

Subtracting Eq. (2.55b) from Eq. (2.55a), we get

$$(\hat{l}\hat{m} - \hat{m}\hat{l}) u_s = 0 \qquad ...(2.55c)$$

Since $u_s \neq 0$, Eq. (2.55c) yields

$$[\hat{l}, \hat{m}] = 0 \qquad ...(2.55d)$$

As the commutator vanishes, $\hat{l}$ and $\hat{m}$ commute. Thus compatible observables are represented by commutating operators.

The incompatibility of two observables implies that the corresponding operators do not commute. Thus the noncommutation of the position operator $\hat{x}$ and the momentum operator $\hat{p}$ of a particle, expressed by Eq. (2.55), shows that simultaneous precise measurements of position and momentum are not possible. The appearance of $\hbar$ in Eq. (2.54) suggests that the uncertainty in the simultaneous measurements in x and p is governed by Heisenberg's uncertainty relation. The possible incompatibility of observables is a quantum phenomenon; it does not appear in the classical limit.

Note that the noncommutation of the position and momentum operators occurs when the position and momentum vectors are in the same direction. If they are in different directions, the operators commute. Thus $[\hat{x}, \hat{p}_x] = i\hbar$, but $[\hat{y}, \hat{p}_x] = 0$, $[\hat{x}, \hat{p}_z] = 0$ etc. In general, one can write

$$[\hat{r}_j, \hat{p}_k] = i\hbar \delta_{jk}$$

where the subscripts j and k refer to the x, y, and z components of the position vector $\vec{r}$ and the momentum vector $\vec{p}$, and δ_{jk} is the *Kronecker delta function*:

$$\delta_{jk} = 1 \text{ when } j = k \text{ and } \delta_{jk} = 0 \text{ when } j \neq k.$$

A few properties of Hermitian Operators

(*i*) The sum or difference of two Hermitian operators is a Hermitian operator.

(*ii*) The product of two commuting Hermitian operators is Hermitian. For example, the operators $\hat{x} = x$ and $\hat{p}_x = -i\hbar \dfrac{\partial}{\partial x}$ do not commute. So, the operator $\hat{x}\hat{p}_x$ or $\hat{p}_x\hat{x}$ is not a Hermitian operator.

The quantities whose operators are Hermitian are referred to as *dynamical variables*. $\hat{x}$ and $\hat{p}_x$ being Hermitian, x and p_x are dynamical variables. But the product xp_x is not a dynamical variable since $\hat{x}$ and $\hat{p}_x$ do not commute.

Let $\hat{l}$ and $\hat{m}$ be two commuting Hermitian operators. Consider the integral $\int \psi_1^* \hat{l}\hat{m}\psi_2 \, d\tau$, where ψ_1 and ψ_2 are two arbitrary wavefunctions. Since $\hat{l}$ is a Hermitian operator

$$\int \psi_1^* \hat{l}\hat{m}\psi_2 \, d\tau = \int \psi_1^* \hat{l}(\hat{m}\psi_2) d\tau = \int (\hat{l}\psi_1)^* \hat{m}\psi_2 \, d\tau \qquad \ldots(i)$$

Again, since $\hat{m}$ is a Hermitian operator,

$$\int (\hat{l}\psi_1)^* \hat{m}\psi_2 \, d\tau = \int (\hat{m}\hat{l}\psi_1)^* \psi_2 \, d\tau \qquad \ldots(ii)$$

From Eqs. (*i*) and (*ii*), we obtain

$$\int \psi_1^* \hat{l}\hat{m}\psi_2 \, d\tau = \int (\hat{m}\hat{l}\psi_1)^* \psi_2 \, d\tau \qquad \ldots(iii)$$

If $\hat{l}$ and $\hat{m}$ commute, $\hat{l}\hat{m} = \hat{m}\hat{l}$ or, $(\hat{l}\hat{m})^* = (\hat{m}\hat{l})^*$.

Then Eq. (*iii*) reduces to

$$\int \psi_1^* \hat{l}\hat{m}\psi_2 \, d\tau = \int (\hat{l}\hat{m}\psi_1)^* \psi_2 \, d\tau \qquad \ldots(iv)$$

Comparing with Eq. (2.46) we find from Eq. (*iv*) that $\hat{l}\hat{m}$ is a Hermitian operator. This proves that *if two Hermitian operators commute, then their product is also a Hermitian operator.*

(*iii*) The eigenvalues of a Hermitian operator are real. Observables in quantum mechanics are represented by Hermitian operators.

(*iv*) The Hermitian adjoint or conjugate $\hat{\alpha}^+$ (alpha dagger) of an operator $\hat{\alpha}$ (not necessarily Hermitian) obeys the relationship

$$\int \psi_1^* \hat{\alpha}^+ \psi_2 \, dx = \int \psi_2 \hat{\alpha}^* \psi_1^* \, dx$$

where ψ_1 and ψ_2 are two functions of x.

If $\hat{\alpha} = \hat{\alpha}^+$, then $\hat{\alpha}$ is a Hermitian operator and is said to be self-adjoint in this case.

If $\hat{\alpha}^+ = -\hat{\alpha}$, then $\hat{\alpha}$ is called anti-Hermitian.

(*v*) If $\hat{\alpha}$ is non-Hermitian, $(\hat{\alpha} + \hat{\alpha}^+)$ and $i(\hat{\alpha} - \hat{\alpha}^+)$ are Hermitian operators.

(*vi*) The Hermitian adjoint of $\hat{\alpha}^+$ is $\hat{\alpha}$.

The Schrödinger Wave Equation

Dirac Notation

The integral $\int \psi_1^* \psi_2 \, d\tau$ is written in compact form as $\langle \psi_1 | \psi_2 \rangle$ and called the scalar product of ψ_2 with ψ_1. Thus, the normalization of ψ is written using Dirac notation as

$$\langle \psi | \psi \rangle = 1 \qquad \ldots(2.55e)$$

Similarly, the Hermitian condition can be written as

$$\langle \psi_1 | \hat{l} \psi_2 \rangle = \langle \hat{l} \psi_1 | \psi_2 \rangle \qquad \ldots(2.56)$$

In writing the scalar product of the above type it is preferable to write either side of Eq. (2.56) as $\langle \psi_1 | \hat{l} | \psi_2 \rangle$ which is termed the *matrix element of* $\hat{l}$ taken between ψ_2 and ψ_1. This form of notation leaves open whether $\hat{l}$ is to operate on ψ_1 or ψ_2. Equation (2.56) is then written as

$$\langle \psi_1 | \hat{l} \, \psi_2 \rangle = \langle \psi_1 | \hat{l} | \psi_2 \rangle = \langle \hat{l} \, \psi_1 | \psi_2 \rangle, \qquad \ldots(2.57)$$

where, of course, $\hat{l}$ is Hermitian. Dirac notation reduces the labour of writing the integrals of wave function products.

2.8. Orthogonality of Eigenfunctions

Like the equation of the vibrating string, the heat equation, and the electromagnetic wave equation, the solutions of the Schrödinger wave equation exhibit the property of *orthogonality*. According to this, if u_E and $u_{E'}$ are two eigenfunctions corresponding to two eigenvalues E and E' of the Schrödinger equation, then in one-dimensional form

$$\int_{-\infty}^{\infty} u_{E'}^*(x) \, u_E(x) \, dx = 0, \text{ when } E \neq E'$$

and

$$\int_{-\infty}^{\infty} u_{E'}^*(x) \, u_E(x) \, dx = 1, \text{ when } E = E'$$

The second equation emphasizes that u_E is normalized.

We shall now establish the orthogonality condition by considering the Schrödinger equation in one dimension only.

The time-independent Schrödinger equation for u_E in one dimension is obtained from Eq. (2.37). This is given by

$$-\frac{\hbar^2}{2m} \frac{d^2 u_E(x)}{dx^2} + V(x) \, u_E(x) = E u_E(x) \qquad \ldots(2.58)$$

The wave equation corresponding to $u_{E'}^*$ is similarly written as

$$-\frac{\hbar^2}{2m} \frac{d^2 u_{E'}^*(x)}{dx^2} + V(x) \, u_{E'}^*(x) = E' u_{E'}^*(x) \qquad \ldots(2.59)$$

Multiply Eq. (2.58) by $u_{E'}^*$ and Eq. (2.59) by u_E. Subtract the two equations and integrate the result from $x = -\infty$ to $x = +\infty$. We get

$$-\frac{\hbar^2}{2m} \int_{-\infty}^{\infty} \left[u_{E'}^* \frac{d^2 u_E}{dx^2} - u_E \frac{d^2 u_{E'}^*}{dx^2} \right] dx = (E - E') \int_{-\infty}^{\infty} u_{E'}^* u_E \, dx \qquad \ldots(2.60)$$

Expressing the integrand of the term within the bracket of the left-hand side of Eq. (2.60) by the derivative of $\left(u_{E'}^* \dfrac{du_E}{dx} - u_E \dfrac{du_{E'}^*}{dx} \right)$ we obtain

$$-\frac{\hbar^2}{2m}\left[u_{E'}^* \frac{du_E}{dx} - u_E \frac{du_{E'}^*}{dx}\right]_{-\infty}^{\infty} = (E - E') \int_{-\infty}^{\infty} u_{E'}^* u_E \, dx \qquad \ldots(2.61)$$

For physically well-behaved wavefunctions, both u and $\dfrac{du}{dx}$ approach zero as $x \to \pm\infty$. Therefore

$$(E - E') \int_{-\infty}^{\infty} u_{E'}^* u_E \, dx = 0. \qquad \ldots(2.62)$$

When $E \neq E'$ we obtain

$$\int_{-\infty}^{\infty} u_{E'}^* u_E \, dx = 0, \qquad \ldots(2.63)$$

i.e., the function u_E and $u_{E'}$ are orthogonal.

A set of eigenfunctions, each of which is normalized and orthogonal to each of the others, is said to be an *orthonormal set of functions*. An orthonormal set of energy eigenfunctions may be expressed in three dimensions as

$$\int u_{E'}^*(\vec{r}) \, u_E(\vec{r}) \, d\tau = \delta_{EE'} \qquad \ldots(2.64)$$

where the integration is carried over the common domain of the functions and $\delta_{EE'}$ is called the *Kronecker delta function*. The value of $\delta_{EE'} = 0$ when $E \neq E'$ and $\delta_{EE'} = 1$ when $E = E'$.

Sometimes it is found that linearly independent and orthogonal eigenfunctions belong to the same eigenvalue. This is called *degeneracy* and the corresponding eigenfunctions are said to be *degenerate*. Let $u_1, u_2, \ldots, u_n$ be the degenerate eigenfunctions, each satisfying the time-independent Schrödinger Eq. (2.37) for the same eigenvalue E. Then, direct substitution into Eq. (2.37) shows that

$$u = \sum_{i=1}^{n} c_i u_i$$

where c_i's are constants, also satisfies the time-independent Schrödinger equation.

Observation

It can be readily shown that the orthogonality property holds for any Hermitian operator $\hat{l}$. Let ψ_1 and ψ_2 be two eigenfunctions of $\hat{l}$ with eigenvalues l_1 and l_2, respectively. Then $\hat{l}\psi_2 = l_2 \psi_2$ and $(\hat{l}\psi_1)^* = l_1 \psi_1^*$, and Eq. (2.46) gives

$$l_2 \int \psi_1^* \psi_2 \, d\tau = l_1 \int \psi_1^* \psi_2 \, d\tau$$

or,

$$(l_1 - l_2) \int \psi_1^* \psi_2 \, d\tau = 0$$

As $l_1 \neq l_2$, we get $\int \psi_1^* \psi_2 \, d\tau = 0$, i.e., ψ_1 and ψ_2 are orthogonal.

2.9. Properties of Eigenfunctions and Boundary Conditions

When an eigenfunction is a mathematically well-behaved function, the measurable quantities such as $\langle x \rangle$, $\langle p \rangle$, etc. which are evaluated from the eigenfunction, turn out to be acceptable. In order to be well-behaved, an eigenfunction $u(x)$ and its derivative $\dfrac{du(x)}{dx}$ are required to obey the following properties:

(i) $u(x)$ and $\dfrac{du(x)}{dx}$ must be *finite*,

The Schrödinger Wave Equation

(ii) $u(x)$ and $\dfrac{du(x)}{dx}$ must be *single-valued*, and

(iii) $u(x)$ and $\dfrac{du(x)}{dx}$ must be *continuous*.

Energy eigenfunctions may be localized or nonlocalized. To treat these two cases on the same basis it is required to enclose the particle in a box of arbitrarily large but finite volume. Then it is assumed that the wave functions obey certain *periodic boundary conditions* at the walls of the box. For simplicity, if we consider a one-dimensional case and assume that the particle lies between x_1 and x_2, then the periodic boundary conditions require

(i) $\qquad\qquad\qquad \psi(x_1) = \psi(x_2)$, and

(ii) $\qquad\qquad\qquad \left(\dfrac{\partial \psi}{\partial x}\right)_{x_1} = \left(\dfrac{\partial \psi}{\partial x}\right)_{x_2}$

These boundary conditions provide simplicity of analysis from mathematical viewpoint, and lead to the discrete energy spectrum of the particle.

Observation

Let $\psi_1, \psi_2, ..., \psi_n$ be the eigenfunctions for eigenvalues $E_1, E_2, ..., E_n$, respectively. The wavefunction ψ resulting from the linear superposition of the eigenfunctions is $\psi = \sum_{i=1}^{n} C_i \psi_i$, where C_i's $(i = 1, 2, ..., n)$ are constants. The functions ψ_i's are orthonormal, so that $\int \psi^* \psi \, d\tau = \sum_{i=1}^{n} C_i^* C_i = 1$.

$C_i^* C_i$ is the probability for the ith eigenstate, so that the sum of the probabilities of the particle occupying all the eigenstates is expectedly unity. The coefficient C_i is obtained from $C_i = \int \psi_i^* \psi \, d\tau$.

The expectation or average energy of the particle having the wavefunction ψ is

$$<E> = \int \psi^* \hat{H} \psi \, d\tau = \int (C_1^* \psi_1^* + C_2^* \psi_2^* + ... + C_n^* \psi_n^*)(C_1 E_1 \psi_1 + C_2 E_2 \psi_2 + ... + C_n E_n \psi_n) \, d\tau = C_1^* C_1 E_1 + C_2^* C_2 E_2 + ... + C_n^* C_n E_n,$$

the eigenfunctions being orthonormal. Thus $<E> = \sum_{i=1}^{n} C_i^* C_i E_i$, i.e., $<E>$ is the probability-weighted average of eigenenergies.

2.10. Probability Current Density

The time derivative of the integral of the position probability density P over a finite volume V gives

$$\frac{d}{dt} \int_V P(\vec{r}, t) \, d\tau = \int_V \frac{\partial}{\partial t} (\psi^* \psi) \, d\tau$$

$$= \int_V \left(\psi^* \frac{\partial \psi}{\partial t} + \frac{\partial \psi^*}{\partial t} \psi \right) d\tau \qquad ...(2.65)$$

Substituting for $\dfrac{\partial \psi}{\partial t}$ and $\dfrac{\partial \psi^*}{\partial t}$ from Eqs. (2.15) and (2.28), respectively in Eq. (2.65), we obtain

$$\frac{d}{dt} \int_V P(\vec{r}, t) \, d\tau = \frac{i\hbar}{2m} \int_V \left[\psi^* \nabla^2 \psi - (\nabla^2 \psi^*) \psi \right] d\tau$$

$$= \frac{i\hbar}{2m} \int_V \vec{\nabla} \cdot \left[\psi^* \vec{\nabla} \psi - (\vec{\nabla} \psi^*) \psi \right] d\tau \qquad \ldots(2.66)$$

$$= \frac{i\hbar}{2m} \oint_S \left[\psi^* \vec{\nabla} \psi - (\vec{\nabla} \psi^*) \psi \right]_n dS \qquad \ldots(2.67)$$

where we have applied Gauss's divergence theorem of vector analysis, S being the surface bounding the volume V, and the subscript n indicating the component of the vector within brackets along the outward normal to the surface element dS.

Let us define a vector

$$\vec{J}_p(\vec{r}, t) = -\frac{i\hbar}{2m} \left[\psi^* \vec{\nabla} \psi - (\vec{\nabla} \psi^*) \psi \right] \qquad \ldots(2.68)$$

Then Eq. (2.66) gives $\quad \dfrac{d}{dt} \int_V P \, d\tau = -\int_V \vec{\nabla} \cdot \vec{J}_p \, d\tau$

or,

$$\int_V \left(\vec{\nabla} \cdot \vec{J}_p + \frac{\partial P}{\partial t} \right) d\tau = 0 \qquad \ldots(2.69)$$

Since Eq. (2.69) is valid for any volume V, we must have

$$\vec{\nabla} \cdot \vec{J}_p + \frac{\partial P}{\partial t} = 0 \qquad \ldots(2.70)$$

Equation (2.70) is similar to the *continuity equation* in electricity:

$$\vec{\nabla} \cdot \vec{J} + \frac{\partial \rho}{\partial t} = 0 \qquad \ldots(2.71)$$

In Eq. (2.70), we have $\vec{J}_p$ in place of the electric current density $\vec{J}$, and P in place of the electric charge density ρ. From this analogy, we can interpret the vector $\vec{J}_p(r, t)$ as the *probability current density*. It gives the probability of a particle passing through a unit cross sectional area normal to its direction of motion per unit time. Equation (2.67) shows that the time rate of increase of the position probability in the volume V results from the flow of the probability current *into* V.

Observe that ψ in Eq. (2.68) cannot be real, for then $\psi = \psi^*$ and the probability current density is always zero. Probability being not directly measurable, the probability current density cannot be measured directly.

2.11. Postulates of Quantum Mechanics

In this Chapter, we have discussed the postulates expressing Schrödinger's formulation of quantum mechanics. For ready reference, these postulates are summarized below.

(*i*) A particle making up a dynamical system can be described by a complex wave function $\psi(x, y, z, t)$ consistent with the uncertainty principle where (x, y, z) are the space coordinates of the particle and t is time.

(*ii*) The total energy E of a particle of mass m and momentum p is given by the classical Hamiltonian

$$H = \frac{p^2}{2m} + V(x, y, z)$$

where $V(x, y, z)$ is the potential energy and $p^2/(2m)$ is the kinetic energy of the particle. Dynamical quantities or observables are associated with operators (shown below) in quantum mechanics. These operators are linear and Hermitian.

The Schrödinger Wave Equation

Observable	Operator
x, y or z	x, y or z
$\vec{p}$	$-i\hbar \vec{\nabla}$
E	$i\hbar \dfrac{\partial}{\partial t}$
V	V
H	$-\dfrac{\hbar^2}{2m}\nabla^2 + V$

The wave function ψ satisfies Schrödinger's equation:

$$\left(-\frac{\hbar^2}{2m}\nabla^2 + V\right)\psi = i\hbar \frac{\partial \psi}{\partial t}$$

(iii) The only possible result of measurement of an observable α is the eigenvalue of the operator $\hat{\alpha}$ satisfying the eigenvalue equation.

$$\hat{\alpha}\, u_s = \alpha_s\, u_s,$$

where u_s is the eigenfunction and α_s is the eigenvalue of the operator $\hat{\alpha}$. The Schrödinger equation gives an eigenvalue equation, the eigenvalue of the Hamiltonian operator being the energy E.

(iv) The wave function ψ is a well-behaved function, i.e. ψ and $\nabla\psi$ are finite, continuous and single-valued.

(v) $\psi^* \psi$ (where ψ^* is the complex conjugate of ψ) is the position probability density of the particle. Since the particle must be found somewhere in space, the wavefunction ψ is normalized, i.e., $\int \psi^* \psi \, d\tau = 1$.

(vi) The expectation or the average value $<\alpha>$ of an observable α is

$$<\alpha> = \int \psi^* \hat{\alpha} \psi \, d\tau$$

(vii) If ψ_n's are the eigenfunctions and $|C_n|^2 = C_n C_n^*$ is the probability of finding the particle in the eigenstate ψ_n, the state of the particle is found by the *principle of superposition*:

$$\psi = \sum_n C_n \psi_n$$

$|C_n|^2$ can be related to the orthonormality of the eigenfunctions.

2.12. Worked-Out Problems

1. The ground state and the first excited state wavefunctions of an atom are ψ_0 and ψ_1, respectively, the corresponding energies being E_0 and E_1. If the system has a 40% probability of being found in the ground state and 60% probability of being found in the first excited state,
 (i) what is the wavefunction of the atom?
 (ii) what is the average energy of the atom? (C.U. 1995)

Ans. (i) Let the wavefunction of the atom be $\psi = C_0 \psi_0 + C_1 \psi_1$, where C_0 and C_1 are constants. Since the wavefunctions are orthonormal, we have

$$\int \psi^* \psi \, d\tau = 1, \text{ or } \int (C_0^* \psi_0^* + C_1^* \psi_1^*)(C_0 \psi_0 + C_1 \psi_1) \, d\tau = 1$$

or, $C_0^* C_0 \int \psi_0^* \psi_0 \, d\tau + C_1^* C_1 \int \psi_1^* \psi_1 \, d\tau = C_0^* C_0 + C_1^* C_1 = 1$. Since the probability for the ground state is 0.4 and that for the first excited state is 0.6, we have $C_0^* C_0 = 0.4$ and $C_1^* C_1 = 0.6$. We can thus choose $C_0 = \sqrt{0.4}$ and $C_1 = \sqrt{0.6}$, so that $\psi = \sqrt{0.4}\, \psi_0 + \sqrt{0.6}\, \psi_1$. [Though C_0 and C_1 are, in general, complex, we take them to be real quantities here as the given data are not sufficient to find both the real and the imaginary parts of C_0 and C_1.]

(ii) Since E_0 and E_1 are energy eigenvalues for the eigenfunctions ψ_0 and ψ_1, the average energy of the atom is $E_{av} = 0.4\, E_0 + 0.6\, E_1$.

2. The expectation value of an observable α is

$$\langle \alpha \rangle = \int \psi^* \hat{\alpha} \psi \, d\tau$$

From this expression, show that $\hat{\alpha}$ is a Hermitian operator.

Ans. Since $\langle \alpha \rangle = \int \psi^* (\hat{\alpha}\psi) \, d\tau$, we have $\langle \alpha \rangle^* = \int \psi (\hat{\alpha}\psi)^* \, d\tau$. $\langle \alpha \rangle$ being the expectation value, must be real, *i.e.*, $\langle \alpha \rangle = \langle \alpha \rangle^*$ Therefore,

$$\int \psi^* (\hat{\alpha}\psi) \, d\tau = \int (\hat{\alpha}\psi)^* \psi \, d\tau$$

or, $\hat{\alpha}$ is a Hermitian operator.

3. The operator $\left(x + \dfrac{d}{dx}\right)$ has the eigenvalue α. Derive the corresponding eigenfunction. *(C.U. 2008)*

Ans. If ψ is the desired eigenfunction, we have

$$\left(x + \frac{d}{dx}\right)\psi = \alpha\, \psi$$

or,

$$\frac{d\psi}{dx} = -(x - \alpha)\,\psi$$

or,

$$\frac{d\psi}{\psi} = -(x - \alpha)\, dx$$

Integrating,

$$\ln \frac{\psi}{\psi_0} = -\left(\frac{x^2}{2} - \alpha x\right)$$

where ψ_0 is the integration constant. Thus

$$\psi = \psi_0 \exp\left[-\left(\frac{x^2}{2} - \alpha x\right)\right]$$

4. The wavefunction of a particle moving in a potential-free region is given by $\psi(x) = A \cos kx$, where A and k are real constants. Is ψ an eigenstate of the operators $\hat{H}$, $\hat{p}_x$, and $\hat{p}_x^2$? If so, find the corresponding eigenvalues.

Ans. $\hat{H}\psi = -\dfrac{\hbar^2}{2m}\dfrac{d^2\psi}{dx^2}$, $\quad \hat{p}_x \psi = -i\hbar \dfrac{d\psi}{dx}$,

and $\hat{p}_x^2 \psi = -\hbar^2 \dfrac{d^2\psi}{dx^2}$.

Using $\psi = A \cos kx$, the above equations reduce to

$\hat{H}\psi = \dfrac{\hbar^2 k^2}{2m}\psi$, $\quad \hat{p}_x \psi = i\hbar\, kA \sin kx$,

and $\hat{p}_x^2 \psi = \hbar^2 k^2 \psi$.

Clearly, ψ is an eigenstate of $\hat{H}$ and $\hat{p}_x^2$, but it is not an eigenstate of $\hat{p}_x$.

The eigenvalue of $\hat{H}$ is $\dfrac{\hbar^2 k^2}{2m}$ and that of $\hat{p}_x^2$ is $\hbar^2 k^2$.

The Schrödinger Wave Equation

5. Write an expression for the probability current density and calculate the same for the wavefunction $\psi = e^{ikr}/r$, where $r^2 = x^2 + y^2 + z^2$, and interpret the result. (C.U. 1998, 2000)

Ans. The expression for the probability current density is

$$\vec{J}_p(\vec{r},t) = -\frac{i\hbar}{2m}\left[\psi^* \vec{\nabla}\psi - (\vec{\nabla}\psi^*)\psi\right]$$

Here $\psi = e^{ikr}/r$. Hence $\psi^* = e^{-ikr}/r$.

The wave function has spherical symmetry. So

$$\vec{\nabla}\psi = \vec{a}_r \frac{d\psi}{dr} = \vec{a}_r \left(\frac{ike^{ikr}}{r} - \frac{e^{ikr}}{r^2}\right)$$

$$= \vec{a}_r \frac{e^{ikr}}{r}\left(ik - \frac{1}{r}\right)$$

Similarly, $\vec{\nabla}\psi^* = \vec{a}_r \dfrac{e^{-ikr}}{r}\left(-ik - \dfrac{1}{r}\right)$,

where $\vec{a}_r$ is the unit vector along $\vec{r}$. Hence

$$\vec{J}_p = \frac{-i\hbar}{2m}\left[\frac{\vec{a}_r}{r^2}\left(ik - \frac{1}{r}\right) + \frac{\vec{a}_r}{r^2}\left(ik + \frac{1}{r}\right)\right]$$

$$= \frac{\hbar k}{mr^2}\vec{a}_r$$

Now, $\qquad \vec{\nabla}\cdot\vec{J}_p = \dfrac{1}{r^2}\dfrac{d}{dr}(r^2 J_p) = 0$

Hence $\dfrac{\partial P}{\partial t} = 0$, where P is the position probability density. The result shows that the position probability does not change with time.

6. Examine if the operator $\hat{L}$ is linear in the following examples:

(i) $\hat{L} f(x) = f(x) + x^2$,   (ii) $\hat{L} f(x) = f(5x^2 + 1)$,

(iii) $\hat{L} f(x) = \left(\dfrac{df}{dx}\right)^2$,   (iv) $\hat{L} f(x) = f(-x)$,   (C.U. 2001, 2005)

and (v) $\hat{L} f(x) = f^*(x)$ where * denotes complex conjugation.   (C.U. 2001, 2005)

Ans. Let $f(x) = f_1(x) + f_2(x)$.

(i) Here $\hat{L} f(x) = \hat{L}(f_1 + f_2) = f(x) + x^2 = f_1 + f_2 + x^2$,

and $\hat{L} f_1 + \hat{L} f_2 = (f_1 + x^2) + (f_2 + x^2) = f_1 + f_2 + 2x^2$

Thus $\hat{L}(f_1 + f_2) \neq \hat{L} f_1 + \hat{L} f_2$, and so the operator $\hat{L}$ is not linear.

(ii) Here $\hat{L} f(x) = \hat{L}(f_1 + f_2) = f(5x^2 + 1) = f_1(5x^2 + 1) + f_2(5x^2 + 1) = \hat{L}f_1 + \hat{L}f_2$.

Also, $\hat{L}[cf(x)] = cf(5x^2 + 1) = c\,\hat{L}f(x)$

where c is an arbitrary complex constant. Hence $\hat{L}$ is a linear operator.

(iii) Here $\hat{L} f(x) = \hat{L}(f_1 + f_2) = \left(\dfrac{df}{dx}\right)^2 = \left(\dfrac{df_1}{dx} + \dfrac{df_2}{dx}\right)^2$

and $\hat{L} f_1 + \hat{L} f_2 = \left(\dfrac{df_1}{dx}\right)^2 + \left(\dfrac{df_2}{dx}\right)^2$

Hence $\hat{L}(f_1 + f_2) \neq \hat{L} f_1 + \hat{L} f_2$, and so $\hat{L}$ is not a linear operator.

(iv) $\hat{L} f(x) = \hat{L}(f_1 + f_2) = f_1(-x) + f_2(-x)$

Also, $\hat{L} f_1 + \hat{L} f_2 = f_1(-x) + f_2(-x)$

Therefore, $\hat{L} f(x) = \hat{L} f_1(x) + \hat{L} f_2(x)$

Again, $\hat{L}[cf(x)] = cf(-x) = c\hat{L} f(x)$, where c is a complex constant. So, $\hat{L}$ is a linear operator.

(v) $\hat{L} f(x) = \hat{L}(f_1 + f_2) = f^*(x) = f_1^*(x) + f_2^*(x) = \hat{L} f_1 + \hat{L} f_2$

Again $\hat{L}[cf(x)] = c^* f^*(x) = c^* \hat{L} f(x)$, where c is a complex constant. Since $\hat{L}[cf(x)] \neq c\hat{L} f(x)$, $\hat{L}$ is not a linear operator.

7. Verify that $[AB, C] = [A, C] B + A [B, C]$ where A, B, C are three operators. (C.U. 2002)

Ans. We have
$$[AB, C] = ABC - CAB \quad \ldots(i)$$
and
$$[A, C] B + A [B, C]$$
$$= (AC - CA) B + A (BC - CB)$$
$$= ACB - CAB + ABC - ACB$$
$$= ABC - CAB \quad \ldots(ii)$$

From (i) and (ii) we obtain
$$[AB, C] = [A, C] B + A [B, C], \text{ proved}.$$

8. Prove that the z-component of the orbital angular momentum operator $\hat{L}_z = -i\hbar \dfrac{\partial}{\partial \phi}$, where ϕ is the azimuthal angle, is Hermitian.

Ans. If ψ_1 and ψ_2 are two wave functions we have

$$\int_0^{2\pi} \psi_1^* \hat{L}_z \psi_2 \, d\phi = \int_0^{2\pi} \psi_1^* \left(-i\hbar \dfrac{\partial}{\partial \phi}\right) \psi_2 \, d\phi$$

$$= -i\hbar \int_0^{2\pi} \psi_1^* \dfrac{\partial \psi_2}{\partial \phi} \, d\phi$$

$$= -i\hbar \left[\psi_1^* \psi_2 \big|_0^{2\pi} - \int_0^{2\pi} \psi_2 \dfrac{\partial \psi_1^*}{\partial \phi} d\phi \right] \text{ (integrating by parts)}$$

The wavefunctions being single-valued, we have $\psi_1^*(2\pi) = \psi_1^*(0)$, $\psi_2(2\pi) = \psi_2(0)$. Hence the first term on the right-hand side of the above equation is zero, giving

$$\int_0^{2\pi} \psi_1^* \hat{L}_z \psi_2 \, d\phi = \int_0^{2\pi} \psi_2 \left(i\hbar \dfrac{\partial}{\partial \phi}\right) \psi_1^* \, d\phi$$

$$= \int_0^{2\pi} \psi_2 \hat{L}_z^* \psi_1^* \, d\phi.$$

This agrees with the condition for an operator to be Hermitian. Hence $\hat{L}_z$ is Hermitian.

9. Verify that $\hat{Q} = \dfrac{1}{2}(\hat{x}\,\hat{p}_x + \hat{p}_x\,\hat{x})$ is a Hermitian operator. (C.U. 2003, 2008)

The Schrödinger Wave Equation

Ans. If ψ_1 and ψ_2 are two wave functions, we have

$$\int_{-\infty}^{\infty} \psi_1^* \hat{Q} \psi_2 \, dx = \frac{1}{2} \int_{-\infty}^{\infty} \psi_1^* \left[-i\hbar x \frac{\partial \psi_2}{\partial x} - i\hbar \frac{\partial}{\partial x}(x\psi_2) \right] dx$$

$$= -\frac{i\hbar}{2} \int_{-\infty}^{\infty} \psi_1^* \left[\psi_2 + 2x \frac{\partial \psi_2}{\partial x} \right] dx$$

$$= -\frac{i\hbar}{2} \left[2 \int_{-\infty}^{\infty} (\psi_1^* x) \frac{\partial \psi_2}{\partial x} dx + \int_{-\infty}^{\infty} \psi_1^* \psi_2 \, dx \right]$$

$$= -\frac{i\hbar}{2} \left[2 (\psi_1^* x) \psi_2 \Big|_{-\infty}^{\infty} - 2 \int_{-\infty}^{\infty} \left(\psi_1^* + x \frac{\partial \psi_1^*}{\partial x} \right) \psi_2 \, dx + \int_{-\infty}^{\infty} \psi_1^* \psi_2 \, dx \right] \quad \text{(integrating by parts)}$$

Since the wavefunctions vanish at infinity, the first term on the right-hand side is zero. So,

$$\int_{-\infty}^{\infty} \psi_1^* \hat{Q} \psi_2 \, dx = \frac{i\hbar}{2} \left[\int_{-\infty}^{\infty} \psi_1^* \psi_2 \, dx + 2 \int_{-\infty}^{\infty} (\psi_2 x) \frac{\partial \psi_1^*}{\partial x} dx \right]$$

$$= \frac{i\hbar}{2} \int_{-\infty}^{\infty} \left[\psi_1^* \psi_2 \, dx + 2x \psi_2 \frac{\partial \psi_1^*}{\partial x} dx \right]$$

$$= \frac{i\hbar}{2} \int_{-\infty}^{\infty} \psi_2 \left[\psi_1^* + 2x \frac{\partial \psi_1^*}{\partial x} \right] dx$$

$$= \int_{-\infty}^{\infty} \psi_2 \hat{Q}^* \psi_1^* \, dx.$$

Hence $\hat{Q}$ is a Hermitian operator.

10. A particle moving in a one-dimensional space has the wavefunction

$$\psi(x, t) = 0, \qquad x < 0$$

$$= Ax \exp(-Bx) \exp\left(\frac{ict}{\hbar}\right), x > 0.$$

where A, B, c are positive constants. Explain whether the particle is bound.

Ans. For $x = -\infty$, $\psi(x, t) = 0$.

Also, $\lim_{x \to \infty} \psi(x, t) = \lim_{x \to \infty} Ax \, e^{-Bx} e^{ict/\hbar} = 0$.

Since the wavefunction vanishes at $x = \pm \infty$, the particle is in a bound state.

11. Normalize the one-dimensional wavefunction

$$\psi(x) = A e^{-\alpha x} \qquad \text{for } x > 0$$
$$= A e^{\alpha x} \qquad \text{for } x < 0$$

where α is a positive constant. *(Rohilkhand 1999)*

Ans. The normalization condition is

$$\int_{-\infty}^{\infty} \psi^* \psi \, dx = 1$$

i.e.,

$$\int_{-\infty}^{0} A^2 e^{2\alpha x} \, dx + \int_{0}^{\infty} A^2 e^{-2\alpha x} \, dx = 1$$

or
$$A^2 \left(\int_{-\infty}^{0} e^{2\alpha x} dx + \int_{0}^{\infty} e^{-2\alpha x} dx \right) = 1$$

or
$$A^2 \left(\frac{e^{2\alpha x}}{2\alpha} \bigg|_{-\infty}^{0} + \frac{e^{-2\alpha x}}{-2\alpha} \bigg|_{0}^{\infty} \right) = 1$$

or $\dfrac{A^2}{\alpha} = 1$ or $A = \sqrt{\alpha}$

So, the normalized wave function is
$$\psi(x) = \sqrt{\alpha}\, e^{-\alpha x} \qquad \text{for } x > 0$$
$$= \sqrt{\alpha}\, e^{\alpha x} \qquad \text{for } x < 0.$$

12. A particle moving one dimensionally is represented by the wavefunction
$$\psi(x) = \left(\frac{\sqrt{2}}{\pi} \right)^{\frac{1}{2}} \frac{x + ix}{1 + ix^2}$$

Find the position probability density where the particle is most likely to be found.

Ans. The position probability density is
$$P = \psi^* \psi = \frac{\sqrt{2}}{\pi} \frac{(x+ix)(x-ix)}{(1+ix^2)(1-ix^2)}$$
$$= \frac{\sqrt{2}}{\pi} \frac{x^2 + x^2}{1 + x^4} = \frac{2\sqrt{2}}{\pi} \frac{x^2}{1 + x^4}$$

The particle is most likely to be found at the point where P is a maximum, i.e., $\dfrac{dP}{dx} = 0$,

or $\dfrac{d}{dx}\left(\dfrac{x^2}{1+x^4} \right) = 0$ or $\dfrac{(1+x^4)(2x) - x^2(4x^3)}{(1+x^4)^2} = 0$

or $1 + x^4 - 2x^4 = 0$ or $x^4 = 1$

or $x = \pm 1$.

13. Find the constant B which makes e^{-ax^2} an eigenfunction of the operator $\left(\dfrac{d^2}{dx^2} - Bx^2 \right)$. What is the corresponding eigenvalue? *(Meerut 1991)*

Ans. The eigenvalue equation of an operator $\hat{l}$ is $\hat{l}\psi = l\psi$, where l is the eigenvalue and ψ is the eigenfunction.

Here $\hat{l} = \dfrac{d^2}{dx^2} - Bx^2$ and $\psi = e^{-ax^2}$

So, $\hat{l}\psi = \left(\dfrac{d^2}{dx^2} - Bx^2 \right) e^{-ax^2} = \dfrac{d^2(e^{-ax^2})}{dx^2} - Bx^2 e^{-ax^2}$

$$= -2a\,\dfrac{d}{dx}\left(x e^{-ax^2}\right) - Bx^2 e^{-ax^2} = -2a\left(e^{-ax^2} - 2ax^2 e^{-ax^2}\right) - Bx^2 e^{-ax^2}$$

$$= (4a^2 x^2 - 2a - Bx^2)\, e^{-ax^2}$$

For e^{-ax^2} to be an eigenfunction, the eigenvalue $(4a^2 - B) x^2 - 2a$ must be independent of x.

So, $\qquad 4a^2 - B = 0 \qquad$ or $\qquad B = 4a^2$

Then $\qquad \hat{l}\psi = -2ae^{-ax^2} = -2a\psi.$

Thus the eigenvalue of $\hat{l}$ is $-2a$.

QUESTIONS

1. (a) Derive the one dimensional form of the Schrödinger wave equation for a free particle of mass m. What would be the form of this equation when the particle is subjected to a force F?

   (b) In what respect does the Schrödinger equation differ from the classical wave equation?

   (Burd. U. 1995)

   (c) Define position probability density and probability current density.

2. (a) Give a physical interpretation of the wavefunction. State and explain the normalization of the wavefunction.

   (b) Argue that $\psi^*(x, t) \psi(x, t)$ must be real, and either positive or zero.

3. Define the expectation value of a dynamical quantity. Write down the expectation values of the energy and momentum of a particle.

4. State Ehrenfest's theorem. Use this theorem to show that classical mechanics agrees with quantum mechanics so far as the expectation values are concerned.

5. Show that

$$\frac{d\langle x \rangle}{dt} = \frac{\langle p_x \rangle}{m} \quad \text{and} \quad \frac{d}{dt}\langle p_x \rangle = \langle F_x \rangle,$$

   where the symbols have their usual meanings. **(C.U. 1997, 2000, 2005)**

6. Derive Schrödinger's (i) time-independent and (ii) time-dependent equations for matter waves. Give a physical interpretation of the wave equation. **(C.U. 1983)**

7. Solve the Schrödinger wave equation for a particle of mass m and potential energy $V(r)$ and derive the eigenvalue equation. What are the eigenfunction and the eigenvalue? Define the stationary state of a particle.

8. (a) Write down the Hamiltonian for a free particle. (b) In quantum mechanics what is the operator corresponding to the Hamiltonian? (c) What is an eigenfunction of this operator? Write down an eigenfunction. (d) What is the time-dependence of the eigenfunction? (e) What can you say about the position and the momentum of a particle described by such a wave function? **(C.U. 1983)**

9. (a) Define observables. What is an operator? What are Hermitian and Hamiltonian operators? Describe briefly the mathematical properties of operators.

   (b) Why is an observable in quantum mechanics represented by a Hermitian operator? Examine whether the operator $\dfrac{i\partial}{\partial x}$ is Hermitian or not. **(Burd. U. 1993)**

   (c) Prove that the Hamiltonian operator is Hermitian. **(C.U. 2002)**

10. Do the one-dimensional position and momentum operators commute? If not, obtain the relevant relationship and discuss its significance. **(C.U. 1987)**

11. Write a short note on 'Dirac notation'.

12. What is an operator? What is a wavefunction? What is the physical significance of normalising a wavefunction? Write the time-independent Schrödinger equation. If p be the linear momentum associated with the position coordinate q, show that the operator $pq - qp = \hbar/i$. **(C.U. 1990)**

13. Show that compatible observables can be represented by commutating operators. What is the implication of noncommutation of two operators?

14. Let L be the total angular momentum of a particle and L_x, L_y, L_z be its three cartesian components. $\hat{L}^2$ commutes with each of $\hat{L}_x, \hat{L}_y$ and $\hat{L}_z$. But $\hat{L}_x, \hat{L}_y$ and $\hat{L}_z$ do not commute with each other. What conclusions can you draw about simultaneous precise measurements of L^2, L_x, L_y and L_z?

(C.U. 1994)

(**Ans.** L^2 and any one of L_x, L_y and L_z can be simultaneously measured precisely. But any two of L_x, L_y and L_z cannot be simultaneously measured precisely.)

15. (a) Write the time-independent Schrödinger equation. Show that this equation is linear, i.e., if ψ_1 and ψ_2 are two solutions, $(a\psi_1 + b\psi_2)$ is also a solution. (C.U. 1993)

(b) Write the time-dependent Schrödinger equation. (C.U. 2000)

(c) Write time-dependent Schrödinger's equation in one dimension. Show that the general solution of this equation is $\psi(x, t) = \phi(x) e^{-iEt/\hbar}$. (N. Beng. U. 2001)

16. Define probability current density and relate it to the position probability density.

17. (a) Evaluate the commutation $\left[x, \dfrac{\partial}{\partial x} \right]$. (C.U. 1996) [**Ans.** -1]

(b) Show that (i) $x^n \hat{p}_x - \hat{p}_x x^n = ni\hbar\, x^{n-1}$, (ii) $[f(x), \hat{p}_x] = i\hbar \dfrac{df}{dx}$,

(iii) $[\hat{x}, \hat{p}_x^n] = i\hbar n \hat{p}_x^{n-1}$ and $[\hat{p}_x, x^n] = -i\hbar n x^{n-1}$. (C.U. 1999)

18. State the postulates of Schrödinger's formulation of quantum mechanics.

19. What is the significance of Hermitian operators in quantum mechanics? Mention any three of their properties.

20. (a) "The probability current density cannot be directly measured." Why?

(b) The probability current density for one-dimensional motion is given by:

$$J_x = \dfrac{i\hbar}{2m} \left(\psi \dfrac{\partial \psi^*}{\partial x} - \psi^* \dfrac{\partial \psi}{\partial x} \right)$$

From this expression, show that ψ cannot be real. (cf. C.U. 2000)

(c) For the wave function $\psi = A \exp i(ax - \omega t)$, find the probability current density. (A = const.)

[**Ans.** $\hbar a |A|^2 /m$]

21. Show that the eigenvalues of a Hermitian operator are real. (C.U. 2001)

22. What do you mean by normalized, orthogonal and orthonormal wave functions? Give one example of each. (Purvanchal Univ. 2001)

23. Find the expectation value of the position of a particle whose normalized wave function is

$$\psi(x) = N \exp[-x^2/2a + ikx]$$

(Purvanchal Univ. 2002) [**Ans.** 0]

24. Show that the function $\psi(x) = Ae^{ikx} + Be^{-ikx}$ is an eigenfunction of $\hat{p}^2$ where p is the momentum of the particle. (N. Beng. U. 2001)

25. A system has two possible energy values E_0 and $2E_0$, and at a certain instant, it is in a state with the expectation energy $\dfrac{3}{2} E_0$. Obtain the wave function for this state, given that ψ_0 and ψ_1 are the wavefunctions for the two possible stationary states. What is the wave function after the elapse of a time interval t? (C.U. 2002)

[**Ans.** $\dfrac{1}{\sqrt{2}} (\psi_0 + \psi_1), \dfrac{1}{\sqrt{2}} (\psi_0 e^{-iE_0 t/\hbar} + \psi_1 e^{-i2E_0 t/\hbar})$]

The Schrödinger Wave Equation

26. Show that the eigenfunctions of a Hermitian operator belonging to different eigenvalues are orthogonal. (C.U. 2002, 07)

27. Prove the following identities in commutation
    (i) $[A, BC] = [A, B] C + B [A, C]$
    (ii) $[A + B, C] = [A, C] + [B, C]$
    (iii) $[A, B + C] = [A, B] + [A, C]$
    (iv) $[A, [B, C]] + [B, [C, A]] + [C, [A, B]] = 0$ (C.U. 2006)
    where A, B and C are operators.

28. The wave function of a particle in a stationary state with an energy E_0 at time $t = 0$ is $\psi(x)$. After how much minimum time will the wave function be again $\psi(x)$? (C.U. 2003) (Ans. $2\pi\hbar/E_0$)

29. What are the properties of a well-behaved wave-function? Which of the following wavefunctions is well-behaved and why? (i) $\psi(x) = Ae^{x^2}$, (ii) $\psi(x) = Axe^{-x^2}$ (A = constant, and $-\infty < x < \infty$). (C.U. 2003) [Ans. (ii)]

30. Find the commutator $[\hat{A}, \hat{B}]$ where $\hat{A} = x^3$ and $\hat{B} = x\dfrac{d}{dx}$. (Meerut 2000) [Ans. $-3x^3$]

31. Given an operator $\hat{\alpha} = -\dfrac{d^2}{dx^2}$, and two functions $f_1 = \sin x$ and $f_2 = \sin 2x$ ($0 \leq x \leq 2\pi$). Show that f_1 and f_2 are eigenfunctions of the given operator. Normalize the functions and find the eigenvalues. Also show that f_1 and f_2 are orthogonal. (K.U. 2004)

    [Ans. $\dfrac{1}{\sqrt{\pi}}, \dfrac{1}{\sqrt{\pi}}, 1, 4$]

32. If a system has two eigenstates ψ_1 and ψ_2 with eigenvalues E_1 and E_2, under what condition will the linear combination $(a\psi_1 + b\psi_2)$ be also an eigenstate? (K.U. 2004) [Ans. $E_1 = E_2$]

33. Find the expectation value of the position and of the momentum of a particle represented by a suitably normalized wave function

    $$\Psi(x) = A\exp\left(-\dfrac{x^2}{a^2} + ik_0 x\right), \quad (-\infty < x < \infty)$$

    where A is the normalization constant. [Ans. 0, $\hbar k_0$] (C.U. 2007)

34. Calculate the position probability density and the probability current density for the wave function

    $$\psi(x) = \left(\dfrac{a}{\sqrt{\pi}}\right)^{1/2} \exp\left(-\dfrac{1}{2}a^2 x^2 + ikx\right),$$ where a and k are constants. (cf. C.U. 2008)

    $$\left[\text{Ans. } \dfrac{a}{\sqrt{\pi}} \exp(-a^2 x^2), \dfrac{a\hbar k}{\sqrt{\pi}\, m} \exp(-a^2 x^2)\right]$$

35. Show that, if the one-dimensional potential energy $V(x)$ is an even function of x, the eigenfunctions are either even or odd functions of x.

36. Prove that the expectation value of the square of an observable α cannot be negative. (C.U. 2008)

    [**Hint.** With $\psi = \sum\limits_{i=1}^{n} C_i \psi_i$, where ψ_i is the eigenfunction for the eigenvalue α_i^2 of the observable α^2,

    $<\alpha^2> = \sum\limits_{i=1}^{n} C_i C_i^* \alpha_i^2$. As $C_i C_i^* \not< 0$ and $\alpha_i^2 \not< 0$, the result follows.]

Chapter 3

Free Particles and Wave Packets

3.1. THE FREE PARTICLE

A free particle is defined as one which is subject to no forces of any kind, and so moves in a region of constant potential. We shall examine the wave function for a free particle by limiting ourselves to a one-dimensional geometry in which the particle motion is confined to the x-axis only. For simplicity of analysis, we shall consider the potential to be zero in which case the time-independent Schrödinger equation [Eq. (2.37)] reduces to

$$\frac{d^2 u}{dx^2} + k^2 u(x) = 0, \qquad \ldots(3.1)$$

where

$$k = \sqrt{\frac{2mE}{\hbar^2}} \qquad \ldots(3.2)$$

The solutions of Eq. (3.1) are of the form

$$u(x) = A e^{ikx}, \qquad \ldots(3.3)$$

where A is an arbitrary constant. The stationary states are given by [see Eq. (2.38)]

$$\psi(x, t) = u(x) e^{-iEt/\hbar}$$
$$= A e^{i(kx - \omega t)}, \qquad \ldots(3.4)$$

where

$$\omega = \frac{E}{\hbar} = \frac{\hbar k^2}{2m}$$

Equation (3.3) gives the form of a free particle eigenfunction corresponding to the eigenvalue E. Equation (3.4) represents a wave travelling in the positive x-direction and can hence be applied for a particle moving in the positive x-direction. The momentum operator $-i\hbar \dfrac{\partial}{\partial x}$ operating on the wave function $\psi(x, t)$ gives

$$-i\hbar \frac{\partial \psi}{\partial x} = -i\hbar \frac{\partial}{\partial x}\left[A e^{i(kx - \omega t)} \right]$$
$$= \hbar k A e^{i(kx - \omega t)}$$
$$= \hbar k \psi$$
$$= \sqrt{2mE}\, \psi, \qquad \ldots(3.5)$$

the last step following from Eq. (3.2). Equation (3.5) shows that the wavefunction $\psi(x, t)$ is an eigenfunction of the momentum operator, and $\sqrt{2mE}$ is the eigenvalue of this operator. In fact, $p = \sqrt{2mE}$ is the momentum of a particle moving in the positive x-direction with energy E in a region of zero potential. Thus the momentum of the particle is precisely defined.

The probability density for the particle is $\psi^*\psi = A^2$, i.e., a constant independent of x. Consequently, the particle is equally likely to be found anywhere and the uncertainty in its position Δx is

Free Particles and Wave Packets

infinite. According to the uncertainty principle, the uncertainty in position Δx and the uncertainty in momentum Δp are related by $\Delta x \Delta p \geq \hbar$. Hence when $\Delta x = \infty$, Δp must be zero. Thus precise values of p given by $p = \hbar k = \sqrt{2mE}$ are possible, as seen in the preceding paragraph.

Clearly, the time-independent Schrödinger equation for a free particle leads to the de Broglie relation $\lambda = \dfrac{2\pi}{k} = \dfrac{2\pi\hbar}{p} = \dfrac{h}{p}$.

A problem appears in normalising the wavefunction given by Eq. (3.4). We have

$$\int_{-\infty}^{\infty} \psi\psi^* \, dx = A^2 \int_{-\infty}^{\infty} dx = 1, \quad \ldots(3.6)$$

since the particle must be somewhere in the space. Equation (3.6) shows that A must be zero as $\int_{-\infty}^{\infty} dx$ is infinite. This difficulty arises since we are considering an ideal case of infinite length. In practice, the particle is confined to a finite length so that normalisation is possible.

For a particle moving in the negative x-direction, the solution of Eq. (3.1) is Ae^{-ikx} having the momentum eigenvalue $-\hbar k = -\sqrt{2mE}$. Thus, for momenta $+\hbar k$ and $-\hbar k$, we have the same energy eigenvalue E. So, each energy eigenvalue of a free particle is *doubly degenerate*.

3.2. PARTICLE CONFINED IN AN ENCLOSURE

Consider a particle of mass m moving freely along the x-axis between two rigid walls situated at $x = 0$ and $x = L$. It is prohibited from being found anywhere outside this region. As before, we assume that the particle moves in a region of zero potential. The time-independent Schrödinger equation for the problem is given by Eq. (3.1). For convenience the solution to this equation is written in the form

$$u(x) = A \cos kx + B \sin kx, \quad \ldots(3.7)$$

where A and B are real constants. The boundary conditions for the problem are

$$u(0) = 0 \quad \ldots(3.8a)$$

and

$$u(L) = 0. \quad \ldots(3.8b)$$

To satisfy the condition (3.8 a) we must have $A = 0$. Hence

$$u(x) = B \sin kx. \quad \ldots(3.9)$$

Using the condition (3.8 b) we obtain from Eq. (3.9)

$$B \sin kL = 0 \quad \ldots(3.10)$$

We cannot make $B = 0$ as the wavefunction will then vanish. Hence Eq. (3.10) is satisfied only when

$$kL = n\pi, \quad \ldots(3.11)$$

where $n = 1, 2, 3,..$ etc. We exclude the value $n = 0$ for then k and hence $u(x)$ become zero. A particle with $k = 0$, *i.e.* $E = 0$ cannot therefore exist in the region considered. The solution of the problem is therefore

$$u_n(x) = B \sin k_n x, \quad \ldots(3.12)$$

where

$$k_n = \dfrac{n\pi}{L}, \, n = 1, 2, 3... \quad \ldots(3.12a)$$

The number n is called the *quantum number*. The allowed values of k are discrete and are given by Eq. (3.12 a). The allowed values of energy are given by

$$E_n = \dfrac{\hbar^2 k_n^2}{2m} = \dfrac{\pi^2 \hbar^2 n^2}{2mL^2}; \, n = 1, 2, 3... \quad \ldots(3.13)$$

The subscript n is used to emphasize that the allowed values of energy E are discrete. In other words, the energy of the particle is *quantised*. The ground state of the particle is given by $n = 1$. In this state the energy of the particle is the lowest. Denoting the ground state energy by E_1 the nth energy level is found from Eq. (3.13) to be given by

$$E_n = n^2 E_1, \qquad n = 2, 3, 4... \qquad ...(3.14)$$

The energy level diagram for the particle is shown in Fig. 3.1.

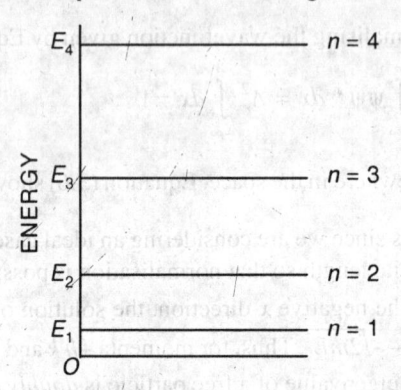

Fig. 3.1. Energy level diagram of a particle confined in a region of zero potential.

The nth state is an eigenstate of the total energy belonging to the eigenvalue E_n. This is because

$$\hat{H} u_n = \hat{T} u_n$$
$$= -\frac{\hbar^2}{2m} \frac{\partial^2 u_n}{\partial x^2} = \frac{\hbar^2 k_n^2}{2m} u_n$$
$$= E_n u_n. \qquad ...(3.15)$$

The nth state has the momentum of magnitude $p_n = \hbar k_n = \dfrac{\pi \hbar n}{L}$ ...(3.16)

The momentum can be both positive and negative as the particle travels back and forth between the walls, being reflected at the ends. This can be visualized by noting that the time-independent wave function of Eq. (3.12) can be expressed as $u_n(x) = \dfrac{B}{2i} \left(e^{in\pi x/L} - e^{-in\pi x/L} \right)$.

The first term within parentheses on the right-hand side represents a plane wave with $k_n = n\pi x/L$ propagating along the positive x-direction, and the second term represents a plane wave of equal amplitude with $k_n = -n\pi x/L$ going in the opposite direction. The superposition of the two waves in the region $0 < x < L$ gives standing waves.

The nth state is not an eigenstate of momentum, for $\hat{p} u_n = -i\hbar \dfrac{\partial u_n}{dx} = -i\hbar B k_n \cos k_n x$. This is expected since the momentum having two possible directions, is not precisely defined for the state.

Note that when L is large, the separation between the discrete energy and momentum values is small, and the states may be considered *quasicontinuous*.

We shall now normalize the wavefunction given by Eq. (3.12) for $0 < x < L$. Note that $u_n(x) = 0$ for $x < 0$ and $x > L$. The probability that the particle lies somewhere in the region $0 < x < L$ is unity, i.e.

$$\int_0^L u_n^2(x) \, dx = 1$$

or,

$$B^2 \int_0^L \sin^2 \frac{n\pi}{L} x \, dx = 1 \text{ or, } B^2 \frac{L}{2} = 1 \text{ or, } B = \sqrt{\frac{2}{L}} \qquad ...(3.17)$$

Free Particles and Wave Packets

Hence the normalised wavefunction $u_n(x)$ is given by

$$u_n(x) = \sqrt{\frac{2}{L}} \sin \frac{n\pi}{L} x \qquad ...(3.18)$$

The wavefunctions for $n=1$, $n = 2$ and $n = 3$ are plotted in Fig. 3.2.

The probability density for the location of the particle is given by

$$P(x) = u_n^2(x) = \frac{2}{L} \sin^2 \frac{n\pi x}{L}. \qquad ...(3.19)$$

$P(x)$ is a maximum when $\dfrac{n\pi x}{L} = \dfrac{\pi}{2}, \dfrac{3\pi}{2}, \dfrac{5\pi}{2}$, etc., i.e. $x = \dfrac{L}{2n}, \dfrac{3L}{2n}, \dfrac{5L}{2n}$ etc. Thus for the state $n = 1$, the most probable position of the particle is at $x = \dfrac{L}{2}$. For $n = 2$, the most probable positions occur at $x = \dfrac{L}{4}$ and $x = \dfrac{3L}{4}$. For $n = 3$, these positions are given by $x = \dfrac{L}{6}, x = \dfrac{3L}{6}$ and $x = \dfrac{5L}{6}$. The variation of $P(x)$ with x for $n = 1, 2$, and 3 is shown in Fig. 3.3.

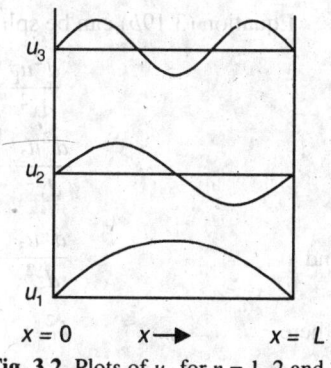

Fig. 3.2. Plots of u_n for $n = 1, 2$ and 3.

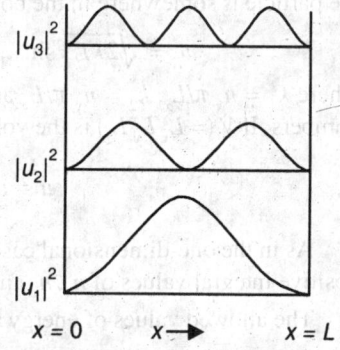

Fig. 3.3. Plots of $|u_n|^2$ for $n = 1, 2$ and 3.

According to classical ideas, the probability of finding the particle anywhere in a small region dx within the interval $0 < x < L$ is the same everywhere and is equal to $\dfrac{dx}{L}$. Thus the probability density is simply $1/L$. This is because classically the particle travels with a uniform speed from one wall to the other and is perfectly reflected at the walls. The quantum mechanical probability density is not a constant as shown in Fig. 3.3 and differs from the classical result.

Note that the ground-state energy E_1 [$= \pi^2 \hbar^2 / (2mL^2)$] agrees with Heisenberg's uncertainty principle so far as the order of magnitude is concerned. The uncertainty in position is $\Delta x \sim L$, so that the uncertainty in momentum Δp_x is at least $\hbar/\Delta x$, i.e., $\Delta p_x \sim \hbar/L$. The minimum kinetic energy is therefore $\Delta p_x^2/(2m) = \hbar^2/(2mL^2)$, which is in conformity with E_1 in order of magnitude.

Extension to three-dimensional box

We now consider that the particle is in a three-dimensional box bound by rigid walls. The particle cannot escape from the box, and its potential energy in the box is zero. The box is defined by $0 \le x \le L_x, 0 \le y \le L_y$ and $0 \le z \le L_z$.

Let the solution of the time-independent Schrödinger equation (2.37) be written as

$$u(\vec{r}) = u_1(x) u_2(y) u_3(z)$$

Substituting into Eq. (2.37) and noting that $V = 0$ in the box, we get

$$u_2 u_3 \frac{d^2 u_1}{dx^2} + u_3 u_1 \frac{d^2 u_2}{dy^2} + u_1 u_2 \frac{d^2 u_3}{dz^2} + k^2 u_1 u_2 u_3 = 0 \qquad ...(3.19a)$$

where $k^2 = 2mE/\hbar^2$. Dividing Eq. (3.19a) by $u_1 u_2 u_3$ we obtain

$$\frac{1}{u_1} \frac{d^2 u_1}{dx^2} + \frac{1}{u_2} \frac{d^2 u_2}{dy^2} + \frac{1}{u_3} \frac{d^2 u_3}{dz^2} + k^2 = 0 \qquad ...(3.19b)$$

Equation (3.19b) can be split up into three equations :

$$\frac{d^2 u_1}{dx^2} + k_1^2 u_1 = 0 \qquad \ldots(3.19c)$$

$$\frac{d^2 u_2}{dy^2} + k_2^2 u_2 = 0 \qquad \ldots(3.19d)$$

and

$$\frac{d^2 u_3}{dz^2} + k_3^2 u_3 = 0 \qquad \ldots(3.19e)$$

where

$$k_1^2 + k_2^2 + k_3^2 = k^2 = \frac{2mE}{\hbar^2} \qquad (3.19f)$$

Putting the appropriate boundary conditions and employing the normalization requirement that the particle is somewhere in the box, we find as in the one-dimensional case

$$u_1 = \sqrt{2/L_x} \sin k_1 x, \quad u_2 = \sqrt{2/L_y} \sin k_2 y \text{ and } u_3 = \sqrt{2/L_z} \sin k_3 z$$

where $k_1 = n_x \pi/L_x$, $k_2 = n_y \pi/L_y$ and $k_3 = n_z \pi/L_z$. n_x, n_y and n_z are integers representing quantum numbers. If $V (= L_x L_y L_z)$ is the volume of the box, the wave functions are

$$u = u_1 u_2 u_3 = \sqrt{\frac{8}{V}} \sin\left(\frac{n_x \pi}{L_x} x\right) \sin\left(\frac{n_y \pi}{L_y} y\right) \sin\left(\frac{n_z \pi}{L_z} z\right)$$

As in the one-dimensional case, $n_x = n_y = n_z = 0$ is not acceptable since this leads to $u = 0$. Only positive integral values of n_x, n_y and n_z are allowed.

The allowed values of energy are

$$E = \frac{\pi^2 \hbar^2}{2m} \left[\left(\frac{n_x}{L_x}\right)^2 + \left(\frac{n_y}{L_y}\right)^2 + \left(\frac{n_z}{L_z}\right)^2\right],$$

where m is the mass of the particle. Different integral values of n_x, n_y and n_z give different values of E. If the box is a cube of side L, we have for the energy

$$E = \frac{\hbar^2 \pi^2}{2m L^2} (n_x^2 + n_y^2 + n_z^2)$$

Note that each value of E is now obtained from three different sets of values of (n_x, n_y, n_z), for example, $(1, 2, 2)$, $(2, 1, 2)$ and $(2, 2, 1)$. The three wave functions are associated with the same energy here, so that the energy level has a three-fold degeneracy.

3.3. Wave Packets

In Sec. 3.1 we have seen that a wavefunction of the form $Ae^{ik'x}$ represents a state in which the position of the particle is completely undefined. However, in a practical experiment, the position of the particle would not be completely undefined. Therefore, we shall consider the effect of localizing the particle by modifying the amplitude of the wavefunction.

Suppose that an amplitude proportional to $e^{-a^2 x^2/2}$ where a is a real constant is introduced into the wavefunction. The resultant wavefunction is termed a *wave packet* and it assumes the form given by

$$\psi(x, 0) = A e^{-a^2 x^2/2} e^{ik'x} \qquad \ldots(3.20)$$

In this case x varies from $-\infty$ to $+\infty$ and $\psi(x, 0)$ is not an eigenfunction of the energy.

The function $\psi(x, 0)$ is normalised by choosing the constant $A = \left(\dfrac{a}{\sqrt{\pi}}\right)^{1/2}$:

Free Particles and Wave Packets 57

$$\int_{-\infty}^{\infty} \psi^*(x, 0) \psi(x, 0) \, dx = \frac{a}{\sqrt{\pi}} \int_{-\infty}^{\infty} e^{-a^2 x^2} \, dx = 1 \qquad \ldots(3.21)$$

The position probability density given by $|\psi|^2$ is then

$$|\psi|^2 = \frac{a}{\sqrt{\pi}} e^{-a^2 x^2} \qquad \ldots(3.22)$$

Evidently, $|\psi|^2$ has the Gaussian form as shown in Fig. 3.4.

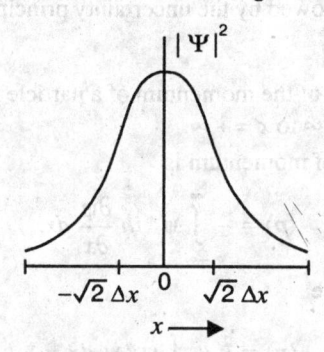

Fig. 3.4. Position probability density against x for a wave packet at $t = 0$.

The uncertainty Δx of the position of the particle is defined as the root-mean-square deviation of the values of x from the average or the expectation value. Therefore

$$(\Delta x)^2 = \langle (x - \langle x \rangle)^2 \rangle = \langle x^2 \rangle$$
$$- \langle 2x \langle x \rangle \rangle + \langle \langle x \rangle^2 \rangle$$
$$= \langle x^2 \rangle - \langle x \rangle^2 \qquad \ldots(3.23)$$

Now

$$\langle x \rangle = \frac{a}{\sqrt{\pi}} \int_{-\infty}^{\infty} x e^{-a^2 x^2} \, dx = 0$$

and

$$\langle x^2 \rangle = \frac{a}{\sqrt{\pi}} \int_{-\infty}^{\infty} x^2 e^{-a^2 x^2} \, dx = \frac{1}{2a^2} \qquad \ldots(3.24)$$

Thus,

$$\Delta x = \{\langle x^2 \rangle - \langle x \rangle^2\}^{1/2} = \frac{1}{a\sqrt{2}} \qquad \ldots(3.25)$$

It is evident from Fig. 3.4 that the maximum probability density occurs at $x = \langle x \rangle = 0$ and it falls to $\frac{1}{e}$ of its maximum value at $x = \pm \sqrt{2} \, \Delta x$.

The expectation value $\langle p \rangle$ of momentum is given by

$$\langle p \rangle = -\frac{i\hbar a}{\sqrt{\pi}} \int_{-\infty}^{\infty} (ik' - a^2 x) e^{-a^2 x^2} \, dx = \hbar k', \qquad \ldots(3.26)$$

The quantity $\langle p^2 \rangle$ is

$$\langle p^2 \rangle = -\frac{\hbar^2 a}{\sqrt{\pi}} \int_{-\infty}^{\infty} (-a^2 - k'^2 - 2ia^2 k' x + a^4 x^2) e^{-a^2 x^2} \, dx$$

$$= \hbar^2 \left(k'^2 + \frac{a^2}{2} \right) \qquad \ldots(3.27)$$

The uncertainty Δp in momentum of the particle is therefore

$$\Delta p = + \{(p^2) - (p)^2\}^{1/2} = \frac{\hbar a}{\sqrt{2}} \qquad \ldots(3.28)$$

Thus the value of the product of the two uncertainties is given by

$$\Delta x \Delta p = \frac{\hbar}{2} \qquad \ldots(3.29)$$

This is the minimum value allowed by the uncertainty principle.

3.4. Worked-out Problems

1. Find the expectation value of the momentum of a particle free to move in a one-dimensional space of zero potential from $x = -\infty$ to $x = +\infty$.

Ans. The expectation value of momentum is

$$\langle p \rangle = -\int_{-\infty}^{\infty} \psi^* i\hbar \frac{\partial \psi}{\partial x} dx.$$

We have $\psi = A e^{i(kx - \omega t)}$. Hence

$$\langle p \rangle = \hbar k \int_{-\infty}^{\infty} \psi^* \psi \, dx = \hbar k$$

since

$$\int_{-\infty}^{\infty} \psi^* \psi \, dx = 1.$$

2. Show that the solutions to the time-independent Schrödinger equation for a particle moving one dimensionally in a region of zero potential between two rigid walls at $x = -\frac{L}{2}$ to $x = +\frac{L}{2}$ are given by $u_n(x) = C_n \cos k_n x$, where $k_n = \frac{n\pi}{L}$ ($n = 1, 3, 5, \ldots$ etc.) and $u_n(x) = C_n \sin k_n x$, where $k_n = \frac{n\pi}{L}$ ($n = 2, 4, 6, \ldots$ etc.).

Ans. We know that for the particle confined between $x' = 0$ and $x' = L$ the solutions are

$$u_n(x') = C_n \sin k_n x'$$

where

$$k_n = \frac{n\pi}{L} \quad (n = 1, 2, 3, \ldots \text{ etc.}).$$

To obtain the results for the particle confined between $x = -\frac{L}{2}$ to $x = +\frac{L}{2}$ we make a coordinate transformation by putting $x = x' - \frac{L}{2}$.

Therefore

$$u_n(x) = C_n \sin k_n \left(x + \frac{L}{2} \right)$$

$$= C_n \sin k_n x \cos k_n \frac{L}{2} + C_n \cos k_n x \sin k_n \frac{L}{2}.$$

When n is odd, i.e., $n = 1, 3, 5$, etc.,

$$\cos k_n \frac{L}{2} = \cos \frac{n\pi}{2} = 0 \text{ and } \left| \sin k_n \frac{L}{2} \right| = \left| \sin \frac{n\pi}{2} \right| = 1$$

Hence the solution is

$$u_n(x) = C_n \cos k_n x.$$

Free Particles and Wave Packets

Similarly when n is even, *i.e.* $n = 2, 4, 6, \ldots$ etc.,

$$\left|\cos k_n \frac{L}{2}\right| = \left|\cos \frac{n\pi}{2}\right| = 1$$

and

$$\sin k_n \frac{L}{2} = \sin \frac{n\pi}{2} = 0$$

Hence the wave function is

$$u_n(x) = C_n \sin k_n x.$$

3. Show that the wavefunctions given by Eq. (3.18) are orthonormal.

Ans. Let $u_n(x)$ and $u_m(x)$ be the wavefunctions corresponding to the eigenvalues E_n and E_m, respectively, n and m being integers.

Then

$$u_n(x) = \sqrt{\frac{2}{L}} \sin \frac{n\pi}{2} x \text{ and } u_m(x) = \sqrt{\frac{2}{L}} \sin \frac{m\pi}{2} x.$$

We have

$$\int_0^L u_m^* u_n \, dx = \frac{2}{L} \int_0^L \sin \frac{m\pi}{2} x \sin \frac{n\pi}{2} x \, dx$$

$$= 0, \text{ when } n \neq m$$

and

$$\int_0^L u_m^* u_n \, dx = 1, \text{ when } n = m$$

Hence the wavefunctions are orthonormal.

4. Use the wavefunction given by Eq. (3.18) to find the expectation values of x and p for the particle moving between $x = 0$ and $x = L$. Explain your result physically.

Ans. We have $\langle x \rangle = \int_0^L u_n^* x u_n \, dx = \frac{2}{L} \int_0^L x \sin^2 \frac{n\pi}{L} x \, dx = \frac{2}{L} \times \frac{L^2}{4} = \frac{L}{2}$

and

$$\langle p \rangle = -\int_0^L u_n^* i\hbar \frac{\partial u_n}{\partial x} dx = \frac{-2i\hbar}{L} \frac{n\pi}{L} \int_0^L \sin \frac{n\pi}{L} x \cos \frac{n\pi}{L} x \, dx = 0$$

The expectation value of x is $L/2$, *i.e.*, at the centre of the region. This result is expected physically since the particle moves freely between $x = 0$ and L. The expectation value of momentum is zero because the particle moves back and forth between $x = 0$ and $x = L$ and reverses its momentum upon reflection at the ends. The magnitude of the momentum is such that $p^2 = 2mE$ but the momentum is equally likely to be positive and negative. Hence the measurement of momentum is expected to average out to zero.

5. A particle of mass m confined between $0 \leq x \leq L$ has a probability 64% of occupying the ground state and a probability 36% of occupying the first excited state. What is the average energy of the particle?

Ans. The average energy of the particle is $<E> = 0.64 \, E_1 + 0.36 \, E_2$, where $E_1 = \hbar^2 \pi^2/(2mL^2)$ is the ground state energy, and $E_2 = 4\hbar^2 \pi^2/(2mL^2)$ is the first excited state energy. Therefore, $<E> = \hbar^2 \pi^2 (0.64 + 4 \times 0.36)/(2mL^2) = 1.04 \, \hbar^2 \pi^2/(mL^2)$.

6. The eigenvalue equation for the momentum operator is $(\hbar/i)(\partial \psi / \partial x) = \lambda \psi$. Solve this equation and hence show that for ψ to be a physically admissible eigenstate, the eigenvalue λ must be real. (C.U. 1996)

Ans. The given equation can be written as

$$\frac{d\psi}{\psi} = \frac{i\lambda}{\hbar} dx$$

Integrating we have,

$$\psi = \psi_0 e^{i\lambda x/\hbar}, \qquad \ldots(i)$$

where ψ_0 is the integration constant. For ψ to be a physically admissible eigenstate of momentum, the uncertainty in position must be infinite. So ψ must extend from $-\infty$ to $+\infty$ and be well-behaved everywhere. This is possible only when λ in Eq. (i) is real.

7. A particle having zero potential energy is confined between two rigid walls at $x = 0$ and $x = L$. Show that for a very large quantum number, the probability of finding the particle in any small interval Δx between x and $x + \Delta x$ is independent of x.

Ans. The probability of finding the particle between x and $x + \Delta x$ is

$$\int_x^{x+\Delta x} |u_n(x)|^2 \, dx = \frac{2}{L} \int_x^{x+\Delta x} \sin^2 \frac{n\pi x}{L} \, dx$$

$$= \frac{1}{L} \int_x^{x+\Delta x} \left(1 - \cos \frac{2n\pi x}{L}\right) dx$$

$$= \frac{\Delta x}{L} - \frac{1}{2n\pi} \left[\sin \frac{2n\pi}{L}(x+\Delta x) - \sin \frac{2n\pi x}{L}\right]$$

$$= \frac{\Delta x}{L} - \frac{1}{n\pi} \cos \frac{n\pi}{L}(2x + \Delta x) \sin \frac{n\pi \Delta x}{L}.$$

For a very large quantum number n, the last term on the right-hand side is negligible, so that the probability is given by the first term $\Delta x/L$, which is independent of x.

8. Show, by solving Schrödinger's equation, that a free particle cannot have negative energy.

Ans. Schrödinger's equation (in time-independent form) for a free particle moving along the x-axis is

$$-\frac{\hbar^2}{2m} \frac{d^2 u}{dx^2} = Eu$$

where E is the energy of the particle. Writing $\alpha^2 = -2mE/\hbar^2$, we have

$$\frac{d^2 u}{dx^2} = \alpha^2 u \qquad \ldots(i)$$

If $E < 0$, $\alpha^2 > 0$, i.e., α is real. The solution of (i) is

$$u(x) = A e^{\alpha x} + B e^{-\alpha x}$$

where A and B are constants. As $x \to \infty$, $A e^{\alpha x} \to \infty$, and as $x \to -\infty$, $B e^{-\alpha x} \to \infty$. Hence $u(x)$ is not finite and well-behaved everywhere for $E < 0$. Consequently, the free particle cannot have negative energy.

9. A particle of mass m and charge e moves in a circular orbit of radius r. If x is the distance round the orbit,

(a) show that the functions

$$\psi = \frac{1}{\sqrt{2\pi r}} \exp(\pm inx/r),$$

where n is zero or a positive integer, are solutions of Schrödinger's equation with energy $n^2\hbar^2/(2mr^2)$.

(b) Also show that the orbiting particle has a magnetic dipole moment

$$\mu = \pm \frac{ne\hbar}{2m}.$$

Ans. (a) One-dimensional Schrödinger's equation is

$$\frac{d^2 \psi}{dx^2} + k^2 \psi = 0, \text{ where } k = \sqrt{\frac{2mE}{\hbar^2}}$$

Free Particles and Wave Packets 61

The solution is $\psi = Ae^{\pm ikx}$, where A is a constant.
The normalization condition gives

$$\int_0^{2\pi r} |\psi|^2 \, dx = 2\pi r A^2 = 1 \text{ or, } A = \frac{1}{\sqrt{2\pi r}}.$$

Also, since x and $x + 2\pi r$ are the same point, we have

$$e^{\pm ikx} = e^{\pm ik(x+2\pi r)}, \text{ i.e., } e^{\pm i2\pi kr} = 1$$

or, $$2\pi kr = 2n\pi$$

or, $$k = \frac{n}{r}.$$

Therefore,

$$\psi = \frac{1}{\sqrt{2\pi r}} e^{\pm inx/r} \text{ proved.}$$

The energy is

$$E_n = \frac{\hbar^2 k^2}{2m} = \frac{\hbar^2 n^2}{2mr^2}.$$

(b) The magnetic dipole moment is

$$\mu = \text{area} \times \text{current} = \pm \pi r^2 \cdot \frac{e}{\tau},$$

where τ is the period of revolution. If v is the velocity of the particle, then

$$\frac{1}{\tau} = \frac{v}{2\pi r} = \frac{p}{2\pi rm} = \frac{\hbar k}{2\pi rm} = \frac{n\hbar}{2\pi r^2 m}$$

where p is the momentum. Hence

$$\mu = \pm \pi r^2 e \times \frac{n\hbar}{2\pi r^2 m} = \pm \frac{ne\hbar}{2m}, \text{ proved.}$$

10. Can you observe the energy states of a ball of mass 10 g moving in a box along its length which is 10 cm? Given $h = 6.6 \times 10^{-34}$ J.s. *(Kanpur 1987)*

Ans. The energy states are given by

$$E_n = \frac{\pi^2 \hbar^2 n^2}{2mL^2} = \frac{n^2 h^2}{8mL^2}$$

Here $h = 6.6 \times 10^{-34}$ J.s, $m = 10$ g $= 10^{-2}$ kg, and $L = 10$ cm $= 0.1$ m

So, $$E_n = \frac{n^2 \times (6.6 \times 10^{-34})^2}{8 \times 10^{-2} \times (0.1)^2} \text{ J} = \frac{n^2 \times (6.6 \times 10^{-34})^2}{8 \times 10^{-4} \times 1.6 \times 10^{-19}} \text{ eV}$$
$$= n^2 \times 3.4 \times 10^{-45} \text{ eV}$$

The difference between two successive energy levels is

$$\Delta E_n = [(n+1)^2 - n^2] \times 3.4 \times 10^{-45} \text{ eV} = (2n+1) \times 3.4 \times 10^{-45} \text{ eV}$$

For $n = 1, 2, 3, \ldots$, this difference is too small to be observed. So, an energy continuum will be found.

11. A particle of mass m, confined to a one-dimensional zero potential region $0 \le x \le L$, has the normalized wavefunction

$$\psi(x, 0) = 2\sqrt{\frac{2}{5L}} \left(1 + \cos\frac{\pi x}{L}\right) \sin\frac{\pi x}{L}$$

at time $t = 0$. (i) What is the wavefunction at a subsequent time $t = t_0$? (ii) What is the average energy at $t = 0$ and $t = t_0$?

Ans. (i) The energy eigenfunctions are

$$\psi_n = \sqrt{\frac{2}{L}} \sin \frac{n\pi x}{L}$$

and the energy eigenvalues are

$$E_n = \frac{n^2 \pi^2 \hbar^2}{2mL^2}; \quad n = 1, 2, 3,\dots$$

The wavefunction at time t can be written as

$$\psi(x, t) = \sum_n C_n(0)\, \psi_n(x, 0)\, \exp\left(-\frac{iE_n t}{\hbar}\right)$$

Given,

$$\psi(x, 0) = 2\sqrt{\frac{2}{5L}} \left(1 + \cos\frac{\pi x}{L}\right) \sin\frac{\pi x}{L}$$

$$= 2\sqrt{\frac{2}{5L}} \sin\frac{\pi x}{L} + \sqrt{\frac{2}{5L}} \sin\frac{2\pi x}{L}$$

Hence,

$$C_1(0) = \frac{2}{\sqrt{5}},\ C_2(0) = \frac{1}{\sqrt{5}},\ C_n(0) = 0 \text{ for } n \neq 1, 2.$$

Therefore,

$$\psi(x, t_0) = 2\sqrt{\frac{2}{5L}} \exp\left(-\frac{i\pi^2 \hbar t_0}{2mL^2}\right) \sin\frac{\pi x}{L} + \sqrt{\frac{2}{5L}} \exp\left(-\frac{i 2\pi^2 \hbar t_0}{mL^2}\right) \sin\frac{2\pi x}{L}$$

(ii) The average energy is

$$\langle E \rangle = \sum_n |C_n(0)|^2 E_n = \frac{4}{5} E_1 + \frac{1}{5} E_2$$

$$= \frac{4\pi^2 \hbar^2}{5mL^2}.$$

The average energy is independent of t.

QUESTIONS

1. (a) What is a free particle?
   (b) A particle is free to move in a one-dimensional space from $x = -\infty$ to $x = +\infty$. (i) Write the momentum operator in Schrödinger's representation and find its eigenfunction. (ii) What is the probability of finding the particle in a small region dx? Explain your result. (iii) What is the physical significance of normalizing a wavefunction? Can you normalize the above wavefunction? (iv) Do you get discrete energy levels for such a particle? Explain. (C.U. 1984)
   (c) Write the eigenvalue equation for the energy of a free particle in one dimension and show that each eigenvalue is doubly degenerate. (CU. 1999, 2002)
2. A particle of mass m is free to move in a force-free region in one dimension between two rigid walls situated at $x = -\frac{L}{2}$ and $x = +\frac{L}{2}$.
   (a) Find the eigenfunctions and eigenvalues of the Hamiltonian.
   (b) Sketch the wavefunction of the ground state and the first excited state.
   (c) Find the ratio of the energy eigenvalues corresponding to the first excited state and the ground state. (cf. C.U. 1982)

Free Particles and Wave Packets

3. A particle is free to move one dimensionally in a region of zero potential between two rigid walls at $x = 0$ and $x = a$. (i) Find the position probability density in the nth state. Does it agree with the result expected classically? (ii) Show that the nth state is an eigenstate of energy. Is it also an eigenstate of momentum? (iii) If E_n is the energy for the nth state and ΔE_n is the energy separation between the $(n + 1)$ th and the nth state, show that

$$\Delta E_n/E_n = (2n + 1) / n^2$$

4. (a) What is a wave packet?

   (b) Consider a one-dimensional wave packet given by $\psi(x, 0) = Ae^{-\alpha^2 x^2/2} e^{ik'x}$. Normalize the wavefunction and establish the uncertainty relation for this wave packet.

   (c) Show that for a one-dimensional wave packet [$\psi(x)$ vanishes as $x \to \pm\infty$], $\langle p \rangle$ is real.

   (C.U. 2000)

5. Find the lowest energy of an electron moving in a one-dimensional force free region of length 2 Å.

   [Ans. 9.375 eV]

6. A particle of mass m and zero potential energy can move back and forth freely along the x-axis from $x = -a/2$ to $x = +a/2$ and cannot go outside this region. (i) Show that the wavefunction for the lowest energy state of the particle is $\psi(x, t) = \sqrt{\dfrac{2}{a}} \cos \dfrac{\pi x}{a} e^{-iEt/\hbar}$, where E represents the energy of the particle. (ii) Establish the uncertainty relation between momentum and position of the particle. (iii) Explain why the uncertainty in momentum appears.

[Hint. Evaluate $(\Delta x)^2 = \langle x^2 \rangle - \langle x \rangle^2$ and $(\Delta p)^2 = \langle p^2 \rangle - \langle p \rangle^2$

Show that $\langle x^2 \rangle = \dfrac{a^2}{2\pi^2}\left(\dfrac{\pi^2}{6} - 1\right) = 0.033 a^2$, $\langle x \rangle = 0$, $\langle p^2 \rangle = \left(\dfrac{\pi\hbar}{a}\right)^2$, and $\langle p \rangle = 0$. Hence $\Delta x \Delta p = 0.18 a \dfrac{\pi \hbar}{a}$

$= 0.57 \hbar$, which is the uncertainty relation. Δp is a measure of fluctuations about the average $\langle p \rangle = 0$ that are found in observing the momentum of the particle. The fluctuations appear because both the directions of momentum are possible, i.e., the particle is sometimes observed with momentum $p = +\sqrt{2mE}$ and sometimes with momentum $p = -\sqrt{2mE}$.]

7. For a free particle show that Schrödinger's time-independent wave equation leads to the de Broglie relation $\lambda = h/p$.   (C.U. 1985)

8. The wavefunction of a particle is $\psi(x) = (2/L)^{1/2} \sin(\pi x/L)$ for $0 \le x \le L$, and $\psi(x) = 0$ elsewhere. What is the probability of finding the particle in the region (i) $0 \le x \le L/4$, (ii) $0 \le x \le L/2$, and (iii) $0 \le x \le 3L/4$? What is the average position of the particle?

   [Ans. $(\pi - 2)/(4\pi)$, $1/2$, $(3\pi + 2)/(4\pi)$, $L/2$]

9. (a) A particle of mass m and zero potential energy moves to-and-fro along the x-axis between two rigid walls at $x = 0$ and $x = L$. Obtain the energy eigenvalues. (b) Show that the lowest energy level agrees with the uncertainty principle in order of magnitude. (c) How are the eigenfunctions and eigenvalues modified when the particle is confined in a three-dimensional box?

10. The wavefunction of a free particle moving along the x-axis is given by $\psi = Ae^{ikx} + Be^{-ikx}$. What is the probability of the particle moving in the (i) positive x-direction and (ii) negative x-direction?

    [Hint. The required probabilities are $|J_x|/(|J_x| + |J_{-x}|)$ and $|J_{-x}|/(|J_x| + |J_{-x}|)$, where J_x and J_{-x} are the probability current densities for forward and backward waves, respectively.]

    [Ans. $A^2/(A^2 + B^2)$, $B^2/(A^2 + B^2)$]

11. The wavefunction of a particle is $\psi(x) = A \exp(-\alpha x^2)$, $(-\infty < x < \infty)$ where A and α are constants. Normalize the wavefunction. What is the probability of finding the particle in the region $0 < x < \infty$?

    (Burd. U. 1995) [Ans. $(2\alpha/\pi)^{1/4}$, $1/2$]

12. Set up the time-independent Schrödinger equation for the three-dimensional potential $V(x, y, z) = 0$ for $0 < x < L$, $0 < y < L$, $0 < z < L$; $V(x, y, z) = \infty$ elsewhere. Write the appropriate boundary conditions. Obtain the energy eigenvalues and the corresponding eigenfunctions.   (C.U. 1997)

13. The normalized wavefunction $\psi(x) = \left(\dfrac{30}{a}\right)^{1/2} \left(\dfrac{x}{a}\right)\left(1 - \dfrac{x}{a}\right)$ describes a particle at $t = 0$ in a rigid box $(0 < x < a)$. (*i*) Calculate the expectation value of energy. (*ii*) Comparing this energy with the energy levels in the box, find the state in which the particle is most likely to be found. (*iii*) Determine the probability of obtaining the lowest energy level from energy measurements.

[**Hint.** (*iii*) Writing $\psi(x) = \Sigma\, C_i\, u_i(x)$, we have $C_i = \int_0^a \psi(x)\, u_i(x)\, dx$. The probability of occupancy of the level i is C_i^2.] [**Ans.** $\dfrac{5\hbar^2}{ma^2}$, $n = 1$, 0.999]

14. Show explicitly that for a particle in a one-dimensional box of length L, the eigenfunctions belonging to the eigenvalues E_1 and E_2 are orthogonal $(E_1 \neq E_2)$ **(N. Beng. U. 2001)**

Chapter 4

Application of Quantum Mechanics to Some Potential Problems

In this chapter we shall study solutions of the time-independent Schrödinger equation when a particle passes from one region of constant potential into a number of adjacent regions each of which is characterized by a different constant potential. The potentials are assumed to change in a discontinuous manner while going from one region to the next. Also we shall confine our attention to a one-dimensional geometry for the simplicity of the analysis.

4.1. PARTICLE AT A RECTANGULAR POTENTIAL BARRIER

Consider the simplest case of a particle moving from region 1 of constant potential V_1 to region 2 of constant potential V_2 as shown in Fig. 4.1. If u_1 and u_2 denote respectively the energy eigenfunctions in regions 1 and 2, we have

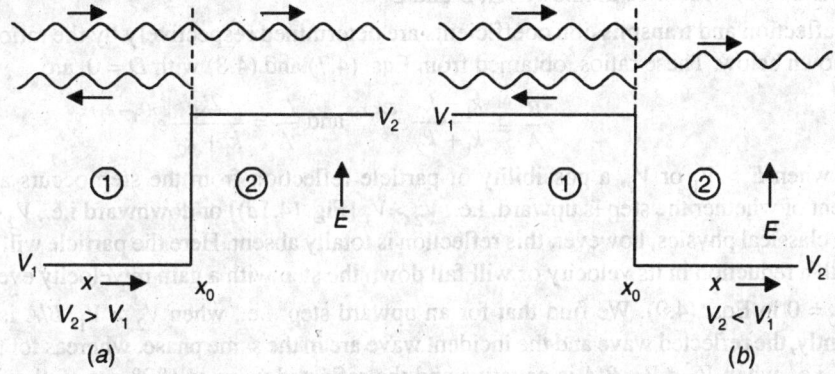

Fig. 4.1. A particle with an energy E is incident on a potential step, where $E > V_1$ or V_2

$$\left(-\frac{\hbar^2}{2m}\frac{d^2}{dx^2} + V_1\right)u_1 = Eu_1, \qquad \ldots(4.1)$$

$$\left(-\frac{\hbar^2}{2m}\frac{d^2}{dx^2} + V_2\right)u_2 = Eu_2, \qquad \ldots(4.2)$$

where E represents the total energy of the particle and m its mass. Let us now consider two cases:

Case 1 : $E > V_1$ or V_2

Under this condition the solutions of Eqs. (4.1) and (4.2) are

$$u_1 = Ae^{ik_1x} + Be^{-ik_1x} \qquad \ldots(4.3)$$

and

$$u_2 = Ce^{ik_2x} + De^{-ik_2x}, \qquad \ldots(4.4)$$

where

$$k_1 = \left[\frac{2m(E-V_1)}{\hbar^2}\right]^{1/2} \qquad \ldots(4.5)$$

and

$$k_2 = \left[\frac{2m(E-V_2)}{\hbar^2}\right]^{1/2} \qquad \ldots(4.6)$$

Equations (4.3) and (4.4) represent plane waves for nonlocalized particles and contain all the basic information. Here we shall not consider localization or normalization since they can be accomplished by constructing a wave packet.

Since $V(x)$ is finite everywhere we must keep E, u_1 and u_2 finite; otherwise, the energies and the position probability densities will be infinite. These conditions require that $\dfrac{d^2 u_1}{dx^2}$ and $\dfrac{d^2 u_2}{dx^2}$ in Eqs. (4.1) and (4.2) must be finite everywhere. As a consequence, $\dfrac{du}{dx}$ and u must be continuous. Applying these conditions at $x = x_0$ we obtain from Eqs. (4.3) and (4.4)

$$A e^{ik_1 x_0} + B e^{-ik_1 x_0} = C e^{ik_2 x_0} + D e^{-ik_2 x_0} \qquad \ldots(4.7)$$

and

$$ik_1 \left(A e^{ik_1 x_0} - B e^{-ik_1 x_0} \right) = ik_2 \left(C e^{ik_2 x_0} - D e^{-ik_2 x_0} \right) \qquad \ldots(4.8)$$

In Eq. (4.3) the terms $A e^{ik_1 x}$ and $B e^{-ik_1 x}$ denote respectively the incident and the reflected waves in region 1. Similarly, the quantity $C e^{ik_2 x}$ is the transmitted wave in region 2 moving away from the potential discontinuity at $x = x_0$. Note that in this region there cannot be any reflected wave since this region extends to infinity without any reflecting discontinuity. Therefore, $D = 0$. Thus we are left with Eqs. (4.7) and (4.8) with three unknowns A, B and C.

The reflection and transmission coefficients are determined respectively by the ratios B/A and C/A, as shown below. These ratios, obtained from Eqs. (4.7) and (4.8) with $D = 0$, are

$$\frac{B}{A} = \frac{k_1 - k_2}{k_1 + k_2} e^{i 2 k_1 x_0} \text{ and } \frac{C}{A} = \frac{2 k_1}{k_1 + k_2} e^{i(k_1 - k_2) x_0} \qquad \ldots(4.9)$$

Thus when $E > V_1$ or V_2, a possibility of particle reflection from the step occurs and this is independent of whether the step is upward, i.e., $V_2 > V_1$ [Fig. (4.1a)] or downward i.e., $V_2 < V_1$ [Fig. (4.1b)]. In classical physics, however, this reflection is totally absent. Here the particle will climb the barrier with a reduction in its velocity or will fall down the step with a gain in velocity every time.

Let $x_0 = 0$ in Eqs. (4.9). We find that for an upward step, i.e., when $V_2 > V_1$, B/A is positive. Consequently, the reflected wave and the incident wave are in the same phase, whereas for the downward step, i.e., when $V_2 < V_1$, B/A is negative and the reflected wave is 180° out of phase with the incident wave. The first case is similar to the case of light reflected from an interface in passing from a denser medium to a rarer medium, and the second case is similar to the reflection of light in passing from a rarer to a denser medium.

The probability current density for the incident wave is

$$J_I = \frac{i\hbar}{2m} \left(\psi_I \frac{d\psi_I^*}{dx} - \psi_I^* \frac{d\psi_I}{dx} \right)$$

where ψ_I is the wavefunction for the incident wave. Since $\psi_I = A e^{ik_1 x}$, we have

$$J_I = \frac{i\hbar}{2m} |A|^2 (-ik_1 - ik_1) = \frac{\hbar k_1}{m} |A|^2$$

Since the reflected wave is $\psi_R = B e^{-ik_1 x}$, the probability current density for the reflected wave is

$$J_R = \frac{i\hbar}{2m} \left(\psi_R \frac{d\psi_R^*}{dx} - \psi_R^* \frac{d\psi_R}{dx} \right) = -\frac{\hbar k_1}{m} |B|^2$$

The transmitted wave is $\psi_T = C e^{ik_2 x}$. So, the probability current density for the transmitted wave is

$$J_T = \frac{i\hbar}{2m} \left(\psi_T \frac{d\psi_T^*}{dx} - \psi_T^* \frac{d\psi_T}{dx} \right) = \frac{\hbar k_2}{m} |C|^2$$

Application of Quantum Mechanics to Some Potential Problems

The ratio $|J_R/J_I|$ is referred to as the *reflection coefficient*, R. Thus

$$R = \left|\frac{J_R}{J_I}\right| = \left|\frac{B}{A}\right|^2 = \left(\frac{B}{A}\right)\left(\frac{B}{A}\right)^* = \left(\frac{k_1 - k_2}{k_1 + k_2}\right)^2 \quad (4.9a)$$

using Eq. (4.9). The ratio $|J_T/J_I|$ is termed the *transmission coefficient*, T. Using Eq. (4.9) we obtain

$$T = \left|\frac{J_T}{J_I}\right| = \left|\frac{C}{A}\right|^2 \frac{k_2}{k_1} = \left(\frac{C}{A}\right)\left(\frac{C}{A}\right)^* \frac{k_2}{k_1} = \frac{4k_1 k_2}{(k_1 + k_2)^2} \quad (4.9b)$$

From Eqs. (4.9a) and (4.9b), we find that

$$R + T = 1$$

This shows that the probability is conserved at the interface between the regions 1 and 2. The particle is either reflected or transmitted; it cannot remain on the interface.

From Eq. (4.9) we have $|B/A| < 1$ and $|C/A| > 1$ when $V_2 > V_1$, for then $k_1 > k_2$. Hence the amplitude of the reflected wave is less and that of the transmitted wave is greater than the amplitude of the incident wave. The kinetic energy of the particle is larger in region 1 than in region 2, and the de Broglie wavelength is shorter in region 1 then.

Case 2 : $V_1 < E < V_2$

This situation is depicted in Fig. 4.2. Classically this means that a particle is incident on a potential step with a kinetic energy that is insufficient to climb the step. In region 1, k_1 is real [see Eq. 4.5] and so plane wave solutions are possible and a reflection probability can be considered. In region 2, however k_2 is imaginary as is evident from Eq. (4.6). So we have in region 2

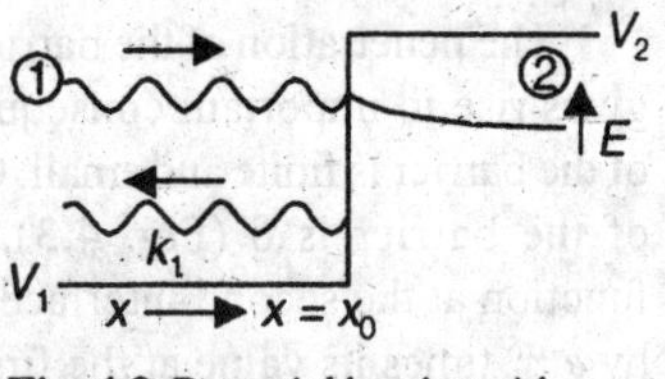

$$u_2 = Ce^{-\alpha x}, \quad \ldots(4.10)$$

where

$$\alpha = \left[\frac{2m(V_2 - E)}{\hbar^2}\right]^{1/2} \quad \ldots(4.11)$$

Fig. 4.2. Potential barrier with an incident particle of energy E, where $V_1 < E < V_2$.

Note that D is assumed zero again; otherwise, the quantity $De^{+\alpha x}$ would produce divergence at infinity. Since u and $\frac{du}{dx}$ are continuous at $x = x_0$ we obtain

$$Ae^{ik_1 x_0} + Be^{-ik_1 x_0} = Ce^{-\alpha x_0} \quad \ldots(4.12)$$

and

$$ik_1\left(Ae^{ik_1 x_0} - Be^{-ik_1 x_0}\right) = -C\alpha e^{-\alpha x_0} \quad \ldots(4.13)$$

Solving Eqs. (4.12) and (4.13) we get

$$\frac{B}{A} = -\frac{\alpha + ik_1}{\alpha - ik_1} e^{i2k_1 x_0} \quad \ldots(4.14)$$

From Eq. (4.14) we find that $|B/A|^2 = 1$. That is, like the classical case, all the incident particles are reflected. However, in this case C/A has the nonzero value given by

$$\frac{C}{A} = \frac{2ik_1}{ik_1 - \alpha} e^{ik_1 x_0} e^{\alpha x_0}$$

Therefore,

$$u_2 = \frac{2ik_1 A}{ik_1 - \alpha} e^{ik_1 x_0} e^{-\alpha(x - x_0)} \quad \ldots(4.15)$$

Evidently, the particle penetrates the reflecting wall, and the amount of penetration depends on α. In the region $x > x_0$ the probability density is $u_2^* u_2$. From Eq. (4.15) we obtain

$$u_2^* u_2 = \frac{4k_1^2 AA^*}{\alpha^2 + k_1^2} e^{-2\alpha(x-x_0)} \qquad \ldots(4.15a)$$

It appears from Eq. (4.15a) that the probability of finding the particle with a coordinate $x > x_0$ becomes appreciable in a region starting at $x = x_0$ and extending over a *penetration distance* d_p given by $d_p = 1/\alpha$. Obviously, the penetration distance is the depth in the region 2 where the wave function drops to $1/e$ of its value at the interface. We have

$$d_p = \frac{1}{\alpha} = \frac{\hbar}{\sqrt{2m(V_2 - E)}}. \qquad \ldots(4.16)$$

The barrier neither stores particles nor it allows transmission of particles through it since the reflection coefficient $|B/A|^2 = 1$. The penetration of the barrier can be thought of as yielding a dwell time on the barrier, which results from the complex nature of B/A (phase shift due to reflection is neither zero nor 180°).

QUANTUM MECHANICAL TUNNELING

The penetration of the barrier by the incident wave gives rise to important consequences when the width of the barrier is finite and small. Consider that the width of the barrier is δ (Fig. 4.3). Therefore the wave function at the second interface $(x_0 + \delta)$ will be given by $e^{-\alpha\delta}$ times its value at the first interface (x_0).

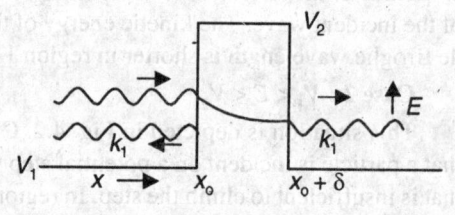

Fig. 4.3. Particle incident on a barrier $(E < V_2)$.

From the second interface the wave can come out as a propagating wave again. This phenomenon, referred to as *quantum mechanical tunneling*, has significant practical consequences.

We shall now tackle the algebra of the tunneling problem by setting $x_0 = 0$ and derive an expression for the transmission probability through the barrier of width δ.

Let u_1, u_2 and u_3 be the eigenfunctions corresponding to the regions $x < 0$, $0 < x < \delta$ and $x > \delta$, respectively. Solving the time-independent Schrödinger equation for the three regions we get

$$u_1 = A_1 e^{ik_1 x} + A_2 e^{-ik_1 x}; (x \leq 0) \qquad \ldots(4.17)$$

$$u_2 = A_3 e^{\alpha x} + A_4 e^{-\alpha x}; (0 \leq x \leq \delta) \qquad \ldots(4.18)$$

$$u_3 = A_5 e^{ik_1 x}; (x \geq \delta) \qquad \ldots(4.19)$$

where

$$k_1 = \left[\frac{2m}{\hbar^2}(E - V_1)\right]^{1/2}$$

and

$$\alpha = \left[\frac{2m}{\hbar^2}(V_2 - E)\right]^{1/2}$$

The continuity of u and $\dfrac{du}{dx}$ at $x = 0$ and at $x = \delta$, when applied to Eqs. (4.17) to (4.19), gives

$$A_1 + A_2 = A_3 + A_4 \qquad \ldots(4.20)$$

$$ik_1(A_1 - A_2) = \alpha(A_3 - A_4) \qquad \ldots(4.21)$$

$$A_3 e^{\alpha\delta} + A_4 e^{-\alpha\delta} = A_5 e^{ik_1 \delta} \qquad \ldots(4.22)$$

$$\alpha(A_3 e^{\alpha\delta} - A_4 e^{-\alpha\delta}) = ik_1 A_5 e^{ik_1 \delta} \qquad \ldots(4.23)$$

From the relations (4.20) and (4.21) we get

$$(\alpha + ik_1) A_1 + (\alpha - ik_1) A_2 = 2\alpha A_3 \quad \ldots(4.24)$$

and
$$(\alpha - ik_1) A_1 + (\alpha + ik_1) A_2 = 2\alpha A_4 \quad \ldots(4.25)$$

Similarly, from Eqs. (4.22) and (4.23) we can write

$$2\alpha A_3 e^{\alpha\delta} = (\alpha + ik_1) A_5 e^{ik_1\delta} \quad \ldots(4.26)$$

and
$$2\alpha A_4 e^{-\alpha\delta} = (\alpha - ik_1) A_5 e^{ik_1\delta} \quad \ldots(4.27)$$

Putting the values of $2\alpha A_3$ and $2\alpha A_4$ from Eqs. (4.24) and (4.25) into Eqs. (4.26) and (4.27) respectively, and eliminating A_2 from the resulting equations, we get

$$\frac{A_5}{A_1} = \frac{e^{-ik_1\delta}}{\cosh(\alpha\delta) + \frac{i(\alpha^2 - k_1^2)}{2k_1\alpha}\sinh(\alpha\delta)} \quad \ldots(4.28)$$

Since the incident wave is $A_1 e^{ik_1 x}$ and the transmitted wave is $A_5 e^{ik_1 x}$ the probability current density for the incident wave is $J_I = \hbar k_1 |A_1|^2 / m$ and that for the transmitted wave is $J_T = \hbar k_1 |A_5|^2 / m$. Therefore the transmission coefficient is

$$T = \frac{J_T}{J_I} = \left|\frac{A_5}{A_1}\right|^2 = \frac{1}{1 + \sinh^2(\alpha\delta)\left[1 + \frac{(\alpha^2 - k_1^2)^2}{4\alpha^2 k_1^2}\right]}$$

$$= \left[1 + \frac{(V_2 - V_1)^2 \sinh^2 \alpha\delta}{4(E - V_1)(V_2 - E)}\right]^{-1} \quad \ldots(4.29)$$

In Eq. (4.29), $V_1 < E < V_2$. If the energy of the particle is greater than the barrier height, i.e., when $E > V_2$ it can be similarly shown that

$$\left|\frac{A_5}{A_1}\right|^2 = \left[1 + \frac{(V_2 - V_1)^2 \sin^2 k_2\delta}{4(E - V_1)(E - V_2)}\right]^{-1} \quad \ldots(4.30)$$

where
$$k_2 = \left[\frac{2m(E - V_2)}{\hbar^2}\right]^{1/2} \quad \ldots(4.31)$$

A plot of the transmission coefficient $\left|\frac{A_5}{A_1}\right|^2$ against $\left(\frac{E - V_1}{V_2 - V_1}\right)$ is shown in Fig. (4.4). It is obvious from the figure that even when $E < V_2$ there is transmission through the barrier in violation of the classical dynamics which predicts complete reflection from the barrier and no transmission through it.

Note that the transmission coefficient is oscillatory when $E > V_2$. The transmission is total for several incident energy values and the transmission drops off for energies slightly larger or slightly less. Maximum transmission occurs when $k_2\delta = n\pi$, i.e. when the width of the barrier is equal to an integral number of half wavelengths. This type of constructive interference provides the physical basis for thin-film optical filters.

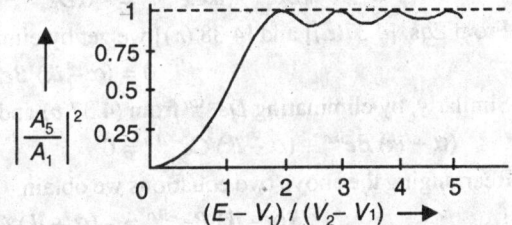

Fig. 4.4. Transmission probability as a function of energy.

Quantum mechanical tunneling is responsible for the operation of tunnel diodes and electron emission through thin insulating films. Tunneling also offers an explanation for Zener breakdown in semiconductor junction diodes.

Let $V_1 = 0$ in Eq. (4.29). If $\alpha\delta \gg 1$, then $\sinh^2 \alpha\delta \simeq \frac{1}{4} \exp(2\alpha\delta) \gg 1$. In this case, Eq. (4.29) reduces to (vide worked-out problem no. 2)

$$T = \frac{16E(E - V_2)}{V_2^2} \exp(-2\alpha\delta) \qquad ...(4.31a)$$

4.2. Particle In A One-Dimensional Square Well

An important new feature of quantum systems results when we study the behaviour of a particle in a finite square-well potential. The square-well potential is defined by

$$V(x) = V \text{ for } x < -a \text{ or } x > a$$
$$= 0 \text{ for } -a < x < a \qquad ...(4.32)$$

The one-dimensional square-well potential is shown in Fig. 4.5.

Consider first the bound states for which the total energy E of the particle is less than the depth of the potential well V, i.e. $E < V$. For convenience, we denote the regions $x < -a$, $-a < x \leq a$ and $x > a$ by 1, 2 and 3 and the energy eigenfunctions in these regions by u_1, u_2 and u_3, respectively. The solutions of the time-independent Schrödinger equations give

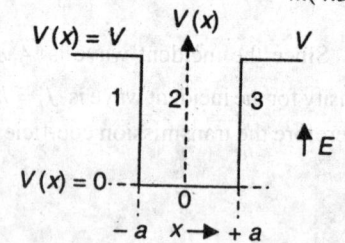

Fig. 4.5. One-dimensional square-well potential.

$$u_1 = Ae^{\alpha x} \qquad ...(4.33)$$
$$u_2 = Be^{ikx} + Ce^{-ikx} \qquad ...(4.34)$$

and
$$u_3 = De^{-\alpha x}, \qquad ...(4.35)$$

where we have incorporated the boundary conditions, namely, the wave functions become zero as $x \to \pm\infty$. The quantities α and k are given by

$$\alpha = \left[\frac{2m(V - E)}{\hbar^2}\right]^{1/2}; \; k = \left[\frac{2mE}{\hbar^2}\right]^{1/2} \qquad ...(4.36)$$

By applying the conditions that u and $\frac{du}{dx}$ are continuous at $x = -a$ and $x = +a$ we get

$$Ae^{-\alpha a} = Be^{-ika} + Ce^{ika} \qquad ...(4.37a)$$
$$Be^{ika} + Ce^{-ika} = De^{-\alpha a} \qquad ...(4.37b)$$
$$\alpha Ae^{-\alpha a} = ikBe^{-ika} - ik Ce^{ika} \qquad ...(4.38a)$$
$$ikBe^{ika} - ikCe^{-ika} = -\alpha De^{-\alpha a} \qquad ...(4.38b)$$

From Eqs. [4.37(a)] and [4.38(a)] we get by eliminating $Ae^{-\alpha a}$

$$0 = (\alpha - ik) Be^{-ika} + (\alpha + ik) Ce^{ika} \qquad ...(4.39a)$$

Similarly, by eliminating $De^{-\alpha a}$ from (4.37 b) and (4.38 b) we get

$$(\alpha + ik) Be^{ika} + (\alpha - ik) Ce^{-ika} = 0 \qquad ...(4.40a)$$

Rearranging the above two equations we obtain

$$(\alpha - ik) Be^{-ika} = -(\alpha + ik) Ce^{ika} \qquad ...(4.39b)$$
$$(\alpha + ik) Be^{ika} = -(\alpha - ik) Ce^{-ika} \qquad ...(4.40b)$$

Multiplying the left-hand sides and the right-hand sides of Eqs. (4.39b) and (4.40b) we obtain

$$(\alpha^2 + k^2) B^2 = (\alpha^2 + k^2) C^2 \qquad ...(4.41)$$

Hence $B^2 = C^2$

or, $B = \pm C$ ...(4.42)

Application of Quantum Mechanics to Some Potential Problems

Therefore, the solutions inside the square-well are either of the following forms:

(i) $$e^{ikx} + e^{-ikx} = 2\cos kx, \qquad \ldots(4.43)$$

which are even functions of x.

(ii) $$e^{ikx} - e^{-ikx} = i2\sin kx, \qquad \ldots(4.44)$$

which are odd functions of x.

The equations in (4.37) and (4.38) form a set of homogeneous equations in the coefficients A, B, C and D. In order that they will have nonzero solutions, the determinant of the coefficients must vanish. These equations are linearly dependent and only give ratios of the coefficients. The actual magnitudes of the coefficients may be obtained by normalization:

$$\int_{-\infty}^{\infty} u^* u \, dx = \int_{-\infty}^{-a} u_1^* u_1 \, dx + \int_{-a}^{a} u_2^* u_2 \, dx + \int_{a}^{\infty} u_3^* u_3 \, dx = 1 \qquad \ldots(4.45)$$

This is a laborious calculation and we shall not perform it, particularly because the exact knowledge of these coefficients will not be required in the following discussion.

Using the relation (4.42) in either Eq. (4.39 b) or (4.40 b) we get

$$\frac{\alpha - ik}{\alpha + ik} = \mp e^{2ika} \qquad \ldots(4.46)$$

Rationalizing the left-hand side of the above equation and equating the real and imaginary parts from both sides we find if $B = C$,

$$\cos 2ka = -\frac{\alpha^2 - k^2}{\alpha^2 + k^2} \qquad \ldots(4.47a)$$

and

$$\sin 2ka = \frac{2\alpha k}{\alpha^2 + k^2} \qquad \ldots(4.47b)$$

If, on the other hand, $B = -C$ we get

$$\cos 2ka = \frac{\alpha^2 - k^2}{\alpha^2 + k^2} \qquad \ldots(4.48a)$$

and

$$\sin 2ka = -\frac{2\alpha k}{\alpha^2 + k^2} \qquad \ldots(4.48b)$$

Consider the case $B = C$ first when u_2 is an even function of x. We obtain from Eq. (4.47 a):

$$\tan^2 ka = \frac{\alpha^2}{k^2} \qquad \ldots(4.49)$$

or,

$$\tan ka = \pm \frac{\alpha}{k}$$

In order to satisfy Eq. (4.47b) we must retain only the plus sign. Therefore

$$k \tan ka = \alpha \qquad \ldots(4.49a)$$

If

$$\mu^2 = \frac{2mV}{\hbar^2}, \qquad \ldots(4.50)$$

then from (4.36) we get $\alpha^2 = \mu^2 - k^2$. Equation (4.49a) therefore reduces to

$$k \tan ka = \sqrt{\mu^2 - k^2} \qquad \ldots(4.51)$$

Similarly, when $B = -C$, u_2 is an odd function of x and we obtain

$$-k \cot ka = +\sqrt{\mu^2 - k^2} \qquad \ldots(4.52)$$

Equations (4.51) and (4.52) are satisfied only for a certain set of discrete values of k and hence of energy E. This is shown in Fig. 4.6. Here a graphical solution of the transcendental Eqs. (4.51) and (4.52) are given. The points of intersection of the circle $y = \sqrt{\mu^2 - k^2}$ with the plots of $y = k \tan ka$ and $y = -k \cot ka$ are the allowed values of k. Denoting them by k_n we obtain the energy eigenvalues

$$E_n = \frac{\hbar^2 k_n^2}{2m} \quad (n = 1, 2, ...). \quad ...(4.53)$$

Observe that there are only a finite number of intersection points. For a very shallow well $V \to 0$ and therefore μ, the radius of circle, becomes also very small. Consequently, there will be one and only one energy level in the range $0 < E < V$. The number of energy states increases for a deeper well.

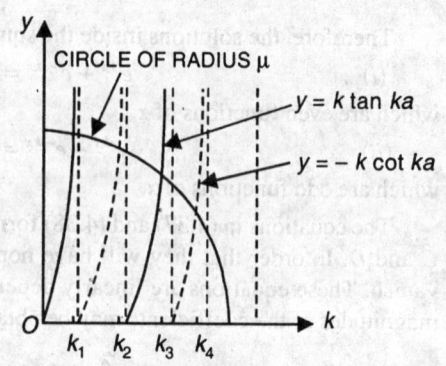

Fig. 4.6. Graphical solution of Eqs. (4.51) and (4.52).

It is seen that for $n = 1, 3, 5, ...$ the wave function is an even function of x whereas for $n = 2, 4, 6, ...$ it becomes an odd function of x, as shown in Fig. 4.7. In fact, it can be shown quite generally that when the potential $V(x)$ is an even function of x, the eigenfunctions must be either even or odd functions of x. Such wavefunctions are said to possess even or odd *parity*.

Let us now consider the case when $E > V$. This gives the solutions u_1 and u_3 oscillatory in nature like the solution u_2 inside. In this case it will be seen that the

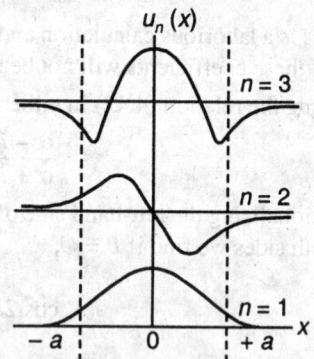

Fig. 4.7. Wave function as a function of x_1 for $n = 1, 2,$ and 3.

Schrödinger equation along with the boundary conditions will be satisfied for any value of energy E and a continuum of states or energy will be obtained. This is evident from the following analysis.

We know that when $V(x)$ is an even function of x, the eigenfunctions become either even or odd functions of x. Considering the *even eigenfunctions* we can write

$$\begin{aligned} u_3(x) &= A_1 \cos k_0 x + A_2 \sin k_0 x & (x > a) \\ u_1(x) &= u_3(-x) & (x < -a) \\ u_2(x) &= A_0 \cos kx & (-a < x < a) \end{aligned} \quad ...(4.54)$$

where k_0 is now given by

$$k_0 = \left[\frac{2m(E-V)}{\hbar^2}\right]^{1/2} \quad ...(4.55)$$

Applying continuity of u and $\frac{du}{dx}$ at $x = a$, we obtain

$$A_0 \cos ka = A_1 \cos k_0 a + A_2 \sin k_0 a,$$

and
$$-k A_0 \sin ka = -k_0 A_1 \sin k_0 a + k_0 A_2 \cos k_0 a$$

Dividing these equations by A_0 and solving them for A_1/A_0 and A_2/A_0, we obtain

$$\frac{A_1}{A_0} = \cos ka \cos k_0 a + \frac{k}{k_0} \sin ka \sin k_0 a,$$

and
$$\frac{A_2}{A_0} = \cos ka \sin k_0 a - \frac{k}{k_0} \sin ka \cos k_0 a \quad ...(4.56)$$

Thus it is always possible to obtain perfectly good values for A_1 and A_2 in terms of A_0, where the

Application of Quantum Mechanics to Some Potential Problems 73

value of A_0 is fixed by normalisation requirements. This means that a solution of u_1, u_2 and u_3 satisfying all the boundary conditions can be obtained for any value of E or k chosen initially. Thus we see that for $E < V$ discrete energy levels will occur, and for $E > V$ there will be a continuum of energy states (Fig. 4.8).

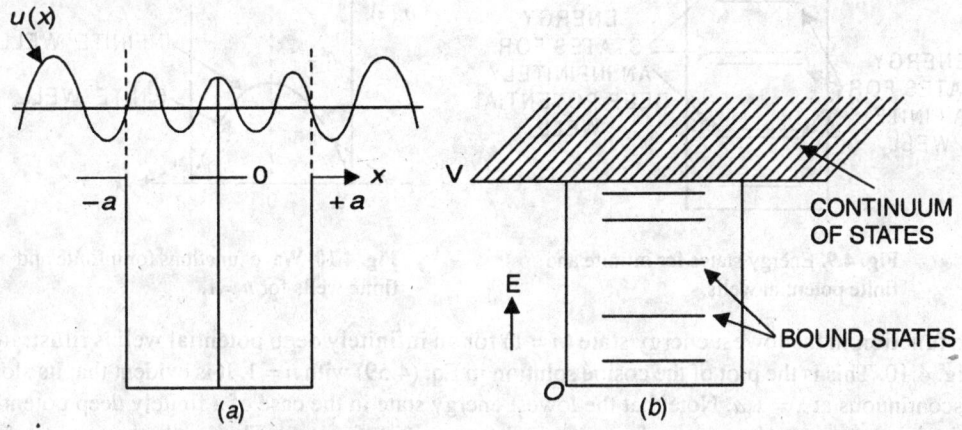

Fig. 4.8. (*a*) Variation of eigenfunction; (*b*) Energy states for $E > V$ and $E < V$.

4.3. Particle In An Infinitely Deep Potential Well

In order to obtain the general feature of the eigenenergies and to evolve a useful artifice for identifying the bound states for a square-well potential of finite depth it is essential to consider the limiting case when $V \to \infty$. We note from Eqs. (4.36) that as $V \to \infty$, $\alpha \to \infty$. It is evident from Eqs. (4.33) and (4.35) that the wave functions outside the potential well become zero in this case. The wave function u_2 inside the well, on the other hand, is non-zero and is given by

$$u_2 = B' \cos kx + C' \sin kx, \qquad \ldots(4.56)$$

where

$$k = \left(\frac{2mE}{\hbar^2}\right)^{1/2}$$

Since $u = 0$ at $x = \pm a$ we obtain

$$B' \cos ka + C' \sin ka = B' \cos ka - C' \sin ka = 0 \qquad \ldots(4.57)$$

which is possible, if $\qquad B' \cos ka = C' \sin ka = 0 \qquad \ldots(4.58)$

This means that the solutions are divided into two classes:

(*i*) $\qquad \cos ka = 0$ and $C' = 0$; and

(*ii*) $\qquad \sin ka = 0$ and $B' = 0$.

The complete set of the eigenfunction u_2 is therefore given by

$$u_2 = \left(\frac{1}{a}\right)^{1/2} \cos\left(\frac{n\pi x}{2a}\right); \quad n = 1, 3, 5, \ldots$$

or, $\qquad u_2 = \left(\frac{1}{a}\right)^{1/2} \sin\left(\frac{n\pi x}{2a}\right); \quad n = 2, 4, 6, \ldots \qquad \ldots(4.59)$

where the normalization condition $\int_{-a}^{+a} u_2^2 \, dx = 1$ has been used.

The eigenstates E are

$$E = \frac{\hbar^2 k^2}{2m} = \frac{\hbar^2 n^2 \pi^2}{2m(2a)^2}, \qquad \ldots(4.60)$$

The discrete k values occur at $ka = \frac{n\pi}{2}$; therefore from Fig. 4.6 we see that the energy eigenvalues for an infinite well are higher than those for a finite well. This is indicated in Fig. 4.9. The

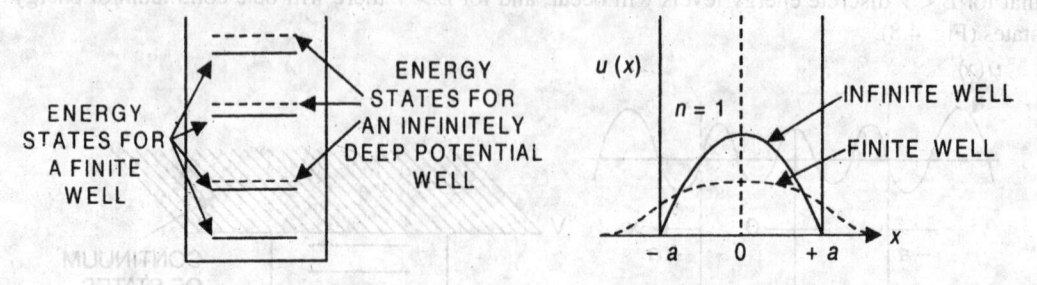

Fig. 4.9. Energy states for infinite and finite potential wells.

Fig. 4.10. Wave functions for infinite and finite wells for $n = 1$.

wave function in the lowest energy state ($n = 1$) for an infinitely deep potential well is illustrated in Fig. 4.10. This is the plot of the cosine solution in Eq. (4.59) with $n = 1$. It is evident that its slope is discontinuous at $x = \pm a$. Note that the lowest energy state in the case of a finitely deep potential well is also represented by an even function similar to a cosine function. The next higher energy state is an odd function like a sine function, and so on. Therefore the infinitely deep potential well provides a useful means for identifying and labeling solutions for a finitely deep well.

The probability density function u^*u for the first few eigenfunctions of the infinite square well are shown in Fig. 4.11. The dotted curves are the predictions of classical mechanics.

It is seen that the quantum mechanical probability density oscillates more and more as n increases. In the limit of n tending to infinity the quantum mechanical probability density will approach the probability density that is predicted by classical mechanics. The classical behaviour regarding the energy continuum can be drawn by considering the fractional separation of the energy eigenvalues as n approaches infinity.

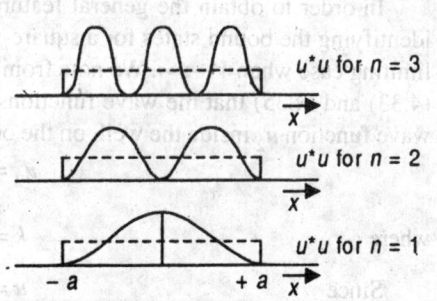

Fig. 4.11. Probability densities for $n = 1, 2,$ and 3 for infinite square well.

We note from the square-well potential problem of finite depth that in the bound state the particle may be found in the classically forbidden region beyond the boundaries of the well with a finite probability. Thus if the walls of the well are of finite height and thickness it is possible for the particle to escape from the well by tunneling. Such a tunneling process can explain the spontaneous α decay of radioactive nuclei, as discussed in the following section.

4.4. Emission Of α– Particles From A Radioactive Element

Elements having atomic number $Z \geq 81$ usually undergo spontaneous disintegration with the emission of α-particles, β-particles and γ-rays. This phenomenon is known as *natural radioactivity*. An α-particle is made up of two neutrons and two protons. The potential energy of an α-particle emitted from a radioactive element of atomic number Z is given by

$$V(r) = \frac{(Z-2)e}{4\pi\epsilon_0 r} 2e = \frac{2e^2(Z-2)}{4\pi\epsilon_0 r}, \qquad \ldots(4.61)$$

where r is the distance of the α-particle from the residual nucleus and ϵ_0 is the free space permittivity ($\epsilon_0 = 8.854 \times 10^{-12}$ F/m). Note that Eq. (4.61) is valid for distances greater than r_0, the effective radius of the nucleus (Fig. 4.12). When the α-particle is inside the nucleus, *i.e.*, $r \leq r_0$, a strong force of attraction between the α-particle and the nucleus comes into play and exists over a short distance r_0 from the centre of the nucleus. This means that for $r \leq r_0$ the potential energy of the α-particle is negative as shown in Fig. 4.12.

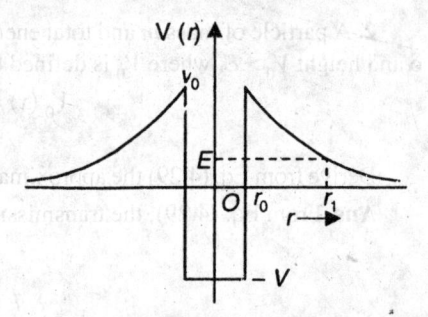

For a 5.3 MeV α-particle emitted from $_{84}Po^{210}$ nucleus, we obtain from Eq. (4.61) (by putting $V = 5.3$ MeV and $Z = 84$) $r = r_1 = 4.45 \times 10^{-14}$m. Also, with $r = r_0 = 10^{-14}$m, we get from Eq. (4.61), $V(r_0) = 23.58$ MeV. Therefore the width of the barrier is $\delta = r_1 - r_0 = 3.45 \times 10^{-14}$m. Evidently an α-particle which is inside the nucleus, according to classical mechanics, cannot come out.

Fig. 4.12. Variation of potential energy of the α-particle with its distance r from the nucleus.

Quantum mechanics, on the other hand, tells us that it is possible for the α-particle to come out of the barrier by tunneling. The relevant theory was put forward in 1928 by Gamow, Condon and Gurney. According to them, the position of the α-particle is represented by its wave function ψ as shown in Fig. 4.13. The wavefunction is oscillatory inside the well, drops rapidly in amplitude during its passage through the barrier, and again becomes oscillatory but with a much reduced amplitude outside the barrier. The nonzero value of the wave function outside the barrier predicts that there is a finite chance of finding the particle outside the nucleus.

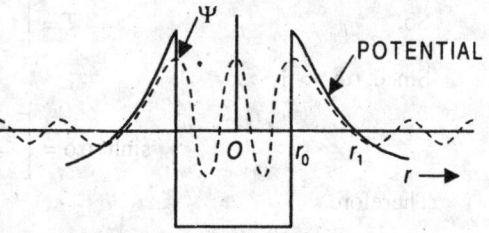

Fig. 4.13. Wave function of the α-particle as a function of r.

Let us find the transmission coefficient using Eq. (4.31a). Here $\alpha\delta = \frac{\delta}{\hbar}\sqrt{2m(V_0 - E)}$

$$= \frac{3.45 \times 10^{-14}}{1.054 \times 10^{-34}} \times [2 \times 4 \times 1.67 \times 10^{-27} \times (23.58 - 5.3) \times 1.6 \times 10^{-13}]^{1/2} = 64.7.$$ So, Eq. (4.31 a) yields for the transmission coefficient of the α-particle through the barrier

$$T = \frac{16 \times 5.3 \times 18.28}{(23.58)^2} \exp(-129.4)$$

$$= 1.77 \times 10^{-56}$$

This is an order-of-magnitude calculation, since the barrier is not strictly rectangular.

4.5. WORKED-OUT PROBLEMS

1. A spherical dust particle of radius $r = 10^{-5}$m, density $\rho = 10^4$ kg/m^3, and moving at a velocity $v = 10^{-2}$ m/s, encounters a potential step of height equal to twice the kinetic energy of the particle. Estimate the penetration distance d_p of the particle inside the step.

Ans. The mass m of the particle is

$$m = \frac{4}{3}\pi r^3 \rho = \frac{4}{3}\pi \times 10^{-15} \times 10^4 \text{ kg} = 4.2 \times 10^{-11} \text{ kg}.$$

The kinetic energy of the particle before impinging is

$$E = \frac{1}{2}mv^2 = \frac{1}{2} \times 4.2 \times 10^{-11} \times 10^{-4} = 2.1 \times 10^{-15} \text{ J}$$

According to the problem, $V_2 - E = 2.1 \times 10^{-15}$ J, where V_2 is the step height. Therefore, using Eq. (4.16) we get

$$d_p = \frac{\hbar}{\sqrt{2m(V_2 - E)}} = \frac{1.054 \times 10^{-34}}{\sqrt{2 \times 4.2 \times 10^{-11} \times 2.1 \times 10^{-15}}} = 2.5 \times 10^{-22} \text{ m}$$

2. A particle of mass m and total energy E, is incident on a rectangular potential barrier of width δ and height $V_0 > E$, where V_0 is defined by

$$V_0(x) = V_0 \text{ for } 0 < x < \delta$$
$$= 0 \text{ for } x < 0 \text{ or } x > \delta.$$

Derive from Eq. (4.29) the approximate expression for the transmission coefficient when $\alpha\delta \gg 1$.

Ans. From Eq. (4.29), the transmission coefficient is

$$T = \left(\frac{A_5}{A_1}\right)^2 = \left[1 + \frac{(V_2 - V_1)^2 \sinh^2(\alpha\delta)}{4(E - V_1)(V_2 - E)}\right]^{-1}$$

For the present problem $V_2 = V_0$ and $V_1 = 0$.

So,
$$T = \left[1 + \frac{V_0^2 \sinh^2 \alpha\delta}{4E(V_0 - E)}\right]^{-1}$$

Since, $\alpha\delta \gg 1$,

$$\sinh^2 \alpha\delta = \left[\frac{e^{\alpha\delta} - e^{-\alpha\delta}}{2}\right]^2 \approx \frac{1}{4} e^{2\alpha\delta}$$

Therefore,

$$T \approx \left[1 + \frac{V_0^2 e^{2\alpha\delta}}{16 E(V_0 - E)}\right]^{-1} \approx \left[\frac{V_0^2 e^{2\alpha\delta}}{16 E(V_0 - E)}\right]^{-1}$$

$$\approx \frac{16 E}{V_0}\left(1 - \frac{E}{V_0}\right) e^{-2\alpha\delta},$$

where
$$\alpha = \left[\frac{2m(V_0 - E)}{\hbar^2}\right]^{1/2}$$

3. Calculate the transmission coefficient for an electron of total energy 2 eV incident upon a rectangular potential barrier of height 4 eV and width 10^{-9}m.

Ans. We know $m = 9.11 \times 10^{-31}$ kg, $e = 1.6 \times 10^{-19}$ coulomb, and $\hbar = 1.054 \times 10^{-34}$ J.s.

Given, $\delta = 10^{-9}$m, $V_0 = 4 \times 1.6 \times 10^{-19}$ J, and $E = 2 \times 1.6 \times 10^{-19}$ J

Hence,
$$\alpha\delta = \left[\frac{2 \times 9.11 \times 10^{-31} \times 2 \times 1.6 \times 10^{-19}}{(1.054 \times 10^{-34})}\right]^{1/2} \times 10^{-9}$$

$$= (52.52 \times 10^{18})^{1/2} \times 10^{-9} = 7.244.$$

Thus $\alpha\delta \gg 1$ can be assumed for this problem, and from problem 2, we can write

$$T \approx \frac{16 \times 2}{4}\left(1 - \frac{2}{4}\right) e^{-14.488} = 2 \times 10^{-6}$$

4. Calculate the lowest energy of an electron confined to move in a one-dimensional potential well of width 1 Å and of infinite depth.

Ans. We have from Eq. (4.60) for the lowest energy ($n = 1$) of an electron in an infinite potential well of width $2a$

$$E = \frac{\hbar^2 \pi^2}{2m(2a)^2}.$$

Given, $2a = 1\text{Å} = 1 \times 10^{-10}$m. Also, $m = 9.11 \times 10^{-31}$ kg and $\hbar = 1.054 \times 10^{-34}$ J.s. Substituting the values we find

$$E = \frac{(1.054 \times 10^{-34})^2 \times (3.142)^2}{2 \times 9.11 \times 10^{-31} \times (10^{-10})^2} = 6 \times 10^{-18} \text{ J}$$

$$= \frac{6 \times 10^{-18}}{1.6 \times 10^{-19}} \text{ eV} = 37.5 \text{ eV}.$$

5. The quantum mechanical transmission coefficient of an α-particle through a nuclear potential barrier is 2.5×10^{-24}. Taking the velocity of the α-particle and the nuclear radius as 1.7×10^7 m/s and 10^{-14}m, respectively, calculate the mean lifetime of α-decay. Why does the α-emission half-lives of natural radioelements differ by large amounts?

Ans. The transmission coefficient (T) gives the probability of tunneling through the potential barrier in each collision with the barrier wall. The value $T = 2.5 \times 10^{-24}$ shows that the α-particle has 2.5 chances in 10^{24} collisions with the barrier to come out of the nucleus. The number of collisions of the α-particle with the barrier in one second is

$$n = \frac{\text{velocity of the particle}}{\text{nuclear diameter}} = \frac{1.7 \times 10^7}{2 \times 10^{-14}} = 8.5 \times 10^{20}.$$

So, the probability of the α-particle to escape in 1s is

$$P = nT = 8.5 \times 10^{20} \times 2.5 \times 10^{-24} = 2.12 \times 10^{-3}.$$

The mean lifetime of α-decay is

$$\tau = \frac{1}{P} = \frac{10^3}{2.12} = 472 \text{ s} = 7 \text{ min } 52 \text{ s}.$$

Since the transmission coefficient T decreases exponentially with $2\alpha\delta$ [see Eq. (4.31a)], a small change in $\alpha\delta$ (due to slight changes in the barrier height and width) causes a large change in T and hence in τ. This explains the wide variation in the α-emission half-lives of natural radioelements.

QUESTIONS

1. A particle of mass m and total energy E moves from a region of constant potential V_1 to a region of constant potential V_2. Derive expressions for the reflection and transmission coefficients when
   (i) $E > V_1$ or V_2, and
   (ii) $V_1 < E < V_2$.

2. What is quantum mechanical tunneling? An electron of mass m and total energy E is incident on a rectangular potential barrier of height V_0, where $V_0 > E$ and of width δ. Derive an expression for the transmission coefficient.

3. Derive an expression for the transmission coefficient of a particle through a rectangular potential barrier for an energy less than the barrier height. Discuss qualitatively how α-decay is explained in the light of the above derivation. (C.U. 1987, 99)

4. A particle is confined to an infinitely deep potential well. Derive an expression for the energy eigenstates of the particle. Sketch the eigenfunction for the lowest energy state of the particle in this case.

5. Show that the quantization law $E = \dfrac{\hbar^2 n^2 \pi^2}{2m(2a)^2}$ for an infinitely deep potential well of width $2a$ can be obtained directly from the de Broglie relation $p = h/\lambda$, by fitting an integral number of half de Broglie wavelengths $\lambda/2$ into the width $2a$ of the well.

   [**Hint.** By plotting the first few eigenfunctions of an infinitely deep potential well it can be seen that the following relation is obeyed: $\dfrac{n\lambda}{2} = 2a$, where $n = 1, 2, 3, \ldots\ldots$

   So, $\lambda = \dfrac{2 \times 2a}{n}$. Now $p = \dfrac{h}{\lambda} = \dfrac{hn}{2 \times 2a}$.

Within the well, the potential energy of the particle is zero, and the energy is entirely kinetic.

Hence $E = \dfrac{p^2}{2m} = \dfrac{h^2 n^2}{4 \times 2m \times (2a)^4} = \dfrac{\pi^2 \hbar^2 n^2}{2m(2a)^2}.$

6. Show that the fractional difference ΔE_n in the energy between adjacent eigenvalues for a particle in an infinite rectangular potential well is given by $\Delta E_n = \dfrac{2n+1}{n^2} E_n$. Using this expression comment on the classical limit of the system.

7. A proton of energy 3 MeV is incident on a rectangular potential barrier of height 10 MeV and thickness 10^{-14}m. Calculate the transmission coefficient of the proton. (mass of the proton = 1.67×10^{-27} kg) [Ans. 3×10^{-5}]

8. Compute the zero-point energy ($n=1$) of a neutron in a nucleus by considering as if it were in an infinite square well potential of width equal to the nuclear diameter of 10^{-14} m. (Assume, mass of the neutron = 1.675×10^{-27} kg.) [Ans. 2.04 MeV]

9. An α-particle of energy 10 MeV is incident on a potential barrier of height 30 MeV. Calculate the width of the potential barrier so that the transmission coefficient is 2×10^{-3}. Assume, mass of the α-particle = 6.69×10^{-27} kg. [Ans. 1.91×10^{-15}m approx]

10. Discuss quantum mechanical tunneling to explain the emission of α-particles from a radioactive nucleus. What is the prediction of classical mechanics in such cases?

11. Find the height of the potential barrier in eV for α-particles emitted from an element (atomic number, $Z = 86$). Assume that the effective nuclear radius r_0 is 10^{-14}m. [Ans. 24.16 MeV]

12. Determine the transmission coefficient for a proton of energy 1 MeV through a 4 MeV high rectangular potential energy barrier of width 10^{-12}cm. [Ans. 0.0015]

13. A particle of mass m moves in a one-dimensional square potential well with infinitely high walls. The width of the well is L. Find the allowed energy values and the eigenfunctions of the particle. Describe the eigenfunctions corresponding to the different eigenvalues diagrammatically. Present also a plot of the probability densities of these states. (C.U. 1985)

14. The height of the potential barrier for alpha particles inside the ^{238}U nucleus is 28 MeV, the effective width being 10^{-12} cm. Calculate the quantum number n for which the energy eigenvalue of an alpha particle inside the ^{238}U nucleus is closest to the kinetic energy of the alpha particle emitted from the ^{238}U nucleus, which is 4 MeV. Given, mass of an alpha particle = 4.0026 u, 1u = 1.66 kg. [Ans. 3]

15. A beam of electrons impinges on an energy barrier of height 0.03 eV and of infinite width. Find the fraction of the number of electrons reflected and transmitted at the barrier if the energy of the impinging electrons is 0.04 eV. (Purvanchal Univ. 2002) [Ans. 1/9, 8/9]

16. An electron with a kinetic energy of 10 eV is moving from left to right along the x-axis. The potential energy is $V = 0$ for $x < 0$ and $V = 20$ eV for $x > 0$ (i) Write Schrödinger's equation for $x < 0$ and $x > 0$. (ii) Show that the reflection coefficient is unity. (iii) Sketch the solutions in the two regions. (iv) What is the wavelength for $x < 0$? (v) Is there any possibility of finding the electron in the region $x > 0$? If yes, how do you physically recoincile your answer with (ii) above?

(C.U. 2005)

Chapter 5

The Harmonic Oscillator

The harmonic oscillator is of tremendous importance in Physics since it provides an excellent model for many actual physical systems of real technical interest. The main reason for this is that almost all systems in stable equilibrium will execute simple harmonic vibration about the point of stable equilibrium for small disturbances from equilibrium. For example, it is used to describe acoustic and thermal properties of solids which arise from atomic vibrations. Similarly, the electromagnetic disturbances may be treated by assuming them to arise from excitations of a set of harmonic oscillators. Besides, many important phenomena in solids such as the lifetimes of excited optical states, the origin of magnetoelastic effects, noise in quantum oscillators, etc. are most successfully discussed in terms of harmonic oscillator eigenstates and eigenvalues.

5.1. HARMONIC OSCILLATOR AND ENERGY EIGENFUNCTIONS

A linear harmonic oscillator is a particle which is bound to an equilibrium position by a force that is proportional to the displacement of the particle from that position. For a one-dimensional system this force may therefore be expressed as

$$F = -Ax = -\frac{dV(x)}{dx}, \quad \ldots(5.1)$$

where $V(x)$ is the potential energy when the particle is displaced by an amount x and A is the force constant. The potential energy $V(x)$ in this case then must be a parabola of the form

$$V(x) = \frac{1}{2}Ax^2 \quad \ldots(5.2)$$

Schrödinger's equation for the problem therefore reduces to

$$\left(-\frac{\hbar^2}{2m}\frac{\partial^2}{\partial x^2} + \frac{1}{2}Ax^2\right)\psi(x,t) = i\hbar\frac{\partial}{\partial t}\psi(x,t) \quad \ldots(5.3)$$

where m is the mass of the oscillating particle. Since the potential energy does not contain time explicitly we can solve the equation by the method of separation of variables :

$$\psi(x,t) = u(x)\phi(t) \quad \ldots(5.4)$$

The time-independent Schrödinger equation for the one-dimensional harmonic oscillator is then given by

$$\frac{\hbar^2}{2m}\frac{d^2u}{dx^2} + \left(E - \frac{1}{2}Ax^2\right)u = 0 \quad \ldots(5.5)$$

The time-dependent Schrödinger equation, on the other hand, gives

$$\phi(t) = e^{-iEt/\hbar} \quad \ldots(5.6)$$

In Eqs. (5.5) and (5.6), E is the separation constant and represents the total energy of the system. To solve Eq. (5.5) we introduce the following parameters :

$$\lambda = \frac{2E}{\hbar}\left(\frac{m}{A}\right)^{1/2} = \frac{2E}{\hbar\omega_0}, \quad \ldots(5.7)$$

$$\alpha^4 = \frac{mA}{\hbar^2} \qquad \ldots(5.8)$$

and
$$\xi = \alpha x, \qquad \ldots(5.9)$$

where $\omega_0 = (A/m)^{1/2}$ corresponds to the angular frequency of the respective classical harmonic oscillator. With the help of Eqs. (5.7) to (5.9), Eq. (5.5) can be written as

$$\frac{d^2 u}{d\xi^2} + (\lambda - \xi^2) u = 0 \qquad \ldots(5.10)$$

For larger values of $|\xi|$, i.e., when $\xi^2 \gg \lambda$, Eq. (5.10) reduces to

$$\frac{d^2 u}{d\xi^2} - \xi^2 u \cong 0 \qquad \ldots(5.11)$$

Equation (5.11) is approximately satisfied for large ξ by $u(\xi) = e^{-\xi^2/2}$ since then $\frac{d^2 u}{d\xi^2} = (\xi^2 - 1) \exp(-\xi^2/2) = (\xi^2 - 1) u(\xi)$. That is, $\frac{d^2 u}{d\xi^2} - \xi^2 u(\xi) = 0$, if $\xi^2 \gg 1$. This suggests that the solutions $u(\xi)$ for Eq. (5.10) may be expressed by

$$u(\xi) = H(\xi) e^{-\xi^2/2} \qquad \ldots(5.12)$$

Putting (5.12) in Eq. (5.10) we get

$$\frac{d^2 H}{d\xi^2} - 2\xi \frac{dH}{d\xi} + (\lambda - 1) H = 0 \qquad \ldots(5.13)$$

Assuming a power series for $H(\xi)$:

$$H(\xi) = \sum_{s=0}^{\infty} a_s \xi^s \qquad \ldots(5.14)$$

we obtain
$$\frac{dH}{d\xi} = \sum_{s=1}^{\infty} s a_s \xi^{s-1} \qquad \ldots(5.15)$$

and
$$\frac{d^2 H}{d\xi^2} = \sum_{s=2}^{\infty} s(s-1) a_s \xi^{s-2}.$$

Putting these into Eq. (5.13) gives

$$\sum_{s=0}^{\infty} \left[(s+1)(s+2) a_{s+2} + \{(\lambda - 1) - 2s\} a_s \right] \xi^s = 0 \qquad \ldots(5.16)$$

Equation (5.16) must hold for all values of ξ and hence the coefficient of each power of ξ must vanish separately. We obtain at once a *recursion relation* between a_s and a_{s+2}:

$$a_{s+2} = -\frac{\lambda - 1 - 2s}{(s+1)(s+2)} a_s \qquad \ldots(5.17)$$

Note that the relation (5.17) couples only the powers of ξ differing by 2. The relation allows us to calculate successively the coefficients $a_2, a_4, a_6 \ldots$ in terms of a_0 and the coefficients $a_3, a_5 \ldots$ in terms of a_1. However, a_0 and a_1 are not interrelated. Therefore, a_0 and a_1 remain as the two fundamental arbitrary constants: this is so because a second-order ordinary differential equation must have two arbitrary constants in its solution.

Let us see what happens when the series $H(\xi)$ does not terminate. It is found from Eq. (5.17) that for large values of s the ratio of the coefficients a_{s+2} and a_s reduces to

$$\left. \frac{a_{s+2}}{a_s} \right|_{s \to \infty} \to \frac{2}{s}. \qquad \ldots(5.18)$$

The series expansion of e^{ξ^2} is

$$e^{\xi^2} = 1 + \frac{\xi^2}{1!} + \frac{\xi^4}{2!} + \frac{\xi^6}{3!} + \ldots = \sum_s b_s \xi^\sigma \qquad \ldots(5.19)$$

For large values of ξ, the initial terms of this series give negligible contribution compared with those for which s is large. This is valid also for the series $H(\xi)$. For the series (5.19) the ratio of the coefficients b_{s+2} and b_s is

$$\frac{b_{s+2}}{b_s} = \frac{(s/2)!}{\left(\frac{s}{2}+1\right)!} = \frac{1}{\left(\frac{s}{2}+1\right)} \underset{\sigma \to \infty}{\approx} \frac{2}{s}. \qquad \ldots(5.20)$$

From Eqs. (5.18) and (5.20) we find that the series $H(\xi)$, when not terminated, behaves like e^{ξ^2} for large values of ξ. The solutions obtained from Eq. (5.12) are given by

$$u(\xi) = H(\xi) e^{-\xi^2/2} \equiv e^{\xi^2} e^{-\xi^2/2} = e^{\xi^2/2} \qquad \ldots(5.21)$$

for large values of ξ. As these functions do not remain finite at $x = \pm \infty$, they are not acceptable as wave functions for the physical system.

Thus the series (5.14) must terminate after a finite number of terms. From the recursion relation (5.17) we find that $H(\xi)$ can be terminated at ξ^n by making

$$\lambda = 2n + 1. \qquad \ldots(5.22)$$

This compels a_{n+2} and all successive a_{n+2r} to be zero.

When n is even, the even-subscript series terminates and the odd-subscript series is eliminated by choosing $a_1 = 0$. Similarly, a_0 is put equal to zero so that the series is finite when n is an odd integer. In either case, a physically admissible solution results if, and only if, Eq. (5.22) is satisfied. Using Eq. (5.7) in Eq. (5.22) we get

$$E_n = (n + \frac{1}{2}) \hbar \omega_0, \qquad \ldots(5.23)$$

where n is referred to as the *quantum number* taking values 0, 1, 2 etc. Equation (5.23) shows that the energy eigenvalues are discrete and *nondegenerate* since, associated with each value of n, there is one value of $H(\xi)$ and one value of $u(x)$.

When n is even and $a_1 = 0$ there is one set of solutions, and when n is odd and $a_0 = 0$, we obtain another set of solutions. These sets of solutions are said to be of *even* and *odd parity*.

An interesting feature of this system is that its lowest energy state corresponding to $n = 0$, is characterized not by $E_0 = 0$, but by $E_0 = \frac{\hbar \omega_0}{2}$. Thus, the lowest energy state of the harmonic oscillator is characterised by the presence of a minimum *zero-point* energy $\frac{1}{2} \hbar \omega_0$ we shall see later (Sec. 5.4) that the zero-point energy is consistent with Heisenberg's uncertainty principle. The energy level diagram of the harmonic oscillator is illustrated in Fig. 5.1. The equal spacing of energy levels is a sole property of the parabolic potential well.

It is noteworthy here that in Einstein's theory of specific heat of solids, the atoms of the solid are regarded as independent harmonic oscillators. If the zero-point energy term is not included there, the total vibrational energy in that theory at high temperatures does not agree with the classical result.

5.2. Hermite Polynomials And The Eigenfunctions

Let us choose the arbitrary constant a_0 or a_1, as the case may be, so that the coefficient of the highest power of ξ in the polynomial is $2^n[= 2^{(\lambda-1)/2}]$. The above procedure results in a set of polynomial solutions $H_n(\xi)$ of the form $H_0(\xi) = 1$, $H_1(\xi) = 2\xi$, $H_2(\xi) = 4\xi^2 - 2$, $H_3(\xi) = 8\xi^3 - 12\xi$ etc. These polynomials are called *Hermite polynomials* and Eq. (5.13) is known as the *Hermite equation*.

There are several relations involving Hermite polynomials. To serve our purpose we shall consider only the following relations:

$$\int_{-\infty}^{\infty} e^{-\xi^2} H_n(\xi) H_m(\xi) d\xi = 2^n n! \sqrt{\pi} \delta_{nm}, \quad ...(5.24)$$

where δ_{nm} is Kronecker's delta.

$$\frac{dH_n(\xi)}{d\xi} = 2n H_{n-1}(\xi) \quad ...(5.25)$$

$$2\xi H_n(\xi) = H_{n+1}(\xi) + 2n H_{n-1}(\xi) \quad ...(5.26)$$

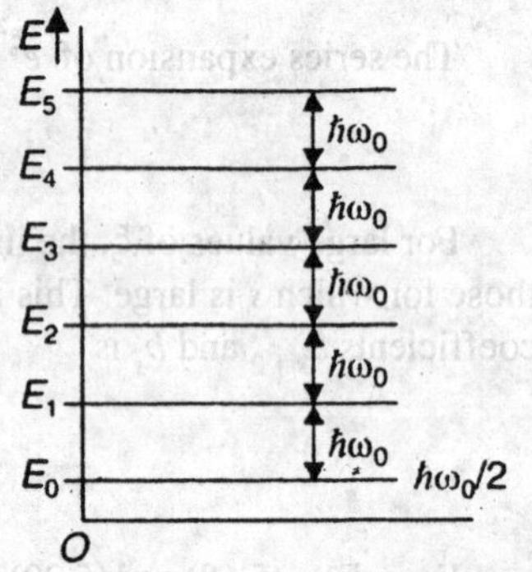

Fig. 5.1. Energy levels of a one-dimensional harmonic oscillator.

Therefore the actual eigenfunctions [from Eq. (5.12)] are given in terms of Hermite polynomials as follows:

$$u_n(x) = A_n H_n(\xi) e^{-\xi^2/2}$$

$$= A_n H_n(\alpha x) e^{-\alpha^2 x^2/2} \quad ...(5.27)$$

In Eq. (5.27), A_n is the normalization constant and is defined from the requirement

$$\int_{-\infty}^{\infty} u_n^*(x) u_n(x) dx = 1$$

or $$\frac{1}{\alpha} \int_{-\infty}^{\infty} A_n^2 e^{-\xi^2} H_n^2(\xi) d\xi = 1$$

Using the relation (5.24) we obtain

$$\frac{A_n^2}{\alpha} \cdot 2^n n! \sqrt{\pi} = 1$$

since $\delta_{nm} = 1$ when $n = m$.

So, $$A_n = \left[\frac{\alpha}{2^n n!} \frac{1}{\sqrt{\pi}} \right]^{1/2} \quad ...(5.28)$$

and $$\int_{-\infty}^{\infty} u_n^*(x) u_m(x) dx = \delta_{nm} \quad ...(5.29)$$

The eigenfunctions $u_n(x)$ therefore form an *orthonormal set*.

The time-dependent wavefunctions $\psi_n(x, t)$ are found from Eq. (5.27) by multiplying by a factor $e^{-iE_n t/\hbar}$ in accordance with Eq. (5.4), i.e.,

$$\psi_n(x, t) = A_n H_n(\alpha x) e^{-\alpha^2 x^2/2} e^{-iE_n t/\hbar} \quad ...(5.30)$$

5.3. Correspondence with Classical Oscillator

We know that the total energy of a classical oscillator is

$$E = \frac{1}{2} m \omega_0^2 a^2, \quad ...(5.31)$$

where a is the amplitude of oscillation.

Hence $$a = \sqrt{2E/(m\omega_0^2)}$$

The Harmonic Oscillator

For a quantum oscillator, $E = (n + \frac{1}{2})\hbar\omega_0$. Therefore, in terms of quantum oscillator energy, the classical oscillator amplitude is

$$a = \sqrt{(2n+1)\hbar/m\omega_0} = \frac{\sqrt{2n+1}}{\alpha} \qquad \ldots(5.32)$$

where α is defined in Eq. (5.8).

The probability $P(x)\,dx$ of observing the classical particle in the region dx is the ratio of the time it spends in this region during one complete cycle to the period of oscillation (T). The time spent dt in dx is given by

$$dt = \frac{2dx}{v(x)}, \qquad \ldots(5.33)$$

where $v(x)$ is the velocity. The numerical factor 2 appears in the right-hand side of Eq. (5.33) since the particle passes through dx twice during a complete cycle.

When $x = a \sin\omega_0 t$, $v(x) = \frac{dx}{dt} = \omega_0 a \cos\omega_0 t = \omega_0 a\sqrt{1 - (x^2/a^2)}$

and

$$P(x)dx = \frac{2dx}{T\omega_0 a\sqrt{1 - (x^2/a^2)}} = \frac{2dx}{2\pi\sqrt{a^2 - x^2}}$$

$$= \frac{dx}{\pi\sqrt{a^2 - x^2}} \qquad \ldots(5.34)$$

This is the classical quantity corresponding to the quantum mechanical counterpart $u^*u\,dx$. Using Eq. (5.32), the probability amplitude for a classical oscillator with energy $(n + \frac{1}{2})\hbar\omega_0$ is therefore given by

$$P(x) = \frac{1}{\pi\sqrt{[(2n+1)/\alpha^2] - x^2}} \qquad \ldots(5.35)$$

$P(x)$ is a maximum (infinite) when $x = \pm\frac{\sqrt{2n+1}}{\alpha} = \pm\frac{\sqrt{2n+1}}{(mA)^{1/4}} \cdot \hbar^{1/2}$

This value of x also determines the limits of the classical motion of the oscillator.

Let us now compare this with the quantum probability amplitude $u_n^* u_n$, where u_n is given by Eq. (5.27).

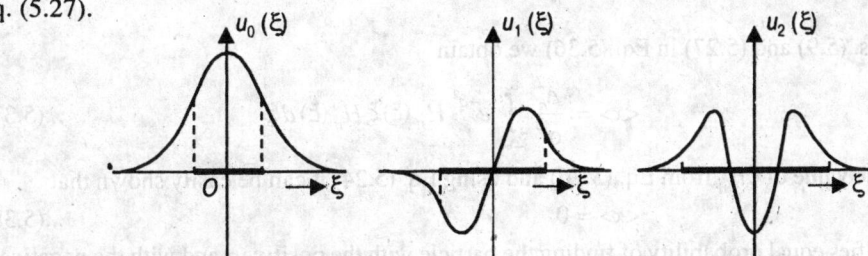

Fig. 5.2. $u_n(\xi)$ versus ξ for the first three harmonic oscillator solutions. The bold lines bounded by the vertical dashed lines mark the classical trajectory.

Figure 5.2 shows the behaviour of the wavefunctions. For the ground state $(n = 0)$, the wavefunction is $u_0 = (\alpha/\sqrt{\pi})^{1/2} \exp(-\xi^2/2)$. For $n = 1$ and $n = 2$, the wavefunctions are respectively $u_1 = (2\alpha/\sqrt{\pi})^{1/2} \xi \exp(-\xi^2/2)$ and $u_2 = [(\alpha/(2\sqrt{\pi})]^{1/2} (2\xi^2 - 1) \exp(-\xi^2/2)$. The wavefunctions are found to penetrate into the region beyond the limits of amplitude of the classical vibration. Also, the wavefunctions become alternately even and odd functions of x as n increases. For $n = 0$ and even, $u_n(-x) = u_n(x)$, *i.e.* the wavefunction is an even function of x and is said to have *even parity*. But, if n is odd, $u_n(-x) = -u_n(x)$, *i.e.*, the wavefunction is an odd function of x and has *odd parity*.

Figure 5.3 shows the plots of the quantum mechanical probability amplitude and the classical probability amplitude; the latter being computed from Eq. (5.35). The plots are shown for $n = 0$ and $n = 12$. [Observe that the curve of quantum-mechanical $u_n^* u_n$ shows $(n + 1)$ oscillatory peaks.] It is evident from the plots corresponding to $n = 0$ that the two probability amplitudes do not show any similarity. The classical probability distribution shows that the probability of finding the particle near the ends of the region of allowed motion is a maximum, while the quantum results predict that the maximum probability of finding the particle occurs at the centre.

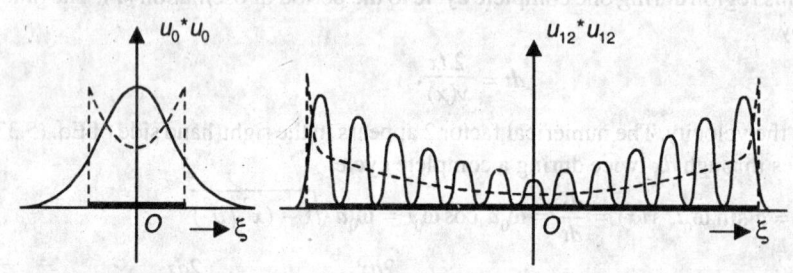

Fig. 5.3. Probability amplitude distributions for $n = 0$ and $n = 12$: solid curve for quantum and dashed curve for classical oscillators.

For $n = 12$ the classical probability and the quantum probability distributions do show some similarity, a similarity which continues to increase with increasing value of n. However, the quantum distribution differs from the classical distribution in that the former possesses nodes and antinodes while the latter is smooth. With increase in n, the nodes become more and more closely spaced. For example, consider a laboratory harmonic oscillator such as a weight of a few grams on a spring. The amplitude of the oscillator is a few centimetres and $n \approx 10^{26}$. The 10^{26} nodes of the quantum probability amplitude distribution would be so closely spaced as to defy observation. Also, an estimate of the probability would give only the mean value of the quantity $u_n^* u_n$, which would be indistinguishable from the classical distribution. Thus for large masses and energies, the classical results agree with the quantum mechanical results; this is an example of the correspondence principle.

5.4. UNCERTAINTY PRINCIPLE AND THE ZERO-POINT ENERGY

The expectation value of x is

$$<x> = \int_{-\infty}^{\infty} u_n^*(x) \, x \, u_n(x) \, dx. \qquad ...(5.36)$$

Using Eqs. (5.9) and (5.27) in Eq. (5.36) we obtain

$$<x> = \frac{A_n^2}{\alpha^2} \int_{-\infty}^{\infty} e^{-\xi^2} H_n(\xi) \xi H_n(\xi) \, d\xi \qquad ...(5.37)$$

Putting the value of ξH_n from Eq. (5.26) and using Eq. (5.24) it can be easily shown that

$$<x> = 0. \qquad ...(5.38)$$

This signifies equal probability of finding the particle with the positive x and with the negative x co-ordinates and that $u_n^*(x) u_n(x)$ is symmetric about the origin.

Now
$$<x^2> = \int_{-\infty}^{\infty} u_n^*(x) \, x^2 \, u_n(x) \, dx$$

$$= \frac{A_n^2}{\alpha^3} \int_{-\infty}^{\infty} e^{-\xi^2} H_n(\xi) \, \xi^2 \, H_n(\xi) \, d\xi \qquad ...(5.39)$$

Applying the recursion relation [Eq. (5.26)] twice we obtain

$$\xi^2 H_n(\xi) = \frac{1}{4} H_{n+2}(\xi) + (n + \frac{1}{2}) H_n(\xi) + n(n-1) H_{n-2}(\xi) \quad ...(5.40)$$

Using Eq. (5.40) in Eq. (5.39) and the orthogonality relation (5.29) we obtain

$$<x^2> = \left(n + \frac{1}{2}\right) \frac{A_n^2}{\alpha^3} \int_{-\infty}^{\infty} e^{-\xi^2} H_n(\xi) H_n(\xi) \, d\xi = \frac{(n + \frac{1}{2})}{\alpha^2}$$

$$= \left(n + \frac{1}{2}\right) \frac{\hbar}{(mA)^{1/2}}. \quad ...(5.41)$$

The expectation value of the x-component of momentum is

$$<p_x> = \frac{\hbar}{i} \int_{-\infty}^{\infty} u_n^*(x) \frac{d}{dx} u_n(x) \, dx$$

$$= \frac{\hbar}{i} A_n^2 \int_{-\infty}^{\infty} e^{-\xi^2/2} H_n(\xi) \frac{d}{d\xi} e^{-\xi^2/2} H_n(\xi) \, d\xi \quad ...(5.42)$$

Again, $\quad \frac{d}{d\xi} e^{-\xi^2/2} H_n(\xi) = -\xi e^{-\xi^2/2} H_n(\xi) + e^{-\xi^2/2} \frac{dH_n}{d\xi} \quad ...(5.43)$

Using Eq. (5.26) and Eq. (5.27) we obtain

$$\frac{d}{d\xi} e^{-\xi^2/2} H_n(\xi) = e^{-\xi^2/2} \left[n H_{n-1}(\xi) - \frac{1}{2} H_{n+1}(\xi) \right] \quad ...(5.44)$$

Putting Eq. (5.44) into Eq. (5.42) and using Eq. (5.24) we find

$$<p_x> = 0. \quad ...(5.45)$$

That is, the probabilities of the particle momentum directed in positive and negative x directions are equal.

Now, $\quad <p_x^2> = A_n^2 \int_{-\infty}^{\infty} u_n^*(x) \left(\frac{\hbar}{i} \frac{d}{dx}\right)^2 u_n(x) \, dx$

$$= -\alpha \hbar^2 A_n^2 \int_{-\infty}^{\infty} e^{-\xi^2/2} H_n(\xi) \frac{d^2}{d\xi^2} e^{-\xi^2/2} H_n(\xi) \, d\xi \quad ...(5.46)$$

Again it can be easily shown that

$$\frac{d^2}{d\xi^2} \left[e^{-\xi^2/2} H_n(\xi) \right] = e^{-\xi^2/2} \left[n(n-1) H_{n-2}(\xi) - (n + \frac{1}{2}) H_n(\xi) + \frac{1}{4} H_{n+2}(\xi) \right]$$

Using Eq. (5.24) and the above relation we obtain from Eq. (5.46)

$$<p_x^2> = \alpha \hbar^2 A_n^2 \int_{-\infty}^{\infty} e^{-\xi^2} H_n(\xi) (n + \frac{1}{2}) H_n(\xi) \, d\xi$$

$$= (n + \frac{1}{2}) \alpha^2 \hbar^2. \quad ...(5.47)$$

Denoting the uncertainties in position and momentum by Δx and Δp_x respectively we get

$$(\Delta x)^2 = <x^2> - <x>^2 = \frac{1}{\alpha^2}\left(n + \frac{1}{2}\right),$$

and

$$(\Delta p_x)^2 = <p_x^2> - <p_x>^2 = (n + \frac{1}{2}) \alpha^2 \hbar^2.$$

Therefore,
$$\Delta x\, \Delta p_x = \left(n + \frac{1}{2}\right)\hbar \qquad ...(5.48)$$

Evidently, $\Delta x\, \Delta p_x$ has the minimum value $\frac{\hbar}{2}$ for $n = 0$. We shall now show that the existence of the zero-point energy $E_0 = \frac{1}{2}\hbar\omega_0$ is in conformity with the uncertainty principle. The total energy is of order

$$E = \frac{(\Delta p_x)^2}{2m} + \frac{1}{2}m\omega_0^2\, (\Delta x)^2$$

where Δp_x and Δx are the uncertainties in momentum and position. By the uncertainty principle,

$$\Delta x\, \Delta p_x = \frac{\hbar}{2}$$

Hence,
$$E = \frac{(\Delta p_x)^2}{2m} + \frac{m\hbar^2 \omega_0^2}{8(\Delta p_x)^2}$$

For E to be a minimum, $\dfrac{dE}{d(\Delta p_x)} = 0$ or $\dfrac{\Delta p_x}{m} - \dfrac{m\hbar^2 \omega_0^2}{4(\Delta p_x)^3} = 0$

or, $\Delta p_x^2 = \frac{1}{2} m\hbar\omega_0$. So, the minimum total energy or the zero-point energy is

$$E_0 = \frac{1}{4}\hbar\omega_0 + \frac{m\hbar^2\omega_0^2}{4m\hbar\omega_0} = \frac{1}{2}\hbar\omega_0.$$

5.5. Raising And Lowering Operators

Let us introduce two operators

$$\hat{a} = \sqrt{\frac{m\omega_0}{2\hbar}}\, \hat{x} + \frac{i}{\sqrt{2m\hbar\omega_0}}\, \hat{p}$$

$$\hat{a}^+ = \sqrt{\frac{m\omega_0}{2\hbar}}\, \hat{x} - \frac{i}{\sqrt{2m\hbar\omega_0}}\, \hat{p} \qquad ...(5.49)$$

Note that $\hat{a}$ and $\hat{a}^+$ are not Hermitian, but $\left(\hat{a}^+\right)^+ = \hat{a}$.

Also,
$$\hat{a}\hat{a}^+ - \hat{a}^+\hat{a} = -\frac{i}{\hbar}(\hat{x}\hat{p} - \hat{p}\hat{x}) = 1 \qquad ...(5.50)$$

since $\hat{x}$ and $\hat{p}$ obey the commutation relation $[\hat{x}, \hat{p}] = i\hbar$.

Equation (5.50) shows that $\hat{a}$ and $\hat{a}^+$ satisfy the commutation relation

$$\left[\hat{a}, \hat{a}^+\right] = 1. \qquad ...(5.51)$$

The operators $\hat{a}$ and $\hat{a}^+$ help us to avoid the position and momentum representations and focus on the Hamiltonian operator $\hat{H}$. This can be appreciated by noting that

$$\hbar\omega_0\, \hat{a}^+\hat{a} = \hbar\omega_0 \left(\sqrt{\frac{m\omega_0}{2\hbar}}\,\hat{x} - \frac{i\hat{p}}{\sqrt{2m\hbar\omega_0}}\right)\left(\sqrt{\frac{m\omega_0}{2\hbar}}\,\hat{x} + \frac{i\hat{p}}{\sqrt{2m\hbar\omega_0}}\right)$$

$$= \frac{\hat{p}^2}{2m} + \frac{1}{2}m\omega_0^2 \hat{x}^2 + \frac{i\omega_0}{2}(\hat{x}\hat{p} - \hat{p}\hat{x})$$

$$= \hat{H} - \frac{1}{2}\hbar\omega_0$$

The Harmonic Oscillator

The Hamiltonian operator $\hat{H}$ of the oscillator can hence be written as

$$\hat{H} = \hbar\omega_0\left(\hat{a}^+\hat{a} + \frac{1}{2}\right) \qquad \ldots(5.52)$$

We can now derive the commutators of the operators $\hat{a}$ and $\hat{a}^+$ with $\hat{H}$:

$$[\hat{H},\hat{a}] = [\hbar\omega_0\,\hat{a}^+\hat{a},\hat{a}] = \hbar\omega_0\,[(\hat{a}^+\hat{a})\hat{a} - \hat{a}(\hat{a}^+\hat{a})]$$

$$= \hbar\omega_0[\hat{a}^+\hat{a} - \hat{a}\hat{a}^+]\hat{a} = \hbar\omega_0\,[\hat{a}^+,\hat{a}]\hat{a}$$

$$= -\hbar\omega_0\,\hat{a} \qquad \ldots(5.53)$$

Similarly, $\qquad [\hat{H},\hat{a}^+] = \hbar\omega_0\,[\hat{a},\hat{a}^+]\hat{a}^+ = \hbar\omega_0\,\hat{a}^+ \qquad \ldots(5.54)$

If u is the eigenfunction of $\hat{H}$ belonging to the energy eigenvalue E of the oscillator, we can write

$$\hat{H}u = Eu \qquad \ldots(5.55)$$

From Eq. (5.53), we obtain

$$\hat{H}(\hat{a}u) - \hat{a}(\hat{H}u) = -\hbar\omega_0\,\hat{a}u$$

or, $\qquad \hat{H}(\hat{a}u) - \hat{a}\,Eu = -\hbar\omega_0\,\hat{a}u$ [using Eq. (5.55)]

or, $\qquad \hat{H}(\hat{a}u) = (E - \hbar\omega_0)\,\hat{a}u \qquad \ldots(5.56)$

So, $\hat{a}u$ is an eigenfunction of $\hat{H}$ with the eigenvalue $(E - \hbar\omega_0)$, whereas u is an eigenfunction of $\hat{H}$ with the eigenvalue E. Therefore, the operator $\hat{a}$ is called a *lowering* or *step-down operator* that decreases the energy eigenvalue by $\hbar\omega_0$.

Repeated application of the lowering operator $\hat{a}$ to eigenfunctions will finally lead to the state of lowest energy, the ground state. If u_0 is the ground-state eigenfunction,

$$\hat{a}u_0 = 0 \qquad \ldots(5.57)$$

since the operation with $\hat{a}$ cannot lower the energy further. From Eq. (5.52), we get

$$\hat{H}u_0 = \hbar\omega_0\left(\hat{a}^+\hat{a} + \frac{1}{2}\right)u_0 = \frac{1}{2}\hbar\omega_0\,u_0 \qquad \ldots(5.58)$$

using Eq. (5.57). Obviously, $\frac{1}{2}\hbar\omega_0$ is the ground-state or zero-point energy.

Again, Eq. (5.54) gives

$$\hat{H}(\hat{a}^+u_0) - \hat{a}^+(\hat{H}u_0) = \hbar\omega_0\,\hat{a}^+u_0$$

or, $\qquad \hat{H}(\hat{a}^+u_0) = \left(\hbar\omega_0 + \frac{1}{2}\hbar\omega_0\right)\hat{a}^+u_0 \qquad \ldots(5.59)$

where we have used Eq. (5.58). Thus, operating on the eigenfunction of energy eigenvalue $\frac{1}{2}\hbar\omega_0$, $\hat{a}^+$ leads to an eigenstate with a higher energy eigenvalue $\frac{1}{2}\hbar\omega_0 + \hbar\omega_0$. Hence $\hat{a}^+$ is called a *raising* or *step-up operator*. The state $\hat{a}^+u_0$ is the first excited state of the oscillator, which can be denoted by u_1. Repeated application of $\hat{a}^+$ will give states of higher excitation with energy eigenvalues

$$E_n = \left(n + \frac{1}{2}\right)\hbar\omega_0, \quad n = 0, 1, 2, \ldots \qquad \ldots(5.60)$$

Thus we find that the energy quantization can be derived from the properties of $\hat{H}, \hat{a}$, and $\hat{a}^+$

Operating on u_0, n times with $\hat{a}^+$, we obtain

$$\hat{H}(\hat{a}^+)^n u_0 = \left(n + \frac{1}{2}\right)\hbar\omega_0 (\hat{a}^+)^n u_0 \qquad ...(5.61)$$

If u_n denotes the normalized eigenfunction with the energy eigenvalue E_n, we can write

$$(\hat{a}^+)^n u_0 = c_n u_n \qquad ...(5.62)$$

where c_n is a constant; it is implied that $c_0 = 1$.

Let us define the operator $\hat{N}$ by

$$\hat{N} = \hat{a}^+ \hat{a} \qquad ...(5.63)$$

Then $\hat{N}u_n = \hat{a}^+ \hat{a} u_n = \left(\dfrac{\hat{H}}{\hbar\omega_0} - \dfrac{1}{2}\right)u_n$ [from Eq. (5.52)]

$= n\, u_n.$

So, $\hat{N}u_n$ has the eigenvalue n for the nth excited energy eigenstate. Hence $\hat{N}$ is termed the *number operator*, giving the number of energy quanta in an eigenstate. As the operator $\hat{a}^+$ increases the number of energy quanta, it is also referred to as the *creation operator*; correspondingly, $\hat{a}$ which decreases the number of energy quanta, is known as the *destruction* or *annihilation operator*.

5.6. WORKED-OUT PROBLEMS

1. A simple harmonic oscillator consisting of a particle of mass m is acted on by a linear restoring force of force constant A. Show that the wave function $\psi(x, t)$ for the lowest energy state ($n = 0$) of this oscillator is given by

$$\psi_0(x, t) = \frac{(mA)^{1/8}}{(\pi\hbar)^{1/4}} e^{-\sqrt{mA}\, x^2/(2\hbar)} e^{-\frac{i}{2}\sqrt{A/m}\, t}$$

Ans. We have from Eq. (5.30) the following expression of $\psi_n(x, t)$ for $n = 0$:

$$\psi_0(x, t) = A_0 H_0(\alpha x) e^{-\alpha^2 x^2/2} e^{-iE_0 t/\hbar} \qquad ...(i)$$

Now, $H_0(\alpha x) = 1$, $\alpha^2 = \dfrac{(mA)^{1/2}}{\hbar}$, and $E_0 = \dfrac{1}{2}\hbar\omega_0' = \left(\dfrac{A}{m}\right)^{1/2}\dfrac{\hbar}{2}$.

Also we have the normalization condition

$$\int_{-\infty}^{\infty} \psi_0^*(x,t)\psi_0(x,t)\, dx = A_0^2 \int_{-\infty}^{\infty} e^{-\alpha^2 x^2}\, dx = 1$$

or, $\qquad 2A_0^2 \displaystyle\int_0^{\infty} e^{-\alpha^2 x^2}\, dx = 1$

since $e^{-\alpha^2 x^2}$ is an even function of x. Using a table of definite integrals we get

$$2A_0^2 \frac{\sqrt{\pi}}{2\alpha} = 1$$

or, $\qquad A_0 = \dfrac{\alpha^{1/2}}{(\pi)^{1/4}} = \dfrac{(mA)^{1/8}}{(\pi\hbar)^{1/4}}$

Therefore, substituting the values of the different parameters we get

$$\psi_0(x, t) = \frac{(mA)^{1/8}}{(\pi\hbar)^{1/4}} e^{-\sqrt{mA}\, x^2/(2\hbar)} e^{-i\sqrt{A/m}\, t/2}$$

2. Find the probability density for the lowest energy state of wave function of the simple harmonic oscillator quoted in Example 1. Comment on the graphical plot of the same.

The Harmonic Oscillator

Ans. We have $\psi(x, t) = A_0 e^{-\alpha^2 x^2/2} e^{-iE_0 t/\hbar}$

where $A_0 = \dfrac{(mA)^{1/8}}{(\pi\hbar)^{1/4}}, \quad \alpha^2 = \dfrac{(mA)^{1/2}}{\hbar}$

and $E_0 = \dfrac{1}{2}\hbar\omega_0 = \dfrac{1}{2}\hbar\sqrt{A/m}$

The probability density as defined in quantum mechanics, is

$$P(x) = \psi^*\psi = u_0^* u_0 = A_0^2 e^{-\alpha^2 x^2}$$

A plot of $P(x)$ against x is shown in Fig. 5.4.

Evidently, quantum mechanics predicts that the particle is most likely to be found near the equilibrium point, *i.e.*, near $x = 0$, since $P(x)$ is maximum at $x = 0$. Also, there is no well defined limits outside which the probability is precisely zero. The probability reduces to $\dfrac{1}{e}$ times the maximum value at $x = \pm\dfrac{1}{\alpha} = \pm\sqrt{\dfrac{2E_0}{A}}$ where E_0 is the zero-point energy $\left(=\dfrac{1}{2}\hbar\omega_0\right)$

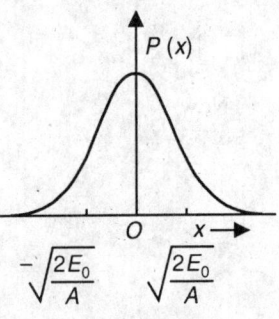

Fig. 5.4.

and A is the force constant. This value of x is the classical oscillator amplitude. The probability of finding the oscillator inside the limits of classical motion is

$$\dfrac{\int_0^{1/\alpha} P(x)\,dx}{\int_0^\infty P(x)\,dx} = \dfrac{\int_0^{1/\alpha} e^{-\alpha^2 x^2}\,dx}{\int_0^\infty e^{-\alpha^2 x^2}\,dx} = \dfrac{\int_0^1 e^{-z^2}\,dz}{\int_0^\infty e^{-z^2}\,dz} \quad \text{(putting } z = \alpha x\text{)}$$

$$= \dfrac{\int_0^1 \left(1 - z^2 + \dfrac{z^4}{2!} - \dfrac{z^6}{3!}....\right)dz}{\dfrac{1}{2}\int_0^\infty p^{-1/2} e^{-p}\,dp} \quad \text{(putting } z^2 = p, \text{ so that } 2z\,dz = dp \text{ or } dz = \dfrac{1}{2\sqrt{p}}dp\text{)}$$

$$= \dfrac{\left[z - \dfrac{z^3}{3} + \dfrac{z^5}{10} - \dfrac{z^7}{42} + \dfrac{z^9}{216} - \dfrac{z^{11}}{1320} + ...\right]_0^1}{\dfrac{1}{2}\Gamma\left(\dfrac{1}{2}\right)}$$

$$= \dfrac{2}{\sqrt{\pi}}\left[1 - \dfrac{1}{3} + \dfrac{1}{10} - \dfrac{1}{42} + \dfrac{1}{216} - \dfrac{1}{1320} + ...\right] \approx \dfrac{2}{1.7724}[0.74673] \approx 0.84$$

Thus the oscillator has approximately 16% probability of lying outside the limits of the classical amplitude.

3. Show by direct substitution that the eigenvalue and the eigenfunction corresponding to $n = 1$ of a simple harmonic oscillator satisfy the time-independent Schrödinger equation.

Ans. The eigenvalue and the eigenfunction corresponding to $n = 1$ are given by

$$E_1 = \dfrac{3}{2}\hbar\omega_0 = \dfrac{3}{2}\hbar\left(\dfrac{A}{m}\right)^{1/2}$$

$$u_1 = 2A_1 \alpha x e^{-\alpha^2 x^2/2},$$

where
$$\alpha = \frac{(mA)^{1/4}}{\sqrt{\hbar}}$$

The time-independent Schrödinger equation is

$$-\frac{\hbar^2}{2m}\frac{d^2u}{dx^2} + \frac{1}{2}Ax^2 u = Eu$$

Now
$$\frac{du_1}{dx} = 2A_1 \alpha e^{-\alpha^2 x^2/2} - 2A_1 \alpha x \frac{\alpha^2}{2} 2x e^{-\alpha^2 x^2/2}$$

$$= 2A_1 \alpha e^{-\alpha^2 x^2/2}\left[1 - \alpha^2 x^2\right]$$

$$\frac{d^2 u_1}{dx^2} = -2A_1 \alpha \frac{\alpha^2}{2} 2x e^{-\alpha^2 x^2/2}[1 - \alpha^2 x^2]$$

$$-2A_1 \alpha e^{-\alpha^2 x^2/2} 2\alpha^2 x$$

$$= -6A_1 \alpha^3 x e^{-\alpha^2 x^2/2} + 2A_1 \alpha^5 x^3 e^{-\alpha^2 x^2/2}$$

$$= 2A_1 \alpha x e^{-\alpha^2 x^2/2}[\alpha^4 x^2 - 3\alpha^2]$$

$$= \alpha^2[\alpha^2 x^2 - 3] u_1$$

Putting into the L.H.S. of the Schrödinger equation we get

$$-\frac{\hbar^2}{2m}\frac{(mA)^{1/2}}{\hbar}\left[\frac{(mA)^{1/2}}{\hbar}x^2 - 3\right]u_1 + \frac{1}{2}Ax^2 u_1$$

$$= \frac{3}{2}\frac{\hbar}{m}(mA)^{1/2} u_1 = \frac{3}{2}\hbar\left(\frac{A}{m}\right)^{1/2} u_1$$

$$= E_1 u_1 = \text{R.H.S.}$$

Thus the eigenvalue and the eigenfunction for $n = 1$, satisfy the time-independent Schrödinger equation.

4. An oscillator consists of a weight of 1 kg at the end of a light rod of length 1 m. If the amplitude of oscillation is 0.1 m, calculate

 (*i*) the frequency of oscillation (ω_0),
 (*ii*) the energy of oscillation (*E*),
 (*iii*) the approximate value of the quantum number (*n*),
 (*iv*) the separation between adjacent energy levels, and
 (*v*) the zero-point energy in electron volt. (*cf. C.U. 1993*)

Ans. (*i*) We know that $\omega_0 = \sqrt{g/l}$, where g is the acceleration due to gravity and l is the length of the rod. Given, $l = 1$ m. Also, $g = 9.8$ m/s². Hence

$$\omega_0 = \sqrt{9.8} = 3.13 \text{ rad/s}$$

(*ii*) The total energy is

$$E = \frac{1}{2}m\omega_0^2 a^2$$

Given, $m = 1$ kg and $a = 0.1$ m

The Harmonic Oscillator 91

So, $E = 0.5 \times 1 \times 9.8 \times 10^{-2}$ J
$= 4.9 \times 10^{-2}$ J

(*iii*) We know that

$$E = (n + \frac{1}{2})\hbar\omega_0,$$

where $\hbar$ is reduced Planck's constant ($\hbar = 1.054 \times 10^{-34}$ J. s), and n is the quantum number.

Thus the quantum number is

$$n = \frac{E}{\hbar\omega_0} - \frac{1}{2} = \frac{4.9 \times 10^{-2}}{1.054 \times 10^{-34} \times 3.13} - \frac{1}{2}$$

$$\approx 1.5 \times 10^{32}$$

(*iv*) The separation between adjacent energy levels is

$$\hbar\omega_0 = 1.054 \times 10^{-34} \times 3.13$$
$$= 3.3 \times 10^{-34} \text{ J}$$

(*v*) The zero-point energy is

$$\frac{1}{2}\hbar\omega_0 = 1.65 \times 10^{-34} \text{ J}$$

$$= \frac{1.65 \times 10^{-34}}{1.6 \times 10^{-19}} \approx 10^{-15} \text{ eV}$$

5. A laboratory oscillator consisting of a mass of 1 gm on a spring exhibits a period of 1 sec. The velocity of the mass when it crosses the zero displacement position is 10 cm/sec.

(*i*) Is this oscillator possibly in an eigenstate of the Hamiltonian ?

(*ii*) What is the approximate value of the quantum number n associated with the energy E of the oscillator?

(*iii*) Has the zero-point energy any significance here ? (*cf. C.U.* 2003, 2005)

Ans. (*i*) & (*ii*) The total energy E of the oscillator is

$$E = \frac{1}{2}mv^2 = \frac{1}{2} \times 10^{-3} \times (10^{-1})^2 = 5 \times 10^{-6} \text{ joule.}$$

Now, the energy of a quantum oscillator with quantum number n is $(n + \frac{1}{2})\hbar\omega_0$. The oscillator in question is in an eigenstate of the Hamiltonian when

$$(n + \frac{1}{2})\hbar\omega_0 = 5 \times 10^{-6}$$

We have $\hbar = 1.054 \times 10^{-34}$ J.s and $\omega_0 = \frac{2\pi}{T} = 2\pi$ rad/s ($\because T = 1$ sec.)

Substituting the values we get

$$n \cong 10^{28}$$

Since n is an integer and very large we conclude that the oscillator is probably in an eigenstate of the Hamiltonian. The value of n is about 10^{28}.

(*iii*) The zero-point energy, *i.e.* $\frac{1}{2}\hbar\omega_0$ is very small compared with the energy $n\hbar\omega_0$ since n is very large ($\approx 10^{28}$). The zero-point energy can thus be neglected and has no significance here.

6. Consider a linear harmonic oscillator for which the total energy is

$$E = \frac{p_x^2}{2m} + \frac{1}{2}m\omega^2 x^2,$$

where the symbols have their usual meanings. The particle is assumed to be confined to a region ~ a. Use the uncertainty principle to obtain the minimum energy (ground state energy) of the oscillator.

(C.U. 1997, 2004, N. Beng. U. 2001).

Ans. The uncertainty in x is $\Delta x = a$. By the uncertainty principle,

$$\Delta x \, \Delta p_x = \hbar/2. \text{ So, } \Delta p_x = \frac{\hbar}{2\Delta x} = \frac{\hbar}{2a}$$

Therefore, $E = \dfrac{(\Delta p_x)^2}{2m} + \dfrac{1}{2}m\omega^2(\Delta x)^2 = \dfrac{\hbar^2}{8ma^2} + \dfrac{1}{2}m\omega^2 a^2$.

For E to be a minimum, $\dfrac{dE}{da} = 0$, i.e.,

$$-\frac{\hbar^2}{4ma^3} + m\omega^2 a = 0 \quad \text{or} \quad a^2 = \frac{\hbar}{2m\omega}$$

So, the minimum total energy or the ground state energy is

$$E_{min} = \frac{\hbar}{8m}(2m\omega) + \frac{1}{2}m\omega^2\left(\frac{\hbar}{2m\omega}\right) = \frac{1}{2}\hbar\omega$$

7. Using the fundamental commutation relation $[\hat{x}, \hat{p}] = i\hbar$, find $[\hat{p}, \hat{H}]$ where $\hat{H}$ is the Hamiltonian operator of a linear harmonic oscillator. Hence determine whether the ground state wavefunction ψ_0 of the Hamiltonian is an eigenstate of $\hat{p}$ or not. (C.U. 2003)

Ans. We have $\hat{p} = -i\hbar\dfrac{\partial}{\partial x}$ and $\hat{H} = -\dfrac{\hbar^2}{2m}\dfrac{\partial^2}{\partial x^2} + \dfrac{1}{2}Ax^2$.

So, $(\hat{p}\hat{H} - \hat{H}\hat{p})\psi = -i\hbar\dfrac{\partial}{\partial x}(\hat{H}\psi) + i\hbar\hat{H}\left(\dfrac{\partial\psi}{\partial x}\right)$

$$= i\hbar\left[-\frac{\partial}{\partial x}\left(-\frac{\hbar^2}{2m}\frac{\partial^2\psi}{\partial x^2} + \frac{1}{2}Ax^2\psi\right) + \left(-\frac{\hbar^2}{2m}\frac{\partial^2}{\partial x^2} + \frac{1}{2}Ax^2\right)\frac{\partial\psi}{\partial x}\right]$$

$$= i\hbar\left[-\frac{1}{2}Ax^2\frac{\partial\psi}{\partial x} - Ax\psi + \frac{1}{2}Ax^2\frac{\partial\psi}{\partial x}\right] = -i\hbar Ax\psi$$

Hence $[\hat{p}, \hat{H}] = -i\hbar Ax = -Ax[\hat{x}, \hat{p}]$

The ground state wavefunction ψ_0 of the Hamiltonian is not an eigenstate of $\hat{p}$ since $\hat{p}$ and $\hat{H}$ do not commute, as shown above.

8. What is the amplitude of the zero-point oscillation of a pendulum whose length is 1 m and the mass of the bob is 1 kg ? Can you detect it ?

Ans. The zero-point energy is $E = \dfrac{1}{2}\hbar\omega = \dfrac{1}{2}m\omega^2 a^2$, where a is the amplitude. For a pendulum,

$$\omega = \sqrt{g/l} = \sqrt{9.8/1} = 3.13 \text{ rad/s.}$$

The Harmonic Oscillator

$$a = \sqrt{\frac{\hbar}{m\omega}} = \sqrt{\frac{1.054 \times 10^{-34}}{1 \times 3.13}} = 0.58 \times 10^{-17}\,m$$

The amplitude is too small to be detected, reflecting that for a classical oscillator, the zero-point energy is of little consequence.

9. Suppose that ψ_0 and ψ_1 are the real, normalized ground and first excited state energy eigenfunctions, respectively, of a linear harmonic oscillator. Consider the wavefunction $\psi = a\psi_0 + b\psi_1$ (where a, b are real numbers) of the oscillator. Show that for the wavefunction ψ the average value of x, i.e. $<x>$ is, in general, not zero. Determine the values of a and b which (i) maximize $<x>$ and (ii) minimize $<x>$.

Ans. The orthonormal requirement is

$$\int_{-\infty}^{\infty} |\psi|^2\, dx = 1 \quad \text{or,} \quad \int_{-\infty}^{\infty} (a\psi_0 + b\psi_1)^2\, dx = 1$$

This gives
$$a^2 + b^2 = 1. \qquad \ldots(i)$$

We have

$$<x> = \int_{-\infty}^{\infty} |\psi|^2 x\, dx = \int_{-\infty}^{\infty} x(a\psi_0 + b\psi_1)^2\, dx$$

$$= a^2 \int_{-\infty}^{\infty} x\psi_0^2\, dx + b^2 \int_{-\infty}^{\infty} x\psi_1^2\, dx + 2ab \int_{-\infty}^{\infty} x\psi_0\psi_1\, dx$$

$$= 2ab \int_{-\infty}^{\infty} x\psi_0\psi_1\, dx = 2ab <\psi_0|x|\psi_1>, \qquad \ldots(ii)$$

since
$$\int_{-\infty}^{\infty} x\psi_0^2\, dx = \int_{-\infty}^{\infty} x\psi_1^2\, dx = 0$$

Equation (ii) shows that $<x>$ is nonzero, in general.

Using Eq. (i), Eq. (ii) can be written as

$$<x> = [1 - (a^2 + b^2 - 2ab)] <\psi_0|x|\psi_1>$$
$$= [1 - (a-b)^2] <\psi_0|x|\psi_1> \qquad \ldots(iii)$$

Since $(a-b)^2$ is a real perfect square, $<x>$ is a maximum in Eq. (iii) when $(a-b)^2 = 0$, or $a = b$. With Eq. (i), we find that if $a = b = 1/\sqrt{2}$, $<x>$ is maximized. Also, if $(a-b)^2$ is a maximum, $<x>$ is a minimum. This happens when $a = -b = 1/\sqrt{2}$.

10. The energy of a linear harmonic oscillator in the third excited state is 0.1 eV. Find the frequency of vibration. *(Rohilkhand 2000)*

Ans. The energy levels of a linear harmonic oscillator are given by

$$E_n = \left(n + \frac{1}{2}\right)\hbar\omega_0 = \left(n + \frac{1}{2}\right)h\nu$$

where ν is the frequency of vibration. For the third excited state $n = 3$. So

$$E_3 = (3 + \frac{1}{2})h\nu = \frac{7}{2}h\nu,$$

Here
$$E_3 = 0.1 \text{ eV} = 0.1 \times 1.6 \times 10^{-19}\text{J} = 1.6 \times 10^{-20} \text{ J}.$$

Hence,
$$1.6 \times 10^{-20} = \frac{7}{2} \times 6.62 \times 10^{-34} \times \nu$$

or,
$$\nu = \frac{2}{7} \times \frac{1.6 \times 10^{-20}}{6.62 \times 10^{-34}} = 6.9 \times 10^{12} \text{ Hz}$$

11. The lowering operator of a linear harmonic oscillator, given by

$$\hat{a} = \sqrt{\frac{m\omega_0}{2\hbar}} x + \sqrt{\frac{\hbar}{2m\omega_0}} \frac{d}{dx}$$

satisfies the relation $\hat{a}u_0 = 0$ where u_0 is the ground-state energy eigenfunction. Hence find the normalized u_0.

Ans. We have
$$\hat{a}u_0 = 0$$

or,
$$\left(\sqrt{\frac{m\omega_0}{2\hbar}} x + \sqrt{\frac{\hbar}{2m\omega_0}} \frac{d}{dx}\right) u_0 = 0$$

Introducing a dimensional variable $\xi = \sqrt{\frac{m\omega_0}{\hbar}} x$,

the above equation reduces to $\left(\xi + \frac{d}{d\xi}\right) u_0 = 0$

or,
$$\frac{du_0}{u_0} = -\xi \, d\xi$$

Integrating,
$$u_0 = A \exp\left(-\frac{\xi^2}{2}\right)$$

where A is the integration constant to be found from the normalization condition

$$\int_{-\infty}^{\infty} |u_0(x)|^2 \, dx = 1.$$

This gives
$$A^2 \int_{-\infty}^{\infty} \exp\left(-\frac{m\omega_0 x^2}{\hbar}\right) dx = 1$$

or,
$$A^2 \sqrt{\frac{\hbar\pi}{m\omega_0}} = 1$$

or,
$$A = \left(\frac{m\omega_0}{\hbar\pi}\right)^{1/4}$$

Thus the normalized u_0 is
$$u_0 = \left(\frac{m\omega_0}{\hbar\pi}\right)^{1/4} \exp\left(-\frac{\xi^2}{2}\right)$$

12. The raising operator of a linear harmonic oscillator is given by

$$\hat{a}^+ = \sqrt{\frac{m\omega_0}{2\hbar}}x - \sqrt{\frac{\hbar}{2m\omega_0}}\frac{d}{dx}$$

The unnormalized ground-state wave function is $u_0 = \exp\left(-\frac{\xi^2}{2}\right)$ where $\xi = \sqrt{\frac{m\omega_0}{\hbar}}x$. Using the property of $\hat{a}^+$, find the normalized wavefunction of the first excited state, u_1.

Ans. We have

$$u_1 = A\hat{a}^+ u_0 = A\left(\sqrt{\frac{m\omega_0}{2\hbar}}x - \sqrt{\frac{\hbar}{2m\omega_0}}\frac{d}{dx}\right)u_0$$

$$= A_1\left(\xi - \frac{d}{d\xi}\right)\exp\left(-\frac{\xi^2}{2}\right) = C\xi\exp\left(-\frac{\xi^2}{2}\right)$$

where A, A_1, and C are constants. The normalization condition is

$$\int_{-\infty}^{\infty} |u_1|^2 \, dx = 1$$

or,
$$2\sqrt{\frac{\hbar}{m\omega_0}} C^2 \int_0^\infty \xi^2 \exp(-\xi^2)d\xi = 1$$

giving
$$C = \sqrt{2}\left(\frac{m\omega_0}{\pi\hbar}\right)^{1/4}$$

Hence the desired normalized wavefunction is

$$u_1 = \sqrt{2}\left(\frac{m\omega_0}{\pi\hbar}\right)^{1/4} \xi \exp(-\xi^2/2)$$

QUESTIONS

1. Discuss the importance of harmonic oscillators in physics.
2. Establish Schrödinger's equation of a linear harmonic oscillator and solve it for different eigenvalues. Discuss the significance of zero-point energy. **(C. U. 1982)**
3. Derive the expressions for the eigenfunctions in terms of Hermite polynomials of a linear harmonic oscillator.
   Show that the eigenfunctions form an orthonormal set.
4. Derive an expression for the probability amplitude for a classical oscillator with energy $(n + \frac{1}{2})\hbar\omega_0$ where the symbols have their usual meanings. Compare this probability with quantum probability.
5. Deduce the position momentum uncertainty relation for a one-dimensional harmonic oscillator. What conclusion can be drawn from this relation ?
6. (a) What is the classical Hamiltonian for a linear harmonic oscillator ?

(b) Write down the Schrödinger equation for this Hamiltonian. What are the eigenvalues and the eigenfunctions of the Hamiltonian? What conditions lead to discrete energy levels? Are all the eigenfunctions orthogonal?

(c) Explain the zero-point energy. How does it reconcile with the classical view point?

(C.U. 1984)

7. Estimate the zero-point energy of the simple harmonic vibrations of an atom. Assume that the restoring force constant A is 10^3 joule/m^2 and the mass of the atom is 10^{-26} kg. [Ans. 0.104 eV]

8. For a linear harmonic oscillator, verify the eigenfunction and the eigenvalue corresponding to $n = 2$ state by direct substitution into the time independent Schrödinger equation.

9. The potential energy $V(x)$ of a linear harmonic oscillator consisting of a particle of mass m and oscillating with an angular frequency ω_0 is

$$V(x) = \frac{1}{2} m \omega_0^2 x^2.$$

(i) Write down the Hamiltonian operator and the time-independent Schrodinger equation for the oscillator.

(ii) The ground state eigenfunction of the Hamiltonian operator is given by

$$u_0 = \left(\frac{\alpha}{\sqrt{\pi}}\right)^{1/2} e^{-\alpha^2 x^2/2},$$

where $\alpha = \left(\frac{m \omega_0}{\hbar}\right)^{1/2}$. Find the energy eigenvalue corresponding to the ground state.

[Ans. $\frac{1}{2}\hbar\omega_0$]

10. (a) Solve the one-dimensional harmonic oscillator problem and show that the boundary conditions lead to the discrete energy spectrum of the system, namely, $E_n = (n + \frac{1}{2})\hbar\omega$; $n = 0, 1, 2, 3$, etc, ω being the characteristic angular frequency.

(b) What is zero-point energy? Show that existence of zero-point energy is in conformity with the uncertainty principle. (C.U. 1986, 99, 2008)

11. The wave function $\psi(x, t)$ for the lowest energy state of a simple harmonic oscillator, consisting of a particle of mass m acted on by a linear restoring force of force constant K, is $\psi(x, t) = A e^{-(\sqrt{mK}/2\hbar)x^2} e^{-(i/2)\sqrt{K/m}\,t}$, where A is a real constant. Verify that $\psi(x, t)$ is a solution of the Schrödinger equation for the appropriate potential. Using $\psi(x, t)$, find the expectation value of x.

(C.U. 1996) [Ans. 0]

12. The lowest-energy eigenfunction of the linear harmonic oscillator is

$$u = \left(\frac{\alpha}{\sqrt{\pi}}\right)^{1/2} e^{-\alpha^2 x^2/2}$$

where $\alpha = \left(\frac{m\omega_0}{\hbar}\right)^{1/2}$, the other symbols having their usual significance. Show that the expectation values of the kinetic and potential energies of the oscillator are the same, each being given by $E_0/2$, where E_0 is the zero-point energy. (Cf. Burd. U. 1996)

13. For the wave function of problem no. 11, find the probability current density. Interpret the result.

[Ans. 0]

14. Explain whether the ground-state wave function of a one-dimensional harmonic oscillator is an eigen function of the momentum operator. (C.U. 2000)

15. Calculate the zero-point energy of a mass of 1.67×10^{-24} gm connected to a fixed point by a spring of force constant 10^4 dyne/cm. (N. Beng. U. 2001) [Ans. 25.5 m eV]

16. Evaluate $\left[\hat{x}, \widehat{H}\right]$ where $\widehat{H}$ is the Hamiltonian for the simple harmonic oscillator.

(C.U. 2002) $\left[\text{Ans. } \dfrac{\hbar^2}{m}\dfrac{\partial}{\partial x}\right]$

17. Prove that the quantum mechanical probability of finding the linear harmonic oscillator in the ground state outside the classical limits of motion, is approximately 16%.

(Purvanchal Univ. 2002)

18. Find the expectation value of energy when the state of the harmonic oscillator is described by the wavefunction $\psi(x,t) = \dfrac{1}{\sqrt{2}}[\psi_0(x,t) + \psi_1(x,t)]$, where $\dfrac{1}{2}\hbar\omega_0$ is the energy for the ground state described by the wavefunction $\psi_0(x,t)$ and $\dfrac{3}{2}\hbar\omega_0$ is the energy for the first excited state described by the wavefunction $\psi_1(x,t)$. **(Agra 1987, Rajasthan 1995)**

[Ans. $\hbar\omega_0$]

19. The wave function of the first excited state of a one-dimensional linear harmonic oscillator has the form $\psi(x) = Ax e^{-\alpha^2 x^2/2}$, where the symbols are usual. Find the expectation value of the momentum operator for this state. Explain your result physically. (C.U. 2004)

[Ans. 0]

20. Define the creation and annihilation operators for a one-dimensional harmonic oscillator and obtain the commutation relation they obey. Find how they commute with the Hamiltonian and give an interpretation. (C.U.2006)

Chapter 6

Spherically Symmetric Potential and the Hydrogen Atom

6.1. Schrödinger's Equation and Its Separation

For a particle of mass m_e, the time-independent Schrödinger equation in three dimensions is

$$\left[-\frac{\hbar^2}{2m_e}\nabla^2 + V(\vec{r})\right]u(\vec{r}) = Eu(\vec{r})$$

For a spherically symmetric potential energy, $V(\vec{r}) = V(r)$, and the Schrödinger equation in spherical coordinates can be written as

$$\frac{-\hbar^2}{2m_e}\left[\frac{1}{r^2}\frac{\partial}{\partial r}\left(r^2\frac{\partial}{\partial r}\right) + \frac{1}{r^2 \sin\theta}\frac{\partial}{\partial \theta}\left(\sin\theta \frac{\partial}{\partial \theta}\right)\right.$$
$$\left. + \frac{1}{r^2 \sin^2\theta}\frac{\partial^2}{\partial \phi^2}\right]u + V(r)u = Eu \qquad ...(6.1)$$

We use the method of separation of variables to solve this equation. Putting $u(r, \theta, \phi) = R(r)Y(\theta, \phi)$ in Eq. (6.1) and dividing the resultant equation by RY we get

$$\frac{1}{R}\frac{d}{dr}\left(r^2 \frac{dR}{dr}\right) + \frac{2m_e r^2}{\hbar^2}[E - V(r)]$$
$$= -\frac{1}{Y}\left[\frac{1}{\sin\theta}\frac{\partial}{\partial \theta}\left(\sin\theta \frac{\partial Y}{\partial \theta}\right) + \frac{1}{\sin^2\theta}\frac{\partial^2 Y}{\partial \phi^2}\right] \qquad ...(6.2)$$

The total derivative $\frac{dR}{dr}$ appears in the left-hand side, but partial derivatives of Y appear in the right-hand side of Eq. (6.2) because R is a function of r only, whereas Y is a function of two variables θ and ϕ.

The left-hand side of Eq. (6.2) depends only on r, whereas the right-hand side depends only on θ and ϕ. Therefore both sides must be equated to a constant, say, λ. This procedure gives a radial equation

$$\frac{1}{r^2}\frac{d}{dr}\left(r^2 \frac{dR}{dr}\right) + \left[\frac{2m_e}{\hbar^2}\{E - V(r)\} - \frac{\lambda}{r^2}\right]R = 0, \qquad ...(6.3)$$

and an angular equation

$$\frac{1}{\sin\theta}\frac{\partial}{\partial \theta}\left(\sin\theta \frac{\partial Y}{\partial \theta}\right) + \frac{1}{\sin^2\theta}\frac{\partial^2 Y}{\partial \phi^2} + \lambda Y = 0. \qquad ...(6.4)$$

The angular equation (6.4) can be further separated by putting $Y(\theta, \varphi) = \Theta(\theta)\Phi(\phi)$ into it, and adopting the same procedure to obtain an equation involving θ only and a second equation in ϕ only.

The ϕ equation is

$$\frac{d^2\Phi}{d\phi^2} + \nu\Phi = 0. \qquad ...(6.5)$$

Spherically Symmetric Potential and the Hydrogen Atom 99

The θ equation is

$$\frac{1}{\sin\theta}\frac{d}{d\theta}\left(\sin\theta\frac{d\Theta}{d\theta}\right)+\left(\lambda-\frac{v}{\sin^2\theta}\right)\Theta=0. \qquad \text{...(6.6)}$$

In Eqs. (6.5) and (6.6), the quantity v is another separation constant.

6.2. Solutions of θ and φ Equations

The general solution of the φ equation (6.5) can be written as

$$\Phi(\phi)=Ae^{i\sqrt{v}\phi}+Be^{-i\sqrt{v}\phi} \qquad v\neq 0 \qquad \text{...(6.7)}$$

and

$$\Phi(\phi)=A+B\phi \qquad v=0$$

The quantity $\Phi(\phi)$ should be continuous and single-valued throughout the domain 0 to 2π. Also, it should have the same value for φ or ($\phi+2q\pi$) where q is an integer. The latter condition requires that v must be equal to the square of an integer. Writing $\sqrt{v}=m$ we obtain from Eq. (6.7)

$$\Phi(\phi)=\frac{1}{\sqrt{2\pi}}e^{im\phi} \qquad \text{...(6.8)}$$

The constant m in Eq. (6.8) is allowed to be a positive or a negative integer, or zero in order to include all physically meaningful solutions. The multiplying constant $\frac{1}{\sqrt{2\pi}}$ is obtained by normalizing Φ over the range 0 to 2π:

$$\int_0^{2\pi}\Phi_m^*\Phi_m\,d\phi=1.$$

By putting $v=m^2$, the θ equation becomes

$$\frac{1}{\sin\theta}\frac{d}{d\theta}\left(\sin\theta\frac{d\Theta}{d\theta}\right)+\left(\lambda-\frac{m^2}{\sin^2\theta}\right)\Theta=0.$$

It is convenient to put $\omega=\cos\theta$ and substitute $\Theta(\theta)=P(\omega)$ in the above equation to obtain

$$\frac{d}{d\omega}\left[(1-\omega^2)\frac{dP}{d\omega}\right]+\left(\lambda-\frac{m^2}{1-\omega^2}\right)P=0 \qquad \text{...(6.9)}$$

Note that the domain of θ is 0 to π. Therefore the domain of ω becomes -1 to $+1$. For the simplest case, *i.e.*, $m=0$, Eq. (6.9) reduces to

$$\frac{d}{d\omega}\left[(1-\omega^2)\frac{dP}{d\omega}\right]+\lambda P=0 \qquad \text{...(6.9a)}$$

Equation (6.9a) is known as *Legendre's equation*. We assume a trial solution in the form of a power series in ω:

$$P(\omega)=\sum_n a_n\omega^n \qquad \text{...(6.10)}$$

We put Eq. (6.10) into Eq. (6.9a) and note that the resulting equation holds for all values of ω between -1 and $+1$. So, the coefficient of ω^l for each l must be separately equated to zero in that equation. This procedure leads to the recursion relation:

$$a_{l+2}=\frac{l(l+1)-\lambda}{(l+1)(l+2)}a_l \qquad \text{...(6.11)}$$

It is evident from Eq. (6.11) that the ratio of the coefficients approaches unity for large l, *i.e.*

$$\frac{a_{l+2}}{a_l}\to 1 \text{ as } l\to\infty \qquad \text{...(6.12)}$$

Obviously, $P(\omega)$ will diverge for $\omega = 1$ unless it is made to terminate after a finite number of terms. The termination condition is found to be

$$\lambda = l(l+1) \qquad ...(6.13)$$

where l is zero or a positive integer. Under this condition the polynomial in ω of order l is referred to as the *Legendre polynomial of the first kind* and is denoted by $P_l(\omega)$. $P_l(\omega)$, when properly normalized, becomes the physically acceptable eigenfunction of the θ equation (6.9) for $m = 0$. Equation (6.9) being a second order differential equation must possess two linearly independent solutions. The second set of solutions, designated as $Q_l(\omega)$, which are polynomials in $(1/\omega)$ of order l do not represent physically acceptable wavefunctions since they have an infinity at $\omega = 0$.

It may be verified by direct substitution that $P_l^m(\omega)$ given by

$$P_l^m(\omega) = (1 - \omega^2)^{|m/2|} \frac{d^{|m|} P_l(\omega)}{d\omega^{|m|}} \qquad ...(6.14)$$

is a solution of the general θ equation, *i.e.*, Eq. (6.9), for arbitrary m. The parameter m is called the 'quantum number' and can assume only integer values as has been imposed by the conditions on the solution of the ϕ equation. As $P_l(\omega)$ is a power series polynomial in ω^n, $\dfrac{d^{|m|} P_l(\omega)}{d\omega^{|m|}}$ will be a power series polynomial in ω. The recursion relation connecting the coefficients of this polynomial is obtained by direct calculation and is given by

$$a_{k+2} = \frac{(k + |m|)(k + |m| + 1) - \lambda}{(k+1)(k+2)} a_k \qquad ...(6.15)$$

The polynomial $P_l^m(\omega)$ will terminate after $(l - |m|)$ terms as must the mth derivative of a polynomial of order l provided

$$[(l - |m|) + |m|][(l - |m|) + |m| + 1] - \lambda = 0 \qquad ...(6.16)$$

The relation (6.16) is equivalent to the condition

$$\lambda = l(l+1), \qquad ...(6.17)$$

i.e., similar to the condition for $m = 0$. From the definition of $P_l^m(\omega)$ we find that $l \geq |m|$ if $P_l^m(\omega)$ is not to be trivially equal to zero. We summarise below the separation constants for θ and ϕ equations :

$$m = 0, \pm 1, \pm 2, ... \pm n$$
$$\lambda = l(l+1) \qquad ...(6.18)$$
$$l \geq |m|$$

$P_l^m(\omega)$ is referred to as the *associated Legendre function*. For $m = 0$, the associated Legendre functions $P_l(\omega)$ reduce to the Legendre polynomials. The first few $P_l(\omega)$ are given by $P_0(\omega) = 1$, $P_1(\omega) = \omega$, $P_2(\omega) = \dfrac{1}{2}(3\omega^2 - 1)$ etc.

Spherical harmonics. The angular part of the complete wavefunction, represented by $Y_{lm}(\theta, \phi)$, is a solution of Eq. (6.4). This is given by

$$Y_{lm}(\theta, \phi) = N_{lm} P_l^m(\cos\theta) \Phi_m(\phi) \qquad ...(6.19)$$

where $\Phi_m(\phi)$ is obtained from Eq. (6.8) and N_{lm} is the normalization constant for the associated Legendre function $P_l^m(\cos\theta)$. $Y_{lm}(\theta, \phi)$ is referred to as *spherical harmonics*.

It may be verified that the functions $Y_{lm}(\theta, \phi)$ are orthogonal :

$$\int_0^\pi \int_0^{2\pi} Y_{lm}(\theta, \phi)^* Y_{l'm'}(\theta, \phi) \sin\theta \, d\theta \, d\phi$$

$$= \int_{-1}^{+1} \int_0^{2\pi} Y_{lm}^* Y_{l'm'} \, d\omega \, d\phi = \delta_{ll'} \delta_{mm'} \qquad ...(6.20)$$

Where $\delta_{ll'}$ and $\delta_{mm'}$ are kronecker's delta functions.

Spherically Symmetric Potential and the Hydrogen Atom 101

In Eq. (6.20) the integrals disappear unless $l = l'$ and $m = m'$. When $m \neq m'$, the ϕ part vanishes whereas for $l \neq l'$ and $|m| = |m'|$, the θ part or the ω part vanishes. The integral

$$\int_{-1}^{+1} P_l^m(\omega) P_l^m(\omega) \, d\omega \qquad ...(6.21)$$

vanishes for $l \neq l'$ and reduces to $\dfrac{2 [(l + |m|)!]}{(2l + 1)[(l - |m|)!]}$

when $l = l'$. Thus

$$N_{lm}^2 = \frac{2l + 1}{2} \frac{(l - |m|)!}{(l + |m|)!}$$

Therefore the normalized spherical harmonics are given by

$$Y_{lm}(\theta, \phi) = \left[\frac{2l + 1}{4\pi} \frac{(l - |m|)!}{(l + |m|)!}\right]^{1/2} P_l^m(\cos\theta) \, e^{im\phi} \qquad ...(6.22)$$

For a central force, where the potential depends on r only, the angular part of the wavefunction is always represented by Eq. (6.22) and is independent of the form of $V(r)$.

6.3. Radial Equation and the Angular Momentum

Substituting $R(r) = \dfrac{\chi(r)}{r}$ and using the relation (6.17) the radial wave equation (6.3) can be written as

$$-\frac{\hbar^2}{2m_e} \frac{d^2\chi}{dr^2} + \left[V(r) + \frac{l(l+1)\hbar^2}{2m_e r^2}\right] \chi = E\chi \qquad ...(6.23)$$

Equation (6.23) for the modified wavefunction χ is similar to the one-dimensional wave equation for a particle in a potential

$$V(r) + \frac{l(l+1)\hbar^2}{2m_e r^2} \qquad ...(6.24)$$

The term $\dfrac{l(l+1)\hbar^2}{2m_e r^2}$, representing additional potential energy, may be found to be related with the angular momentum as follows:

A classical particle of mass m_e moving with an angular velocity ω about an axis is said to possess an angular momentum M about the axis, where

$$M = m_e r^2 \omega,$$

r being the radial distance from the axis, assumed to pass through the origin. In order to keep the particle in this path an inward force

$$m_e \omega^2 r = \frac{M^2}{m_e r^3}$$

is required. This force, termed the 'centripetal force', is supplied by the potential energy and so adds to $V(r)$ an extra 'centrifugal potential energy' of amount $M^2/(2m_e r^2)$. This corresponds to the second term in (6.24) provided we put

$$M = [l(l+1)]^{1/2}\hbar \qquad ...(6.25)$$

Equation (6.25) gives a relation between the quantum number l and the angular momentum M of the particle. We shall consider below the operators corresponding to the three components of the angular momentum vector. We have

$$\vec{M} = \vec{r} \times \vec{p}, \qquad ...(6.26)$$

where $\vec{p}$ is the linear momentum vector. Breaking up into cartesian components we obtain for the operators

$$\hat{M}_x = \hat{y}\hat{p}_z - \hat{z}\hat{p}_y = -i\hbar\left(y\frac{\partial}{\partial z} - z\frac{\partial}{\partial y}\right)$$

$$\hat{M}_y = \hat{z}\hat{p}_x - \hat{x}\hat{p}_z = -i\hbar\left(z\frac{\partial}{\partial x} - x\frac{\partial}{\partial z}\right) \qquad ...(6.27)$$

$$\hat{M}_z = \hat{x}\hat{p}_y - \hat{y}\hat{p}_x = -i\hbar\left(x\frac{\partial}{\partial y} - y\frac{\partial}{\partial x}\right).$$

To transform Eqs. (6.27) into spherical coordinates we note from Fig. 6.1 that

$$x = r\sin\theta\cos\phi$$
$$y = r\sin\theta\sin\phi$$
$$z = r\cos\theta$$

and use proper derivatives. For instance,

$$\frac{\partial}{\partial\phi} = \frac{\partial x}{\partial\phi}\frac{\partial}{\partial x} + \frac{\partial y}{\partial\phi}\frac{\partial}{\partial y} + \frac{\partial z}{\partial\phi}\frac{\partial}{\partial z}$$

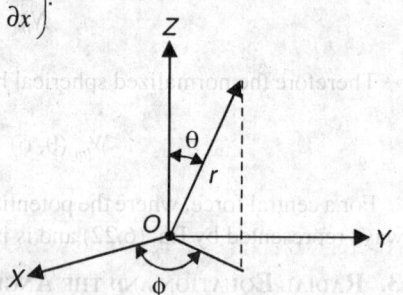

Fig. 6.1. Spherical polar coordinate system.

$$= -r\sin\theta\sin\phi\,\frac{\partial}{\partial x} + r\sin\theta\cos\phi\,\frac{\partial}{\partial y} + 0\,\frac{\partial}{\partial z}$$

$$= -y\frac{\partial}{\partial x} + x\frac{\partial}{\partial y}.$$

Therefore,

$$\hat{M}_z = -i\hbar\frac{\partial}{\partial\phi} \qquad ...(6.28)$$

Adopting the same procedure we find

$$\hat{M}_x = i\hbar\left(\sin\phi\,\frac{\partial}{\partial\theta} + \cot\theta\cos\phi\,\frac{\partial}{\partial\phi}\right) \qquad ...(6.29)$$

$$\hat{M}_y = i\hbar\left(-\cos\phi\,\frac{\partial}{\partial\theta} + \cot\theta\sin\phi\,\frac{\partial}{\partial\phi}\right) \qquad ...(6.30)$$

The operator expressing the square of the total angular momentum is

$$\hat{M}^2 = \hat{M}_x^2 + \hat{M}_y^2 + \hat{M}_z^2$$

$$= -\hbar^2\left[\frac{1}{\sin\theta}\frac{\partial}{\partial\theta}\left(\sin\theta\,\frac{\partial}{\partial\theta}\right) + \frac{1}{\sin^2\theta}\frac{\partial^2}{\partial\phi^2}\right] \qquad ...(6.31)$$

Using Eqs. (6.4), and (6.13), Eq. (6.31) becomes

$$\hat{M}^2 Y_{lm}(\theta,\phi) = l(l+1)\hbar^2 Y_{lm}(\theta,\phi) \qquad ...(6.32)$$

Equation (6.32) shows that $Y_{lm}(\theta,\phi)$ is an eigenfunction of $\hat{M}^2$ with the eigenvalue $l(l+1)\hbar^2$.

Spherically Symmetric Potential and the Hydrogen Atom 103

By using Eqs. (6.8) and (6.28) it can be similarly shown that $\Phi_m(\phi)$ and hence $Y_{lm}(\theta, \phi)$ is an eigenfunction of $\hat{M}_z$ having the eigenvalue $m\hbar$. That is,

$$\hat{M}_z Y_{lm}(\theta, \phi) = m\hbar\, Y_{lm}(\theta, \phi) \qquad ...(6.33)$$

The quantum number l appearing on the right hand side of Eq. (6.32) is termed the *azimuthal* or *orbital angular-momentum quantum number*. The quantum number m appearing on the right-hand side of Eq. (6.33) is defined as the *magnetic quantum number*.

It is clear from the above discussion that $Y_{lm}(\theta, \phi)$ is not an eigenfunction of $\hat{M}_x$ and $\hat{M}_y$ (except for $l = 0$). Also, since $Y_{lm}(\theta, \phi)$ is an eigenfunction of both $\hat{M}_z$ and $\hat{M}^2$ both M_z and M^2 can in general be precisely observed at once. This result can be shown to be in accordance with the uncertainty relation.

Furthermore, the condition $l \geq |m|$, obtained from the θ equation, is found to be physically equivalent to the condition that the z-component of the angular momentum ($m\hbar$) cannot exceed the magnitude of the total angular momentum, $\sqrt{l(l+1)}\,\hbar$. That l can assume only positive (integral) values is understandable since it deals with the *magnitude* of the orbital angular momentum vector. The quantum number m represents the vector component of the angular momentum in some specific direction, and so can be both *positive* and *negative* (integers).

6.4. THE HYDROGEN ATOM

The simplest of all atomic systems in nature is the hydrogen atom. It consists of an electron (mass m_e, charge $-e$) bound electrically to a proton (mass m_p, charge $+e$). Since the proton mass is some 1850 times the electron mass, the proton may be approximately considered stationary at the origin of the coordinate system. The Coulomb force acting on the electron can then be described by a central potential energy given by (in SI units)

$$V(r) = -\frac{e^2}{4\pi\varepsilon_0 r},$$

where ε_0 is the permittivity of free space. Putting this value of $V(r)$ in Eq. (6.23) and treating m_e as the electron mass we see that a stationary state of orbital angular momentum l is associated with a radial wave function $\chi(r)$ satisfying

$$\left[-\frac{\hbar^2}{2m_e}\frac{d^2}{dr^2} + \frac{l(l+1)\hbar^2}{2m_e r^2} - \frac{e^2}{4\pi\varepsilon_0 r}\right]\chi = E\chi \qquad ...(6.34)$$

$-e^2/(4\pi\varepsilon_0 r)$ is the Coulomb potential energy, and $l(l+1)\hbar^2/(2m_e r^2)$ is referred to as the *centrifugal potential energy*. The former is attractive whereas the latter is repulsive.

Let us define two dimensionless quantities ρ and n:

$$\rho \equiv +\left(\frac{-8m_e E}{\hbar^2}\right)^{1/2} r$$

$$n \equiv +\left(-\frac{m_e}{2\hbar^2 E}\right)^{1/2}\frac{e^2}{4\pi\varepsilon_0} \qquad ...(6.34a)$$

The parameter E is negative for a bound state. Therefore, ρ and n are real. Equation (6.34) is then written as

$$\frac{d^2\chi}{d\rho^2} + \left[-\frac{1}{4} + \frac{n}{\rho} - \frac{l(l+1)}{\rho^2}\right]\chi = 0 \qquad ...(6.35)$$

When r or ρ assumes large values, Eq. (6.35) is reduced to the form $\frac{d^2\chi}{d\rho^2} = \frac{\chi}{4}$, which has the solution $\chi \sim e^{-\rho/2}$. Therefore the solution to Eq. (6.35) can be written as $\chi = f(\rho) e^{-\rho/2}$, where $f(\rho)$ is a function whose contribution at large ρ is much smaller compared to $e^{-\rho/2}$.

Putting $\chi = f(\rho) e^{-\rho/2}$ in Eq. (6.35) we get

$$\frac{d^2 f}{d\rho^2} - \frac{df}{d\rho} + \left[\frac{n}{\rho} - \frac{l(l+1)}{\rho^2}\right] f = 0 \qquad ...(6.36)$$

A possible expression for f may be written as

$$f = a_0 \rho^\sigma + a_1 \rho^{\sigma+1} + a_2 \rho^{\sigma+2} + ... \qquad ...(6.36\,a)$$
$$= \rho^\sigma \sum_s a_s \rho^s$$

Substituting this expression into Eq. (6.36), every power of ρ is made separately equal to zero. As a result we find that the terms of the lowest power, $\rho^{\sigma-2}$ cancel, if

$$\sigma(\sigma - 1) = l(l+1) \qquad ...(6.37)$$

Evidently, $\sigma = l + 1$ or $-l$. The first choice *i.e.* $\sigma = l + 1$, is to be accepted as f would become infinite at $r = 0$ for the second. Once σ is ascertained, it may be found that the terms proportional to ρ^{l+s} in Eq. (6.36) cancel, if

$$a_{s+1} = \frac{l + s + 1 - n}{(s+1)(s + 2l + 2)} a_s \qquad ...(6.38)$$

Starting with $s = 0$ we may use Eq. (6.38) to find the functions f which are known as *associated Laguerre polynomials*. Obviously, for large s, $\frac{a_{s+1}}{a_s} = \frac{1}{s}$. The same asymptotic ratio is obtained in the series expansion of e^ρ and we conclude that the behaviour of f is similar to that of e^ρ for large ρ. To make χ bounded as $\rho \to \infty$, the power series in ρ must be finite. The polynomial will terminate after s terms and χ will turn out to be a physically acceptable wave function, if

$$n = l + s + 1 \qquad ...(6.39)$$

As l and s are zero or positive integers, n must be a positive integer equal to or greater than unity. Here n is called the *principal quantum number*. The parameter s which is generally denoted by n' is termed the *radial quantum number*. It is obvious from Eq. (6.39) that n' is redundant when n and l are known. Hence, n' is frequently omitted in the description of a state.

The relation between the eigenenergies of the hydrogen atom and the principal quantum number n is obtained from the definition of n given in Eq. (6.34 a):

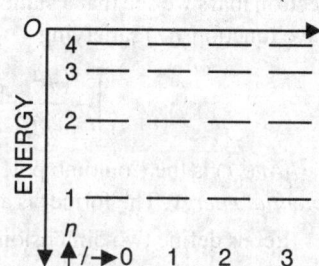

Fig. 6.2. Sketch of energy levels of hydrogen atom.

$$E_n = -\frac{m_e e^4}{32\pi^2 \epsilon_0^2 \hbar^2 n^2} \qquad ...(6.40)$$

The energy levels for $n \leq 4$ of the hydrogen atom are shown in Fig. 6.2. An infinite number of such levels (eigenstates) are possible.

As $n \to \infty$ the states become very crowded and $E_n \to 0$. This zero of the energy occurs at an infinite distance from the nucleus. The negative sign in Eq. (6.40) indicates that as the electron comes closer to the nucleus, its energy decreases because of the attractive Coulomb potential. The energy of each state depends on n and not on l or m.

Note that the eigenfunctions of the hydrogen atom are multiple degenerate. Though the energy depends on the value of n, the specification of the complete eigenfunction $\chi(r)r^{-1} Y_{lm}(\theta, \Phi)$ requires the knowledge of the three quantum numbers n, l and m. For each value of n, there are n possible values of l, where l can be $0, 1, 2, ... (n-1)$. Again, associated with each value of l, there are $(2l+1)$ possible values of m which are $0, \pm 1, \pm 2, ..., \pm l$. Evidently, the total degeneracy of the state of specified n is given by

$$N(n) = \sum_{l=0}^{n-1} (2l+1) = n^2. \qquad ...(6.41)$$

Wavefunction in the ground state

The angular part of the wavefunction is given by Eq. (6.19) while the radial part of the wavefunction is given by $R(r) = \frac{\chi(r)}{r}$. For the ground state $n = 1$, i.e. $l + s + 1 = 1$, from Eq. (6.39). Therefore, $l = 0$, $s = 0$ and $m = 0$ using Eq. (6.18). Hence $Y_{00}(\theta, \phi) = $ constant. The ground-state wavefunction can therefore be written as

$$u(r) = A \frac{\chi(r)}{r} = C \frac{\rho}{r} e^{-\rho/2}, \qquad ...(6.42)$$

where A and C are constants. In arriving at Eq. (6.42) we have used the result that for $l = 0$, $\sigma = 1$ and $\frac{\chi(r)}{r} \cong \frac{\rho}{r} e^{-\rho/2}$, which follows from Eq. (6.36 a). The constant C is determined from the normalisation condition

$$\int u\, u^* \, dv = 1. \qquad ...(6.43)$$

Assuming that the electron is in the volume element between the distances r and $r + dr$ from the nucleus, we have $dv = 4\pi r^2 \, dr$. Hence from Eqs. (6.42) and (6.43) we obtain

$$4\pi C^2 \int_0^\infty \rho^2 e^{-\rho} \, dr = 1 \qquad ...(6.44)$$

The ground state energy (E_1) is obtained by putting $n = 1$ in Eq. (6.40). Thus

$$E_1 = -\frac{m_e e^4}{32\pi^2 \varepsilon_0^2 \hbar^2} \qquad ...(6.45)$$

Using this value in Eq. (6.34a) we get

$$\rho = \frac{2}{a_0} r \qquad ...(6.46)$$

where a_0 is the *first Bohr radius* :

$$a_0 = \frac{4\pi \varepsilon_0 \hbar^2}{m_e e^2} \qquad ...(6.47)$$

Hence Eq. (6.44) can be written as

$$4\pi C^2 \int_0^\infty \frac{4}{a_0^2} r^2 e^{-2r/a_0} \, dr = 1$$

or

$$\frac{16\pi C^2}{a_0^2} \int_0^\infty e^{-2r/a_0} r^2 \, dr = 1 \qquad ...(6.48)$$

Integrating by parts or using gamma functions, it can be shown that

$$\int_0^\infty e^{-2r/a_0} r^2 \, dr = \frac{a_0^3}{4}$$

Therefore,

$$\frac{16\pi C^2}{a_0^2} = \frac{4}{a_0^3}$$

or,

$$C^2 = \frac{1}{4\pi a_0}$$

$$C = \frac{1}{2\sqrt{\pi a_0}}$$

Hence the wavefunction is

$$u(r) = \frac{1}{2\sqrt{\pi a_0}} \frac{2}{a_0} e^{-r/a_0}$$

$$= \frac{1}{\sqrt{\pi}} \left(\frac{1}{a_0}\right)^{3/2} e^{-r/a_0} \qquad \text{...(6.49)}$$

Most probable position of the electron

The most probable position of the electron is given by the value of r for which the probability function is a maximum. The probability dP of locating the electron between r and $r + dr$ is

$$dP = uu^* \, 4\pi r^2 \, dr \qquad \text{...(6.50)}$$

Putting the value of uu^* from Eq. (6.49), we obtain

$$dP = \frac{1}{\pi a_0^3} e^{-2r/a_0} \, 4\pi r^2 \, dr$$

$$= \frac{4}{a_0^3} e^{-2r/a_0} \, r^2 \, dr = P(r) \, dr,$$

where $P(r) = \dfrac{4}{a_0^3} e^{-2r/a_0} r^2$ is termed the probability function. For a maximum, we have

$$\frac{dP(r)}{dr} = 0$$

or,

$$\frac{d}{dr}\left[\frac{4}{a_0^3} e^{-2r/a_0} r^2\right] = 0,$$

or,

$$\frac{4}{a_0^3} \left(2r \, e^{-2r/a_0} - \frac{2r^2}{a_0} e^{-2r/a_0} \right) = 0 \qquad \text{...(6.51)}$$

Equation (6.51) gives $r = a_0$. That is, the probability of finding the electron of the hydrogen atom in the ground state is a maximum at $r = a_0$, where a_0 is the first Bohr radius. Substituting the values $\varepsilon_0 = 8.854 \times 10^{-12}$ F/m, $\hbar = 1.054 \times 10^{-34}$ J.s, $m_e = 9.11 \times 10^{-31}$ kg, and $e = 1.6 \times 10^{-19}$ C in Eq. (6.47) we find $a_0 = 5.29 \times 10^{-11}$ m. This value of a_0 is used as the basic dimensional scale of all atomic systems and processes. The variation of $P(r)$ with r is shown in Fig. 6.3.

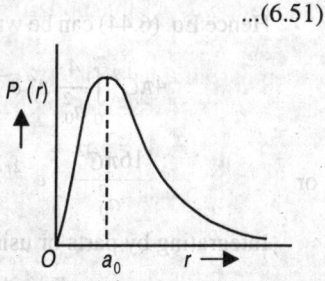

Fig. 6.3. Dependence of $P(r)$ on r.

Spherically Symmetric Potential and the Hydrogen Atom

At $r = a_0$, $u(r)$ reduces to $1/e$ times its value at $r = 0$.

Note. The hydrogen atom has only one orbital electron. As the system tends to remain in a state of minimum energy, this electron normally occupies the ground state. It may move to higher energy levels when excited, such as, by absorbing electromagnetic radiation energy or the kinetic energy of some impinging particles. These excited states of the electron are not stable; the electron has a tendency to come back to its ground state. In doing so, it emits electromagnetic radiation (photons) whose energy is equal to the energy difference between the levels of transition.

We have already shown the wavefunction for the 1s state ($n = 1, l = 0, m = 0$) for the hydrogen atom. The wavefunction for the 2s state ($n = 2, l = 0, m = 0$) is

$$\Psi = \frac{1}{4\sqrt{2\pi}} \left(\frac{1}{a_0}\right)^{3/2} \left(2 - \frac{r}{a_0}\right) \exp(-r/2a_0)$$

For the 2p state with $n = 2, l = 1$ and $m = 0$, the wavefunction is

$$\Psi = \frac{1}{4\sqrt{2\pi}} \left(\frac{1}{a_0}\right)^{3/2} \left(\frac{r}{a_0}\right) \exp(-r/2a_0) \cos\theta,$$

and for the 2p state with $n = 2, l = 1$ and $m = \pm 1$, the function is

$$\Psi_{m=\pm 1} = \mp \frac{1}{8\sqrt{\pi}} \left(\frac{1}{a_0}\right)^{3/2} \left(\frac{r}{a_0}\right) \exp(-r/2a_0) \sin\theta \exp(\pm i\phi),$$

6.5. Worked-Out Problems

1. Calculate, in electron volt, the energies of the first two states corresponding to $n = 1$ and 2 of the hydrogen atom. Also determine the frequency of the emitted radiation and the associated wavelength when transition occurs between these levels.

Ans. The energy corresponding to $n = 1$ is obtained from Eq. (6.40):

$$E_1 = -\frac{m_e e^4}{32\pi^2 \varepsilon_0^2 \hbar^2}$$

$$= \frac{-9.1 \times 10^{-31} \times (1.6 \times 10^{-19})^4}{32 \times \pi^2 \times (8.854 \times 10^{-12})^2 \times (1.055 \times 10^{-34})^2}$$

$$= -2.17 \times 10^{-18} \text{ J} = -13.6 \text{ eV}$$

The electron energy for $n = 2$ is

$$E_2 = \frac{E_1}{4} = -3.4 \text{ eV}$$

The frequency of emitted radiation is

$$\nu = \frac{E_2 - E_1}{h} = \frac{10.2 \times 1.6 \times 10^{-19}}{6.626 \times 10^{-34}} \text{ Hz}$$

$$= 2.463 \times 10^{15} \text{ Hz}$$

The corresponding wavelength is

$$\lambda = \frac{c}{\nu},$$

where c is the velocity of light in free space.

Hence

$$\lambda = \frac{3 \times 10^8}{2.463 \times 10^{15}} = 1.218 \times 10^{-7} \text{ m}$$

$$= 121.8 \text{ nm}$$

2. Compute the expectation value of r for the ground state of a hydrogen atom. The unnormalised ground-state wave function is e^{-r/a_0}, a_0 being the radius of the Bohr orbit. *(C.U. 1986, 2000)*

Ans. Let the ground state wave function be $u(r) = A e^{-r/a_0}$ where the constant A is determined from the normalization condition

$$\int_0^\infty u^*(r)\, u(r)\, 4\pi r^2\, dr = 1$$

Substituting for $u(r)$ we have

$$4\pi A^2 \int_0^\infty r^2 e^{-2r/a_0}\, dr = 1 \qquad \ldots(i)$$

Putting $\dfrac{2r}{a_0} = x$ we obtain from (i)

$$4\pi A^2 (a_0/2)^3 \int_0^\infty x^2 e^{-x}\, dx = 1$$

As $\int_0^\infty x^2 e^{-x}\, dx = \Gamma(3) = 2$ we have

$$A = \frac{1}{\sqrt{\pi}} \left(\frac{1}{a_0}\right)^{3/2}$$

The expectation value of r is given by

$$\langle r \rangle = \int_0^\infty u^*(r)\, r\, u(r)\, 4\pi r^2\, dr$$

$$= 4 \left(\frac{1}{a_0}\right)^3 \int_0^\infty r^3 e^{-2r/a_0}\, dr$$

$$= \frac{4}{a_0^3} \left(\frac{a_0}{2}\right)^4 \int_0^\infty x^3 e^{-x}\, dx, \qquad \left[\text{since } \frac{2r}{a_0} = x\right]$$

$$= \frac{a_0}{4}\, \Gamma(4)$$

$$= \frac{3}{2} a_0$$

[N. B. The expectation value $\langle r \rangle$ is found to be greater than the most probable value of r which is a_0. The reason for this result is the following:

The probability function $P(r) = \dfrac{4}{a_0^3} e^{-2r/a_0} r^2$ increases from zero at $r = 0$ to a maximum at $r = a_0$, and falls (see Fig. 6.3). The variation is asymmetric about $r = a_0$, $P(r)$ remaining finite for large values of r. The measurements on r thus give an expectation value which is greater than the most probable value a_0.**]**

3. Show, on the basis of a simple analysis, that in the ground state of the hydrogen atom the position and the momentum of the electron obey the uncertainty principle. Given: the first Bohr radius, $a_0 = (4\pi \varepsilon_0 \hbar^2)/m_e e^2$ and the ground state energy, $E_1 = -(m_e e^4)/(32\pi^2 \varepsilon_0^2 \hbar^2)$.

Ans. The uncertainty in the position of the electron is (approximately) $\Delta x = a_0$, since the electron is located within a region of size a_0.

Spherically Symmetric Potential and the Hydrogen Atom 109

If p is the magnitude of the linear momentum of the electron, then the momentum components can have values ranging from $-p$ to $+p$, since the momentum can be in any direction. Therefore the uncertainty in any components of linear momentum approximately satisfies the relation $\Delta p = p$.

The total energy of the electron is

$$E_1 = -\frac{m_e e^4}{32\pi^2 \varepsilon_0^2 \hbar^2}$$

$$= \text{kinetic energy} + \text{potential energy}$$

$$\approx \frac{p^2}{2m_e} - \frac{e^2}{4\pi \varepsilon_0 a_0}.$$

The approximate equality sign has been used since the potential energy expression refers to the position of the electron at $r = a_0$ which is uncertain.

So, $\quad \dfrac{p^2}{2m_e} = -\dfrac{m_e e^4}{32\pi^2 \varepsilon_0^2 \hbar^2} + \dfrac{e^2}{4\pi \varepsilon_0 a_0} = -\dfrac{e^2}{8\pi \varepsilon_0 a_0} + \dfrac{e^2}{4\pi \varepsilon_0 a_0}$

$$= \frac{e^2}{8\pi \varepsilon_0 a_0}.$$

Hence $\quad p = \sqrt{m_e e^2 / 4\pi \varepsilon_0 a_0}$.

Thus $\quad \Delta p \, \Delta x \approx p a_0 = \sqrt{m_e e^2 a_0 / 4\pi \varepsilon_0} = \hbar$.

Hence the position and the momentum of the electron obey the uncertainty principle.

4. Write the expression of the operator for the total energy of particle of mass m_e and charge e, the potential energy of which is $-\dfrac{e^2}{4\pi\varepsilon_0 r}$. Verify if the wave function

$$u(r) = \frac{1}{\sqrt{\pi}} \left(\frac{1}{a_0}\right)^{3/2} e^{-r/a_0}$$

is an eigenfunction of the above operator; and, if so, obtain the corresponding eigenvalue $[a_0 = 4\pi \varepsilon_0 \hbar^2 / m_e e^2]$. *(C.U. 1994)*

Ans. The required energy operator is

$$\hat{H} = -\frac{\hbar^2}{2m_e} \nabla^2 + \hat{V}$$

$$= -\frac{\hbar^2}{2m_e} \nabla^2 - \frac{e^2}{4\pi\varepsilon_0 r}.$$

We have

$$\hat{H} u(r) = -\frac{\hbar^2}{2m_e} \frac{1}{r^2} \frac{d}{dr}\left(r^2 \frac{du}{dr}\right) - \frac{e^2 u}{4\pi\varepsilon_0 r}$$

$$= \frac{\hbar^2}{2m_e a_0} \frac{1}{r^2} \frac{d}{dr}(r^2 u) - \frac{e^2 u}{4\pi\varepsilon_0 r}$$

$$= \frac{\hbar^2}{2m_e a_0 r^2}\left(2r - \frac{r^2}{a_0}\right) u - \frac{e^2 u}{4\pi\varepsilon_0 r}$$

Replacing r by its most probable value a_0 on the right–hand side, we get

$$\hat{H}u = \frac{\hbar}{2m_e a_0^2} u - \frac{e^2 u}{4\pi\varepsilon_0 a_0}$$

$$= \left(\frac{\hbar^2}{2m_e} \cdot \frac{m_e^2 e^4}{16\pi^2 \varepsilon_0^2 \hbar^4} - \frac{e^2}{4\pi\varepsilon_0} \cdot \frac{m_e e^2}{4\pi\varepsilon_0 \hbar^2} \right) u \qquad \text{(Substituting for } a_0\text{)}$$

$$= -\frac{m_e e^4}{32\pi^2 \varepsilon_0^2 \hbar^2} u$$

Thus u is an eigenfunction of the total energy operator $\hat{H}$. The eigenvalue is $-m_e e^4/(32\pi^2 \varepsilon_0^2 \hbar^2)$.

5. Is the wavefunction $u(r)$ of Problem No. 4 an eigenfunction of the momentum operator $\hat{p} = -i\hbar \nabla$? (C.U. 1994)

Ans. We have

$$\hat{p}u(r) = -i\hbar \frac{du}{dr} = i\frac{\hbar}{a_0} u(r)$$

Since the multiplying factor of u on the right-hand side is an imaginary quantity, $u(r)$ is not an eigenfunction of $\hat{p}$.

6. Show that only one of the three rectangular Cartesian components of the angular momentum $\vec{M}$ can be precisely measured at any time, but M^2 can be determined simultaneously with any one component of $\vec{M}$

Ans. We have

$$\hat{M}_x = \hat{y}\hat{p}_z - \hat{z}\hat{p}_y, \quad \hat{M}_y = \hat{z}\hat{p}_x - \hat{x}\hat{p}_z \text{ and}$$

$$\hat{M}_z = \hat{x}\hat{p}_y - \hat{y}\hat{p}_x.$$

Therefore
$$[\hat{M}_x, \hat{M}_y] = (\hat{y}\hat{p}_z - \hat{z}\hat{p}_y)(\hat{z}\hat{p}_x - \hat{x}\hat{p}_z)$$
$$- (\hat{z}\hat{p}_x - \hat{x}\hat{p}_z)(\hat{y}\hat{p}_z - \hat{z}\hat{p}_y)$$
$$= (\hat{y}\hat{p}_x - \hat{x}\hat{p}_y)(\hat{p}_z \hat{z} - \hat{z}\hat{p}_z)$$
$$= -\hat{M}_z [\hat{p}_z, \hat{z}] = i\hbar \hat{M}_z,$$

using the commutation relation $[\hat{z}, \hat{p}_z] = i\hbar$.

Similarly, we obtain $[\hat{M}_y, \hat{M}_z] = i\hbar \hat{M}_x$ and $[\hat{M}_z, \hat{M}_x] = i\hbar \hat{M}_y$. As $\hat{M}_x, \hat{M}_y$ and $\hat{M}_z$ do not commute with each other, only one of the three components of $\vec{M}$ can be precisely measured at any time.

Again, proceeding in a similar fashion, we find

$$[\hat{M}^2, \hat{M}_x] = [\hat{M}^2, \hat{M}_y] = [\hat{M}^2, \hat{M}_z] = 0$$

Thus, M^2 can be determined simultaneously with any one of M_x, M_y and M_z.

Spherically Symmetric Potential and the Hydrogen Atom

7. The unnormalized hydrogenic 2p-state wavefunction for $m = 0$ is $(r/a_0) \exp(-r/2a_0) \cos\theta$. Determine the normalization constant. Also, calculate the expectation value of r.

Ans. Let the wavefunction be $u = A\left(\dfrac{r}{a_0}\right) \exp(-r/2a_0) \cos\theta$, where A is the normalization constant. We have

$$\int_{\theta=0}^{\pi} \int_{\phi=0}^{2\pi} \int_{r=0}^{\infty} u\, u^* \, r^2 \sin\theta \, d\theta \, d\phi \, dr = 1$$

So, $2\pi A^2 \displaystyle\int_{\theta=0}^{\pi} \int_{r=0}^{\infty} \left(\dfrac{r}{a_0}\right)^2 \exp(-r/a_0)\, r^2 \cos^2\theta \sin\theta\, d\theta\, dr = 1$

or, $\dfrac{4\pi}{3} A^2 a_0^3 \displaystyle\int_0^{\infty} x^4 e^{-x} dx = 1$ [where $x = r/a_0$]

or, $\dfrac{4\pi}{3} A^2 a_0^3 \, \Gamma(5) = 1$

whence $A = \dfrac{1}{4\sqrt{2\pi}\, a_0^{3/2}}$

The expectation value of r is

$$\langle r \rangle = \int_{\theta=0}^{\pi} \int_{\phi=0}^{2\pi} \int_{r=0}^{\infty} u^*\, r\, u \, r^2 \sin\theta \, d\theta \, d\phi \, dr$$

$$= \dfrac{1}{32\pi a_0^3} \times \dfrac{4\pi}{3} a_0^4 \int_0^{\infty} x^5 e^{-x} dx$$

$$= \dfrac{a_0}{24} \Gamma(6) = 5 a_0$$

8. Show that the effective potential energy of an electron in the hydrogen atom, consisting of the Coulomb and the centrifugal parts, is a minimum at $r = l(l+1) a_0$. Also show that the minimum effective potential energy is negative.

Ans. The effective potential energy is

$$V = \dfrac{l(l+1)\hbar^2}{2 m_e r^2} - \dfrac{e^2}{4\pi\varepsilon_0 r}$$

The first term on the r.h.s. is the centrifugal part and the second term is the Coulomb part. V is a minimum when

$$\dfrac{dV}{dr} = 0$$

or, $-\dfrac{l(l+1)\hbar^2}{m_e r^3} + \dfrac{e^2}{4\pi\varepsilon_0 r^2} = 0$

or, $r = r_m = \dfrac{l(l+1)\, 4\pi\varepsilon_0 \hbar^2}{m_e e^2} = l(l+1)\, a_0$, since $a_0 = \dfrac{4\pi\varepsilon_0 \hbar^2}{m_e e^2}$.

The minimum value of V is

$$V_m = \frac{l(l+1)\hbar^2}{2m_e r_m^2} - \frac{e^2}{4\pi\varepsilon_0 r_m}$$

$$= \frac{1}{l(l+1)a_0}\left(\frac{\hbar^2}{2m_e a_0} - \frac{e^2}{4\pi\varepsilon_0}\right)$$

$$= -\frac{e^2}{8\pi l(l+1)a_0 \varepsilon_0},$$

which is negative.

QUESTIONS

1. Write the time-independent Schrödinger equation for a spherically symmetric potential and separate it into three equations.

2. Show that the expression for the complete ground state eigenfunction of the hydrogen atom problem is given by

$$u = \frac{1}{\sqrt{\pi}}\left(\frac{1}{a_0}\right)^{3/2} e^{-r/a_0},$$

where a_0 is the Bohr radius of the hydrogen atom.

3. (a) Write down the radial part of the time-independent Schrödinger equation for the hydrogen atom and hence find the relation between the eigenenergies (E_n) and the principal quantum number (n). **(C.U. 1981, C.U. 1989)**

$$\left[\text{Ans. } E_n = -\frac{m_e e^4}{32\pi^2 \varepsilon_0^2 \hbar^2 n^2}\right]$$

(b) What is the significance of the negative sign in the expression for E_n? Where is the zero of the energy scale located?

4. Write down the Schrödinger wave equation for the hydrogen atom in spherical polar coordinates. Explain the significance of the different quantum numbers. Mention the order of degeneracy of a certain energy state.

5. Show by substitution that the eigenfunction u_{100} and the eigenvalue E_1 satisfy the time-independent Schrödinger equation for the hydrogen atom.

6. (a) Plot the ground-state eigenfunction of the hydrogen atom as a function of the distance r.

   (b) Show that the most probable position of the electron of the hydrogen atom in the ground state is given by the first Bohr radius (a_0). Also calculate the probability that the electron lies in a sphere of radius R. **(cf. C.U. 2003)**

$$\left[\text{Ans. } 1 - e^{-2R/a_0}\left(1 + \frac{2R}{a_0} + \frac{2R^2}{a_0^2}\right)\right]$$

7. Explain the significance of the various energy levels of the hydrogen atom which has only one orbital electron.

8. The wave function for the ground state of the hydrogen atom problem is given by $\overline{u}(r) = C(\rho/r)e^{-\rho/2}$, where C is a constant and $\rho = 2r/a_0$, a_0 being the first Bohr radius. Determine the constant C from the normalization condition of the wave function.

9. A particle moves in a central field given by the potential $V(r)$. Obtain the differential equation for the radial part of the wave function. Show that this can be transformed into a form which is formally the

Spherically Symmetric Potential and the Hydrogen Atom 113

same as the Schrödinger equation for a one-demensional motion in the region $0 \leq r \leq \infty$ with an effective potential

$$V_{eff}(r) = V(r) + \frac{\hbar^2 \lambda}{2m_e r^2},$$

where the symbols have the usual meanings, λ being a constant. (C.U. 1986)

10. Obtain an expression for the magnitude of the linear momentum of the electron of the hydrogen atom in the ground state. Given, the ground state energy $E_1 = -(m_e e^4)/(32\pi^2 \varepsilon_0^2 \hbar^2)$. Assume that the electron is at the first Bohr radius. [Ans. $m_e e^2/(4\pi \varepsilon_0 \hbar)$]

11. Show that, if we choose z-axis as the polar axis, the z-component of the angular momentum operator is $\hat{M}_z = \frac{\hbar}{i} \frac{\partial}{\partial \phi}$, where ϕ is the azimuthal angle. Hence show that the z-component of the angular momentum is quantised. (Assume the corresponding eigenfunction.) (C.U. 1992)

12. Determine the expectation value of the potential energy of the electron at the ground state of the hydrogen atom. The unnormalised ground state wave function is e^{-r/a_0}, where a_0 is the radius of the Bohr orbit. (Agra 1995, C.U. 1997) [Ans. $-e^2/(4\pi\varepsilon_0 a_0)$]

13. The unnormalized 2s-state wavefunction of the hydrogen atom is $(2 - r/a_0) \exp(-r/2a_0)$. Find the normalization constant and the expectation value $<r>$. [Ans. $1/(4\sqrt{2\pi} a_0^{3/2})$, $6a_0$]

14. Find the expectation value of the kinetic energy of the electron at the 1s state of the hydrogen atom.

[Ans. $\dfrac{m_e e^4}{32 \pi^2 \varepsilon_0^2 \hbar^2}$]

15. (a) Evaluate the commutator brackets $\left[\hat{M}_x, \hat{M}_y\right]$ and $\left[\hat{x}, \hat{M}_z\right]$.

(b) Find the degeneracy of an arbitrary stationary state of a hydrogen atom. Is it an eigenstate of momentum ? (C.U. 2000)

16. The unnormalized wave function for the ground state of the hydrogen atom is $\psi = e^{-r/a_0}$ where a_0 is a constant. Find the expectation value of the Coulomb force on an electron in this state.

(C.U. 2002) [Ans. $- e^2/(2\pi \varepsilon_0 a_0^2)$]

17. The wave function of the 1s electron in the hydrogen atom is $\psi = \left(\pi a_0^3\right)^{-1/2} \exp(-r/a_0)$. Show that the average value of $1/r$ for this electron would be $1/a_0$. (C.U. 2008)

18. The hydrogen atom wave functions are ψ_{nlm} for the quantum numbers n, l, and m. The energy eigenvalues are $E_n = C/n^2$, where C is a constant. What is the expectation value of the energy in a state

$$\phi = \frac{1}{\sqrt{6}} \left(\sqrt{2}\, \psi_{100} - \psi_{210} + \sqrt{3}\, \psi_{211}\right) ?$$

(c.f. C.U. 2008)

[Ans. $5C/12$]

STATISTICAL MECHANICS

Chapter 1

Basic Concepts

1.1. MICROSCOPIC AND MACROSCOPIC SYSTEMS

A system of atomic dimensions or of a smaller size is called a *microscopic* (*i.e.*, small-scale) *system*. A molecule is an example of such a system. A *macroscopic* (*i.e.*, large-scale) *system* is one which is large enough to be observable in the ordinary sense. Such a system contains many atoms or molecules. Solids, liquids, or gases of our daily experience are macroscopic systems. Parameters which characterise the system as a whole, and not the individual particles which make up the system, are termed *macroscopic parameters*. Pressure, volume, temperature, electrical resistivity, etc. are examples of macroscopic parameters. When the macroscopic parameters of an isolated system do not change with time, the system is said to be in *equilibrium*.

Statistical mechanics aims at studying the macroscopic parameters of a system in equilibrium from a knowledge of the microscopic properties of its constituent particles using the laws of mechanics. This is different from the approach of *thermodynamics* which aims at studying a macroscopic system in equilibrium from a macroscopic phenomenological standpoint without considering the microscopic constituents of the system.

Macrostate and Microstate

The macroscopic state or the *macrostate* of a system can be specified by quoting the macroscopic parameters and the energy of the system. The microscopic state or the *microstate* (also called the *quantum state*) of a system of particles can be specified by quoting the state such as the position, momentum, orientation, and configuration of the individual particles. Since a large number of particles constitute the assembly or "ensemble", macro- and microstates can be correlated by applying statistical methods based on the theory of probabilities.

For a given macrostate, a system can have many possible microstates. For example, consider a system of three particles with three energy levels 0, E and $2E$. For the macrostate characterized by the total energy $3E$ of the system, there are two microstates : (*i*) one particle in each of the three energy levels 0, E and $2E$; and (*ii*) all the three particles in the same energy level E. Each of these two microstates leads to the same macrostate with the total energy $3E$ of the system of particles.

A system that does not interact and exchange energy with any other system is called an *isolated system*. A microstate occupied by a system without affecting the macroscopic description is called an *accessible state* of the system. By definition, an isolated system is in equilibrium if the probability of its occupying each accessible state is independent of time. A fundamental postulate of equilibrium statistical mechanics, referred to as the *postulate of equal a priori probabilities*, asserts that an isolated system in equilibrium is found with equal probability in each of its accessible states.

A system for which quantum mechanical effects are unimportant is called a *classical system*.

1.2. CALCULATION OF PROBABILITIES

Suppose we have N identifiable particles to be distributed among n containers such that there are N_1 particles in the first container, N_2 in the second, and so on. To find the number of ways in which this can be done, we begin with two containers. N number of particles are distributed in these two containers such that the first contains N_1 particles, and the second N_2 particles. The number of such distributions is clearly given by

$$^NC_{N_1} = \frac{N!}{N_1!(N-N_1)!} = \frac{N!}{N_1!N_2!} \qquad ...(1.1)$$

since
$$N = N_1 + N_2$$

Suppose now that the second container is partitioned into two chambers containing n_1 and n_2 particles. Then

$$n_1 + n_2 = N_2 = N - N_1. \qquad ...(1.2)$$

The number of independent ways of obtaining the above distribution in the chambers of the second container is, from Eq. (1.1),

$$^{N_2}C_{n_1} = \frac{N_2!}{n_1!n_2!} \qquad ...(1.3)$$

We may visualize the system as one with three containers, the first container having N_1 particles the second having n_1 particles, and the third n_2 particles. The total number of ways in which this arrangement can be achieved, will be given by the product

$$^NC_{N_1} \cdot {}^{N_2}C_{n_1} = \frac{N!}{N_1!N_2!} \cdot \frac{N_2!}{n_1!n_2!} = \frac{N!}{N_1!n_1!n_2!} \qquad ...(1.4)$$

since for each arrangement given by Eq. (1.1), there are $^{N_2}C_{n_1}$ ways of arranging particles in the second and the third containers. We may subdivide the third container into two chambers as before, and proceed in a similar way to cover the case of n containers. Finally, replacing n_1, n_2, etc., respectively by N_2, N_3, etc., we find that the total number of independent ways of arranging N particles among n containers such that there are N_1 particles in the first, N_2 in the second, etc., will be

$$W = \frac{N!}{N_1!N_2!N_3!...N_n!} = \frac{N!}{\prod_{i=1}^{n} N_i!} \qquad ...(1.5)$$

where Π indicates the extended product. W represents the number of *microstates* for the arrangement.

Now, the *mathematical probability* of an event is defined as the ratio of the actual number of cases in which the event occurs to the total number of cases. Thus the probability of the distribution of $N_1, N_2,..., N_n$ particles in containers 1, 2,..., n will be given by dividing W by the *total number of microstates*, *i.e.* the total number of ways of arranging N particles among n containers considering all possible distributions. This latter factor is n^N, since each of the N particles can be in any of the n containers, *i.e.*, each particle can be arranged in n ways. The probability is consequently

$$P_r = \frac{W}{n^N} = \frac{N!}{\prod_{i=1}^{n} N_i!} n^{-N} \qquad ...(1.6)$$

When N is large, Eq. (1.5) can be put in a simpler form by using *Stirling's theorem*. This theorem states that when N is a very large number $\ln N! \cong N(\ln N - 1)$

[Stirling's theorem can be proved by noting that

$$\ln N! = \ln 1 + \ln 2 + \ln 3 + ... + \ln x + ... + \ln N,$$

and
$$\ln N^N = \ln N + \ln N + \ln N + ... + \ln N + ... + \ln N,$$

Hence
$$\ln\left(\frac{N!}{N^N}\right) = \ln\frac{1}{N} + \ln\frac{2}{N} + ... + \ln\frac{x}{N} + ... + \ln\frac{N}{N}$$

$$= \sum_{x=1}^{N} (\ln x - \ln N)$$

Basic Concepts

Replacing the summation by integration, we get

$$\ln\left(\frac{N!}{N^N}\right) = \int_{x=1}^{N} (\ln x - \ln N)\, dx$$

$$= [x \ln x - x]_1^N - [x \ln N]_1^N$$
$$= (N \ln N - N + 1) - (N \ln N - \ln N)$$
$$= \ln N - N + 1$$

Therefore,

$$\ln N! = N \ln N + \ln N - N + 1 \cong N(\ln N - 1)$$

neglecting $\ln N$ and 1 respectively with respect to $N \ln N$ and N. This last step is justified since N is very large.]

From Eq. (1.5) we obtain

$$\ln W = N(\ln N - 1) - \sum_{i=1}^{n} N_i (\ln N_i - 1) = N \ln N - \sum_{i=1}^{n} N_i \ln N_i \qquad ...(1.7)$$

1.3. Phase Trajectory

For the one-dimensional motion of a classical particle in the x-direction, the dynamical state of the particle is specified by its position coordinate x and the momentum p_x. The p_x–x plane is called the *phase plane*, and any point on this plane is referred to as a *phase point*. The locus of the phase point on the phase plane, i.e., the curve showing the variation of p_x with x, is known as the *phase trajectory* of the particle.

Consider a free particle of mass m and energy E moving to-and-fro along the x-axis in the force-free region between two rigid walls at $x = 0$ and $x = L$. For

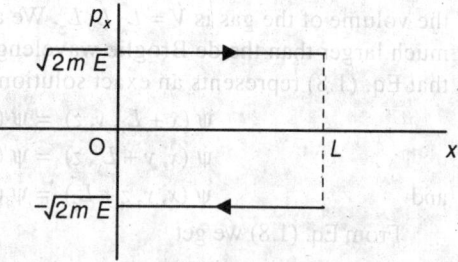

Fig. 1.1. Phase trajectory of a free particle moving between $x = 0$ and $x = L$.

motion along the positive x-direction, $p_x = \sqrt{2mE}$ = constant. Thus the phase trajectory is a line segment parallel to the x-axis at a distance $\sqrt{2mE}$ above it between $x = 0$ and $x = L$ (Fig. 1.1). For motion in the reverse direction, $p_x = -\sqrt{2mE}$, so that the phase trajectory is a line segment parallel to the x-axis at a distance $\sqrt{2mE}$ below it between $x = L$ and $x = 0$ (Fig. 1.1).

1.4. Statistics of An Assembly of Particles

A. Phase space and density of states

Consider a system of noninteracting particles as that forming a perfect monatomic gas. The dynamical condition of a particle is specified by its three position coordinates x, y, z, and three corresponding momenta p_x, p_y, p_z. The specification of these values for all the particles gives the microstate of the system. The set of quantities (x, y, z, p_x, p_y, p_z) gives the coordinates of a point in a six-dimensional space, called the *phase space*.

It is possible to divide the phase space into small cells of equal size by taking fixed intervals $\delta x, \delta y$ and δz for the position coordinates, and $\delta p_x, \delta p_y$ and δp_z for the momentum components. Let $\delta x\, \delta p_x = h_0$ where h_0 is a small constant having the dimensions of angular momentum. The volume of each small cell of the phase space is then given by $\delta x\, \delta y\, \delta z\, \delta p_x\, \delta p_y\, \delta p_z = h_0^3$. Identifying δx as the uncertainty in x and δp_x as that in p_x, we find from Heisenberg's uncertainty principle that h_0 is at least the reduced Planck's constant $\hbar$. Considering all these small cells extending over the entire phase space, we can map out completely the dynamical state of the gas since the whole volume and the whole range of momenta between minus infinity and plus infinity are thus covered.

Quantum mechanically, the wave function $\psi(x, y, z, t)$ of a particle is given by

$$\psi(x, y, z, t) = A e^{i(\vec{k}\cdot\vec{r} - \omega t)} \qquad ...(1.8)$$

which represents a plane wave propagating in the direction of the wave vector $\vec{k}$ and which has a constant amplitude A. $\vec{r}$ is the position vector, and ω is the angular frequency related to the energy of the particle E by

$$E = \hbar\omega \qquad \ldots(1.9)$$

where $\hbar$ is the reduced Planck's constant. The wave vector $\vec{k}$ and the momentum $\vec{p}$ are related by

$$\vec{p} = \hbar \vec{k} \qquad \ldots(1.10)$$

We also have

$$E = \frac{p^2}{2m} = \frac{\hbar^2 k^2}{2m} \qquad \ldots(1.11)$$

Since the wave function ψ has to satisfy the boundary conditions, all possible values of $\vec{k}$ or $\vec{p}$ are not allowed. The energy of the particle is thus quantized.

To make the situation simple, we consider the gas to be enclosed in a rectangular parallelepiped with sides parallel to the x, y, z axes. If the lengths of the sides of the container are $L_x, L_y,$ and L_z, the volume of the gas is $V = L_x L_y L_z$. We assume that the linear dimensions of the container are much larger than the de Broglie wavelength $\lambda = 2\pi/k$. The boundary conditions should be such that Eq. (1.8) represents an exact solution. Hence these conditions are

$$\psi(x + L_x, y, z) = \psi(x, y, z),$$
$$\psi(x, y + L_y, z) = \psi(x, y, z), \qquad \ldots(1.12)$$
and
$$\psi(x, y, z + L_z) = \psi(x, y, z).$$

From Eq. (1.8) we get

$$\psi(x, y, z) = A e^{i\vec{k}\cdot\vec{r}} = A e^{i(k_x x + k_y y + k_z z)} \qquad \ldots(1.13)$$

To satisfy (1.12) we must have

$$k_x = \frac{2\pi n_x}{L_x},$$
$$k_y = \frac{2\pi n_y}{L_y}, \qquad \ldots(1.14)$$
and
$$k_z = \frac{2\pi n_z}{L_z},$$

where n_x, n_y and n_z are integers (positive or negative) or zero.

The discrete or the quantized particle energies are given by

$$E = \frac{\hbar^2}{2m}(k_x^2 + k_y^2 + k_z^2) \qquad \ldots(1.15)$$

$$= \frac{2\pi^2 \hbar^2}{m}\left(\frac{n_x^2}{L_x^2} + \frac{n_y^2}{L_y^2} + \frac{n_z^2}{L_z^2}\right) \qquad \ldots(1.16)$$

The allowed values of $p_x, p_y,$ and p_z are obtained from Eq. (1.10) and (1.14) as

$$p_x = \frac{h n_x}{L_x}$$

$$p_y = \frac{h n_y}{L_y} \qquad \ldots(1.17)$$

Basic Concepts

and
$$p_z = \frac{h n_z}{L_z}$$

where $h = 2\pi\hbar$ = Planck's constant. For a macroscopic volume L_x, L_y, L_z are large so that the allowed momentum components given by Eq.(1.17) are closely spaced. Thus for any small change dp_x of p_x, there are many allowed states of the particle corresponding to different values of n_x. The number Δn_x of the integers n_x for which p_x lies between p_x and $p_x + dp_x$ is obtained from Eq. (1.17):

$$\Delta n_x = \frac{L_x}{h} dp_x \qquad \ldots(1.18)$$

The number of states for which the momentum components lie between p_x and $p_x + dp_x$, p_y and $p_y + dp_y$, p_z and $p_z + dp_z$ is equal to the product

$$\Delta n_x \, \Delta n_y \, \Delta n_z = \frac{L_x L_y L_z}{h^3} dp_x \, dp_y \, dp_z$$

$$= \frac{V}{h^3} dp_x \, dp_y \, dp_z \qquad \ldots(1.19)$$

To find the number of quantum states $g(E)\,dE$ within the energy values E and $E + dE$, we have to consider the volume of the momentum space between concentric spheres of radii p and $p + dp$. From Eq. (1.19) this is found to be

$$g(E)\,dE = \frac{V}{h^3} 4\pi p^2 \, dp \qquad \ldots(1.20)$$

Clearly $4\pi V p^2 \, dp$ is the volume of the phase space occupied by a particle in a container of volume V and with momentum magnitude lying between p and $p + dp$, corresponding to energy between E and $E + dE$. Equation (1.20) reveals that division of the phase space into cells of volume h^3 **gives the number of quantum states or microstates for the particle.**

Using Eq. (1.11), we may write Eq. (1.20) in the form

$$g(E)\,dE = \frac{2\pi V (2m)^{3/2}}{h^3} E^{1/2} \, dE. \qquad \ldots(1.21)$$

The function $g(E)$ gives the number of states per unit energy range and is referred to as the *density of states*.

B. Distribution function

Let N denote the total number of particles in an assembly. Also, let $N_1, N_2, \ldots, N_n$ be the number of particles with energies $E_1, E_2, \ldots, E_n$, respectively, and g_i be the number of quantum states for the energy level E_i. The quantity g_i is called the *degeneracy* of the energy level E_i. The ensemble of these degenerate states is called the *microcanonical ensemble*. The degree of degeneracy of a level is called its *statistical weight*. We shall find the number of ways W in which the particles are distributed among the quantum states.

If the particles are *identifiable*, the number of ways in which the groups of $N_1, N_2, \ldots, N_n$ particles can be chosen from N particles is given by [see Eq. (1.5)]

$$W_1 = \frac{N!}{\prod_{i=1}^{n} N_i!} \qquad \ldots(1.22)$$

n being the number of energy levels. Now N_i particles are to be accommodated in g_i states, each of which has the same *a priori* probability of being occupied. Each of the N_i particles may thus occupy any of the g_i states. The number of ways in which N_i particles are arranged in g_i states is therefore given by $g_i^{N_i}$. Considering the different values of i, the total number of arrangements will be

$$W_2 = \prod_{i=1}^{n} g_i^{N_i}. \qquad \ldots(1.23)$$

The required number of ways W is hence obtained as

$$W = W_1 W_2 = \frac{N!}{\prod_{i=1}^{n} N_i!} \prod_{i=1}^{n} g_i^{N_i} \qquad ...(1.24)$$

The quantity W is called the *thermodynamic probability* for the system. Assuming that N is very large, we shall apply Stirling's theorem to obtain from Eq. (1.24)

$$\ln W = N(\ln N - 1) + \sum_{i=1}^{n} N_i \ln g_i - \sum_{i=1}^{n} N_i (\ln N_i - 1)$$

$$= N \ln N + \sum_{i=1}^{n} N_i \ln g_i - \sum_{i=1}^{n} N_i \ln N_i \qquad ...(1.25)$$

Since the system is in *equilibrium*, the parameters $N_1, N_2, ..., N_n$ are such that W is a maximum subject to the restriction that the total number of particles N and the total energy U are constants. For mathematical convenience, we shall maximise $\ln W$ given by Eq. (1.25), rather than W itself, with respect to $N_1, N_2, ..., N_n$ satisfying that

$$N = \sum_{i=1}^{n} N_i = \text{constant} \qquad ...(1.26)$$

and

$$U = \sum_{i=1}^{n} N_i E_i = \text{constant}. \qquad ...(1.27)$$

Since all microstates are equally probable, the maximum W, *i.e.* the largest number of microstates gives the *most probable distribution* of the particles in the quantum states.

Since the particles are noninteracting, the total energy U of the system is taken equal to the sum of the separate energies of the particles. From Eq. (1.25) we have for the maximum W

$$\delta \ln W = \sum_{i=1}^{n} (\ln g_i - \ln N_i - 1) \, \delta N_i = 0 \qquad ...(1.28)$$

Also, from Eqs. (1.26) and (1.27), we get

$$\delta N = \sum_{i=1}^{n} \delta N_i = 0, \text{ and } \delta U = \sum_{i=1}^{n} E_i \, \delta N_i = 0 \qquad ...(1.29)$$

Combining (1.28) and (1.29) by *Lagrange's method of undetermined multipliers*, we obtain

$$\sum_{i=1}^{n} (\ln g_i - \ln N_i + \alpha + \beta E_i) \, \delta N_i = 0 \qquad ...(1.29a)$$

Since the variations are quite arbitrary, we write

$$\ln g_i - \ln N_i + \alpha + \beta E_i = 0 \qquad ...(1.30)$$

where α and β are constants. Hence

$$f(E_i) = \frac{N_i}{g_i} = e^{\alpha + \beta E_i} \qquad ...(1.31)$$

The quantity $f(E_i)$ defined by the ratio of the number of particles N_i distributed in g_i states to the number of states g_i, is called the *distribution function*. It gives the average number of particles per quantum state of the system.

The particular form of the distribution function given by (1.31) is obtained from the classical assumption of identifiable particles and without using Pauli's exclusion principle; it is referred to as the *Maxwell-Boltzmann distribution function*.

Basic Concepts

1.5. ENTROPY

We shall now show that the number of possible arrangements W given in Sec. 1.4 is related to the entropy of the system and shall also find the value of the constant β.

Let a small amount of heat δQ be added to the system. This will result in a change in the internal energy of the system, which from the first law of thermodynamics, is given by

$$\delta U = \delta Q - P\delta V \qquad ...(1.32)$$

where $P\delta V$ is the work done by the system. Since $U = \sum_i N_i E_i$, we have

$$\delta U = \sum_i E_i \delta N_i + \sum_i N_i \delta E_i. \qquad ...(1.33)$$

The change δE_i in the energy level E_i is possible only if the volume changes. Hence

$$\sum_i N_i \delta E_i = \sum_i N_i \frac{\partial E_i}{\partial V} \delta V = - P\delta V. \qquad ...(1.34)$$

where we have used the relation

$$P = -\frac{\partial U}{\partial V}$$

From Eqs. (1.32), (1.33), and (1.34) we find that

$$\delta Q = \sum_i E_i \delta N_i. \qquad ...(1.35)$$

If the heat δQ is added reversibly, W is always a maximum since the system is always in thermal equilibrium. Keeping the total number of particles constant, we get with the aid of Eqs. (1.28) and (1.29a)

$$\delta \ln W = -\beta \sum_i E_i \delta N_i = -\beta \delta Q. \qquad ...(1.36)$$

Since $\delta \ln W$ is a perfect differential, $\beta \delta Q$ must also be so. But δQ is not a perfect differential; there is an integrating factor $1/T$, T being the absolute temperature. Thus

$$\beta = -\frac{1}{k_B T} \qquad ...(1.37)$$

where k_B is a constant.

From Eq. (1.36) we now obtain

$$k_B \, \delta \ln W = \frac{\delta Q}{T} = \delta S, \qquad ...(1.38)$$

where δS is the change in entropy. Integration of Eq. (1.38) gives

$$S = k_B \ln W + \text{constant}, \qquad ...(1.39)$$

The value of the constant k_B, determined from comparison with experiments, is seen to be the Boltzmann's constant, i.e. $k_B = 1.38 \times 10^{-23}$ J/K.

The relation between S and $\ln W$, given by Eq. (1.39), is known as the *Planck relation* of entropy. Earlier, Boltzmann found a similar relation, but he used the probability P_r in place of W. Since P_r and W are proportional, the two relations are not basically different. A simple deduction of the *Boltzmann relation* is given below.

Since a system, left to itself, attains a state of maximum entropy and also of maximum probability, entropy must be a function of the probability of the state. Thus let $S = f(P_r)$ where P_r is the probability. Consider two independent systems X and Y having entropies S_1 and S_2, and probabilities P_{r1} and P_{r2}, respectively. Then the entropy of the combined system of X and Y will be $S_1 + S_2$, but the probability of the combined system will be P_{r1}, P_{r2}. This is because the entropy is additive while probability is multiplicative. Thus

$$S_1 + S_2 = f(P_{r1}) + f(P_{r1}) = f(P_{r1} P_{r2}) \qquad ...(1.40)$$

Equation (1.40) can be satisfied if we choose
$$S = k_B \ln P_r + \text{const.,} \qquad \text{...(1.41)}$$
which is Boltzmann's relation connecting entropy and probability.

1.6. Perfect Gas Law

We shall apply here Eq. (1.41) to obtain the perfect gas law. We note that the probability of finding a particle within a portion V of a large volume V_0 available to it is V/V_0. If there are N number of particles, then the probability P_r of all the particles lying in the volume V is $(V/V_0)^N$. Hence from Eq. (1.41) we obtain for the entropy

$$\begin{aligned} S &= k_B \ln (V/V_0)^N + \text{const.} \\ &= k_B N [\ln V - \ln V_0] + \text{const.} \end{aligned} \qquad \text{...(1.42)}$$

From thermodynamics, we write
$$dU + PdV = T\,dS \qquad \text{...(1.43)}$$

For a perfect gas, $(\partial U / \partial V)_T = 0$. Hence we obtain from Eq. (1.43)

$$\frac{P}{T} = \left(\frac{\partial S}{\partial V}\right)_T = \frac{Nk_B}{V} \qquad \text{...(1.44)}$$

where Eq. (1.42) has been used. Therefore,
$$PV = Nk_B T \qquad \text{...(1.45)}$$
which is the perfect gas relation. It is clear that the constant k_B in Eq. (1.41) must be the Boltzmann constant.

1.7. Maxwell-Boltzmann Statistics of a System of Particles

The Maxwell-Boltzmann statistics (abbreviated *MB statistics*) giving the distribution function (1.31) is applicable to an ensemble of particles forming a dilute gas. It may be applied, for example, to the free electrons in the conduction band of a semiconductor at ordinary temperatures and for not too high doping.

From the definition of the distribution function, we note that the number of particles N_i occupying the g_i states is given by

$$N_i = g_i f(E_i) \qquad \text{...(1.46)}$$

If the energy levels are very closely spaced, then g_i may be replaced by $g(E)dE$ where $g(E)$ is the density of states [see Eq. (1.21)], and N_i by $N(E)\,dE$, the number of particles with energies between E and $E + dE$. Hence

$$N(E)\,dE = f(E)\,g(E)\,dE \qquad \text{....(1.47)}$$

Equation (1.47) is perfectly general and holds for any form of the distribution function and the density of states. We shall consider here an electron gas system obeying the MB statistics. Since there are two allowed momentum states for the two spin orientations of an electron, the right-hand side of Eq. (1.21) will have to be multiplied by 2, giving for this system

$$g(E)dE = \frac{8\sqrt{2}\pi V m^{3/2}}{h^3} E^{1/2}\,dE. \qquad \text{...(1.48)}$$

Using the MB form of the distribution function, given by Eq. (1.31), and substituting $\beta = -1/k_B T$ we obtain from Eq. (1.47)

$$N(E)\,dE = e^\alpha e^{-E/k_B T} g(E)\,dE \qquad \text{...(1.49)}$$

The constant α is to be determined from the total number of particles N. We have

$$N = \int_0^\infty N(E)\,dE = e^\alpha \int_0^\infty g(E) e^{-E/k_B T}\,dE,$$

Basic Concepts

so that

$$e^\alpha = \frac{N}{\int_0^\infty g(E)e^{-E/k_BT}\,dE} = \frac{Nh^3}{8\sqrt{2}\pi V m^{3/2}} \Bigg/ \int_0^\infty E^{1/2} e^{-E/k_BT}\,dE, \qquad \ldots(1.50)$$

where Eq. (1.48) has been used. The limits of the integral are taken to be zero and infinity, since the energy of the particles, being kinetic, is positive. Although the maximum energy of the particles is very large, say, E_m, the upper limit of the integral is shifted from E_m to infinity for convenience. As the integrand is exponentially damped, the contribution to the integral from E_m to infinity is negligible.

To evaluate the integral in Eq. (1.50) we introduce a dimensionless variable $x = E/k_BT$. Then

$$\int_0^\infty E^{1/2} e^{-E/k_BT}\,dE = (k_BT)^{3/2} \int_0^\infty x^{1/2} e^{-x}\,dx$$

$$= (k_BT)^{3/2}\,\Gamma(3/2) = (k_BT)^{3/2}\,\frac{\sqrt{\pi}}{2} \qquad \ldots(1.51)$$

Substituting this value into Eq. (1.50) we obtain

$$e^\alpha = \frac{N}{2V}\left(\frac{h^2}{2\pi m k_BT}\right)^{3/2} \qquad \ldots(1.52)$$

Finally, we write for the Maxwell-Boltzmann distribution function for our system

$$f(E) = e^\alpha\, e^{-E/k_BT} = \frac{N}{2V}\left(\frac{h^2}{2\pi m k_BT}\right)^{3/2} e^{-E/k_BT} \qquad \ldots(1.53)$$

A plot of $f(E)$, given by Eq. (1.53), is shown in Fig. 1.2 for two different temperatures.

The distribution function is sometimes expressed in the form

$$f(E) = \frac{N}{F}e^{-E/k_BT} \qquad \ldots(1.53a)$$

where

$$F = \frac{2V}{h^3}(2\pi m k_BT)^{3/2} = \int_0^\infty g(E)e^{-E/k_BT}\,dE \qquad \ldots(1.53b)$$

For discrete energy states, we have

$$F = \sum_i g_i e^{-E_i/k_BT} \qquad \ldots(1.53c)$$

where the summation replaces the integration.

The quantity F represents the sum of the *Boltzmann factor* e^{-E/k_BT} over all the accessible states, and is called the "partition function" or "sum-over-states"; the German name is "Zustandsumme". The *importance* of the partition function is that it determines the distribution function $f(E)$ [see Eq. (1.53a)]. So the quantities which are determined by the distribution function are all linked with the partition function. The distribution function proportional to the Boltzmann factor e^{-E/k_BT} is known as the *canonical distribution*.

Various thermodynamic quantities of a system are shown to be given by the partition function (see Sec. 1.8).

Equipartition of energy

The total number of independent quantities required to specify the configuration and position of a system is called the number of *degrees of freedom* of the system. This number is determined by the possibilities of motion of the parts of the system. In the case of a gas particle, if we regard it as a point, its position is specified by the coordinates x, y, z. Thus each gas particle will have three degrees of freedom. If there are N number of gas particles in the system, the number of degrees of freedom of the system will be $3N$. The *law of equipartition of energy* states that the thermal equilibrium energy of the gas particles is uniformly distributed among the various degrees of freedom and for each of them it is $\frac{1}{2} k_B T$. We shall establish this law here using the MB statistics.

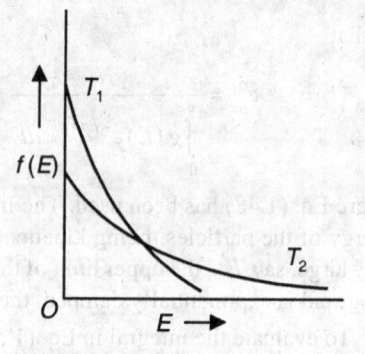

Fig. 1.2. Schematic plots of the MB distribution function for two temperatures T_1 and T_2 ($T_1 < T_2$).

The kinetic energies of a particle associated with its motion along the x, y, and z directions are $p_x^2/2m$, $p_y^2/2m$, and $p_z^2/2m$, respectively. Here p_x, p_y and p_z are the x, y, and z components of the momentum p. The average energy of the particle for its motion along any of the three directions is

$$\bar{E}_j = \frac{\iiint (p_j^2/2m) f(E) dp_x dp_y dp_z}{\iiint f(E) dp_x dp_y dp_z} \quad (j \equiv x, y, z) \qquad \ldots(1.54)$$

where $dp_x dp_y dp_z$ is a small volume element in the momentum space and the integrals extend over the entire momentum space.

Using the Maxwell-Boltzmann form of the distribution function $f(E)$ and noting that $E = (p_x^2 + p_y^2 + p_z^2)/(2m)$ we obtain from Eq. (1.54)

$$\bar{E}_x = \frac{\int_{-\infty}^{\infty} (p_x^2/2m) e^{-p_x^2/2mk_B T} dp_x \int_{-\infty}^{\infty} e^{-p_y^2/2mk_B T} dp_y \int_{-\infty}^{\infty} e^{-p_z^2/2mk_B T} dp_z}{\int_{-\infty}^{\infty} e^{-p_x^2/2mk_B T} dp_x \int_{-\infty}^{\infty} e^{-p_y^2/2mk_B T} dp_y \int_{-\infty}^{\infty} e^{-p_z^2/2mk_B T} dp_z}$$

$$= \frac{\int_{-\infty}^{\infty} (p_x^2/2m) e^{-p_x^2/2mk_B T} dp_x}{\int_{-\infty}^{\infty} e^{-p_x^2/2mk_B T} dp_x}$$

$$= \frac{\int_0^{\infty} (p_x^2/2m) e^{-p_x^2/2mk_B T} dp_x}{\int_0^{\infty} e^{-p_x^2/2mk_B T} dp_x} \qquad \ldots(1.55)$$

Substituting $x = p_x^2/2mk_B T$ we obtain from (1.55)

$$\bar{E}_x = k_B T \frac{\int_0^{\infty} x^{1/2} e^{-x} dx}{\int_0^{\infty} x^{-1/2} e^{-x} dx} = k_B T \frac{\Gamma(3/2)}{\Gamma(1/2)} = \frac{1}{2} k_B T \qquad \ldots(1.56)$$

since $\Gamma(3/2) = (1/2) \Gamma(1/2)$

Thus the mean kinetic energy for motion along the x-direction is $\frac{1}{2} k_B T$. The same value of

Basic Concepts

$\frac{1}{2}k_BT$ is also obtained for $\bar{E}_y$ and $\bar{E}_z$. Thus the law of equipartition of energy is seen to be satisfied. The average energy per particle considering the three degrees of freedom is $\frac{3}{2}k_BT$.

Average speed, root-mean-square speed, and most probable speed

Since the energy E is kinetic, we have $E = \frac{1}{2}mv^2$, where v is the speed of the particle. Converting from E to v, the number of particles with speed between v and $v + dv$ is found from Eqs. (1.48), (1.49) and (1.52) to be

$$N(v)dv = N(E)dE = \left(\frac{2}{\pi}\right)^{1/2} N \left(\frac{m}{k_BT}\right)^{3/2} e^{-mv^2/2k_BT} v^2 dv \qquad ...(1.57)$$

The average thermal speed of a particle is

$$\bar{v} = \int_0^\infty v N(v) dv / \int_0^\infty N(v) dv = \sqrt{2/m} \int_0^\infty E^{1/2} N(E) dE / \int_0^\infty N(E) dE$$

$$= \left(\frac{2k_BT}{m}\right)^{1/2} \int_0^\infty x e^{-x} dx / \int_0^\infty x^{1/2} e^{-x} dx \quad [\text{where } x = E/k_BT]$$

$$= \left(\frac{2k_BT}{m}\right)^{1/2} \Gamma(2)/\Gamma(3/2) = \left(\frac{8k_BT}{\pi m}\right)^{1/2} \qquad ...(1.58)$$

The root-mean-square (abbreviated rms) speed of a system of particles is that speed whose square is the average of the squares of the speeds of the particles.

Denoting the rms speed by v_{rms} we have

$$v_{rms}^2 = \int_0^\infty v^2 N(v) dv / \int_0^\infty N(v) dv = \frac{2}{m} \int_0^\infty E N(E) dE / \int_0^\infty N(E) dE$$

$$= \frac{2}{m} \times \frac{3}{2} k_BT = \frac{3k_BT}{m} \quad \text{or} \quad v_{rms} = \sqrt{3k_BT/m} \qquad ...(1.59)$$

The function $N(v) = \left(\frac{2}{\pi}\right)^{1/2} N \left(\frac{m}{k_BT}\right)^{3/2} e^{-mv^2/2k_BT} v^2$ is plotted against v in Fig. 1.3. The curve, called the *velocity distribution curve*, rises with v, attains a maximum, and then falls due to the predominance of the exponential term. The peak of the curve broadens and shifts to a higher value of v as the temperature rises. With increasing temperature, there are more particles with higher values of v, i.e. higher kinetic energy.

The most probable speed v_{mp} of a particle is that value of the speed v for which the number of particles with speed between v and $v + dv$, is a maximum, i.e., $N(v)$ is a maximum.

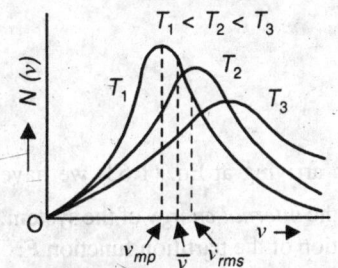

Fig. 1.3. Variation of $N(v)$ with v for three different temperatures T_1, T_2 and T_3.

For a maximum, $\dfrac{dN(v)}{dv} = 0$

or, $2v e^{-mv^2/2k_BT} - v^2 \dfrac{mv}{k_BT} e^{-\frac{mv^2}{2k_BT}} = 0$ or, $v^2 = v_{mp}^2 = \dfrac{2k_BT}{m}$

or, $v_{mp} = \sqrt{2k_BT/m}$ ...(1.60)

From Eqs. (1.58), (1.59) and (1.60) we find that $v_{rms}/\bar{v} = \sqrt{3\pi/8} = 1.085$ and $\bar{v}/v_{mp} = 2/\sqrt{\pi} = 1.128$. Thus $v_{rms} > \bar{v} > v_{mp}$, as shown in Fig. 1.2. Note, however, that the differences between v_{rms}, $\bar{v}$, and v_{mp} are not large. Clearly, most of the particles have speeds near the peak of the distribution curve. There are very few particles with very high or very low speeds.

1.8. THERMODYNAMIC QUANTITIES FROM PARTITION FUNCTION

A *thermal reservoir* or a *heat bath* is defined as a body of such a large heat capacity that its temperature does not change when heat is added to or removed from it. The addition of the heat δQ to a heat bath increases its internal energy by δU if its volume V is constant. If the consequent increase in entropy is δS, then $\delta Q = T\delta S = \delta U$. Since $\delta S = k_B \delta (\ln W)$ where W is the number of accessible quantum states of the heat bath, we have

$$\frac{1}{k_B T} = \left(\frac{\partial (\ln W)}{\partial U}\right)_V$$

This is the defining equation of the temperature T of the heat bath.

An ensemble of particles making up a system in thermal equilibrium with a heat bath is called a *canonical ensemble*. The partition function of a canonical ensemble is given by

$$F = \sum_{i=1}^{n} e^{-E_i/(k_B T)} \qquad \ldots(1.61)$$

where the particles are distributed over nondegenerate quantum states with energies $E_1, E_2, ..., E_n$, and T is the temperature of the heat bath with which the system is in thermal equilibrium.

The entropy of the system is

$$S = k_B \ln W \qquad \ldots(1.62)$$

where W is the number of arrangements for the most probable distribution. For nondegenerate levels, we have from Eq. (1.25) for the most probable distribution

$$\ln W = N \ln N - \sum_{i=1}^{n} N_i \ln N_i$$

$$= N \ln N - \sum_{i=1}^{n} \frac{N}{F} e^{-E_i/(k_B T)} \left(\ln N - \ln F - \frac{E_i}{k_B T}\right)$$

$$= N \ln N - N \ln N + N \ln F + \frac{1}{k_B T}\sum_{i=1}^{n} N_i E_i$$

$$= N \ln F + \frac{U}{k_B T} \qquad \ldots(1.63)$$

In arriving at Eq. (1.63) we have used the relationships $N_i = \frac{N}{F} e^{-E_i/(k_B T)}$ and $U = \Sigma N_i E_i$, U being the *internal energy* of the system. Using Eq. (1.63) in Eq. (1.62), one can express the entropy as a function of the partition function F:

$$S = k_B N \ln F + \frac{U}{T} \qquad \ldots(1.64)$$

The internal energy is

$$U = \sum_{i=1}^{n} N_i E_i = \frac{N}{F} \sum_{i=1}^{n} E_i e^{-E_i/(k_B T)} = \frac{N}{F} k_B T^2 \frac{dF}{dT}$$

$$= N k_B T^2 \frac{d(\ln F)}{dT} \qquad \ldots(1.65)$$

Basic Concepts

Putting Eq. (1.65) into Eq. (1.64), S can be expressed as a function of F and T only.

All the other thermodynamic quantities can be found in terms of S and U, and they can be obtained in terms of F and T. For example, the *Helmholtz free energy* is written from Eq. (1.64) as

$$A = U - TS = -Nk_B T \ln F. \qquad ...(1.66)$$

Since F is determined by the energy levels of the system, the thermodynamic properties depend on these levels and temperature.

For a small reversible change in A, keeping the number of particles constant, we have

$$dA = dU - SdT = -PdV - SdT \qquad ...(1.67)$$

So,

$$P = -\left(\frac{\partial A}{\partial V}\right)_T \qquad ...(1.68)$$

and

$$S = -\left(\frac{\partial A}{\partial T}\right)_V \qquad ...(1.69)$$

Using Eq. (1.66), we can obtain the pressure P and the entropy S as

$$P = Nk_B \left(\frac{\partial (T \ln F)}{\partial V}\right)_T \qquad ...(1.70)$$

$$S = Nk_B \left(\frac{\partial (T \ln F)}{\partial T}\right)_V \qquad ...(1.71)$$

Equation (1.68) allows us to obtain the *isothermal compressibility* K in terms of F:

$$\frac{1}{K} = -V\left(\frac{\partial P}{\partial V}\right)_T = V\left(\frac{\partial^2 A}{\partial V^2}\right)_T \qquad ...(1.72)$$

The heat capacity at constant volume can be expressed from Eq. (1.69) as

$$C_V = T\left(\frac{\partial S}{\partial T}\right)_V = -T\left(\frac{\partial^2 A}{\partial T^2}\right)_V \qquad ...(1.73)$$

The *chemical potential* μ is defined as

$$\mu = \left(\frac{\partial A}{\partial N}\right)_{T,V} = -T\left(\frac{\partial S}{\partial N}\right)_{U,V} \qquad ...(1.74)$$

Clearly, μ is also determined by the partition function.

1.9. Approach to Equilibrium

Suppose that we have two systems A and B which are not in thermodynamic equilibrium with each other. Let the systems be in weak thermal contact with each other to allow energy transfer from one to the other. As the two systems are thermally isolated from the rest of the universe, their total energy $U_T (= U_A + U_B)$ is constant. Since the thermal contact is weak, the entropy of the two systems is the sum of the entropies of each system, i.e.

$$S(U_T) = S_A(U_A) + S_B(U_T - U_A) \qquad ...(1.75)$$

If A and B are in thermal equilibrium, the entropy of the combined system is a maximum. When they are not in thermal equilibrium, A and B evolve to maximize the entropy. The rate of change of entropy with time is obtained from Eq. (1.75) as

$$\frac{dS}{dt} = \frac{dU_A}{dt}\left(\frac{\partial S_A}{\partial U_A} - \frac{\partial S_B}{\partial U_B}\right) = \frac{dU_A}{dt}\left(\frac{1}{T_A} - \frac{1}{T_B}\right) \qquad ...(1.76)$$

By Clausius' principle,

$$\frac{dS}{dt} \geq 0. \qquad ...(1.77)$$

Equation (1.76) shows that when $T_A = T_B$, $\dfrac{dS}{dt} = 0$, or S is a constant. Here the system is in *thermal equilibrium*. If $T_A > T_B$, we have $\dfrac{dU_A}{dt} < 0$, implying that A loses energy. Thus, heat is transferred from the hotter system to the colder one.

Let us now consider a situation where a movable piston separates two systems in thermal equilibrium. Suppose that the movement of the piston increases the volume of the system A and decreases the volume of the system B such that the total volume V_T ($=V_A + V_B$) remains constant. The total entropy is now written as

$$S(V_T) = S_A(V_A) + S_B(V_T - V_A) \qquad \text{...(1.78)}$$

Hence

$$\dfrac{dS}{dt} = \dfrac{dV_A}{dt}\left(\dfrac{\partial S_A}{\partial V_A} - \dfrac{\partial S_B}{\partial V_B}\right) = \dfrac{dV_A}{dt}\left(\dfrac{P_A}{T_A} - \dfrac{P_B}{T_B}\right) \geq 0 \qquad \text{...(1.78a)}$$

Here $T_A = T_B$ as A and B are in thermal equilibrium. If the two pressures P_A and P_B are equal, $dS/dt = 0$ and S is a constant. Here the system is in *mechanical equilibrium*. If $P_A > P_B$, then $\dfrac{dV_A}{dt} > 0$, implying that the side with higher pressure expands.

Finally, we consider the case where the number of particles can change between the interacting systems A and B. Here the interaction is *diffusive* and can be achieved by a membrane separating A and B, allowing particles to pass through. The total number of particles N_T ($= N_A + N_B$) remains constant here. The total entropy is expressed by

$$S(N_T) = S_A(N_A) + S_B(N_T - N_A) \qquad \text{...(1.79)}$$

So,

$$\dfrac{dS}{dt} = \dfrac{dN_A}{dt}\left(\dfrac{\partial S_A}{\partial N_A} - \dfrac{\partial S_B}{\partial N_B}\right) = -\dfrac{dN_A}{dt}\left(\dfrac{\mu_A}{T_A} - \dfrac{\mu_B}{T_B}\right) \geq 0 \qquad \text{...(1.80)}$$

where μ is the chemical potential defined by Eq. (1.74). We assume that the systems are in thermal equilibrium, *i.e.* $T_A = T_B$. If the chemical potentials μ_A and μ_B are equal, then S is a constant and the system is said to be in *chemical equilibrium*. If $\mu_A > \mu_B$, then $dN_A/dt < 0$, showing that the particles escape from the system with the higher chemical potential.

In general, the entropy S is a function of the internal energy U, the volume V, and the number of particles N. Hence

$$dS = \left(\dfrac{\partial S}{\partial U}\right)_{V,N} dU + \left(\dfrac{\partial S}{\partial V}\right)_{U,N} dV + \left(\dfrac{\partial S}{\partial N}\right)_{U,V} dN$$

$$= \dfrac{1}{T} dU + \dfrac{P}{T} dV - \dfrac{\mu}{T} dN \qquad \text{...(1.81)}$$

The term μdN is referred to as the *chemical work*.

1.10. Equipartition Theorem And Its Applications

The *equipartition theorem* or the *law of equipartition of energy* states as follows :

If a system obeying classical statistical mechanics is in equilibrium at the absolute temperature T, each degree of freedom which contributes an independent quadratic term of a coordinate or momentum to its total energy, has an average energy $(1/2) k_B T$.

A degree of freedom implies one of the quadratic terms in the total energy. The equipartition theorem has been established in Sec. 1.7 for a system of gas particles obeying MB statistics. Some applications of this theorem will be considered here.

(i) Harmonic oscillator

A particle of mass m executing a simple harmonic motion along the x-axis has the energy

$$E = \frac{p_x^2}{2m} + \frac{1}{2}\mu x^2,$$

where μ is the spring constant. The first term quadratic in the momentum p_x is the kinetic energy of the particle, and the second term quadratic in the position coordinate x is its potential energy. Because of the two quadratic terms in E, the number of degrees of freedom is two. If the oscillator is in equilibrium with a heat bath at temperature T which is high enough for the particles to be described classically, the equipartition theorem shows that the average energy of the oscillator is $(1/2) k_B T + (1/2) k_B T = k_B T$.

A vibrational mode described by a harmonic oscillator has clearly the mean energy $k_B T$.

(ii) Specific heat of a monatomic ideal gas

For a monatomic gas particle the kinetic energy in three-dimensions is

$$E = \frac{1}{2m}(p_x^2 + p_y^2 + p_z^2)$$

There are three quadratic terms, one for each degree of freedom. The mean energy for each quadratic term is, by the equipartition theorem, $(1/2) k_B T$. The average energy of the particle is therefore $(3/2) k_B T$. Considering one mole of the gas which contains N_A molecules, where N_A is Avogadro's number, the average energy of the gas is

$$U = N_A \left(\frac{3}{2} k_B T\right) = \frac{3}{2} RT \qquad \ldots(1.82)$$

where $R (= N_A k_B)$ is the universal gas constant. The molar specific heat C_V at constant volume is thus given by

$$C_V = \left(\frac{\partial U}{\partial T}\right)_V = \frac{3}{2} R.$$

Since $C_P - C_V = R$, the molar specific heat at constant pressure is $C_P = (5/2) R$.

(iii) Specific heat of diatomic and triatomic gases

For a diatomic particle there are three degrees of freedom for translational motion. The particle has one vibrational mode with two degrees of freedom. The orientation of the molecule is specified by two angles. Consequently there are two conjugate momenta. The rotational energy is quadratic in these momenta, and so the molecule has two rotational degrees of freedom. The mean energy of the molecule at a high temperature T is, by the equipartition theorem, $(7/2) k_B T$. The molar specific heat of a diatomic gas at constant volume is thus $(7/2) R$. The molar specific heat at constant pressure is $(9/2) R$.

A triatomic molecule has a centre of mass that can move in three directions, each with a conjugate momentum. The kinetic energy of translational motion is quadratic in these momenta. So the translational motion has three degrees of freedom. In the absence of an axis of symmetry, three angles are required to specify the orientation of the molecule, and so there are three conjugate momenta. The rotational energy is quadratic in each of these three conjugate momenta, giving three rotational degrees of freedom. A triatomic molecule has three atoms and hence there are nine kinetic energy terms quadratic in momenta. Six of them have already been considered; so we are left with three vibrational modes each with a potential and a kinetic energy term. Hence there are six degrees of freedom for vibration. The total of twelve degrees of freedom gives a mean energy of $6 k_B T$ for the molecule at a high temperature T. The molar specific heat of a triatomic gas is therefore $6R$.

If the triatomic molecule has an axis of symmetry, two angles are needed to specify the orientation of the molecule; so there are two ratational degrees of freedom. The number of translational degrees of freedom is three as before. Consequently, there must be four vibrational modes giving eight degrees of freedom. The total of thirteen degrees of freedom results in a molar specific heat of $13R/2$ at constant volume. Observe that the symmetry of the molecule influences the specific heat.

Validity of the equipartition theorem

We shall now consider the limitations of the equipartition theorem. As the temperature is lowered, the vibrational modes first disappear, next the rotational motion stops, and finally the translational motion slows down as the system liquifies and stops when it freezes.

The classical concepts leading to the equipartition theorem has a fundamental quantum mechanical limitation imposed by Heisenberg's uncertainty principle. Let us consider a particle of the system and assume that s_0 be the distance in which the particle is localized. If the typical momentum of the particle is p_0, then Heisenberg's uncertainty principle would be unimportant and the classical description will apply when $s_0 p_0 \gg \hbar$. For a vibrating atom, the average kinetic energy $\overline{p_x^2}/(2m)$ would be $(1/2) k_B T$, by the equipartition theorem. So, the momentum of the atom is typically of the order of

$$p_0 \simeq \sqrt{\overline{p_x^2}} = \sqrt{m k_B T}.$$

The mean potential energy of the vibrating atom is $\frac{1}{2} \mu \overline{x^2} = \frac{1}{2} k_B T$, by the equipartition theorem. The typical value of the mean amplitude of vibration of the atom is thus of the order of

$$s_0 \simeq \sqrt{\overline{x^2}} = \sqrt{\frac{k_B T}{\mu}}.$$

Heisenberg's uncertainty principle is unimportant if

$$s_0 p_0 = \sqrt{\frac{m}{\mu}} k_B T \gg \hbar$$

or,
$$k_B T \gg \hbar \omega \qquad \ldots(1.83)$$

where $\omega (= \sqrt{\mu/m})$ is the angular frequency of oscillation of the atom. The classical ideas are valid if the temperature is large enough to satisfy Eq. (1.83).

Note that $\hbar \omega$ is the separation between the allowed discrete energy levels of a quantum oscillatior. When Eq. (1.83) is satisfied, the mean energy of the system is very high, and the spacing $\hbar \omega$ between the levels about the mean energy is small compared to the thermal energy $k_B T$. The discreteness of the energy levels is now of no consequence and the energy can be considered to be quasicontinuous. The classical description and the equipartition theorem will now apply.

For monatomic gas particles having translational motion, s_0 would represent the mean separation between the particles. Then quantum mechanical effects are unimportant when $s_0 \gg \hbar/p_0$, i.e. $s_0 \gg \lambda_0$, where $\lambda_0 (= h/p_0)$ is the mean de Broglie wavelength of the gas particles. The mean volume available to a gas particle is $(4/3) \pi s_0^3$, and the total volume of the gas containing N particles, is $(4/3) \pi s_0^3 N = V$

So,
$$s_0 = \left(\frac{3V}{4\pi N}\right)^{1/3} \qquad \ldots(1.84)$$

Allowing for three translational degrees of freedom, the mean energy of a gas particle at temperature T is $p_0^2/(2m) = (3/2) k_B T$. So, $p_0 = \sqrt{3 m k_B T}$ and $\lambda_0 = h/\sqrt{3 m k_B T}$. Hence the validity criterion of the equipartition theorem is

$$s_0 \gg \lambda_0$$

or,
$$\left(\frac{V}{N}\right)^{1/3} \gg \left(\frac{4\pi}{3}\right)^{1/3} \frac{h}{\sqrt{3 m k_B T}} \qquad \ldots(1.85)$$

Basic Concepts

Thus classical ideas are valid when the concentration (N/V) of the gas particles is small, the temperature T is sufficiently high, and the particle mass m is not too small. In a nutshell, the classical description holds quite well for *a dilute gas of heavy particles at high temperatures*.

1.11. Gibbs Paradox

For an ideal gas containing N particles at temperature T, the internal energy is $U = (3/2) N k_B T$, and so its entropy, by Eq. (1.64), is

$$S = N k_B \left(\ln F + \frac{3}{2} \right). \qquad \ldots(1.86)$$

The partition function F can be obtained from Eq. (1.53b) by dropping the factor 2 for spin. Thus

$$F = \frac{V}{h^3} (2\pi m k_B T)^{3/2} = V \left(\frac{2\pi m k_B}{h^2} \right)^{3/2} T^{3/2} \qquad \ldots(1.87)$$

Putting this expression for F in Eq. (1.86) we get

$$S = N k_B \left[\ln V + \frac{3}{2} \ln T + \sigma \right] \qquad \ldots(1.88)$$

where
$$\sigma = \frac{3}{2} \ln \left(\frac{2\pi m k_B}{h^2} \right) + \frac{3}{2}.$$

As $T \to 0$, Eq. (1.88) shows that $S \to -\infty$, which contradicts the third law of thermodynamics. However, at such low temperatures, the classical ideas on the basis of which Eq. (1.88) is derived, break down, and quantum concepts must be applied.

Another serious drawback of Eq. (1.88) is that it does not permit the entropy to behave as an extensive quantity. If the size of the system is multiplied by a scale factor, say, α, then all the thermodynamic quantities must be scaled by the same factor α to behave properly as extensive quantities. If V and N are multiplied by α, the average energy U given by Eq. (1.65) is properly multiplied by the same factor α, but the entropy S given by Eq. (1.88) is not increased by α due to the term $N \ln V$.

Suppose that a partition divides the vessel containing the gas into two parts. If S_1 and S_2 are the entropies of the two parts, then the total entropy should be $S = S_1 + S_2$. If the two parts are equal, then $S_1 = S_2$ and S should be $2S_1$. If each part contains N_1 particles in a volume V_1, then by Eq. (1.88) we obtain

$$S_1 = S_2 = N_1 k_B \left[\ln V_1 + \frac{3}{2} \ln T + \sigma \right]$$

For the entire unpartitioned system Eq. (1.88) gives

$$S = 2N_1 k_B \left[\ln (2V_1) + \frac{3}{2} \ln T + \sigma \right]$$

so that

$$S - 2S_1 = 2N_1 k_B [\ln (2V_1) - \ln V_1] = 2 N_1 k_B \ln 2$$

which shows that S differs from the expected value of $2S_1$.

This apparent paradox is known as the *Gibbs paradox*. The root of the paradox lies in treating the gas particles as distinguishable particles. If one considers the indistinguishability of the particles, the Gibbs paradox is removed. To see this, we refer to Eq. (1.71) which can be written as

$$S = N k_B \left[\ln F + T \frac{\partial}{\partial T} \ln F \right] = k_B \left[\ln F^N + T \frac{\partial}{\partial T} \ln F^N \right]$$

This shows that the single-particle partition function F is multiplied N times to give the N-particle partition function as F^N. If the particles are indistinguishable, an interchange of two particles does not give any new quantum state. Hence the $N!$ possible arrangements of the particles among themselves do not yield distinct situations, and the number of distinct states summed in F^N is thus too large by $N!$. Therefore, the partition function for N particles would be $F_N = F^N/N!$. Then

$$S = k_B \left[\ln\left(\frac{F^N}{N!}\right) + T \frac{\partial}{\partial T} \ln\left(\frac{F^N}{N!}\right) \right] = k_B \left[\ln F^N + T \frac{\partial}{\partial T} F^N - \ln N! \right]$$

$$= N k_B \left[\ln V + \frac{3}{2} \ln T + \sigma \right] - k_B (N \ln N - N)$$

using Stirling's formula : $\ln N! = N \ln N - N$. Therefore,

$$S = N k_B \left[\ln \frac{V}{N} + \frac{3}{2} \ln T + \sigma_0 \right] \qquad ...(1.89)$$

where

$$\sigma_0 = \sigma + 1 = \frac{3}{2} \ln\left(\frac{2\pi m k_B}{h^2}\right) + \frac{5}{2}.$$

The entropy S given by Eq. (1.89) behaves properly and is scaled by α when both V and N are multiplied by α. Therefore, the Gibbs paradox does not arise. Equation (1.89) is known as *Sackur-Tetrode formula*.

Note that since P and U are determined by the derivative of $\ln F$ with respect to V and T, respectively, the results (1.70) and (1.65) for these quantities are left unchanged by the extra term $N!$ in the N-particle partition function. But the expression for S, *i.e.* Eq. (1.71) contains $\ln F + T \frac{\partial \ln F}{\partial T}$ and is thus changed by this extra term. The same comment applies to the free energy given by Eq. (1.66). It depends on the partition function itself rather than its derivative, and so behaves properly like an extensive quantity only if Eq. (1.89) for S is used. So, for an ideal gas

$$A = U - TS = \frac{3}{2} N k_B T - N k_B T \left[\ln \frac{V}{N} + \frac{3}{2} \ln T + \sigma_0 \right]$$

$$= - N k_B T \left[\ln \frac{V}{N} + \frac{3}{2} \ln T + \sigma_0 - \frac{3}{2} \right] \qquad ...(1.90)$$

The chemical potential μ of an ideal gas can be found from Eqs. (1.74) and (1.89). Since $U = \frac{3}{2} N k_B T$, Eq. (1.89) gives

$$S = N k_B \left[\ln \frac{V}{N} + \frac{3}{2} \ln\left(\frac{4\pi m U}{3 h^2 N}\right) + \frac{5}{2} \right].$$

Hence

$$\mu = -T \left(\frac{\partial S}{\partial N}\right)_{U,V} = -k_B T \left[\ln \frac{V}{N} + \frac{3}{2} \ln\left(\frac{4\pi m U}{3 h^2 N}\right) \right]$$

$$= -k_B T \left[\ln \frac{V}{N} + \frac{3}{2} \ln\left(\frac{2\pi m k_B T}{h^2}\right) \right] \qquad ...(1.91)$$

1.12. LIMITATIONS OF MB STATISTICS

(*i*) The MB statistics holds only for a system of distinguishable particles forming a dilute gas when the mean potential energy due to the mutual interaction between the particles is negligible compared to their mean kinetic energy.

Basic Concepts

(*ii*) The MB statistics does not give correctly the entropy of an ideal gas. This leads to Gibbs paradox and predicts the behaviour of the entropy at absolute zero of temperature at variance with the third law of thermodynamics.

(*iii*) If MB statistics is applied to the 'electron gas' in metals, several discrepancies appear between theory and observation. For example, the velocity distribution of photoelectrons and the thermionic emission current density, predicted by MB statistics, do not agree with experimental results. Furthermore, experimental results show that the high-temperature value of the molar specific heat of a metal is $3R$ which, from Debye's theory, is the lattice contribution alone. The electronic contribution to the specific heat of a metal thus turns out to be very small. However, MB statistics predicts that the contribution of the free electrons in metals to the heat capacity would be $3R/2$, a value much greater than that expected.

(*iv*) If MB statistics is applied to a 'photon gas', it fails to account for the experimentally observed energy density of radiation at different portions of the frequency scale, which follows Planck's law of radiation.

The above shortcomings of the classical or MB statistics have been removed by the *quantum statistics* which includes *Fermi-Dirac* (or *FD*) *statistics* and *Bose-Einstein* (or *BE*) *statistics*. We shall discuss FD and BE statistics in the next chapter. In quantum statistics, the particles are taken to be *indistinguishable*, whereas they are *distinguishable* in the classical MB statistics. We shall see later that when the number of particles is much less than the number of available states, both FD and BE distribution functions reduce to MB distribution.

1.13. Worked-out Problems

1. An electron gas obeys the Maxwell Boltzmann statistics. Calculate the average thermal energy (in eV) of an electron in the system at room temperature, *i.e.* 300 K.

Ans. The average thermal energy is given by $\frac{3}{2} k_B T$. Substituting the values $k_B = 1.38 \times 10^{-23}$ J/K and $T = 300$ K we obtain

$$\frac{3}{2} k_B T = \frac{3 \times 1.38 \times 10^{-23} \times 300}{2} \text{ J}$$

$$= \frac{3 \times 1.38 \times 10^{-23} \times 300}{2 \times 1.6 \times 10^{-19}} \text{ eV} = 0.039 \text{ eV}$$

2. In an ideal gas obeying MB statistics there are N number of particles at a temperature T. Find the internal energy of the gas and the heat capacity at constant volume.

Ans. The average internal energy per particle is $\frac{3}{2} k_B T$. Therefore the total internal energy of the gas is $U = \frac{3}{2} N k_B T$.

The heat capacity of the gas at constant volume is given by

$$C_v = \left(\frac{\partial U}{\partial T} \right)_v = \frac{3}{2} N k_B$$

3. Six distinguishable particles are distributed over three nondegenerate levels of energies 0, ε, and 2ε. Calculate the total number of microstates of the system. Find the total energy of the distribution for which the probability is a maximum. (*C.U.* 1993)

Ans. Since the levels are nondegenerate, there is only one state associated with each energy. Since each of the 6 particles can be in any of the 3 energy levels, the total number of ways of arranging the particles, i.e., the total number of microstates is $3^6 = 729$. Let the number of particles in the three energy states be N_1, N_2 and N_3, respectively, where $N_1 + N_2 + N_3 = 6$, the total number of particles. As the particles are distinguishable, the number of ways of choosing N_1, N_2 and N_3 particles from 6 particles is

$$W = \frac{6!}{N_1! N_2! N_3!}$$

W gives the thermodynamic probability. It is a maximum when $N_1! N_2! N_3!$ is a minimum. By inspection we find that $N_1! N_2! N_3!$ is a minimum when $N_1 = N_2 = N_3 = 2$. The corresponding total energy of the distribution is

$$0 \times N_1 + \varepsilon \times N_2 + 2\varepsilon \times N_3 = 2\varepsilon + 4\varepsilon = 6\varepsilon$$

4. A number of identifiable particles are distributed in a two-level system having energies E and $2E$, respectively. Determine the average energy for the most probable distribution, if the degeneracy of each level is the same. (*cf. Burd. U.* 1995)

Ans. Here $g_1 = g_2 = g$ (say), and the most probable distribution is given by Eq. (1.31) with $\beta = -1/(k_B T)$. So, the number of particles in the energy state E is $N_1 = g e^\alpha \cdot e^{-E/k_B T}$ and that in the energy state $2E$ is $N_2 = g e^\alpha e^{-2E/(k_B T)}$. Hence, the average energy is

$$\bar{E} = \frac{EN_1 + 2EN_2}{N_1 + N_2} = \frac{E e^{-E/k_B T} + 2E e^{-2E/k_B T}}{e^{-E/k_B T} + e^{-2E/k_B T}}$$

5. Five identifiable particles are distributed in three nondegenerate levels with energies 0, E, and $2E$. Determine the most probable distribution for a total energy $3E$.

Ans. As the levels are nondegenerate, there is only one state for each energy. Let the number of particles occupying the three energy states be N_1, N_2, and N_3, respectively, where $N_1 + N_2 + N_3 = 5$, the total number of particles. As the particles are identifiable, the number of ways of choosing the particles is $W = \dfrac{5!}{N_1! N_2! N_3!}$

The energy of the system of $0 N_1 + E N_2 + 2E N_3 = 3E$ (given). Hence

$$N_2 + 2N_3 = 3 \qquad \ldots (1)$$

The most probable distribution is the one in which W is a maximum subject to the constraint given by Eq. (*i*). Thus,

if $N_2 = 1$, $N_3 = \dfrac{3-1}{2} = 1$, and $N_1 = 5 - (N_2 + N_3) = 5 - 2 = 3$;

if $N_2 = 3$, $N_3 = \dfrac{3-3}{2} = 0$, and $N_1 = 5 - 3 = 2$.

No other distributions are possible.

For $N_1 = 3$, $N_2 = 1$, and $N_3 = 1$, $W = \dfrac{5!}{3! 1! 1!} = 20$

For $N_1 = 2$, $N_2 = 3$, and $N_3 = 0$, $W = \dfrac{5!}{2! 3! 0!} = 10$

So, the most probable distribution is $N_1 = 3$, $N_2 = 1$ and $N_3 = 1$.

6. In a system of gas particles, obeying MB statistics, calculate the mean square deviation of the particle velocity from the average velocity. Hence find the parameters determining the broadness of the velocity distribution.

Ans. The deviation of the particle velocity v from the average velocity $<v>$ is $\delta = v - <v>$, where the angular brackets $<\ >$ denote the average. The mean square deviation is

Basic Concepts

$$\langle\delta^2\rangle = \langle(v-\langle v\rangle)^2\rangle = \langle v^2 - 2v\langle v\rangle + \langle v\rangle^2\rangle$$
$$= \langle v^2\rangle - 2\langle v\rangle\langle v\rangle + \langle v\rangle^2 = \langle v^2\rangle - 2\langle v\rangle^2$$
$$+ \langle v\rangle^2 = \langle v^2\rangle - \langle v\rangle^2 = v_{rms}^2 - \langle v\rangle^2$$
$$= \frac{k_B T}{m}\left(3 - \frac{8}{\pi}\right)$$

The quantity $\langle\delta^2\rangle$ is a measure of the broadness of the velocity distribution: the larger the $\langle\delta^2\rangle$, the broader the velocity distribution. Thus the parameters determining the broadness of the velocity distribution are the temperature T and the particle mass m. The distribution is broader for a larger T and a smaller m.

7. In a system of particles, 20% of the particles are moving with a speed of 2m/s, 30% with a speed of 4m/s, and 50% with a speed of 3m/s. calculate the average speed, rms speed, and the most probable speed. If the mass of each particle is 2 gm find the mean kinetic energy of a particle.

Ans. The average speed is
$$\bar{v} = 0.2 \times 2 + 0.3 \times 4 + 0.5 \times 3 = 3.1 \text{ m/s.}$$
The rms speed is
$$v_{rms} = (0.2 \times 2^2 + 0.3 \times 4^2 + 0.5 \times 3^2)^{1/2} = 3.18 \text{ m/s.}$$
Since 50% of the particles have the speed 3 m/s, the most probable speed is
$$v_{mp} = 3 \text{ m/s}$$
The mean kinetic energy of a particle of mass $m \ (= 2 \times 10^{-3}$ kg) is
$$\frac{1}{2} m v_{rms}^2 = \frac{1}{2} \times 2 \times 10^{-3} \times 10.1 \text{ J} = 10.1 \text{ mJ}$$

8. Show that the half-width of Maxwell's velocity distribution curve is approximately $\sqrt{2k_B T/m}$.

Ans. Let, on the two sides of the most probable speed (v_{mp}), v_1 and $v_2 \ (> v_1)$ be the speeds where the function $N(v)$ drops to half its peak value. Then the speed difference $(v_2 - v_1)$ is called the half-width of the velocity distribution curve of Fig. 1.3. At $v = v_{mp}$, $N(v) = N(v_{mp}) = $ maximum. If, at $v = v_{mp} + \delta v$, $N(v) = N(v_{mp} + \delta v) = \frac{1}{2} N(v_{mp})$, then using the expression for $N(v)$ we obtain

$$e^{-m(v_{mp} + \delta v)^2/2k_B T}(v_{mp} + \delta v)^2 = \frac{1}{2} e^{-mv_{mp}^2/2k_B T} v_{mp}^2$$

or, $\exp\left[-\frac{m v_{mp}^2}{2k_B T}\left(2 + \frac{\delta v}{v_{mp}}\right)\frac{\delta v}{v_{mp}}\right]\left[1 + 2\frac{\delta v}{v_{mp}} + \left(\frac{\delta v}{v_{mp}}\right)^2\right] = \frac{1}{2}$

Expanding the exponential, retaining the terms up to the second order in $\delta v/v_{mp}$, and noting that $v_{mp}^2 = 2k_B T/m$, we get

$$\left[1 - \left(2 + \frac{\delta v}{v_{mp}}\right)\frac{\delta v}{v_{mp}} + \frac{1}{2}.4\left(\frac{\delta v}{v_{mp}}\right)^2\right]\left[1 + 2\frac{\delta v}{v_{mp}} + \left(\frac{\delta v}{v_{mp}}\right)^2\right] = \frac{1}{2}$$

or, $1 + 2\frac{\delta v}{v_{mp}} + \left(\frac{\delta v}{v_{mp}}\right)^2 - \left(2 + \frac{\delta v}{v_{mp}}\right)\frac{\delta v}{v_{mp}} - 4\left(\frac{\delta v}{v_{mp}}\right)^2 + \frac{1}{2}.4\left(\frac{\delta v}{v_{mp}}\right)^2 = \frac{1}{2},$

neglecting terms higher than the second in $\delta v/v_{mp}$.

So, $1 - 2\left(\frac{\delta v}{v_{mp}}\right)^2 = \frac{1}{2}$

or, $\delta v = \pm v_{mp}/2$.

Thus $v_1 = v_{mp} - v_{mp}/2 = v_{mp}/2$, and $v_2 = v_{mp} + v_{mp}/2 = 3v_{mp}/2$.

Since v_1 and v_2 are equally removed from v_{mp}, the velocity distribution curve is practically symmetric about the peak between v_1 and v_2. (This is an approximation. Actually $(v_2 - v_{mp})$ is slightly greater than $(v_{mp} - v_1)$ and the distribution curve is asymmetric). The half-width is

$$v_2 - v_1 = v_{mp} = \sqrt{2k_BT/m}, \text{ proved. (Actually } v_2 - v_1 = 1.15\, v_{mp})$$

Clearly the half-width is larger (i.e., the distribution is broader) for a larger T and a smaller m. The same conclusion regarding the broadness of the velocity distribution is arrived at in problem no. 6 above.

9. A system containing two spin $\frac{1}{2}$ particles, stationary in space, is placed in an external magnetic field $\vec{B}$. Each particle has a magnetic moment μ which can be aligned either parallel or antiparallel to $\vec{B}$. What are the possible microstates and macrostates of the system ? (C.U. 2001)

Ans. A microstate is specified by the state of each particle; the latter is given by the quantum number m which has two values $+\frac{1}{2}$ and $-\frac{1}{2}$. The macrostate, i.e. the state of the ensemble is specified by the total magnetic moment and the total energy.

Each particle has a magnetic moment μ when aligned parallel to $\vec{B}$, and $-\mu$ when aligned antiparallel to $\vec{B}$. The quantum numbers of the two particles are denoted by m_1 and m_2. A particle has energy $-\mu B$ when its spin points along $\vec{B}$, and energy μB when its spin points opposite to $\vec{B}$. In the following table we list the possible microstates of the system together with the total magnetic moment and total energy, which characterize the system as a whole.

Microstates	Quantum numbers		Total magnetic moment	Total energy
	m_1	m_2		
1	$\frac{1}{2}$	$\frac{1}{2}$	2μ	$-2\mu B$
2	$\frac{1}{2}$	$-\frac{1}{2}$	0	0
3	$-\frac{1}{2}$	$\frac{1}{2}$	0	0
4	$-\frac{1}{2}$	$-\frac{1}{2}$	-2μ	$2\mu B$

Clearly, there are four microstates labeled 1, 2, 3 and 4. The number of macrostates is three : one macrostate is specified by the total magnetic moment 2μ and the total energy $-2\mu B$; the other two macrostates are characterized by 0, 0 and -2μ, $2\mu B$, respectively, the first quantity giving the total magnetic moment and the second one the total energy.

10. A system has two non-degenerate energy levels $E_1 = 0$ and $E_2 = 0.1$ eV. What is the temperature at which the probability of the system occupying the higher energy level is 0.25 ?

Ans. The partition function is

$$Z = e^{-E_1/k_BT} + e^{-E_2/k_BT} = e^0 + e^{-0.1 \times 1.6 \times 10^{-19}/k_BT}$$

$$= 1 + e^{-1.6 \times 10^{-20}/1.38 \times 10^{-23}T} = 1 + e^{-1159.4/T}$$

The probability of the system being in the higher energy level is

$$p = \frac{e^{-E_2/k_BT}}{Z} = 0.25$$

Basic Concepts   139

or, $\dfrac{e^{-1159.4/T}}{1+e^{-1159.4/T}} = \dfrac{1}{4}$ or, $1 - 3\, e^{-1159.4/T} = 0$

or, $e^{1159.4/T} = 3$ or, $\dfrac{1159.4}{T} = \ln 3$

or, $T = \dfrac{1159.4}{\ln 3} = 1055\text{K}$

11. A system has two energy levels of energy 0 and 100 k_B with degeneracies of 2 and 3, respectively. Determine the partition function and the average energy at a temperature of 100 K.

Ans. The partition function is

$$Z = 2e^0 + 3e^{-100k_B/100k_B} = 2 + 3e^{-1} = 3.104$$

The average energy is

$$\bar{E} = \dfrac{0 \times 2e^0 + 100k_B \times 3e^{-1}}{Z} = \dfrac{100 \times 1.38 \times 10^{-23} \times 1.104}{3.104}\,\text{J}$$

$$= \dfrac{49.08 \times 10^{-23}}{1.6 \times 10^{-19}}\,\text{eV} = 3.068\,\text{meV}$$

12. The single-particle partition function of a system of N distinguishable particles is $F = CVT^{3/2}$ where C is a constant. Calculate the internal energy and the pressure of the system.

Ans. The internal energy is

$$U = Nk_B T^2 \dfrac{d}{dT}(\ln F)$$

and the pressure is

$$P = Nk_B T \dfrac{\partial}{\partial V}(\ln F)$$

Now, $\ln F = \ln C + \ln V + \dfrac{3}{2}\ln T$

So, $U = Nk_B T^2 \cdot \dfrac{3}{2T} = \dfrac{3}{2}Nk_B T$

and $P = Nk_B T \dfrac{1}{V} = \left(\dfrac{N}{V}\right)k_B T.$

13. A system of particles occupying single-particle states and obeying MB statistics is in equilibrium at absolute temperature T. The population of the energy 2.3 meV is 63% and that for the energy 11.5 meV is 21%. What is the value of T?

Ans. If N_1 is the population of the energy $E_1 = 2.3$ meV and N_2 is that for the energy $E_2 = 11.5$ meV, we have

$$\dfrac{N_2}{N_1} = e^{-(E_2 - E_1)/k_B T} = \dfrac{21}{63} = \dfrac{1}{3}$$

So, $-\dfrac{E_2 - E_1}{k_B T} = \ln(1/3)$

or, $\quad -\dfrac{(11.5 - 2.3) \times 10^{-3} \times (1.6 \times 10^{-19})}{1.38 \times 10^{-23} \times T} = \ln(1/3)$

or, $\quad T = -\dfrac{9.2 \times 1.6}{0.138 \ln(1/3)} = 97.1 K$

14. A flux of 10^{12} neutrons/m² emerges each second from a port in a nuclear reactor. If these neutrons have a Maxwell-Boltzmann energy distribution corresponding to $T = 300$ K, calculate the density of neutrons in the beam. (Mass of a neutron = 1.67×10^{-27} kg). *(C.U. 2003)*

Ans. The number of neutrons per m³ with velocities between v_x and $v_x + dv_x$, v_y and $v_y + dv_y$, and v_z and $v_z + dv_z$ is $Ae^{-E/k_BT} dv_x\, dv_y\, dv_z$, where A = a constant. The number density of neutrons is

$$n = A \int_{-\infty}^{\infty} \int_{-\infty}^{\infty} \int_{-\infty}^{\infty} e^{-E/k_BT} dv_x\, dv_y\, dv_z \qquad ...(i)$$

where $E = \dfrac{1}{2} m (v_x^2 + v_y^2 + v_z^2)$, m being the neutron mass. The neutrons emerging each second through the port with velocities between v_x and $v_x + dv_x$, v_y and $v_y + dv_y$, and v_z and $v_z + dv_z$, will be $Ae^{-E/k_BT} v_x\, dv_x\, dv_y\, dv_z$. All the emerging neutrons will have v_x between 0 and ∞, v_y between $-\infty$ and $+\infty$, and v_z between $-\infty$ and $+\infty$. This number is

$$n_s = A \int_{0}^{\infty} \int_{-\infty}^{\infty} \int_{-\infty}^{\infty} e^{-E/k_BT} v_x\, dv_x\, dv_y\, dv_z \qquad ...(ii)$$

Substituting for A from (*i*) in (*ii*) gives

$$n_s = \dfrac{n}{2} \dfrac{\int_0^{\infty} e^{-mv_x^2/(2k_BT)} v_x\, dv_x}{\int_0^{\infty} e^{-mv_x^2/(2k_BT)} dv_x} \qquad ...(iii)$$

Let $\quad x = mv_x^2/(2k_BT)$, so $dx = \dfrac{m}{k_BT} v_x\, dv_x$.

So, $\quad dv_x = \dfrac{k_BT}{mv_x} dx = \dfrac{k_BT}{m} \left(\dfrac{m}{2k_BT}\right)^{1/2} x^{-1/2} dx = \left(\dfrac{k_BT}{2m}\right)^{1/2} x^{-1/2} dx.$

Hence (*iii*) yields

$$n_s = n \left(\dfrac{k_BT}{2m}\right)^{1/2} \dfrac{\int_0^{\infty} e^{-x} dx}{\int_0^{\infty} x^{-1/2} e^{-x} dx} = n \left(\dfrac{k_BT}{2\pi m}\right)^{1/2}$$

Thus $\quad n = n_s \left(\dfrac{2\pi m}{k_BT}\right)^{1/2} = 10^{12} \left(\dfrac{2\pi \times 1.67 \times 10^{-27}}{1.38 \times 10^{-23} \times 300}\right)^{1/2} = 1.59 \times 10^9$ m⁻³,

The density is $1.67 \times 10^{-27} \times 1.59 \times 10^9 = 2.66 \times 10^{-18}$ kg/m³.

15. A system of N particles obeying MB statistics possesses three energy levels $E_1 = 0$, $E_2 = \epsilon$, and $E_3 = 10\epsilon$. Find the temperature below which only the levels E_1 and E_2 are occupied. What is the average energy $\bar{E}$ of the system at temperature T? Find also the specific heat per mole, C_v.

Basic Concepts

Ans. If N_1, N_2, and N_3 are the number of particles populating the levels E_1, E_2, and E_3 respectively, we have $N_1 + N_2 + N_3 = N$. By MB statistics, $N_2/N_1 = \exp(-\epsilon/k_BT)$, and $N_3/N_1 = \exp(-10\epsilon/k_BT)$. Hence

$$N_1 = N_3 \exp(10\epsilon/k_BT) \text{ and } N_2 = N_1 \exp(-\epsilon/k_BT) = N_3 \exp(9\epsilon/k_BT). \text{ So,}$$

$$N_3 = \frac{N}{1 + \exp(9\epsilon/k_BT) + \exp(10\epsilon/k_BT)}$$

When $N_3 < 1$, the level E_3 is not occupied. Thus, if T_C is the temperature below which only E_1 and E_2 are populated, we have

$$N_3 = \frac{N}{1 + \exp(9\epsilon/k_BT_C) + \exp(10\epsilon/k_BT_C)} = 1$$

If $N \gg 1$, we obtain

$$N \simeq \exp(10\epsilon/k_BT_C) \quad \text{or,} \quad \ln N \simeq \frac{10\epsilon}{k_BT_C}$$

or,
$$T_C \simeq \frac{10\epsilon}{k_B \ln N}.$$

The average energy of the system of particles is

$$\overline{E} = \frac{E_1 N_1 + E_2 N_2 + E_3 N_3}{N_1 + N_2 + N_3} = \frac{\epsilon[\exp(-\epsilon/k_BT) + 10\exp(-10\epsilon/k_BT)]}{1 + \exp(-\epsilon/k_BT) + \exp(-10\epsilon/k_BT)}$$

The molar specific heat is

$$C_V = N_A \frac{\partial \overline{E}}{\partial T}$$

$$= R\epsilon^2 \beta^2 \frac{\exp(-\beta\epsilon) + 100\exp(-10\beta\epsilon) + 81\exp(-11\beta\epsilon)}{[1 + \exp(-\beta\epsilon) + \exp(-10\beta\epsilon)]^2}$$

where $\beta = 1/k_BT$, N_A is Avogadro's number, and $R (= N_A k_B)$ is the universal gas constant.

QUESTIONS

1. (a) What do you mean by macroscopic and microscopic systems? Give examples. **(C.U. 1996)**
   (b) Explain micro and macro states. **(C.U. 1999, 2005)**
   (c) Explain : accessible states, and the postulate of equal a priori probabilities.

2. (a) Distinguish between mathematical probability and thermodynamic probability.
   (b) What is phase space? Considering a free particle in a container of dimensions l_x, l_y, and l_z, show that the number of quantum states can be obtained by dividing the phase space into cells of volume h^3 where h is Planck's constant.
   (c) What do you mean by the phase trajectory for one-dimensional motion of a particle?

3. (a) Define the terms "density of states" and "distribution function". Comment on their usefulness in determining the properties of a system of particles.
   (b) Show that the member of ways in which N classical noninteracting particles can be distributed in energy states so as to assign N_i particles to the energy E_i is $N! \prod_i g_i^{N_i} / \prod_i N_i!$ where g_i is the degeneracy of the energy state ϵ_i. **(C.U. 1992)**

4. What is the relation between entropy and the probability of state? Establish this relationship using the additive property of entropy and the multiplication property of probability.

5. Write the form of the Maxwell-Boltzmann distribution function. Using this function, obtain the law of equipartition of energy for a system of ideal **gas particles**. What is "partition function"?

6. (a) Obtain an expression for the density of states of a two-dimensional system of free electrons contained in an enclosure of area A and dimensions L_x and L_y. [Ans. $4\pi \, Am/h^2$]

   (b) Consider a free particle inside a three-dimensional box of side L with energy between E and $E + dE$. Calculate the number of microstates. (C.U. 2008)

7. Derive an expression for the average thermal speed of a particle of mass m in an ideal Boltzmann gas at a temperature T. What is the value of this speed for an electron system obeying MB statistics at 300 K? [Ans. $\sqrt{8k_B T/\pi m}$, 1.08×10^5 m/s]

8. Obtain the root-mean-square speed and the most probable speed of a particle in an ideal Boltzmann gas. [Ans. $\sqrt{3k_B T/m}$, $\sqrt{2k_B T/m}$]

9. For a system of gas particles obeying MB distribution function, show that $v_{rms} > \bar{v} > v_{mp}$, where v_{rms} is the rms speed, $\bar{v}$ is the average speed, and v_{mp} is the most probable speed of a particle. Also, plot the velocity distribution curve and show how the curve is altered as the temperature increases.

10. Starting from the MB velocity distribution of a system of gas particles, find the number of particles having kinetic energy between E and $E + dE$. How does the energy distribution curve differ from the velocity distribution curve?

    [Ans. $N(E)\, dE = \dfrac{2N}{\sqrt{\pi}} \cdot \dfrac{E^{1/2}}{(k_B T)^{3/2}} \cdot e^{-E/k_B T}\, dE$. The energy distribution curve, *i.e.*, the plot of $N(E)$ versus E, initially rises with E as $E^{1/2}$, attains a maximum, and then falls. But the velocity distribution curve, *i.e.*, the plot of $N(v)$ versus v initially rises with v as v^2, attains a maximum, and then falls.]

11. What are the limitations of MB statistics?

12. Consider a system of 8 nonineracting distinguishable particles distributed over two nondegenerate energy states $+E$ and $-E$. What is the maximum entropy of the system? Find also the corresponding total energy of the distribution. (cf. Burd. U. 1993) [Ans. 5.86×10^{-23} J/K, 0]

13. N distinguishable particles are distributed in two energy states 0 and E. Find the heat capacity of the system.

    $$\text{Ans. } \dfrac{NE^2 \exp(-E/k_B T)}{k_B T^2 [1 + \exp(-E/k_B T)]^2}$$

14. (a) A linear harmonic oscillator moves with a constant energy along the x-axis. What will be the phase trajectory? (C.U. 1998)

    (b) How is the phase trajectory modified if the oscillations are damped?

    [Hint. Let m, ω, and a be the mass, the angular frequency, and the amplitude of the oscillator. If p_x be the linear momentum at position x, show that $p_x^2/(m^2 \omega^2 a^2) + x^2/a^2 = 1$. Hence $p_x - x$ plot, *i.e.*, the phase trajectory is an ellipse. If the oscillations are damped, the energy decreases with time, and so the ellipse will gradually shrink and spiral towards the origin.]

15. A particle of mass m is thrown vertically upwards from the surface of the earth with an energy E, so that it comes back to its starting point after describing the vertical path. If g is the acceleration due to gravity and p_x is the momentum of the particle at a height x, obtain the phase trajectory.

    [Ans. A portion of the parabola $p_x^2 = 2mE - 2m^2 gx$]

16. A system of distinguishable particles has two nondegenerate single-particle energy states 0 and E. At what temperature would the probability of a particle occupying the excited state be half that of occupying the ground state? (C.U. 1999) [Ans. $E/k_B \ln 2$]

17. Give the expression for the partition function for a system of distinguishable particles distributed in the three nondegenerate states having energies 0, E, and $3E$. The system is in thermal equilibrium at temperature T, and the degeneracy of each state is g.

    [Ans. $g(1 + e^{-E/k_B T} + e^{-3E/k_B T})$]

Basic Concepts

18. A classical particle of mass m is free to move in a cube of side l. If its energy is less then or equal to E, find the volume of the phase space avialable to it. **(C. U. 2001)**

$$\left[\text{Ans. } \frac{8\sqrt{2}}{3} \pi l^3 (mE)^{3/2} \right]$$

19. A system consists of three spin $\frac{1}{2}$ particles, fixed in space, and placed in an external magnetic field $\vec{B}$. Each particle has a magnetic moment μ which can be either parallel or anti-parallel to $\vec{B}$. Find the number of microstates and macrostates of the system. **[Ans. 8, 4]**

20. A system has three levels of energy 0, 50 k_B, and 100 k_B with degeneracies of 1, 2, and 3, respectively. Find the relative population of each level, and the mean energy at a temperature of 300 K.

[Ans. 0.206; 0.350, 0.444, 61.5 k_B]

21. The single-particle partition function for a system of N gas particles obeying MB statistics is F. Derive the relations expressing the pressure and the internal energy of the gas in terms of F.

22. (a) State the equipartition theorem, and applying it determine the molar specific heat of monatomic, diatomic, and triatompric gases.

    (b) When is the equipartition theorem not applicable ?

23. What is Gibbs paradox? Why does it arise and how is it avoided?

24. Using the Sackur-Tetrode formula for entropy derive expressions for the Helmholtz free energy and the chemical potential of an ideal gas.

25. Consider N spin – 1/2 particles, each with magnetic moment μ, in an external magnetic field B such that N_1 particles are parallel and $N_2 (= N - N_1)$ particles are antiparallel to the field. If E is the energy of the system of particles, show that the possible number of configurations is

$$\frac{N!}{\left[\frac{N + E/(\mu B)}{2}\right]! \left[\frac{N - E/(\mu B)}{2}\right]!} \qquad \text{(C.U. 2007)}$$

[**Hint.** Note that $E = \mu B (N_2 - N_1) = \mu B (N - 2N_1)$. Hence find N_1, and then $N_2 (= N - N_1)$. Put these values in $W = \dfrac{N!}{N_1! N_2!}$ where W is the desired number of configurations.]

Chapter 2

Fermi-Dirac and Bose-Einstein Statistics

2.1. FERMI-DIRAC STATISTICS

The Maxwell-Boltzmann distribution function, obtained in Sec. 1.4, suffers from the following two major objections :

First, the particles are assumed to be identifiable although in actual practice electrons or other elementary particles are indistinguishable. Second, any number of particles was allowed to occupy the same quantum state while many particles, particularly electrons, obey Pauli's exclusion principle which does not allow a quantum state to accept more than one particle.

These objections are removed in the distribution function obtained by Fermi and Dirac. This distribution function, known as the *Fermi-Dirac* (abbreviated *FD*) distribution function, gives the statistical behaviour of free electrons in metals and heavily doped semiconductors. Many electrical and thermal properties of solids which the classical statistics failed to explain, could be understood on the basis of the FD statistics. The particles that obey the FD statistics are sometimes called *fermions*. We shall derive below the Fermi-Dirac distribution function.

Consider a system of N *indistinguishable noninteracting* particles obeying the *Pauli exclusion principle*. Let $N_1, N_2, ..., N_n$ be the number of particles in the system with energies $E_1, E_2, ..., E_n$ respectively, and let g_i represent the multiplicity or the degeneracy of the energy level E_i. Since the particles are all indistinguishable, there is only one way of choosing $N_1, N_2, ..., N_n$ particles from the assembly of N particles. The number of ways in which N_i particles are arranged in g_i quantum states (or the number of *microstates* which are equally *a priori* probable) is given by

$$^{g_i}C_{N_i} = \frac{g_i!}{N_i!(g_i - N_i)!} \qquad ...(2.1)$$

Equation (2.1) results from the fact that the particles are indistinguishable and that each quantum state can accommodate only one particle in accordance with Pauli's exclusion principle.

The total number of ways W of distributing $N_1, N_2, ... N_n$ particles in n energy levels is the product of the terms given by Eq. (2.1) over all the levels, *i.e.*

$$W = \prod_{i=1}^{n} \frac{g_i!}{N_i!(g_i - N_i)!} \qquad ...(2.2)$$

where Π denotes the product. We assume that $g_i \gg 1$, $N_i \gg 1$ and $(g_i - N_i) \gg 1$. Application of Stirling's theorem in Eq. (2.2) yields

$$\ln W = \sum_{i=1}^{n} [g_i (\ln g_i - 1) - N_i (\ln N_i - 1) - (g_i - N_i)\{\ln (g_i - N_i) - 1\}]$$

$$= \sum_{i=1}^{n} [g_i \ln g_i - N_i \ln N_i - (g_i - N_i) \ln (g_i - N_i)], \qquad ...(2.3)$$

Since the system is in equilibrium, to find the most probable distribution, W or $\ln W$ must be *maximised* subject to the restrictions that the total number of particles N and the total energy U are constants, *i.e.*

$$N = \sum_{i=1}^{n} N_i = \text{constant}, \qquad \qquad ...(2.4)$$

and
$$U = \sum_{i=1}^{n} N_i E_i = \text{constant}, \qquad \qquad ...(2.5)$$

As in Sec. 1.3 we apply Lagrange's method of undetermined multipliers to obtain from Eqs. (2.3), (2.4) and (2.5)

$$\sum_{i=1}^{n} [\ln(g_i - N_i) - \ln N_i + \alpha + \beta E_i] \, \delta N_i = 0,$$

or
$$\ln(g_i - N_i) - \ln N_i + \alpha + \beta E_i = 0, \qquad \qquad ...(2.6)$$

as the variations are arbitrary. Here α and β are constants. Equation (2.6) readily gives

$$f(E_i) = \frac{N_i}{g_i} = \frac{1}{1 + e^{-\alpha - \beta E_i}}, \qquad \qquad ...(2.7)$$

which is the FD distribution function.

As in Sec. 1.4, β may be shown to be given by

$$\beta = -\frac{1}{k_B T}, \qquad \qquad ...(2.8)$$

where k_B is the Boltzmann constant and T is the absolute temperature. The quantity α is usually expressed in the form

$$\alpha = \frac{E_F}{k_B T}, \qquad \qquad ...(2.9)$$

where E_F is termed the *Fermi energy* or the *Fermi level* of the system. The FD distribution function then takes the form

$$f(E_i) = \frac{1}{1 + e^{(E_i - E_F)/k_B T}} \qquad \qquad ...(2.10)$$

Since only one particle may occupy a quantum state, $f(E_i)$ for FD statistics is the *probability* that a quantum state of energy E_i is occupied. The plot of $f(E_i)$ against E_i/E_F is shown in Fig. 2.1. At $T = 0$, $f(E_i) = 1$ for $E_i < E_F$ and $f(E_i) = 0$ for $E_i > E_F$. Thus at absolute zero of temperature, $f(E_i)$ is a step function, as shown in Fig. 2.1. Here the probability of occupation of all states with energies less than E_F is unity, and that of states with energies higher than E_F is zero. Thus *at absolute zero of temperature, the Fermi level represents the highest occupied energy level.*

At $T > 0$, $f(E_i)$ is close to unity for $E_i \ll E_F$ and approaches zero for $E_i \gg E_F$. The variations of $f(E_i)$ for two different temperatures T_1 and T_2 are shown in Fig. 2.1. If the temperature is not very large, $f(E_i)$ varies rapidly from about unity to about zero over an energy range of a few times $k_B T$ around E_F. At a nonzero temperature, Eq. (2.10) shows that $f(E_i) = 1/2$ at $E_F = E_i$. Thus *the Fermi level is that energy level for which the probability of occupation at $T > 0$ is 1/2.*

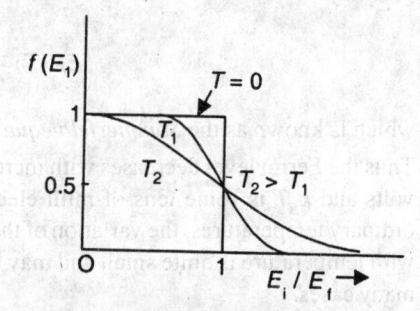

Fig. 2.1. Plot of the Fermi distribution function $f(E_i)$ against normalised energy (E_i/E_F) at three different temperatures.

At low temperatures, when the $f(E_i)$ is the nearly a step function, the distribution function is said to be strongly *degenerate*. At very high temperatures, when the step-like character is lost, it is said to be nearly *nondegenerate*.

Determination of the Fermi level

We shall now apply the FD statistics to an electron gas and find the Fermi level of the system. When the energy levels are closely spaced, the number of particles with energies between E and $E + dE$ is given by Eq. (1.47). Using the density of states function represented by Eq. (1.48) and the FD form for the distribution function $f(E)$ we get from Eq. (1.47)

$$N(E)dE = f(E)g(E)dE = \frac{8\sqrt{2}\pi V m^{3/2}}{h^3} \cdot \frac{E^{1/2}dE}{1 + e^{(E - E_F)/k_B T}} \qquad \ldots(2.11)$$

The total number of electrons is given by

$$N = \int_0^\infty N(E)dE = \int_0^\infty f(E)g(E)dE$$

$$= \frac{8\sqrt{2}\pi V m^{3/2}}{h^3} \int_0^\infty \frac{E^{1/2}dE}{1 + e^{(E - E_F)/k_B T}} \qquad \ldots(2.12)$$

Equation (2.12) is used to determine the Fermi level from known values of N or N/V, the number of particles per unit volume. If $T \neq 0$, the integral in Eq. (2.12) cannot be evaluated analytically and numerical methods have to be applied. For a given concentration of particles, satisfaction of Eq. (2.12) requires that E_F must be a function of temperature.

At $T = 0$, however, using the step-like property of $f(E)$ a simple expression for $E_F(0)$, the Fermi level at absolute zero of temperature, can be found from Eq. (2.12). Since at $T = 0$, the highest occupied energy is $E_F(0)$ and $f(E) = 1$ for $E < E_F(0)$ and $f(E) = 0$ for $E > E_F(0)$, we obtain from Eq. (2.12)

$$N = \frac{8\sqrt{2}\pi V m^{3/2}}{h^3} \int_0^{E_F(0)} E^{1/2} dE = \frac{16\sqrt{2}\pi V m^{3/2}}{3h^3} [E_F(0)]^{3/2},$$

or

$$E_F(0) = \frac{h^2}{8m} \left(\frac{3N}{\pi V} \right)^{2/3} \qquad \ldots(2.13)$$

It will be shown in Sec. 2.2 that at temperatures for which $k_B T \ll E_F$ the variation of E_F with T is given by

$$E_F(T) = E_F(0) \left[1 - \frac{\pi^2}{12} \frac{(k_B T)^2}{E_F^2(0)} \right] \qquad \ldots(2.14)$$

which is known as the *Sommerfeld equation*.

Thus the Fermi level decreases with increasing temperature. For metals, since $E_F(0)$ is a few electron volts and $k_B T$ is some tens of milli electron volts at ordinary temperatures, the variation of the Fermi level with temperature is quite small and may be ignored in many cases.

Some features of the FD statistics

The distribution of the number of particles $N(E)$ as a function of energy for a three dimensional electron gas obeying FD statistics, as given by Eq. (2.11), is shown in Fig. 2.2. Since $N(E)$ is the product of $g(E)$ and $f(E)$, and $g(E)$ varies as $E^{1/2}$, the step-like behaviour of $f(E)$ at $T = 0$ shows that at this temperature $N(E)$ rises parabolically from zero at $E = 0$ and

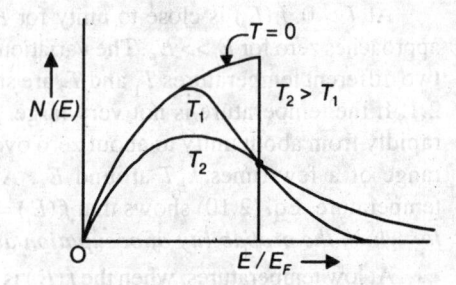

Fig. 2.2. Plot of $N(E)$ against E/E_F at three different temperatures.

falls abruptly to zero as E increases above E_F. As the temperature increases, the abrupt fall of $N(E)$ is smoothened out.

It is instructive to find the internal energy U of an electron gas obeying FD statistics at 0 K. We have

$$U = \int_0^{E_F(0)} EN(E)\,dE = \frac{8\sqrt{2}\pi V m^{3/2}}{h^3} \int_0^{E_F(0)} E^{3/2} f(E)\,dE$$

$$= \frac{8\sqrt{2}\pi V m^{3/2}}{h^3} \cdot \frac{2}{5} \left[E_F(0)\right]^{5/2} = \frac{3}{5} N E_F(0) \qquad \ldots(2.15)$$

where Eq. (2.13) has been used. The internal energy per particle is thus $(3/5) E_F(0)$. Note that classically the internal energy per particle at temperature T is $(3/2) k_B T$.

When the energy is much larger than the Fermi energy, the term unity in the denominator of Eq. (2.10) may be neglected compared to the exponential term, so that the distribution function reduces to

$$f(E) \cong e^{E_F/k_B T}\, e^{-E/k_B T} \qquad \ldots(2.16)$$

where E_i has been replaced by E. If

$$E - E_F \gg k_B T, \qquad \ldots(2.17)$$

i.e., if E_F is many $k_B T$ units less than any energy of a particle in the system, then Eq. (2.16) may be applied. Note that (2.16) is the MB distribution function of Eq. (1.53) with $\alpha = E_F / k_B T$. Thus when (2.17) is satisfied, the FD distribution function can be approximated by the MB function. In other words, the 'exponential tail' of the FD distribution function is the MB function.

As the temperature is increased, the Fermi level decreases and may become negative at sufficiently high temperatures. Since the minimum value of E is zero, (2.17) is clearly satisfied at high temperatures whereby the FD and the MB distribution functions become virtually the same.

The reduction of the FD distribution to the MB distribution at high temperatures can be physically understood as follows. When T is large, the particles are thermally excited to a great extent and are distributed over a wide range of energy values. The number of particles for any energy value is then much smaller than the number of available quantum states. Therefore, it is unlikely that more than one particle will occupy the same quantum state, and the Pauli exclusion principle becomes unimportant. The indistinguishability of the particles also loses its significance; indeed it is found that when $g_i \gg N_i$, Eqs. (2.6) and (1.30) are nearly the same. Consequently there is practically no difference between the MB and the FD distributions.

Equation (2.13) shows that $E_F(0)$ is inversely proportional to the particle mass m and directly proportional to the two-thirds power of the particle number density N/V. Therefore, $E_F(0)$ will be much smaller than $k_B T$ at a lower temperature for a dilute gas of heavy particles. This shows why at normal temperatures ordinary gaseous substances follow the MB statistics instead of the FD statistics. For a dense gas of light particles, such as the free electrons in metals, $E_F(0)$ is very large and is never much less than $k_B T$ for physically attainable temperatures. Such systems therefore obey the FD statistics. In semiconductors, when the doping is not large, the particle number density is small and the system may be treated classically using the MB distribution function.

The quantity $\exp(\alpha) = \exp(E_F / k_B T)$ is called the *degeneracy parameter*, D. Since α can have any value between $-\infty$ and $+\infty$, the degeneracy parameter lies between 0 and ∞. If $0 < D \le \exp(1)$, the system is said to be nondegenerate. On the other hand, if $\exp(1) \le D < \infty$, the system is degenerate. The transition from nondegeneracy to degeneracy is roughly marked by $D = \exp(1)$.

Among the various useful applications of the FD statistics in metals are the evaluation of the electronic contribution to the specific heat and the derivation of Richardson's equation for the thermionic emission current. We shall consider the former in the following section. The latter is discussed elsewhere*.

2.2. Specific Heat Of Conduction Electrons In Metals

The heat capacity of a metal is the sum of the contributions due to the lattice and the free electrons. The lattice contribution is given by the Debye theory [See Sec. 5.3, Solid State Physics]. If the free electrons are assumed to follow the MB statistics, as is done in the classical theory of Drude and Lorentz, the electronic contribution to the heat capacity would be $(3/2) Nk_B$. Here N is the number of atoms in the metal and it is assumed that there is one free electron per atom. Experimental results show that the high-temperature value of the heat capacity is almost equal to $3 Nk_B$ which, according to the Debye theory, is the lattice contribution alone. The electronic contribution should therefore be much smaller than that predicted by the classical theory. Application of the FD statistics leads to this desired result and gives an excellent agreement between theory and experiment.

In order to estimate the heat capacity of free electrons obeying the FD statistics, the first-order dependence of the Fermi energy on temperature must be known. To accomplish this, we introduce a function $\lambda(E)$ such that $\lambda(0) = 0$, and evaluate the integral

$$I = \int_0^\infty f(E) \frac{\partial \lambda(E)}{\partial E} dE, \qquad \ldots(2.18)$$

where $f(E)$ is the FD distribution function. Integrating by parts, we obtain

$$I = [f(E)\lambda(E)]_0^\infty - \int_0^\infty \lambda(E) \frac{\partial f}{\partial E} dE$$

$$= -\int_0^\infty \lambda(E) \frac{\partial f}{\partial E} dE. \qquad \ldots(2.19)$$

Expanding $\lambda(E)$ in a Taylor's series about $E = E_F$, we get

$$\lambda(E) = \lambda(E_F) + (E - E_F)\left(\frac{\partial \lambda}{\partial E}\right)_{E_F} + \frac{1}{2}(E - E_F)^2 \left(\frac{\partial^2 \lambda}{\partial E^2}\right)_{E_F} + \ldots \qquad \ldots(2.20)$$

Writing

$$A_i = -\frac{1}{i!} \int_0^\infty (E - E_F)^i \frac{\partial f}{\partial E} dE \qquad \ldots(2.21)$$

Equation (2.19) can be put in the form

$$I = A_0 \lambda(E_F) + A_1 \left(\frac{\partial \lambda}{\partial E}\right)_{E_F} + A_2 \left(\frac{\partial^2 \lambda}{\partial E^2}\right)_{E_F} + \ldots \qquad \ldots(2.22)$$

It is easy to show that

$$\frac{\partial f}{\partial E} = -\frac{f(1-f)}{k_B T} \qquad \ldots(2.23)$$

---

* "Electronics : Fundamentals and Applications" by D. Chattopadhyay and P.C. Rakshit (New Age International, New Delhi).

Fermi-Dirac and Bose-Einstein Statistics

Figure 2.2 A shows the plot of $-(\partial f/\partial E)$ versus E. For convenience f is also plotted against E in Fig. 2.2A. $\partial f/\partial E$ is significant for energies a few $k_B T$ about $E = E_F$ and is zero for energies far removed from E_F since either f or $(1-f)$ will be zero for such energies. If $E_F \gg k_B T$, the lower limit of the integral in Eq. (2.21) may therefore be shifted to $-\infty$. Then we obtain

$$A_0 = -\int_{-\infty}^{\infty} \frac{\partial f}{\partial E} dE = [-f]_{-\infty}^{\infty} = 1. \qquad \ldots(2.24)$$

$$A_1 = -\int_{-\infty}^{\infty} (E - E_F)\frac{\partial f}{\partial E} dE. \qquad \ldots(2.25)$$

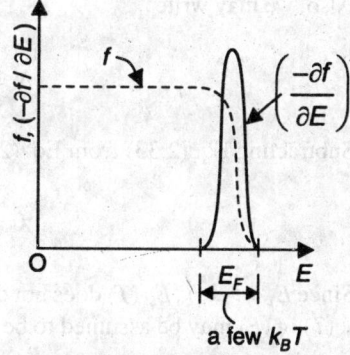

Fig. 2.2 A. Plots of $-(\partial f/\partial E)$ (solid curve) and f (broken curve) against E for $T \neq 0$.

Now $\partial f/\partial E$ is an even function of $(E - E_F)$ so that the integrand is an odd function of $(E - E_F)$. The integral must therefore vanish, giving

$$A_1 = 0 \qquad \ldots(2.26)$$

Again,

$$A_2 = -\frac{1}{2}\int_{-\infty}^{\infty} (E - E_F)^2 \frac{\partial f}{\partial E} dE$$

$$= \frac{(k_B T)^2}{2} \int_{-\infty}^{\infty} \frac{x^2 e^x}{(1+e^x)^2} dx, \qquad \ldots(2.27)$$

where we have put $x = (E - E_F)/k_B T$. From standard tables the value of the integral in Eq. (2.27) is found to be $\pi^2/3$, whence

$$A_2 = \frac{\pi^2}{6}(k_B T)^2 \qquad \ldots(2.28)$$

Substitution of the values of A_0, A_1 and A_2 from Eqs. (2.24), (2.26) and (2.28) respectively into Eq. (2.22) gives to the first order

$$I = \int_0^{\infty} f(E) \frac{\partial \lambda}{\partial E} dE$$

$$= \lambda(E_F) + \frac{\pi^2}{6}(k_B T)^2 \left(\frac{\partial^2 \lambda}{\partial E^2}\right)_{E_F} \qquad \ldots(2.29)$$

Let us take

$$\lambda(E) = \int_0^E g(E) dE, \qquad \ldots(2.30)$$

where $g(E)$ is the density of states function:

$$g(E) dE = \frac{8\sqrt{2}\pi V m^{3/2}}{h^3} E^{1/2} dE. \qquad \ldots(2.31)$$

Then, using Eq. (2.29) we get

$$N = \int_0^{\infty} f(E) g(E) dE$$

$$= \int_0^{E_F(T)} g(E) dE + \frac{\pi^2}{6}(k_B T)^2 \left(\frac{\partial g}{\partial E}\right)_{E_F} \qquad \ldots(2.32)$$

Also, we may write

$$N = \int_0^{E_F(0)} g(E)\,dE. \qquad ...(2.33)$$

Subtracting Eq. (2.33) from Eq. (2.32) we obtain

$$0 = \int_{E_F(0)}^{E_F(T)} g(E)\,dE + \frac{\pi^2}{6}(k_B T)^2 \left(\frac{\partial g}{\partial E}\right)_{E_F} \qquad ...(2.34)$$

Since $E_F \gg k_B T$, $E_F(T)$ does not differ much from $E_F(0)$, so that in the small range from $E_F(0)$ to $E_F(T)$, $g(E)$ may be assumed to be a constant. Equation (2.34) therefore yields

$$g(E_F)[E_F(0) - E_F(T)] = \frac{\pi^2}{6}(k_B T)^2 \left(\frac{\partial g}{\partial E}\right)_{E_F}. \qquad ...(2.35)$$

Using Eq. (2.31), we get from Eq. (2.35)

$$E_F(T) = E_F(0) - \frac{\pi^2}{12} \frac{(k_B T)^2}{E_F(T)} \qquad ....(2.36)$$

Since $E_F(T)$ is nearly equal to $E_F(0)$, the quantity $E_F(T)$ in the second term on the right-hand side of Eq. (2.36) may be replaced by $E_F(0)$, giving

$$E_F(T) = E_F(0)\left[1 - \frac{\pi^2}{12}\left(\frac{k_B T}{E_F(0)}\right)^2\right]. \qquad ...(2.37)$$

which is the *Sommerfeld equation*.

To determine the electronic heat capacity, we take

$$\lambda(E) = \int_0^E E g(E)\,dE. \qquad ...(2.38)$$

Then

$$\frac{\partial \lambda}{\partial E} = E g(E)$$

and

$$\frac{\partial^2 \lambda}{\partial E^2} = \frac{\partial}{\partial E}[E g(E)] \qquad ...(2.39)$$

Using these values in Eq. (2.29) we get

$$\int_0^\infty f(E)\frac{\partial \lambda}{\partial E}\,dE = \int_0^\infty E g(E) f(E)\,dE = U, \qquad ...(2.40)$$

and

$$U = \int_0^{E_F(T)} E g(E)\,dE + \frac{\pi^2}{6}(k_B T)^2 \left[\frac{\partial}{\partial E}\{E g(E)\}\right]_{E_F} \qquad ...(2.41)$$

where U is the internal energy. With the aid of Eq. (2.31) we obtain from Eq. (2.41)

$$U = \int_0^{E_F(0)} E g(E)\,dE + \int_{E_F(0)}^{E_F(T)} E g(E)\,dE + \frac{\pi^2}{6}(k_B T)^2 \cdot \frac{3}{2} g(E_F)$$

$$\approx U_0 + [E_F(T) - E_F(0)] E_F(0) g[E_F(0)] + \frac{\pi^2}{4}(k_B T)^2 g[E_F(0)] \qquad ...(2.42)$$

where U_0 is the internal energy at absolute zero of temperature. In deriving Eq. (2.42), the approximations already indicated in connection with Eq. (2.35) have been used. Substituting for $E_F(T)$ from Eq. (2.37) into Eq. (2.42) we get

$$U = U_0 + \frac{\pi^2}{6} (k_B T)^2 g[E_F(0)] \qquad ...(2.43)$$

From Eqs. (2.31) and (2.13) we get

$$g[E_F(0)] = \frac{3N}{2E_F(0)} = \frac{3N}{2k_B T_F}, \qquad ...(2.44)$$

where T_F is the *Fermi temperature* given by $k_B T_F = E_F(0)$.

Using Eq. (2.44), we obtain from Eq. (2.43)

$$U = U_0 + \frac{\pi^2}{4} \frac{N k_B T^2}{T_F} \qquad ...(2.45)$$

The electronic heat capacity at constant volume is given by

$$C_V^{(e)} = \frac{\partial U}{\partial T} = \frac{\pi^2}{2} N k_B \left(\frac{T}{T_F} \right) \qquad ...(2.46)$$

For $E_F(0) \approx 5$ eV one finds that at $T = 300$ K, $C_V^{(e)} \approx N k_B / 40$, which is a small fraction of the classical value of $3 N k_B / 2$. The use of the FD statistics thus removes the specific heat difficulty of the classical theory. The physical reason behind the much smaller heat capacity of the FD free-electron system is the following : Thermal excitation of electrons takes place only if they find unoccupied states within a few times $k_B T$ in energy. This is possible only for the electrons in the vicinity of the Fermi level. The electrons with energies much less than the Fermi energy do not find such unoccupied states and hence cannot be thermally excited. Thus only a small fraction of the total number of electrons effectively contributes to the heat capacity.

Equation (2.46) shows that the electronic specific heat rises linearly with T. At low temperatures, the specific heat due to the lattice vibrations is proportional to T^3. The total specific heat of a metal at low temperatures should therefore be given by

$$C_V = AT + BT^3, \qquad ...(2.47)$$

where A and B are constants. At sufficiently low temperatures, the first term predominates. Thus the electronic specific heat can be determined from low temperature experiments on metals. From Eq. (2.47) we write

$$\frac{C_V}{T} = A + BT^2 \qquad ...(2.48)$$

A plot of C_V/T against T^2 should therefore be a straight line whose intercept on the vertical axis gives the constant A (Fig. 2.3). The experimental data give such plots, which shows that the temperature dependences given by (2.47) are correct.

2.3. BOSE-EINSTEIN STATISTICS

The Bose-Einstein (abbreviated BE) statistics gives the statistical behaviour of an ensemble of indistinguishable particles which do not obey the Pauli exclusion principle. Phonons and photons may be treated with the help of the BE statistics. The particles that obey the BE statistics are sometimes called *bosons*.

The problem here is to distribute N_i identical particles among the g_i quantum states of the

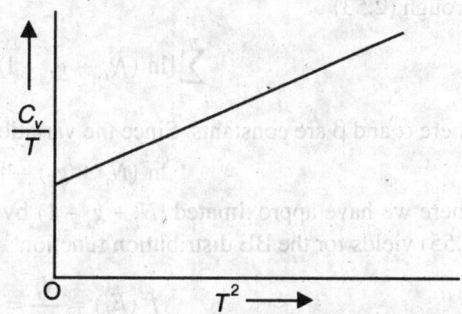

Fig. 2.3. Plot of C_V/T against T^2 at low temperatures.

energy level E_i keeping in mind that there is no restriction regarding the number of particles that may occupy a quantum state.

Let us consider the following array :

$$|\cdot\cdot|\cdot|\cdot\cdot\cdot\cdot| \quad |\cdot\cdot\cdot|\cdot|\cdot\cdot\cdot|$$

where a line denotes a quantum state, and the particles contained in a quantum state are shown by points on the right of the quantum state. In the array there are $g_i + N_i$ objects consisting of the quantum states and the particles. Keeping the first quantum state, *i.e.* the first line fixed in its place, the remaining $(g_i + N_i - 1)$ objects can be arranged in $(g_i + N_i - 1)!$ ways. However, among these the number of ways of permuting the quantum states among themselves, *i.e.* $(g_i - 1)!$, and the number of ways of permuting the particles among themselves, *i.e.* $N_i!$, cannot be regarded as independent arrangements. Therefore, the number of ways in which N_i particles can be distributed in g_i states (or the number of *microstates* which are equally *a priori* probable) is given by

$$A = \frac{(N_i + g_i - 1)!}{N_i!(g_i - 1)!} \qquad ...(2.49)$$

The total number of ways W of distributing $N_1, N_2, ..., N_n$ particles in n energy levels is the product of the terms represented by (2.49) over all the levels, *i.e.*

$$W = \prod_{i=1}^{n} \frac{(N_i + g_i - 1)!}{N_i!(g_i - 1)!} \qquad ....(2.50)$$

Applying Stirling's theorem for large enough N_i and g_i in Eq. (2.50) we get

$$\ln W = \sum_{i=1}^{n} [(N_i + g_i - 1)\{\ln(N_i + g_i - 1) - 1\} - N_i(\ln N_i - 1)$$

$$- (g_i - 1)\{\ln(g_i - 1) - 1\}]$$

$$= \sum_{i=1}^{n} [(N_i + g_i - 1)\ln(N_i + g_i - 1) - N_i \ln N_i$$

$$- (g_i - 1)\ln(g_i - 1)] \qquad ...(2.51)$$

As the system is in equilibrium, for the most probable distribution, W or $\ln W$ has to be maximised under the conditions that the total number of particles N and the total energy U are constants, *i.e.*

$$N = \sum_{i=1}^{n} N_i = \text{constant}, \qquad ...(2.52)$$

and

$$U = \sum_{i=1}^{n} N_i E_i = \text{constant}, \qquad ...(2.53)$$

As before, we apply Lagrange's method of undetermined multipliers and obtain from Eqs. (2.51) through (2.53)

$$\sum_{i=1}^{n} [\ln(N_i + g_i - 1) - \ln N_i + \alpha + \beta E_i]\delta N_i = 0, \qquad ...(2.54)$$

where α and β are constants. Since the variations are arbitrary, we obtain from Eq. (2.54)

$$\ln(N_i + g_i) - \ln N_i + \alpha + \beta E_i = 0 \qquad ...(2.55)$$

where we have approximated $(N_i + g_i - 1)$ by $(N_i + g_i)$ assuming large enough $(N_i + g_i)$. Equation (2.55) yields for the BE distribution function

$$f(E_i) = \frac{N_i}{g_i} = \frac{1}{e^{-\alpha - \beta E_i} - 1} \qquad ...(2.56)$$

Fermi-Dirac and Bose-Einstein Statistics

At very high temperatures, the particles are distributed over a wide energy range so that the number of particles in any energy range would be much smaller than the number of quantum states in that range. That is, $g_i \gg N_i$. In this case, Eq. (2.55) reduces to

$$\ln g_i - \ln N_i + \alpha + \beta E_i = 0 \qquad \text{...(2.57)}$$

which is the same as Eq. (1.30) for a Maxwell-Boltzmann system. Thus at high temperatures the BE distribution function reduces to the MB distribution function given by

$$f(E_i) = e^{\alpha + \beta E_i} \qquad \text{...(2.58)}$$

It is then easily appreciated that the quantity β must be identified as

$$\beta = -\frac{1}{k_B T} \qquad \text{...(2.59)}$$

Equation (2.56) is now written as

$$f(E_i) = \frac{1}{e^{-\alpha} e^{E_i/k_B T} - 1} \qquad \text{...(2.60)}$$

The quantity α can be determined in terms of the number of particles in the system as in the case of MB and FD systems. It is seen that at high temperatures α must be a very large negative number in order that Eq. (2.60) may reduce to the form of Eq. (2.58). As the temperature approaches zero, $\alpha \to 0$ and Eq. (2.60) reduces to

$$f(E_i) = \frac{1}{e^{E_i/k_B T} - 1} \qquad \text{...(2.61)}$$

It is readily seen from Eq. (2.61) that at the absolute zero of temperature, the particles tend to occupy the lowest energy state. This phenomenon is known as the *Bose condensation*. It is believed that Bose condensation is responsible for the superfluid state of liquid helium.

In some cases, the number of particles in the system does not remain constant. In this situation $\alpha = 0$ and the BE distribution function is also given by (2.61). It is noted from (2.61) that if $E_i \gg k_B T$, the BE distribution reduces to the MB distribution. The function $f(E_i)$, as given by Eq. (2.61), is plotted against E_i for two different temperatures in Fig. 2.4.

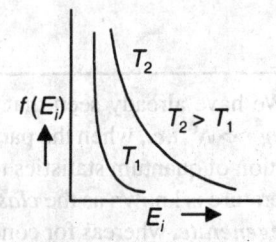

Fig. 2.4. Plot of $f(E_i)$ versus E_i at two different temperatures T_1 and T_2.

Comparison of MB, BE, and FD distribution functions

We have already discussed in detail the Maxwell-Boltzmann, Bose-Einstein, and the Fermi-Dirac systems. For ready reference, the basic postulates of the three statistics are listed in Table 2.1. The salient features of the three distribution functions are given in Table 2.2.

Table 2.1 : Postulates of MB, BE, and FD statistics

MB	BE	FD
(1) The particles of the system in equilibrium are distinguishable.	The particles of the system in equilibrium are indistinguishable.	The Particles of the system in equilibrium are indistinguishable.
(2) Pauli's exclusion principle does not apply.	Pauli's exclusion principle is not obeyed.	Pauli's exclusion principle applies.

(3) The system is isolated, so that the total number of particles is a constatnt. That is, $\Sigma \, \delta N_i = 0$	If the system is isolated, so that the total number of particles is a constant, we have $\Sigma \, \delta N_i = 0$	The system being isolated, the total number of particles is a constant, i.e., $\Sigma \, \delta N_i = 0$
(4) The particles are noninteracting, so that the total energy of the system is a constant. That is, $\Sigma E_i \, \delta N_i = 0$	The particles are noninteracting, so that the total energy of the system is a constant. So, $\Sigma E_i \, \delta N_i = 0$	The particles being noninteracting, the total energy is a constant, i.e., $\Sigma E_i \, \delta N_i = 0$

Table 2.2 : Comparison of MB, BE, and FD distribution functions

MB	BE	FD
(1) Holds for distinguishable particles; approximations of BE and FD distributions at $E \gg k_B T$	Helds for indistinguishable particles not obeying Pauli's exclusion principle	Holds for indistinguishable particles obeying the exclusion principle
(2) $f(E) = e^{\alpha} e^{-E/k_B T}$	$f(E) = \dfrac{1}{e^{-\alpha} e^{-E/k_B T} - 1}$	$f(E) = \dfrac{1}{e^{(E-E_F)/k_B T} + 1}$
(3) Applies to common gases at normal temperatures.	Applies to photon gas, phonon gas. Particles with intergral or zero spin [e.g., π meson (also called pion)] obey BE statistics.	Applies to electron gas in metals. Particles having half-integral spin [e.g., protons, neutrons, electrons, neutrinos, positrons, muons (also called μ mesons)] obey FD statistics.

We have already seen that both FD and BE distribution functions reduce to MB distribution when $g_i \gg N_i$, i.e., when the particle number is quite small, and when the temperature T is high. The reduction of quantum statistics to the MB form at sufficiently low concentration or sufficiently high temperature is known as the *classical limit of quantum statistics*. A gas in the classical limit is called *nondegenerate*, whereas for concentrations and temperatures where FD or BE distribution functions must be used, the gas is called *degenerate*.

2.4. SYMMETRY PROPERTIES

Let us consider a system of N identical gas particles occupying a volume V. Suppose that the space coordinates and the spin coordinate, if any, of the ith particle be collectively labelled by q_i. A quantum state of the particle is specified by the momentum components and the direction of spin orientation. We use the index r_i to denote this quantum state. The set of quantum numbers $r_1, r_2, ..., r_N$ give the state of the ensemble, *i.e.*, the whole gas. The wavefunction ψ of the gas in this state can be expressed as $\psi_{(r_1, r_2, ..., r_N)} (q_1, q_2, ..., q_n)$.

(*a*) For the classical case of MB statistics, the particles are distinguishable, and any number of particles can occupy the same single-particle state r. Upon interchange of two particles, no symmetry requirements are imposed on the wavefunction.

(*b*) For particles obeying BE statistics or bosons, the quantized spin angular momentum $L_s = s\hbar$ of each particle is characterized by zero or integral values of the spin quantum number s, *i.e.*, $s = 0, 1, 2, ...$. The quantum-mechanical symmetry requirement is that the complete wavefunction ψ must be *symmetric* for bosons. That is, an interchange of two particles does not alter the wavefunction. For example, an interchange of two particles i and j gives

$$\psi (..., q_i, q_j, ...) = \psi (..., q_j, q_i, ...) \qquad ...(2.61A)$$

Fermi-Dirac and Bose-Einstein Statistics

where, for simplicity, we have dropped the subscript $(r_1, r_2, ..., r_N)$. The state of the ensemble is thus left unaltered by the particle interchange, the particles being indistinguishable. There is no restriction on the number of particles in a single-particle state r.

(c) For FD particles or fermions, the spin quantum number s takes half-integral values, *i.e.*, $s = \frac{1}{2}, \frac{3}{2}, \frac{5}{2}, ...$ Here the quantum-mechanical symmetry requirement is that the complete wavefunction ψ be *antisymmetric*, *i.e.*, ψ must change sign upon interchange of two particles. Thus an interchange of particles i and j gives

$$\psi(..., q_i, q_j, ...) = -\psi(..., q_j, q_i, ...) \qquad ...(2.61B)$$

As before, such an interchange does not change the state of the ensemble, the particles being indistinguishable. Let us interchange two particles i and j in the same single-particle state r. This gives

$$\psi(..., q_i, q_j, ...) = \psi(..., q_j, q_i, ...) \qquad ...(2.61C)$$

Equations (2.61B) and (2.61C) yield

$$\psi = 0 \text{ (if particles } i \text{ and } j \text{ occupy the same quantum state)} \qquad ...(2.61D)$$

So, FD statistics rules out the existence of two (or more) particles in the same single-particle state. That is, FD statistics obeys Pauli's exclusion principle.

2.5. Planck's Law of Radiation

An elegant deduction of Planck's law follows from the BE statistics. We shall give this deduction below.

We consider a black-body chamber containing electromagnetic radiation in thermal equilibrium at temperature T. Quantum mechanically the radiation is looked upon as an assembly of photons which are indistinguishable particles. The total number of photons is unspecified since the walls of the chamber can readily absorb and emit photons. Hence the photon distribution function is given by Eq. (2.61) with E_i representing the energy of a photon in state i. When applied to photons, Eq. (2.61) is referred to as the *Planck distribution*.

The electric field $\vec{\xi}$ in the electromagnetic radiation satisfies the wave equation

$$\nabla^2 \vec{\xi} = \frac{1}{c^2} \frac{\partial^2 \vec{\xi}}{\partial t^2}, \qquad ...(2.62)$$

where c is the velocity of the photons.

The solution of this equation has the form

$$\vec{\xi} = \vec{\xi}_0 \exp[i(\vec{k} \cdot \vec{r} - \omega t)] \qquad ...(2.63)$$

where $\vec{\xi}_0$ is a constant, and $\vec{k}$ is the wavevector given by

$$k = \frac{\omega}{c} \qquad ...(2.64)$$

Now, the energy E of a photon and its momentum $\vec{p}$ are written as

$$E = \hbar\omega \qquad ...(2.65)$$

and

$$\vec{p} = \hbar \vec{k} \qquad ....(2.66)$$

Using Eq. (2.64) we obtain

$$p = \frac{\hbar\omega}{c} \qquad ...(2.67)$$

As Maxwell's equation $\nabla \cdot \vec{\xi} = 0$ must be satisfied, Eq. (2.63) yields $\vec{k} \cdot \vec{\xi} = 0$. Thus $\vec{\xi}$ is perpendicular to the direction of propagation determined by the wavevector $\vec{k}$. For each $\vec{k}$, two possible components of $\vec{\xi}$, transverse to $\vec{k}$, can therefore be specified. In other words, for each $\vec{k}$, two photons corresponding to two directions of polarization of $\vec{\xi}$ are possible.

The values of $\vec{k}$ that are allowed can be determined from boundary conditions. Taking the enclosure in the shape of a parallelepiped of edge lengths L_x, L_y, and L_z, we may proceed as in Sec. 1.3. Imposing boundary conditions as given by Eq. (1.12) we obtain the values of $\vec{k}$ expressed by Eq. (1.14). The number of states per unit volume for photons with a given direction of polarization and momentum values lying between p and $p + dp$ is obtained from Eq. (1.20). This is

$$g(E)\, dE = \frac{4\pi p^2}{h^3} dp \qquad \ldots(2.68)$$

The number of photons in this range is

$$N(E)\, dE = f(E)\, g(E)\, dE = \frac{4\pi p^2}{h^3} \frac{dp}{e^{E/k_B T} - 1} \qquad \ldots(2.69)$$

where Eq. (2.61) is used with the subscript i omitted.

Using Eqs. (2.65) and (2.67) we express Eq. (2.69) as

$$N(E)\, dE = \frac{1}{2\pi^2 c^3} \cdot \frac{\omega^2 d\omega}{e^{\hbar\omega/k_B T} - 1} \qquad \ldots(2.70)$$

This has to be multiplied by 2 to obtain the number of photons per unit volume having both directions of polarization and angular frequency lying between ω and $\omega + d\omega$. Let $\rho_\omega\, d\omega$ be the *energy density* (*i.e.*, the energy per unit volume) of photons with both directions of polarization in the angular frequency range between ω and $\omega + d\omega$. Then

$$\rho_\omega\, d\omega = [2N(E)\, dE]\, \hbar\omega = \frac{\hbar}{\pi^2 c^3} \cdot \frac{\omega^3 d\omega}{e^{\hbar\omega/k_B T} - 1} \qquad \ldots(2.71)$$

which is *Planck's law* for black body radiation.

In terms of the requency $\nu = \omega/(2\pi)$, Eq. (2.17) is written as

$$\rho_\nu\, d\nu = \frac{8\pi h}{c^3} \frac{\nu^3 d\nu}{e^{h\nu/k_B T} - 1} \qquad \ldots(2.72)$$

where $h\, (= 2\pi\hbar)$ is Planck's constant and $\rho_\nu\, d\nu$ is the energy density of radiation with frequencies between ν and $\nu + d\nu$. One can also write Planck's law in terms of the wavelength $\lambda\, (= c/\nu)$. Noting that $d\nu = |c\, d\lambda / \lambda^2|$, we have from Eq. (2.72)

$$\rho_\lambda\, d\lambda = \frac{8\pi hc}{\lambda^5} \frac{d\lambda}{e^{ch/\lambda k_B T} - 1} \qquad \ldots (2.73)$$

where $\rho_\lambda\, d\lambda$ is the energy density for wavelengths between λ and $\lambda + d\lambda$.

Figure 2.5 depicts the plot of ρ_λ versus λ for three temperatures T_1, T_2 and T_3. The function ρ_λ attains a maximum at $\lambda = \lambda_m$ which increases as T decreases.

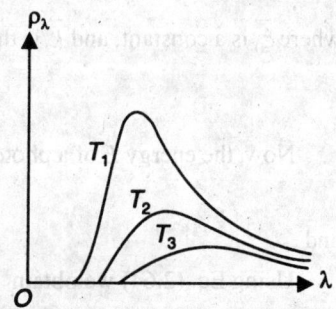

Fig. 2.5. ρ_λ versus λ for three temperatures T_1, T_2, and T_3 ($T_1 > T_2 > T_3$).

2.6. Deductions From Planck's Law of Radiation

Planck's law can explain the experimental data on the spectral distribution of radiation from a black body. The law is quite general, and a number of classical laws of radiation can be derived from it, as shown below.

(i) Wien's displacement law

To find the wavelenth λ_m at which ρ_λ is a maximum, we put $d\rho_\lambda / d\lambda = 0$ and thus obtain from Eq. (2.73)

$$-\frac{5}{\lambda^6 (e^{ch/\lambda k_B T} - 1)} + \frac{e^{ch/\lambda k_B T}}{\lambda^5 (e^{ch/\lambda k_B T} - 1)^2} \frac{ch}{\lambda^2 k_B T} = 0$$

or,
$$\frac{x}{5} = \frac{e^x - 1}{e^x} = 1 - e^{-x} \quad \ldots (2.74)$$

where
$$x = \frac{ch}{\lambda k_B T} \quad \ldots (2.75)$$

By inspection we find that the transcendental equation (2.74) has the solution $x \simeq 5$. A numerical or a graphical solution is found by plotting $y = x/5$ and $y = 1 - e^{-x}$ (Fig. 2.6). The point of intersection gives the solution $x = 4.965$ or $\frac{hc}{\lambda k_B T} = 4.965$,

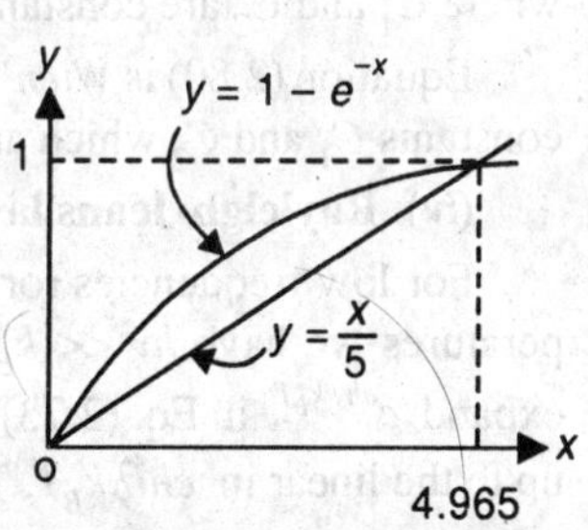

Fig. 2.6. Graphical solution of Eq. (2.74)

or
$$\lambda T = \lambda_m T = \frac{hc}{4.965 \, k_B}$$

or
$$\lambda_m T = 0.2896 \times 10^{-2} \text{ m.K} = \text{constant} \quad \ldots (2.76)$$

which is *Wien's displacement law*, stated as follows :-

When the temperature of a black body is raised, the position of maximum emission is displaced to shorter waves such that the product $\lambda_m T$ remains constant.

Application of Wien's displacement law

Wien's displacement law can be used to estimate the temperature of stars. From the emission spectrum of a star, λ_m can be found. Assuming that the star behaves as a black body, its temperature can be determined from Eq. (2.76). For example, from the solar spectrum, we get $\lambda_m = 475.3$ nm which is on the short wavelength side of visible light. Hence, the temperature of the sun is $T = 0.2896 \times 10^{-2} / (475.3 \times 10^{-9}) = 6093$ K.

Conversely, if the temperature of the black body is known, the wavelength of maximum emission can be obtained from Eq. (2.76).

(ii) Stefan-Boltzmann law

The total energy radiated per unit volume is

$$U = \int_0^\infty \rho_v \, dv = \frac{8\pi h}{c^3} \int_0^\infty \frac{v^3 dv}{e^{hv/k_B T} - 1}$$

Putting $x = hv / k_B T$, so that $dv = (k_B T / h) \, dx$, we have

$$U = \frac{8\pi (k_B T)^4}{c^3 h^3} \int_0^\infty \frac{x^3 dx}{e^x - 1}$$

The integral has the value $\pi^4/15$, giving

$$U = \alpha T^4 \quad \ldots (2.77)$$

where
$$\alpha = \frac{8\pi^5 k_B^4}{15 c^3 h^3} = \text{constant} \quad \ldots (2.78)$$

Equation (2.77) is *Stefan-Boltzmann law*, stating that *the total energy emitted by a black body is proportional to the fourth power of its absolute temperature.*

Stefan's constant is given by

$$\sigma = \frac{\alpha c}{4} = \frac{2\pi^5 k_B^4}{15 c^2 h^3} \qquad (2.79)$$

(iii) Wien's law

For high frequencies (or short wavelengths) and low temperatures such that $h\nu \gg k_B T$ or $hc/\lambda k_B T \gg 1$, one can neglect unity compared to $e^{ch/\lambda k_B T}$ in the denominator on the right-hand side of Eq. (2.73). Then Planck's law simplifies to

$$\rho_\lambda \, d\lambda = C_1 \lambda^{-5} e^{-C_2/\lambda T} \, d\lambda \qquad \ldots (2.80)$$

where C_1 and C_2 are constants : $C_1 = 8\pi hc$ and $C_2 = hc/k_B$.

Equation (2.80) is *Wien's law of radiation*. Wien stated it as an empirical law with two adjustable constants C_1 and C_2 which are so chosen as to fit the experimental data at short wavelengths.

(iv) Rayleigh-Jeans law

For low frequencies (or long wavelengths) and high temperatures we have $h\nu \ll k_B T$ or $hc/\lambda k_B T \ll 1$. We can then expand $e^{ch/\lambda k_B T}$ in Eq. (2.73) in a power series and retain terms up to the linear in $ch/\lambda k_B T$. Thus we obtain

$$\rho_\lambda \, d\lambda = \frac{8\pi k_B T}{\lambda^4} \, d\lambda \qquad \ldots (2.81)$$

Equation (2.81) is *Rayleigh-Jeans law* which states that *the energy density of radiation is inversely proportional to the fourth power of wavelength.*

Figure 2.7 compares Planck's, Wien's and Rayleigh-Jeans radiation laws. As expected, Wien's law agrees with Planck's law for short wavelengths, whereas Rayleigh-Jeans law agrees with Planck's law for long wavelengths. Note that Rayleigh-Jeans law does not give a maximum ρ_λ for finite temperatures at finite wavelengths. But Wien's law predicts a maximum ρ_λ at a wavelength λ_m that can be found by setting $\partial \rho_\lambda / \partial \lambda = 0$ in Eq. (2.80). Thus we obtain $\frac{C_2}{\lambda T} = 5$, or $hc = 5\lambda k_B T$, or

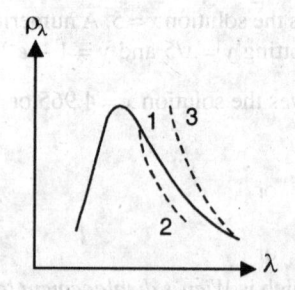

Fig. 2.7. Comparison of the plots of ρ_λ versus λ for Planck's law (curve 1) Wien's law (curve 2), and Rayleigh-Jeans law (curve 3).

$$\lambda T = \lambda_m T = \frac{hc}{5k_B} = 0.2876 \times 10^{-2} \text{ m.K.}, \qquad \ldots (2.82)$$

which compares favourably with Eq. (2.76) predicted by Planck's law.

2.7. WORKED-OUT PROBLEMS

1. Calculate the Fermi energy at 0 K of metallic silver containing one free electron per atom. The density and atomic weight of silver is 10.5 g/cm³ and 108, respectively. **(cf. C.U. 1998)**

Ans. We have

$$E_F(0) = \frac{h^2}{8m} \left(\frac{3N}{\pi V} \right)^{2/3}$$

The number density of free electrons is given by

$$\frac{N}{V} = \frac{6.02 \times 10^{23} \text{ atom/mole}}{108 \text{ g/mole}} \times 10.5 \text{ g/cm}^3$$

Fermi-Dirac and Bose-Einstein Statistics

$$= 5.9 \times 10^{22} \text{ cm}^{-3} = 5.9 \times 10^{28} \text{ m}^{-3}.$$

So, $$E_F(0) = \frac{(6.6 \times 10^{-34} \text{ J.s.})^2}{8 \times 9.1 \times 10^{-31} \text{ kg}} \left(\frac{3 \times 5.9 \times 10^{28} \text{ m}^{-3}}{3.14} \right)^{2/3}$$

$$= 8.8 \times 10^{-19} \text{ J} = \frac{8.8 \times 10^{-19}}{1.6 \times 10^{-19}} \text{ eV} = 5.5 \text{ eV}$$

2. Using the data of prob. 1, determine the internal energy of the electron gas per unit volume at 0K.

Ans. The required internal energy is

$$U = \frac{3}{5} \left(\frac{N}{V} \right) E_F(0) = \frac{3}{5} \times 5.9 \times 10^{28} \times 8.8 \times 10^{-19} \text{ J/m}^3$$

$$= 3.11 \times 10^{10} \text{ J/m}^3.$$

3. Using the data of prob. 1, calculate the molar specific heat due to free electrons at constant volume in metallic silver at 100 K.

Ans. We have $C_V^{(e)} = \frac{\pi^2}{2} N k_B \left(\frac{T}{T_F} \right)$.

Here, $N = 6.02 \times 10^{23}$ atoms / mol, $T = 100$ K, $T_F = \frac{E_F(0)}{k_B}$

$$= \frac{8.8 \times 10^{-19} \text{ J}}{1.38 \times 10^{-23} \text{ J/K}} = 63768 \text{ K}$$

$$C_V^{(e)} = \frac{(3.14)^2}{2} \times 6.02 \times 10^{23} \times 1.38 \times 10^{-23} \times \left(\frac{100}{63768} \right)$$

$$= 0.064 \text{ J K}^{-1} \text{ mol}^{-1}.$$

4. Show that for a two-dimensional electron gas, the number of electrons per unit area is given by

$$n = \frac{4\pi m k_B T}{h^2} \ln (e^{E_F/k_B T} + 1).$$

Ans. The density of states function for a two-dimensional electron gas is given by

$$g(E) dE = \frac{2A}{h^2} (2\pi p dp) = \frac{4\pi m A}{h^2} dE,$$

where A is the area of the two-dimensional container, and m is the electron mass. The number of electrons is

$$N = \int_0^\infty g(E) f(E) dE = \frac{4\pi m A}{h^2} \int_0^\infty \frac{dE}{1 + e^{(E - F_F)/k_B T}}.$$

Therefore, the number of electrons per unit area is given by

$$n = \frac{N}{A} = \frac{4\pi m k_B T}{h^2} \int_0^\infty \frac{dx}{1 + e^{x - \eta}},$$

where $x = E / k_B T$ and $\eta = E_F / k_B T$. The integral is written as

$$I = \int_0^\infty \frac{dx}{1 + e^{x - \eta}} = \int_0^\infty \frac{e^{-x} dx}{e^{-x} + e^{-\eta}}.$$

Substituting $z = e^{-x} + e^{-\eta}$, we get $dz = -e^{-x} dx$

When $x = 0$, $z = 1 + e^{-\eta}$, and when $x = \infty$, $z = e^{-\eta}$. Hence

$$I = \int_{e^{-\eta}}^{1+e^{-\eta}} \frac{dz}{z} = \ln\left(\frac{1 + e^{-\eta}}{e^{-\eta}}\right) = \ln(1 + e^{\eta}),$$

so that
$$n = \frac{4\pi m k_B T}{h^2} \ln(1 + e^{E_F/k_B T}).$$

5. In a system of two particles, each particle can be in any one of three possible quantum states. Find the ratio of the probability that the two particles occupy the same state to the probability that the two particles occupy different states for MB, BE, and FD statistics. **(C.U. 1996)**

Ans. Let r be the desired ratio. If W_1 is the number of ways in which the two particles occupy the same state and W_2 is that in which they occupy different states, then, $r = W_1/W_2$.

For MB statistics, the particles are distinguishable, and $W_1 = 3$. To calculate W_2, we note that one particle can be in any one of the 3 quantum states, and the remaining particle can be in any one of the 2 quantum states that are available to it. So, $W_2 = 3 \times 2 = 6$. Therefore, for MB statistics, $r = W_1/W_2 = 3/6 = 1/2$.

For FD and BE statistics, the particles are indistinguishable. In the case of FD, $W_1 = 0$ since the two particles cannot be in the same quantum state. Here $W_2 = {}^3C_2 = 3$. So, for FD statistics, $r = W_1/W_2 = 0/3 = 0$.

For BE statistics, $W_1 = 3$ and $W_2 = {}^3C_2 = 3$.
So, $r = W_1/W_2 = 3/3 = 1$.

6. Two particles are to be distributed in two nondegenerate energy states. Find the number of distributions according to MB, BE, and FD statistics. Show the distributions diagramatically.

Ans. In MB statistics, the particles are distinguishable and each of the two particles can be in any of the two energy states. So, the number of distributions is $2^2 = 4$.

In BE statistics, the particles are indistinguishable and each particle can occupy any energy state. So, the number of distributions is $\frac{(2 + 2 - 1)!}{2!(2-1)!} = 3$.

In FD statistics, the particles are indistinguishable and not more than one particle can occupy a state. So, the number of distribution is ${}^2C_2 = 1$.

The distributions are diagramatically shown below. In MB statistics the particles being distinguishable, are denoted by A, B. In BE and FD statistics, they are indistinguishable, and are so represented by A.

MB statistics States		BE statistics States		FD statistics States	
1	2	1	2	1	2
A	B	A	A	A	A
AB	–	AA	–		
B	A	–	AA		
–	AB				

7. Show that if f is the FD distribution function, $-(\partial f/\partial E)$ is a maximum at the Fermi level. Also show that $-(\partial f/\partial E)$ is symmetric about the Fermi level.

Ans. We have

Fermi-Dirac and Bose-Einstein Statistics

$$f = \frac{1}{1+e^{(E-E_F)/k_B T}} = \frac{1}{1+e^x},$$

where $\quad x = \dfrac{E-E_F}{k_B T}$, E_F being the Fermi level.

So, $\quad \dfrac{\partial f}{\partial E} = \dfrac{df}{dx}\dfrac{\partial x}{\partial E} = -\dfrac{e^x}{k_B T(1+e^x)^2}$

or, $\quad -\dfrac{\partial f}{\partial E} = \dfrac{1}{k_B T}\dfrac{e^x}{(1+e^x)^2},$ ...(i)

and $\quad -\dfrac{\partial^2 f}{\partial E^2} = -\dfrac{d}{dx}\left(\dfrac{\partial f}{\partial E}\right)\dfrac{\partial x}{\partial E} = \dfrac{1}{k_B T}\dfrac{d}{dx}\left(-\dfrac{\partial f}{\partial E}\right)$

$= \dfrac{1}{(k_B T)^2} \dfrac{(1+e^x)^2 e^x - 2e^x(1+e^x)e^x}{(1+e^x)^4}$

$-\dfrac{\partial f}{\partial E}$ is a maximum when $-\dfrac{\partial^2 f}{\partial E^2} = 0$

or, $\quad (1+e^x)^2 - 2e^x(1+e^x) = 0$
or, $\quad (1+e^x)(1+e^x - 2e^x) = 0$
or, $\quad e^{2x} = 1 = e^0$
or, $\quad x = \dfrac{E-E_F}{k_B T} = 0$, or, $E = E_F$

Thus $-\dfrac{\partial f}{\partial E}$ is a maximum at the Fermi level.

If E is below E_F by the same amount, x is replaced by $-x$, and Eq. (i) gives

$$-\dfrac{\partial f}{\partial E} = \dfrac{1}{k_B T}\dfrac{e^{-x}}{(1+e^{-x})^2} = \dfrac{1}{k_B T}\dfrac{e^x}{(1+e^x)^2}$$

Thus $-\dfrac{\partial f}{\partial E}$ is unchanged upon the reversal of the sign of x. Hence $-(\partial f/\partial E)$ is symmetric about $x = 0$, i.e., about $E = E_F$.

8. Deduce the pressure - volume relationship for a free electron gas obeying FD statistics at 0K. Hence find an expression for the bulk modulus of the gas.

Ans. If U is the internal energy of a system of particles occupying a volume V at a pressure P, we have from thermodynamics $P = -\dfrac{\partial U}{\partial V}$.

For a free electron gas obeying FD statistics at 0K, Eqs. (2.15) and (2.13) give

$$U = \frac{3}{5}NE_F(0) = \left(\frac{3}{5}\right)\left(\frac{h^2}{8m}\right)N\left(\frac{3N}{\pi V}\right)^{2/3}$$

So, $\quad P = -\dfrac{\partial U}{\partial V} = \dfrac{2}{5}\left(\dfrac{h^2}{8m}\right)\left(\dfrac{N}{V}\right)\left(\dfrac{3N}{\pi V}\right)^{2/3} = \left(\dfrac{h^2}{20m}\right)\left(\dfrac{3}{\pi}\right)^{2/3}\left(\dfrac{N}{V}\right)^{5/3}$

which is the desired pressure - volume relationship. It shows that

$$PV = \frac{2}{5}NE_F(0) = \frac{2}{3}U.$$

The bulk modulus is

$$B = -V\frac{\partial P}{\partial V} = \frac{5}{3}\left(\frac{h^2}{20m}\right)\left(\frac{3}{\pi}\right)^{2/3}\left(\frac{N}{V}\right)^{5/3}$$

$$= \frac{5}{3}P = \frac{2}{3}\frac{NE_F(0)}{V} = \frac{10}{9}\frac{U}{V}$$

9. Three spin 1/2 fermions are to be distributed in two levels having energies E_1 and E_2, respectively. Each level can accomodate a maximum of two fermions with opposite spins, +1/2 and –1/2. Determine the number of microstates and macrostates.

Ans. If two particles with opposite spin are accomodated in the energy level E_1, the remaining particle in the level E_2 can have spin either + 1/2 or – 1/2, giving two microstates corrsponding to the macrostate of energy $(2E_1 + E_2)$ (see the following table). Similarly, if two particles of opposite spin occupy the energy level E_2, the remaining particle in the energy level E_1 will have two possible spins +1/2 and –1/2. Thus we get two more microstates for the macrostate of energy $(2E_2 + E_1)$. The interchange of the two particles of opposite spin in the same energy level does not give any new microstate. Therefore, the total number of microstates is 4 and the total number of macrostates is 2 with energies $2E_1 + E_2$ and $2E_2 + E_1$.

Serial number of microstate	Spin of particles in energy level E_1	Spin of particles in energy level E_2	Total energy	Serial number of macrostate
1	$\frac{1}{2}, -\frac{1}{2}$	$\frac{1}{2}$	$2E_1 + E_2$	1
2	$\frac{1}{2}, -\frac{1}{2}$	$-\frac{1}{2}$	$2E_1 + E_2$	
3	$\frac{1}{2}$	$\frac{1}{2}, -\frac{1}{2}$	$E_1 + 2E_2$	2
4	$-\frac{1}{2}$	$\frac{1}{2}, -\frac{1}{2}$	$E_1 + 2E_2$	

QUESTIONS

1. In what way does the Fermi-Dirac distribution differ from the Maxwell-Boltzmann distribution ?
2. What is 'Fermi energy'? Does it depend on temperature ? Define the "degeneracy parameter".
3. Sketch the FD distribution function at the absolute zero of temperature and at a finite nonzero temperature. What do you mean by degenerate and nondegenerate distribution functions ?
4. Indicate how you can determine the Fermi level from a knowledge of the particle density in the system.
5. Show that at high temperatures and low concentrations the FD distribution reduces to the MB distribution. Explain this result physically.
6. (a) Show that the internal energy of a Fermi gas of free particles at absolute zero of temperature is $(3/5) NE_F(0)$. **(cf. C.U. 1994)**

Fermi-Dirac and Bose-Einstein Statistics 163

(b) The specific heat of silver at liquid helium temperatures is represented by $C_v = \gamma T + \alpha T^3$ where γ and α are constants. Explain the origin of the term linear in temperature. **(cf. C.U. 1985)**

(c) Show that the average energy of an electron in a metal at $T = 0$ is $3E_F/5$, where E_F is the Fermi energy. **(C.U. 1997, 2004, 08)**

7. (a) Explain physically why you would expect a lower value of the electronic specific heat in metals when FD statistics is used instead of MB statistics.

(b) What are the symmetry requirements for the total wavefunction for MB, FD, and BE systems?

8. (a) Discuss the difference between the Fermi-Dirac and the Bose-Einstein statistics. What is Bose condensation?

(b) Discuss comparatively the basic postulates of Boltzmann, Bose and Fermi statistics.
(C.U. 1989, 97, 2004)

(c) Name the statistics (BE or FD) obeyed by each of the following particles: neutron, π– meson, muon, photon, neutrino. **(C.U. 1992, 97)**

9. (a) Compare among MB, FD, and BE distribution functions.

(b) Which statistics is obeyed by an atomic nucleus?

[**Ans.** If the nucleus contains an odd number of FD particles, i.e., if the mass number is odd, it obeys FD statistics. If the number is even, BE statistics is obeyed].

(c) Which statistics will apply to (i) deuterons and α-particles, and (ii) He^3 atoms?
[**Ans.** (i) BE, (ii) FD]

10. (a) For a system of noninteracting particles, how does the distribution function differ for particles obeying Bose-Einstein and Fermi-Dirac statistics? Give one example of each of the systems where you would apply the above two distributions.

(b) At $T = 0$, sketch the distribution functions in the above two cases.

(c) In what limit would these distributions tend to the classical Maxwell-Boltzmann distribution?
(C.U. 1982)

11. Starting from Bose-Einstein distribution formula establish Planck's law of black body radiation.
(C.U. 1987, 2000, 2005, 08)

12. A system of identical noninteracting particles obeys Pauli's principle. Obtain the distribution law. Discuss (i) the classical limit and (ii) the $T = 0$ behaviour of the gas. **(C.U. 1985)**

13. Write BE and FD distribution functions (no proof required). What are the basic assumptions used in the derivation of these distribution functions? Sketch the FD distribution function for $T = 0K$ and $T > 0K$. **(C.U. 1995)**

14. Find how the electronic specific heat of a metal behaves as a function of temperature. **(C.U. 1995)**

15. Write notes on:

(a) Specific heat due to electrons in metals at very low temperatures;

(b) Planck's law of black body radiation. **(C.U. 1983)**

16. Calculate the Fermi energy at 0 K of sodium containing one free electron per atom. The density and the atomic weight of sodium is $9.7 \times 10^2 kg/m^3$ and 23, respectively. Also find the Fermi temperature.
[**Ans.** 3.13 eV, 36290 K]

17. Which particles do obey Bose-Einstein statistics? Derive the BE distribution formula. Deduce Planck's radiation formula from BE statistics. **(C.U. 1985)**

18. Derive the Fermi-Dirac distribution formula. Discuss the nature of the distribution function at $T = 0$ and when T is finite. What is Fermi energy? **(C.U. 1986, 99)**

19. At the same temperature, which will exert the greatest and the least pressure–a gas obeying MB statistics, a gas of bosons, and a gas of fermions?

[Hint : A gas of fermions will exert the greatest pressure since the proportion of high-energy particles in the Fermi distribution is larger than that in the other two distributions. A gas of bosons will exert the least pressure since the proportion of low-energy particles in the Bose distribution is larger than that in the other distributions.]

20. The particles of a gas have mass 1.6×10^{-24} gm, the density of the gas is 10^{-4} gm/c.c. and its temperature is 10^3 K. Discuss the energy distribution according to Fermi statistics and illustrate with diagram.
(c.f. C.U. 1989)

[Hint : Calculate $E_F(0)$ and compare it with $k_B T$.]

21. Consider a system of 2 identical particles each of which can be in any one of 3 single particle states. How many states of the system are possible if the particles obey (i) Maxwell-Boltzmann, (ii) Fermi-Dirac, and (iii) Bose-Einstein statistics?
(C.U. 1994)

[Ans. (i) 9, (ii) 3, (iii) 6]

22. Prove that for a system at $T > 0$ K obeying FD statistics, the probability that a level lying ΔE below the Fermi level is unoccupied is the same as the probability of occupation of a level lying ΔE above the Fermi level.
(C.U. 1995)

[Hint. : The probability of occupancy of level E is $f(E)$, whereas the probability of nonoccupancy is $1 - f(E)$].

23. Consider a free electron at the Fermi level in a metal at 0K. Show that the de Broglie wavelength of the electron is $2\left(\dfrac{\pi}{3n}\right)^{1/3}$, where n is the number of free electrons per unit volume in the metal.

[Hint : Use $\lambda = h/p$, $p = \sqrt{2mE_F(0)}$ and Eq. (2.13).]

24. If f is the FD distribution function, show that (i) $\dfrac{\partial f}{\partial E} = -\dfrac{f(1-f)}{k_B T}$, and (ii) $\int_{-\infty}^{\infty}\left(-\dfrac{\partial f}{\partial E}\right) dE = 1$

25. Show that as $T \to 0$, $-\dfrac{\partial f}{\partial E} = \delta(E - E_F)$ where f is the FD distribution funciton and δ is Dirac's delta funciton.

26. (a) Show that Planck's law reduces to Wien's law for $h\nu \gg k_B T$ and to Rayleigh-Jeans law for $h\nu \ll k_B T$.
(C.U. 1998)

(b) Deduce Wien's displacement law and Stefan-Boltzmann law from Planck's law of radiation.

(c) Explain how Wien's displacement law can be used to estimate the temperature of stars.
(Mangalore Univ. 2005)

27. Show that the single-particle occupation number (effective number of particles in a single-particle state) for a fermion at T(K) with an energy within $\pm k_B T$ of the Fermi energy has an approximate range of 0.46.
(C.U. 2000)

[Hint. $f(E_F - k_B T) = \dfrac{1}{1 + (1/e)}$ and $f(E_F + k_B T) = \dfrac{1}{1 + e}$. The desired range is $\Delta f = f(E_F - k_B T) - f(E_F - k_B T) = (e - 1)/(e + 1)$.]

28. A system has nondegenerate single-particle states with 0, 1, 2, 3 energy units. Three particles are to be distributed in these states such that the total energy of the system is 3 units. Find the number of microstates if the particles obey (i) MB statistics, (ii) BE statistics, and (iii) FD statistics.

[Ans. (i) 10, (ii) 3, (iii) 1]

29. Calculate the pressure exerted by an ideal gas of electrons at 0K. What is the physical reason for the nonzero pressure?
(C.U. 2002)

[Hint. : Refer to worked-out problem No. 8. The pressure is non-zero because the energy is finite at 0 K.]

30. Calculate the energy difference $(E - E_F)$ in eV at 300 K where the FD distribution function is (i) 0.95 and (ii) 0.05.

[Ans. (i) -0.076 eV, (ii) $+0.076$ eV]

Chapter 3
Third Law of Thermodynamics

3.1. Introduction

The entropy S of a system can be determined from a knowledge of the macrostate of the system. As the temperature T of the system is lowered and the absolute zero of temperature is approached, the system tends to attain the lowest possible energy E_0, called the *ground-state energy*. With a decrease of the energy E, the entropy of the system decreases and becomes vanishingly small as the system tends to attain its ground-state energy. In symbols, as $T \to 0$, $E \to E_0$ and $S \to 0$. This principle is known as *Nernst's theorem* or *the third law of thermodynamics*. Of course, the system must be in equilibrium as the temperature is decreased. This result has an experimental difficulty since at very low temperatures, the rate of attaining equilibrium is very slow. A useful result can however be obtained at a temperature T_0 which is small but not inconvenient. At this temperature the entropy corresponding to the degrees of freedom not associated with the nuclear spins of the constituent atoms would be negligible. But even at this low temperature, the nuclear spins would be oriented at random owing to the weak interacting forces. Thus as $T \to T_0$, $S \to S_0$ where S_0 is determined by the random orientations of the nuclear spins. Note that S_0 is a constant depending only on the types of atomic nuclei constituting the system but is independent of the system parameters related to the spatial arrangement of the atoms and the interatomic interactions. The process $T \to T_0$ may be denoted by $T \to 0_+$; this indicates an approach towards a limiting temperature, which is very small but large enough for the spins to be randomly oriented. This process is extremely useful from the practical point of view as it avoids working at temperatures which are prohibitively low. As one goes to temperatures much less than T_0, the nuclear spins will deviate from their random configuration.

The entropy behaviour as $T \to 0_+$ makes it possible to state the *third law of thermodynamics* in the following manner :

The limiting property of the entropy S of a system is that as $T \to 0_+$, $S \to S_0$ where S_0 is a constant independent of the parameters of the particular system.

3.2. Consequences of The Third Law

A. Limiting behaviour of thermodynamic systems near absolute zero of temperature :

If C_V and C_P denote the heat capacities of a system at constant volume and at constant pressure, respectively, then one can write

$$C_V = T \left(\frac{\partial S}{\partial T} \right)_V, \qquad ...(3.1)$$

and

$$C_P = T \left(\frac{\partial S}{\partial T} \right)_P. \qquad ...(3.2)$$

According to the third law of thermodynamics, as $T \to 0$, $S \to S_0$, where S_0 is independent of the system parameters. Hence, integrating Eq. (3.1) we obtain

$$S(T) - S(0) = \int_0^T \frac{C_V(T')}{T'} dT' \qquad ...(3.3)$$

Clearly, as $T \to 0$, $C_V(T)$ must approach zero; otherwise, the integral on the right-hand side of (3.3) will diverge. Similar conclusions regarding the behaviour of C_p can be made by working with Eq. (3.2). Thus we conclude that

$$\text{as } T \to 0, \quad C_V \to 0 \text{ and } C_p \to 0 \qquad ...(3.4)$$

The physical reason for this result is that as $T \to 0$, the system attains its ground-state energy, and a further decrease of temperature is not accompanied with a further decrease of energy. At very low temperatures, the difference between C_p and C_V becomes smaller. This result is not conflicting with the relation $C_p - C_V = R$ for one mole of a perfect gas. It simply reflects that as $T \to 0$, the system tends to its ground state and quantum mechanical effects come into play so that the classical equation of state $PV = RT$ is no longer true.

The third law of thermodynamics also predicts a limiting property of the volume coefficient of expansion of a system. This coefficient α is defined by

$$\alpha = \frac{1}{V}\left(\frac{\partial V}{\partial T}\right)_P \qquad ...(3.5)$$

By using the Maxwell relation $(\partial S/\partial P)_T = -(\partial V/\partial T)_P$ we may write

$$\alpha = -\frac{1}{V}\left(\frac{\partial S}{\partial P}\right)_T \qquad ...(3.6)$$

Since the limiting value of the entropy, i.e. S_0, is independent of the system parameters, it must not depend on volume or pressure changes in this limit. Thus as $T \to 0$, $(\partial S/\partial P)_T \to 0$ and $\alpha \to 0$.

B. Application to special systems

The use of the third law of thermodynamics in some particular systems leads to interesting results. We shall discuss one such application below.

Consider the transformation of a solid from one crystalline form to another. An example is the transformation of grey tin into white tin. Grey tin is the stable form at temperatures less than $T_0 = 292K$ and white tin is stable at temperatures higher than T_0. At the transition temperature T_0 the two forms exist in equilibrium with each other, and some amount of heat is absorbed by the tin in transforming from the grey to the white variety. This heat is known as the *heat of transformation* and has the value 535 calories per gram-atom at the transition temperature.

Although white tin is the unstable form below the transition temperature T_0, the rate of transformation into the grey form is very slow with respect to the times of experimental interest. White tin can thus exist down to the lowest temperatures. The specific heats of both grey tin and white tin can therefore be measured at temperatures below T_0.

At the transition temperature the transformation between the white and the grey forms is reversible. Let $S_g(T_0)$ and $S_w(T_0)$ denote the entropies at the transition temperature of one gram-atom of grey tin and white tin, respectively. Then, for the reversible isothermal transition from the grey to the white form, we have

$$S_w(T_0) - S_g(T_0) = \int_{grey}^{white} \frac{dQ}{T} = \frac{Q_0}{T_0}, \qquad ...(3.7)$$

where Q_0 is the heat of transformation per gram atom. If $C_g(T)$ and $C_w(T)$ are the heat capacities per gram-atom of grey and white tin, respectively, then we can write

$$S_g(T_0) = S_g(0) + \int_0^{T_0} \frac{C_g(T)}{T} dT, \qquad ...(3.8)$$

and

$$S_w(T_0) = S_w(0) + \int_0^{T_0} \frac{C_w(T)}{T} dT, \qquad ...(3.9)$$

Third Law of Thermodynamics

Here $S_g(0)$ and $S_w(0)$ are the limiting entropies as $T \to 0$ of grey and white tin, respectively, associated with possible nuclear spin orientations. Since we are considering the same number of the same kind of nuclei, according to the third law of thermodynamics we must have

$$S_g(0) = S_w(0) \qquad ...(3.10)$$

Employing Eqs. (3.8) through Eq. (3.10), we get from Eq. (3.7)

$$Q_0 = T_0 \left[\int_0^{T_0} \frac{C_w(T)}{T} dT - \int_0^{T_0} \frac{C_g(T)}{T} dT \right] \qquad ...(3.11)$$

Using the experimentally measured values of $C_w(T)$ and $C_g(T)$ and evaluating the two integrals numerically, it is found that

$$\int_0^{T_0} \frac{C_w(T)}{T} dT = 12.30 \text{ cal./deg} \qquad ...(3.12)$$

and

$$\int_0^{T_0} \frac{C_g(T)}{T} dT = 10.53 \text{ cal./deg} \qquad ....(3.13)$$

As $T_0 = 292$ K, we get from Eq. (3.11)

$Q_0 = 292 \times (12.30 - 10.53) = 517$ cal

Allowing for the experimental errors, the agreement between this value and the experimental value of Q_0, *i.e.*, 535 calories, is quite good. This can be regarded as a strong evidence in favour of the third law of thermodynamics.

QUESTIONS

1. State and explain the third law of thermodynamics.
2. Show that when $T \to 0$, $C_P \to 0$ and $C_V \to 0$. Explain this result physically.
3. Prove that when the temperature approaches absolute zero, the volume coefficient of expansion of a system also approaches zero.
4. Discuss some consequences of the third law of thermodynamics.
5. Can you give some evidence in support of Nernst's theorem ?
6. Write a note on the third law of thermodynamics and its consequences. (C.U. 1983)

$$S_1(0) = S_2(0) \qquad (3.10)$$

$$Q_0 = 292 \times (12.30 - 10.53) = 516 \text{ cal}$$

SOLID STATE PHYSICS

Chapter 1

Crystals and Their Properties

1.1. STRUCTURE OF SOLIDS

Generally, *a solid is defined as any substance that deforms elastically under small shear stresses*. This definition excludes gases and liquids, for they exhibit only viscous resistance to shear. Viscosity opposes shear deformation but does not generate restoring forces. The branch of physics in which the properties of solids are discussed is known as *Solid State Physics* or the *physics of condensed matter* in a broader sense. A solid can, in general, be divided into two classes : (*i*) *crystalline* and (*ii*) *amorphous*.

A *crystal* is a substance which is formed by regular repetition in three dimensions of identical units, where a unit may contain one or more atoms.[1] Truly speaking, a crystalline solid is composed of an agglomeration of a large number of small crystalline grains. These grains are arranged in a more or less random fashion and joined at interfaces referred to as *grain boundaries*[2]. The resulting substance is termed as *polycrystalline*. However, if the atoms are arranged in a regular manner throughout a sample of macroscopic size, the specimen is termed as *a single crystal*. The branch of science devoted to the study of structures and physical properties of crystalline solids is known as *crystallography*.

An *amorphous solid* is that in which the constituent atoms take their positions in a regular manner over short distances of the order of several interatomic spacings (an interatomic spacing is approximately 3Å in most solids). However, no correlation between the positions of the atoms occurs over distances perhaps, a few hundred interatomic spacings. Glass and plastics are two common examples of amorphous solids.

Single crystals of most materials are generally prepared by growing from a saturated solution or pulling from the melt. Since measurements on single crystals give a maximum amount of information about the basic properties of the material we shall restrict our discussion on single crystal samples.

1.2. CLASSIFICATION OF CRYSTALS

The crystals are commonly classified into *five* types according to the binding forces that are responsible for holding the atoms of the crystal together. These types are : (*i*) *Ionic crystals*, (*ii*) *Covalent crystals*, (*iii*) *Metallic crystals*, (*iv*) *Molecular crystals* and (*v*) *Hydrogen-bonded crystals*.

The crystal binding is measured by the *cohesive energy* which is defined as the energy released when the atoms or the molecules, originally in the gaseous state at very low temperatures, are allowed to condense into the solid phase. Accordingly, this energy is equal to the energy of sublimation to a close approximation. We discuss below each of the different classes of crystals.

---

1. The fact that a macroscopic specimen is crystalline does not necessarily mean that this regular stacking extends throughout the volume.
2. The atomic arrangement within the grains is essentially regular. On the other hand, the grain boundaries show irregularities and are considered as localization of very severe lattice disruption and dislocation. The grains are usually as small as 10^{-6} m in diameter.

(*i*) **Ionic crystal.** An ionic crystal is formed when a strongly electropositive metal is combined with a strongly electronegative atom, such as a halide. Thus NaCl, KBr, LiF etc. form ionic crystals in the solid state. Here the transfer of valence electrons from one atom to the other takes place resulting in a crystal that is composed of positive and negative ions. The electrostatic interaction between these ions becomes the source of cohesive energy. The electronic configuration of the ions is basically an inert gas configuration, the charge distribution of each ion being spherically symmetric. The binding energy of the ionic crystals is relatively high and so they possess fairly high melting and boiling points. Since the valence electrons are tightly bound to the respective ions, the electrons cannot contribute to the charge transport, and so the ionic crystals behave as insulators. However, at high temperatures the ions in these crystals diffuse through the solid. As a result, these crystals conduct electricity at elevated temperatures by diffusion of positive and negative ions. This *ionic conductivity* is many orders of magnitude smaller than the *electronic conductivity* of metals partly because the mass of an ion is about 10^4 times the mass of an electron so that the mobility of an ion is much less than that of an electron. In the case of alkali halides, the alkali ions are much smaller than the halogen ions. Thus the alkali ions are more mobile than the halogen ions, and contribute more to the ionic charge transport in these crystals.

Ordinarily, the ionic crystals are diamagnetic because the electronic configuration is that of closed shells, leaving no unpaired electron spins. However, by virtue of the associated intrinsic magnetic moment the ionic crystals can make a paramagnetic contribution to the susceptibility.

Ionic crystals are usually transparent to visible light and exhibit a strong absorption peak in the infrared region of the spectrum. This absorption is accompanied by the forced vibration of the ions, which is induced by the oscillating electric field of the electromagnetic radiation. A resonant absorption peak results when the frequency of the electromagnetic radiation matches the normal modes of oscillation of the ions.

(*ii*) **Covalent (Homopolar) crystals.** Covalent crystals are those in which the neighbouring atoms share the valence electrons equally. They are different from the ionic crystals in which actual transfer of valence electrons from one atom to another takes place. Thus, in covalent crystals the net charge associated with any atom is zero. Covalent crystals are often formed by the elements of the columns III, IV and V of the periodic table. An example of a covalent crystal is diamond. Here each carbon atom shares its four valence electrons with its four nearest atoms and forms *covalent electron-pair bonds*. The covalent bond is stronger than the bond that exists in ionic crystals. The covalent bonds are strongly directional, *i.e.* the electrons tend to be aligned along the lines joining the adjacent atoms and the lines, in turn, tend to be disposed tetrahedrally about any atom. The characteristic tetrahedral disposition of covalent bonds is shown by diamond or zinc blende structures. According to the Pauli principle, the electron pair of a covalent bond must have antiparallel spin orientations.

Covalent crystals are hard and brittle, and because of their high binding energies they possess high melting and boiling points. Covalent crystals of high purity are insulators at low temperatures since each valence electron is an integral part of a bond and is not free to move through the crystal. However, at sufficiently high temperatures, the thermal energy of vibration of the crystal becomes sufficient to break the covalent bonds and generate free electrons. These electrons are able to move through the solid. The unsaturated bond, left behind, is called *a hole* and it carries an effective positive charge due to the absence of an electron. Also the unsaturated bond may shift about and contribute to an electric current. Thus covalent crystals like Ge, Si and silicon carbide show semiconducting properties at ordinary temperatures. Their conductivity is very sensitive to the presence of a very small amount of impurity atoms, and increases with rise in temperature.

Pure covalent crystals are diamagnetic as there is complete pairing of electron magnetic moments within each bond. However, covalent crystals containing impurities show evidence of paramagnetism owing to either an excess or a deficiency of bonding electrons. The covalent crystals are opaque to shorter wavelengths but transparent to longer-wavelength radiation. The transition is fairly abrupt and occurs at a characteristic wavelength generally in the visible or near infrared region.

(*iii*) **Metallic crystals.** The metallic elements in the free state form metallic crystals. Here the valence electrons are so influenced by the presence of the neighbouring atoms that they behave as free or highly mobile electrons. In other words, there is no tendency for the valence electrons to be localized within any given region of the material. These electrons may be looked upon as being shared by all the atoms. Thus there are about 10^{23} electrons per cm^3 which can contribute to charge transport in metals. The valence electrons in metals are referred to as *free* or *conduction electrons* since they participate in the conduction of electric current. The presence of the large number of conduction electrons accounts for the very high thermal and electrical conductivity of metals. The most characteristic optical properties of metals, namely, the high optical reflection and absorption coefficients are due to its high electrical conductivity.

The interaction of the ion cores with the conduction electrons in alkali metals makes a large contribution to the binding energy. An alkali metal crystal may be thought of as an array of positive ions embedded in a nearly uniform sea of negative charge. In the case of transition metals, additional binding forces arising from interactions among the inner electron shells, are present. Transition group elements have incomplete *d*-electron shells. A covalent bond formed between the *d*-electrons of the unfilled shells gives the transition elements a high binding energy. For example, in iron and tungsten, the inner electronic shells make a large contribution to the binding. As a result, these elements possess high melting points. On the other hand, the binding energy of alkali metals is very low; consequently they have relatively low melting and boiling points.

The spins and magnetic moments of the conduction electrons in metals except in ferromagnetics are evenly paired in the absence of a magnetic field. However, this balance of magnetic moments is disturbed when an external magnetic field is applied. Most metals are paramagnetic, and some are diamagnetic or ferro-magnetics.

(*iv*) **Molecular crystals.** Molecular crystals are those in which the binding arises primarily from dipolar forces between the atoms or the molecules of the crystal. Even if an atom or a molecule has no average dipole moment, it will have in general an instantaneous, fluctuating dipole moment that originates from the instantaneous positions of the electrons in their orbits around the nucleus.

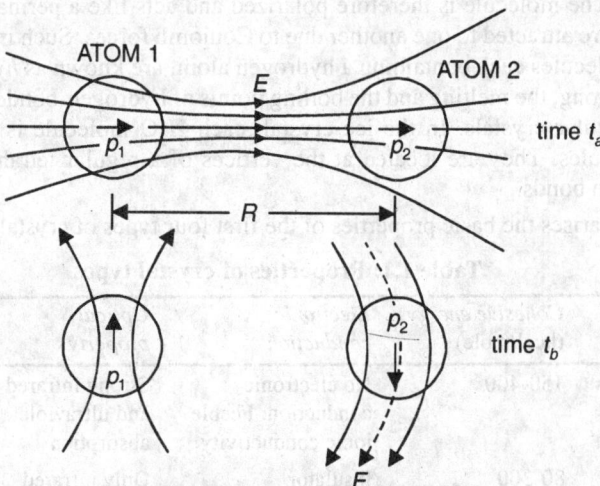

Fig. 1.1. Two atoms separated by a distance R are shown for two times, t_a and t_b. In both the cases, the interaction is attractive.

Figure 1.1 shows two atoms separated by a distance R. An instantaneous dipole moment of magnitude p_1 on atom 1 will produce an electric field (E) varying as p_1/R^3 at the centre of atom 2.

This field in turn will induce at the centre of atom 2 an instantaneous dipole moment given by $p_2 = \alpha E \sim \alpha p_1 / R^3$, where α is the electronic polarizability. The potential energy $U(R)$ of the dipole moment is given by

$$U(R) \cong -p_2 E \qquad \qquad ...(1.1)$$

So that $U(R) \sim \alpha p_1^2 / R^6$ and is negative.

The interaction between p_1 and p_2 is attractive in character and provides the necessary cohesive energy to the molecular crystals. This is termed variously as the *van der Waals interaction*, the *London interaction*, or the *fluctuating dipole interaction*. Since the binding energy falls off as $1/R^6$ these forces are short-range and are usually quite weak. Hence the molecular crystals are characterized by small binding energies, and low melting and boiling points. Crystals of the inert gases such as He, Ne, A etc., in the solid state are usually molecular crystals. Solid oxygen and nitrogen crystals have both molecular and covalent character. In these crystals, the entities are not free atoms, but O_2 or N_2 molecules.

Each molecule in molecular crystals is electrically neutral and interacts with another molecule only weakly. This makes molecular crystals good insulators, showing neither electronic nor ionic conductivity. The magnetic behaviour of the molecular crystals depends on the magnetic properties of the individual molecules. Thus, if the molecules are paramagnetic, the crystal also shows paramagnetism. However, the molecular crystals are in general diamagnetic, as are almost all saturated molecules except oxygen and a few others.

(*v*) **Hydrogen bonded crystals.** In the category of insulators, hydrogen-bonded crystals are separately listed. Because of the large ionization potential, it is very difficult to detach an electron completely from hydrogen, so that it does not act like an alkali metal ion in forming ionic crystals. Also, it can form only one covalent bond via electron sharing, and therefore cannot act as a typical covalent crystal atom which provides four bonds in tetrahedral covalent crystals. Furthermore, the proton having a very minute size gives a structure unachievable with any other positive ion. Ice is an example where hydrogen bonding is important.

When a pair of electrons is shared between a hydrogen atom and a highly electronegative atom (like oxygen) the shared electron pair is attracted more towards the electronegative atom. The electronegative atom thus gets slightly negatively charged and the hydrogen atom gets an equal amount of positive charge. The molecule is therefore polarized and acts like a permanent electric dipole. Many such dipoles are attracted to one another due to Coulomb forces. Such interactions in permanently polarized molecules each containing a hydrogen atom, are known as *hydrogen bonds*. Such bonds being quite strong, the melting and the boiling points of hydrogen-bonded crystals are higher than those of molecular crystals. In the ice crystal, each H_2O molecule is surrounded by four neighbouring molecules. They are located at the vertices of a regular tetrahedron, and are held together by hydrogen bonds.

Table 1.1 summarises the basic properties of the first four types of crystals.

Table 1.1: Properties of crystal types

	Crystal type	Cohesive energy (kcal/mole)	Electronic conduction	Optical property	Magnetic property
(i)	Ionic crystals	150-400	No electronic conduction. Feeble ionic conductivity	Strong infrared and ultraviolet absorption	Diamagnetic
(ii)	Covalent crystals	80-300	Insulator or semiconductor	Only infrared absorption	Diamagnetic
(iii)	Metals	20-200	Conductivity high	Reflection of visible radiation	Para-dia-, or ferromagnetic
(iv)	Molecular crystals	1-20	Very good insulator	Characterised by properties of the molecules	Diamagnetic in general

Crystals and Their Properties

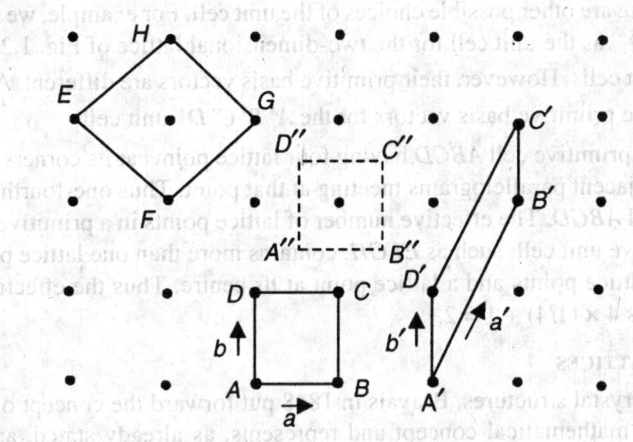

Fig. 1.2. A two-dimensional crystal lattice and unit cells. The dots are lattice points.

1.3. Some Crystallographic Terms

A *space lattice* or simply a *lattice* is defined to be a regular periodic array of geometric points arranged in space such that each point has identical surroundings. The points are known as *lattice points*. A crystal is obtained by placing an atom (in simple structures) or a group of atoms (in more complex structures) at each lattice point in a regular manner. Such an atom or a group of atoms serves as the basic building unit for the entire crystal structure, and is termed the *basis*. Thus,

$$\text{lattice} + \text{basis} \longrightarrow \text{crystal}.$$

Figure 1.2 shows a two-dimensional crystal lattice. Here the parallelogram *ABCD* defined by the vectors $\vec{a}$ and $\vec{b}$ may be selected as a *unit cell* of the lattice. The vectors $\vec{a}$ and $\vec{b}$ are called the *basis vectors*. It is obvious from the figure that all translations of the *ABCD* parallelogram by integral multiples of the basis vectors $\vec{a}$ and $\vec{b}$ along the $\vec{a}$ and $\vec{b}$ directions will result in translating it to an exactly identical region in the crystal. Thus, the whole crystal may be reproduced merely by translating the area *ABCD* along the $\vec{a}$ and $\vec{b}$ directions by suitably multiplying the basis vectors $\vec{a}$ and $\vec{b}$. In other words, each lattice point in the crystal may be described by a vector $\vec{R}$ where $\vec{R} = h\vec{a} + k\vec{b}$; the constants h and k being integers.

By extending the above procedure to *three dimensions* we define the following terms:

Unit Cell. This is a region of the crystal defined by the basis vectors $\vec{a}, \vec{b}$ and $\vec{c}$ such that a translation of this region by any integral multiple of these vectors will result in a similar region of the crystal.

Basis Vectors. Three linearly independent vectors $\vec{a}, \vec{b}$ and $\vec{c}$ used to define a unit cell are called the basis vectors.

Primitive Unit Cell. The smallest unit cell occupying the smallest volume in the given lattice is termed a primitive unit cell.

Primitive Basis Vectors. The three vectors $\vec{a}, \vec{b}$ and $\vec{c}$ required to define a primitive unit cell are called primitive basis vectors.

It is clear from the above definitions that a lattice point in a three-dimensional crystal lattice is given by the vector $\vec{R}$, where

$$\vec{R} = h\vec{a} + k\vec{b} + l\vec{c}, \ (h, k, l \text{ integers})$$

Note that there are other possible choices of the unit cell. For example, we may choose $A'B'C'D'$ or $A''B''C''D''$ as the unit cell for the two-dimensional lattice of Fig. 1.2. These unit cells are also primitive unit cells. However, their primitive basis vectors are different. As shown, the vectors $\vec{a'}$ and $\vec{b'}$ are the primitive basis vectors for the $A'B'C'D'$ unit cell.

Consider the primitive cell $ABCD$ having four lattice points at its corners. Each lattice point is shared by four adjacent parallelograms meeting at that point. Thus one-fourth of each corner point belongs to the cell $ABCD$. The effective number of lattice points in a primitive cell is thus $4 \times (1/4) = 1$. A nonprimitive unit cell, such as $EFGH$, contains more than one lattice point. The cell $EFGH$ has four corner lattice points and a lattice point at its centre. Thus the effective number of lattice points in $EFGH$ is $4 \times (1/4) + 1 = 2$.

1.4. Bravais Lattices

To describe crystal structures, Bravais in 1848 put forward the concept of the space lattice. A space lattice is a mathematical concept and represents, as already stated, an infinite number of points in space such that the arrangement of points about a given point is identical to that about any other point. There are just *fourteen* ways in which the points in space lattices can be arranged in order that all the lattice points have exactly the same surroundings. These fourteen point lattices are called *Bravais lattices* and are shown in Fig. 1.3. A general unit cell for a space lattice is depicted in Fig. 1.2A. It is a parallelepiped defined by the basis vectors $\vec{a}, \vec{b}$, and $\vec{c}$. The shape and size of the unit cell depend on the lengths a, b, c, and the angles α, β, γ between them.

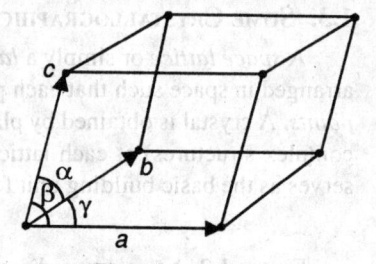

Fig. 1.2A. unit cell

The Bravais lattices can be grouped under *seven* crystal systems, namely, cubic, tetragonal, orthorhombic, monoclinic, triclinic, trigonal, and hexagonal. We are mainly concerned here with the cubic lattice. For the cubic lattice the lengths of the basis vectors a, b and c are equal and the angles α, β, γ between them are each equal to $90°$. For the tetragonal lattice, $a = b \neq c$ and $\alpha = \beta = \gamma = 90°$. For the unit cell of the orthorhombic lattice, $a \neq b \neq c$ and $\alpha = \beta = \gamma = 90°$. The monoclinic lattice is characterized by $a \neq b \neq c$ and $\alpha = \gamma = 90° \neq \beta$; and the triclinic type by $a \neq b \neq c$ and $\alpha \neq \beta \neq \gamma$. The trigonal unit cell has $a = b = c$ and $\alpha = \beta = \gamma < 120°$, $\neq 90°$, whereas the hexagonal unit cell has $a = b \neq c$ and $\alpha = \beta = 90°, \gamma = 120°$.

Quantities specifying the size of a unit cell, e.g., the length a in a cubic crystal, are called *lattice constants* or *lattice parameters*. For a general unit cell, $a, b, c, \alpha, \beta,$ and γ are the lattice parameters.

1.5. Some Crystal Structures

Cubic Lattices. There are three possible cubic lattices : these are *(i) the simple cubic* (sc) [Fig. 1.3 *(i)*], *(ii) the body-centered cubic* (bcc); [Fig. 1.3 *(ii)*], and *(iii) the face-centered cubic* (fcc) [Fig. 1.3 *(iii)*] lattices. The unit cell of the simple cubic lattice of Fig. 1.3 *(i)* is a primitive unit cell since a minimum of one lattice point is contained in each cell. If each lattice point is occupied by a single atom, the primitive cell will contain only one atom. This is explained below.

In a simple cubic cell there are 8 lattice points at the eight corners of the cell. This means that each lattice point is shared equally among 8 unit cells that adjoin at each corner. In other words, there are 8 corner lattice points and 1/8th of each belongs to this particular cell, giving a total of one lattice point per unit cell. Materials with simple cubic structure are very rare. The only known element with this structure is the alpha phase of polonium.

An fcc structure is formed by adding to the simple cubic lattice an extra lattice point at the centre of each square face. Thus, for the fcc lattice in addition to 8 corner lattice points shared equally among 8 cells, there are also 6 face-centre lattice points, each shared between two cells. Therefore, the number of lattice points per cubic cell is $8 (1/8) + 6 (1/2)$, *i.e.* 4. Some elements having the fcc configuration are Al, Ca, Cu, Ni, Pb, Pt etc.

Crystals and Their Properties

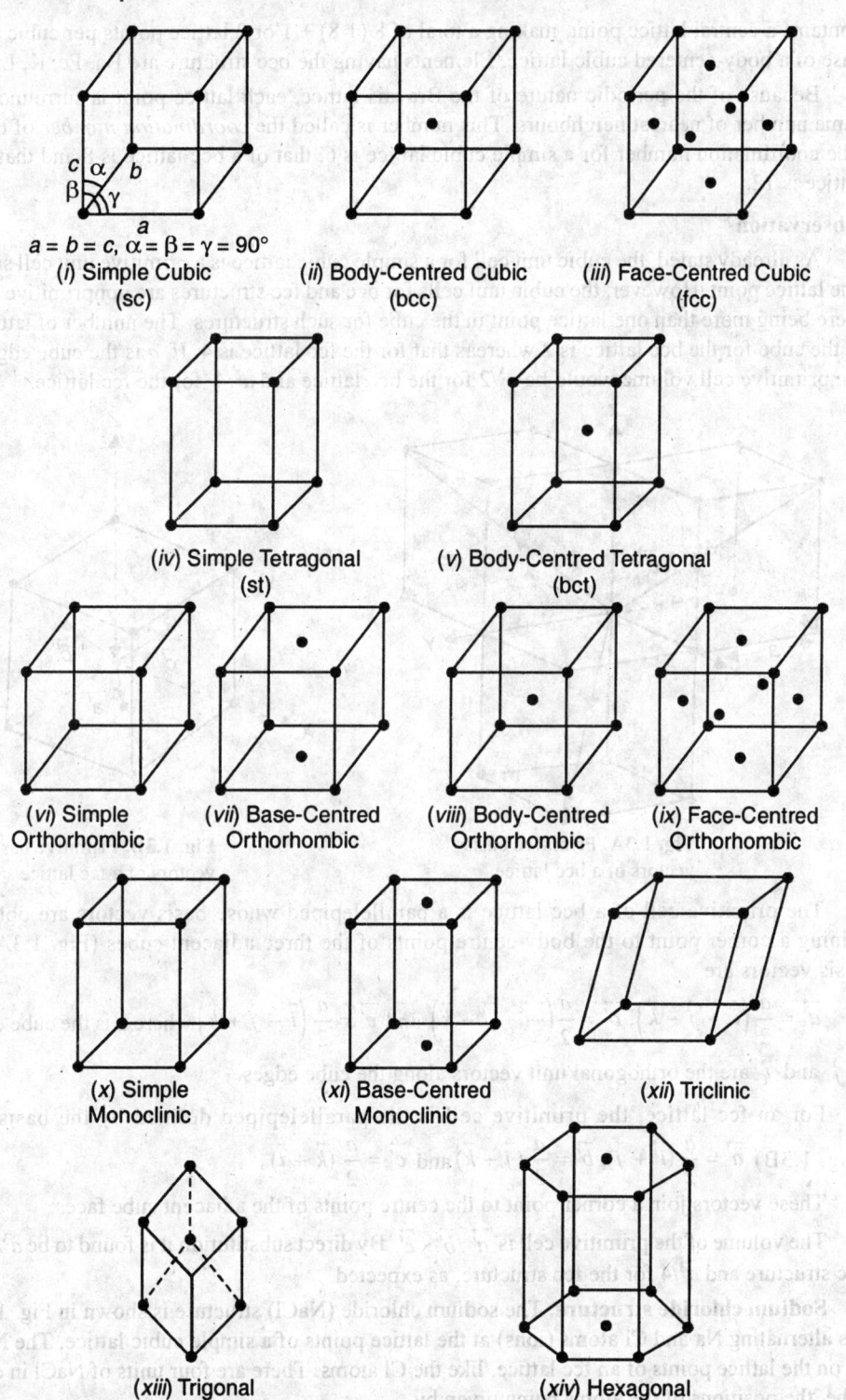

$a = b = c$, $\alpha = \beta = \gamma = 90°$
(i) Simple Cubic
(sc)

(ii) Body-Centred Cubic
(bcc)

(iii) Face-Centred Cubic
(fcc)

(iv) Simple Tetragonal
(st)

(v) Body-Centred Tetragonal
(bct)

(vi) Simple Orthorhombic

(vii) Base-Centred Orthorhombic

(viii) Body-Centred Orthorhombic

(ix) Face-Centred Orthorhombic

(x) Simple Monoclinic

(xi) Base-Centred Monoclinic

(xii) Triclinic

(xiii) Trigonal

(xiv) Hexagonal

Fig. 1.3. Fourteen Bravais lattices.

A bcc structure, on the other hand is formed by adding one lattice point at the centre of each simple cubic lattice. So, in addition to 8 corner lattice points shared equally among 8 cells, each cell contains a central lattice point, making a total of 8 (1/8) + 1 or 2 lattice points per cubic cell in the case of a body-centered cubic lattice. Elements having the bcc structure are Ba, Fe, K, Li, Na etc.

Because of the periodic nature of the Bravais lattice, each lattice point is surrounded by the same number of nearest neighbours. This number is called the *coordination number* of the lattice. The coordination number for a simple cubic lattice is 6, that of a bcc lattice is 8 and that of an fcc lattice is 12.

Observation

As already stated, the cubic unit cell for a simple cubic lattice is a primitive unit cell since it has one lattice point. However, the cubic unit cells for bcc and fcc structures are nonprimitive unit cells, there being more than one lattice point in the cube for such structures. The number of lattice points in the cube for the bcc lattice is 2 whereas that for the fcc lattice is 4. If a is the cube edge, clearly the primitive cell volume would be $a^3/2$ for the bcc lattice and $a^3/4$ for the fcc lattice.

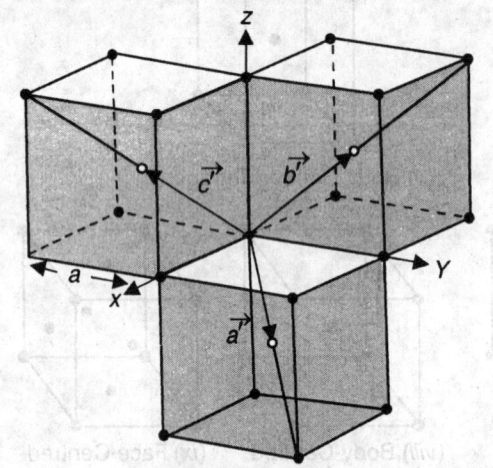

Fig. 1.3A. Primitive basis vectors of a bcc lattice

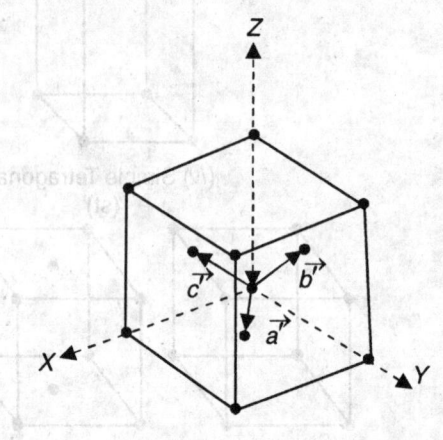

Fig. 1.3B. Primitive basis vectors of a fcc lattice

The primitive cell of a bcc lattice is a parallelepiped whose basis vectors are obtained by joining a corner point to the body-centre points of the three adjacent cubes (Fig. 1.3A). These basis vectors are

$\vec{a'} = \frac{a}{2}(\vec{i} + \vec{j} - \vec{k})$, $\vec{b'} = \frac{a}{2}(-\vec{i} + \vec{j} + \vec{k})$ and $\vec{c'} = \frac{a}{2}(\vec{i} - \vec{j} + \vec{k})$ where a is the cube edge, and $\vec{i}, \vec{j}$ and $\vec{k}$ are the orthogonal unit vectors along the cube edges.

For an fcc lattice, the primitive cell is the parallelepiped defined by the basis vectors (Fig. 1.3B) $\vec{a'} = \frac{a}{2}(\vec{i} + \vec{j})$, $\vec{b'} = \frac{a}{2}(\vec{j} + \vec{k})$ and $\vec{c'} = \frac{a}{2}(\vec{k} + \vec{i})$.

These vectors join a corner point to the centre points of the adjacent cube faces.

The volume of the primitive cell is $\vec{a'} \cdot \vec{b'} \times \vec{c'}$. By direct substitution it is found to be $a^3/2$ for the bcc structure and $a^3/4$ for the fcc structure, as expected.

Sodium chloride structure. The sodium chloride (NaCl) structure is shown in Fig. 1.4 (a). It has alternating Na and Cl atoms (ions) at the lattice points of a simple cubic lattice. The Na atoms lie on the lattice points of an fcc lattice, like the Cl atoms. There are four units of NaCl in each unit cube, the positions of the atoms being given by

Cl - (0, 0, 0); $\left(\frac{a}{2}, \frac{a}{2}, 0\right)$; $\left(\frac{a}{2}, 0, \frac{a}{2}\right)$; $\left(0, \frac{a}{2}, \frac{a}{2}\right)$

$$\text{Na} - \left(\frac{a}{2},\frac{a}{2},\frac{a}{2}\right); \left(0,0,\frac{a}{2}\right); \left(0,\frac{a}{2},0\right); \left(\frac{a}{2},0,0\right)$$

The distance between two successive Cl (or Na) atoms (ions) along the cube edge is the lattice constant a.

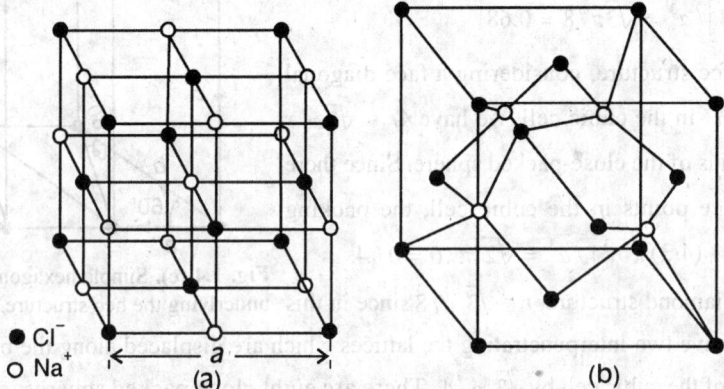

- ● Cl⁻
- ○ Na⁺

Fig. 1.4: (a) NaCl structure, (b) diamond structure.

It is evident from the figure that each atom is surrounded by six atoms of the opposite kind. Crystals whose structures have NaCl arrangement are KCl, PbS, MnO, MgO, etc.

Diamond Structure. A diamond crystal is formed by carbon atoms. The diamond lattice [Fig. 1.4 (b)] consists of two interpenetrating fcc lattices which are displaced along the body diagonal of the cubic cell by one quarter the diagonal length. The coordination number of the diamond lattice is 4. It is not a Bravais lattice since the arrangement of points surrounding a given point is different from that of its nearest neighbours. The elements which crystalize in the diamond structure are diamond, silicon, germanium and grey tin.

Zincblende Structure. Zincblende structure has an equal number of zinc and sulphur ions distributed on a diamond lattice so that each lattice point has four nearest neighbours of opposite kind. Some compounds which crystallize in the zincblende structure are CdS, HgTe, AlP, GaP, GaAs, InAs, InSb etc.

Packing fraction. The packing of lattice points or atoms in a unit cell may be more dense or close in one structure than in the other. One measure of this packing is the coordination number. The higher the coordination number, the more closely packed the structure. The coordination numbers of the simple cubic, fcc, bcc, and diamond structures are respectively 6, 12, 8, and 4. So, of the three cubic Bravais lattices, the fcc is the most dense and the simple cubic is the least dense. The diamond structure is less dense than any of the three cubic Bravais lattices.

Another measure of the packing of atoms in the cell is the *packing fraction*. Suppose that identical solid spheres are so distributed in space that their centres lie on the lattice points of the crystal structure with the neighbouring spheres just touching without overlapping. Such an arrangement of spheres is referred to as a *close-packed arrangement*. *The ratio of the volume of the spheres in the unit cell to the volume of the unit cell is called the packing fraction of the structure.* The packing fraction is also referred to as the *packing efficiency* or the *packing factor*.

For a simple cubic structure of lattice constant a the radius of a close-packed sphere is $a/2$. As there is one lattice point in the cell, the volume of the sphere in the cell is $(4/3)\pi(a/2)^3$. The volume of the cell is a^3. Hence, the packing fraction of the simple cubic lattice is $[(4/3)\pi(a/2)^3]$ $a^3 = \pi/6 = 0.52$.

For the bcc structure, if r is the radius of the close-packed sphere, then considering a body diagonal of length $\sqrt{3}\,a$ in the cubic unit cell, we have $4r = \sqrt{3}\,a$, so that $r = \sqrt{3}\,a/4$. Since there are two lattice points in the unit cell, the volume of the close-packed spheres in the cell is $2 \times (4/3) \pi r^3 = \sqrt{3} \pi a^3 /8$. Hence, the packing fraction is $\left[\sqrt{3} \times \pi a^3 /8\right]/a^3 = \sqrt{3}\pi/8 = 0.68$.

For the fcc structure, considering a face diagonal of length $\sqrt{2} a$ in the cubic cell, we have $4r = \sqrt{2} a$, r being the radius of the close-packed sphere. Since there are four lattice points in the cubic cell, the packing fraction is $[4 \times (4/3) \pi r^3]/a^3 = \sqrt{2} \pi/6 = 0.74$.

For the diamond structure, $r = \sqrt{3} a/8$ since in this structure we have two interpenetrating fcc lattices which are displaced along the body diagonal of length $\sqrt{3} a$ of the cubic cell by $\sqrt{3} a/4$. There are eight close-packed spheres, each of radius r, in the cell so that the packing fraction is $8 \times (4/3) \pi r^3 /a^3 = \sqrt{3}\pi/16 = 0.34$

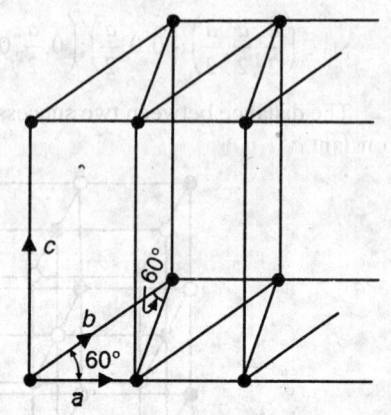

Fig. 1.4 (c). Simple hexagonal Bravais lattice underlying the hcp structure.

Hexagonal Close-Packed Structure. It is not a Bravais lattice but is important since about 30 elements, some of which are Cd, La, Mg, and Zn, crystallize in this form. The basic element of the hexagonal close-packed (hcp) structure is a simple hexagonal Bravais lattice of which the three primitive vectors $\vec{a}, \vec{b}$, and $\vec{c}$ are as shown in Fig. 1.4 (c). $\vec{a}$ and $\vec{b}$ generate a lattice of equilateral triangles in the horizontal plane and $\vec{c}$ stacks the planes a distance c above one another. The hcp structure consists of two such interpenetrating simple hexagonal Bravais lattices, displaced vertically by a distance $c/2$ along the common $\vec{c}$-axis and displaced horizontally such that the points of one lie directly above the centroids of the triangles described by the points of the other. The name 'close-packed' arises since close-packed hard spheres can be stacked in this structure. Close-packed hard spheres are first placed in the triangular lattice of the first layer. In the depressions left between the spheres, the next layer of hard spheres are placed, forming a second triangular layer which is shifted with respect to the first. In the depressions left between the spheres of the second layer, another layer of spheres is placed so that these spheres lie directly over the spheres of the first layer. The process is repeated, the fourth layer lying directly over the second, and so on. The resulting lattice is hcp with $c = \sqrt{8/3} \times a = 1.633a$. The value of $c/a = 1.633$ is ideal and is not achieved if the physical units in the hcp structure are not close-packed spheres. In fact, for most of the materials having this structure c/a departs from this ideal value.

Observations

The operations which transform a crystal to itself are called *symmetry operations*. Such operations are *translation, rotation, reflection,* and *inversion,* or a combination of them.

The translation symmetry operation implies that a lattice point $\vec{r}$ under the lattice translation vector operation $\vec{T}$, gives a point $\vec{r}'$ which is identical to $\vec{r}$, i.e., $\vec{r}' = \vec{r} + \vec{T}$, where $\vec{T} = h\vec{a} + k\vec{b} + l\vec{c}$; h, k, l are integers and $\vec{a}, \vec{b}, \vec{c}$ are primitive basis vectors.

If the rotation of a lattice about an axis by an angle $2\pi/n$ leaves the lattice invariant, the lattice is said to have n-fold rotation symmetry about the axis. Compatible with translational symmetry, n takes the values 1, 2, 3, 4, and 6 only. A two-dimensional square lattice repeats itself with a minimum rotation of 90° about an axis passing through the centre of the square and perpendicular to its plane. Thus the two-dimensional square lattice has 4-fold rotation symmetry. A regular hexagon has correspondingly 6-fold rotation symmetry.

If a plane divides the crystal lattice into two identical parts which are mirror images of each other, the lattice is said to have reflection symmetry. Again, if in a lattice, each point located at $\vec{r}$ relative to a lattice point, has an identical point located at $-\vec{r}$ relative to the same lattice point, the lattice is said to possess inversion symmetry.

The *essential symmetry* of a cubic crystal is that it has a 3-fold rotational symmetry about each of the four body diagonals of the cubic cell. A cubic lattice has also a 4-fold rotational symmetry about any of the cube edge, a reflection symmetry about a cube face, etc. Because of its symmetry, the cubic crystal is referred to as the *isometric crystal*.

1.6. Crystal Planes and Miller Indices

The position and orientation of a crystal plane within the crystal may be determined by any three noncollinear points in the plane. If each of the points lies on a crystal axis, the position and the orientation of the plane may be obtained by specifying the locations of the points along the axes in terms of the lattice constants or the basis vectors. For example, if the lattice points contained in the plane have coordinates (4, 0, 0), (0, 1, 0), and (0, 0, 2) relative to the axis vectors from some origin, the location of the plane within the crystal may be specified by the three numbers 4, 1, 2. However, for structure analysis it is more convenient to specify the orientation of a plane by Miller indices, defined as follows (Fig. 1.5):

(*i*) Take any lattice point in the crystal and form coordinate axes with this lattice point as origin in the directions of the basis vectors.

(*ii*) Determine the intercepts of the plane on the axes. Express these intercepts as multiples of the basis vectors along the axes.

(*iii*) Find the reciprocals of these numbers and reduce them to the smallest three integers h, k, l having the same ratio. Then the Miller indices of the plane are given by (*hkl*).

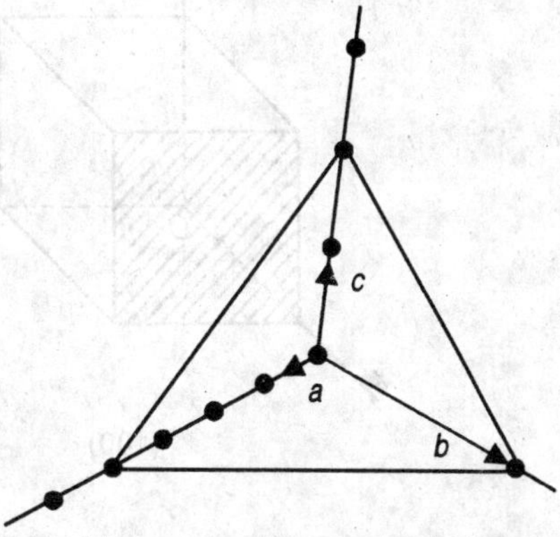

Consider the plane as shown in Fig. 1.5. The intercepts of the plane on the axes as multiples of the basis vectors a, b and c are 4, 1 and 2, respectively. The reciprocals of these intercepts are $\frac{1}{4}, 1$, and $\frac{1}{2}$. The Miller indices are obtained

Fig. 1.5. Determination of Miller indices.

by reducing them to three smallest integers having the same ratio. This is done by multiplying each of the reciprocals by 4 resulting in 1, 4, 2. Therefore the Miller indices are (142).

If a plane is parallel to one or two of the basis vectors $\vec{a}, \vec{b}, \vec{c}$ the corresponding intercept or intercepts will be at infinity and the corresponding Miller indices are zero. If, for example, a plane is parallel to $\vec{c}$ and has intercepts $4a$ and b, the Miller indices of the plane are (140). On the other hand, for a plane parallel to both $\vec{b}$ and $\vec{c}$ and having intercept $4a$, the Miller indices are (100).

It should be noted that the Miller indices do not refer to a particular plane but a set of parallel planes having a specific orientation. This is because the Miller indices refer to the ratio of the reciprocal intercepts and not the individual intercepts. Since parallel planes in a crystal behave identically, it is necessary to specify the orientation of a set of parallel planes rather than locate a particular plane. The *significance* of the Miller indices is that they give such specifications.

For a plane having an intercept on the negative side of an axis the corresponding index is negative. This is indicated by placing a minus sign on the top of the corresponding index :

($h\bar{k}l$). Some important planes in a cubic crystal and their Miller indices are illustrated in Fig. 1.6.

A number of planes whose Miller indices differ by permutation of numbers or of minus signs may be crystallographically equivalent in respect of density of lattice points and interplanar spacing. Consider, for example, a cubic lattice. In this case, the planes obtained by permutations of the three Miller indices among themselves, such as (hkl), (khl), (lhk), etc. are, by symmetry, equivalent. Besides, the planes (hkl), ($\bar{h}kl$), ($hk\bar{l}$), ($h\bar{k}l$) etc. obtained by assigning various combinations of minus signs to the Miller indices for cubic lattices are crystallographically equivalent. It is customary to represent the complete set of the equivalent planes of which (khl) is a member by enclosing the Miller indices in curly brackets: {hkl}. In a cubic lattice, the cube faces are represented by (100), (010), (001), ($\bar{1}$00), (0$\bar{1}$0), and (00$\bar{1}$). Because they are equivalent in the above sense, we can represent the set of cube faces by {100}.

The *indices of a direction* in a crystal can be expressed as a set of the smallest integers which have the same ratios as the components of a vector in the desired direction, written as multiples of the basis vectors $\vec{a}, \vec{b}, \vec{c}$. The direction index of a vector $h\vec{a} + k\vec{b} + l\vec{c}$ is thus [hkl]. The letters h, k and l are integers and have no common factor greater than unity. The square brackets enclosing the three indices denote direction indices. In a cubic crystal, the x-axis is the [100] direction, the $-y$ axis is the [0$\bar{1}$0] direction and the $-z$ axis is the [00$\bar{1}$] direction.

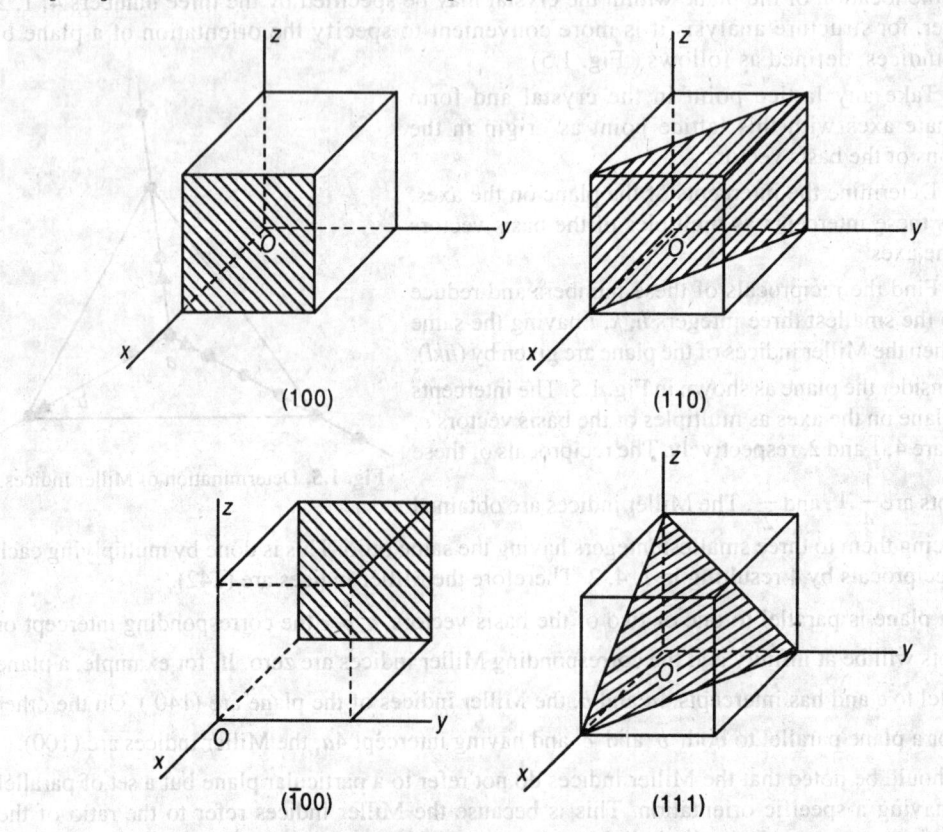

Fig. 1.6. Miller indices of some planes in a cubic crystal.

Like crystal planes with different Miller indices, directions with different direction indices may be crystallographically equivalent. This is because normals to equivalent planes are equivalent directions. The entire set of crystallographically equivalent directions of which [hkl] is a member, is written by enclosing the direction indices in angle brackets, viz. <hkl>. In cubic crystals, a direction with direction indices [hkl] is normal to a plane having Miller indices (hkl), but this is not generally valid for other systems. That is why bcc and fcc structures are referred with respect to the cubic unit cell rather than with respect to their primitive unit cells.

Position of a point in the unit cell

The positions of points in a unit cell are quoted in terms of fractional parts of the basis vector magnitudes along the respective coordinate directions with a corner lattice point as the origin. The coordinates of the centre point of a unit cell, for example, are given by $(\frac{1}{2}, \frac{1}{2}, \frac{1}{2})$. Similarly, the coordinates of the face centres are $(\frac{1}{2}, \frac{1}{2}, 0)$, $(0, \frac{1}{2}, \frac{1}{2})$, $(\frac{1}{2}, 0, \frac{1}{2})$, etc.

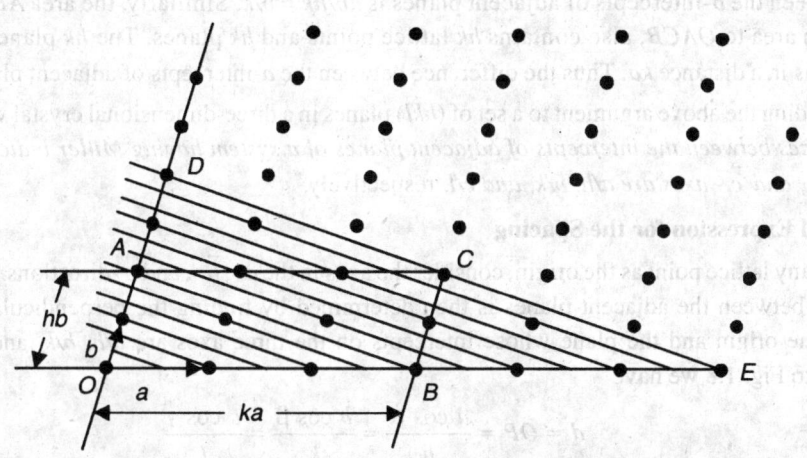

Fig. 1.7. A set of parallel planes in a two-dimensional lattice.

1.7. SPACING BETWEEN ADJACENT PLANES IN THE LATTICE

The crystallographic planes pass through the lattice points in the crystal. Consider a set of parallel crystallographic planes in a two-dimensional crystal, as shown in Fig. 1.7. The system of planes is determined by the intercepts ka and hb along $\vec{a}$ and $\vec{b}$ axes, respectively. Therefore the Miller indices are (hk) [or $(h k 0)$ if the planes are assumed to exist parallel to $\vec{c}$ -axis normal to the plane of the paper in a three-dimensional crystal]. The letters h and k represent the smallest integers possible in the particular situation. In Fig. 1.7 $h = 2$ and $k = 3$; therefore the figure is a representation of the (23) or (230) planes. It is evident from the figure that the intercepts made by adjacent planes on the $\vec{a}$ and $\vec{b}$ axes differ by $a/2$ and $b/3$, respectively. Likewise, for a set of Miller indices (hk), these intercepts would differ by a/h and b/k. This is proved below.

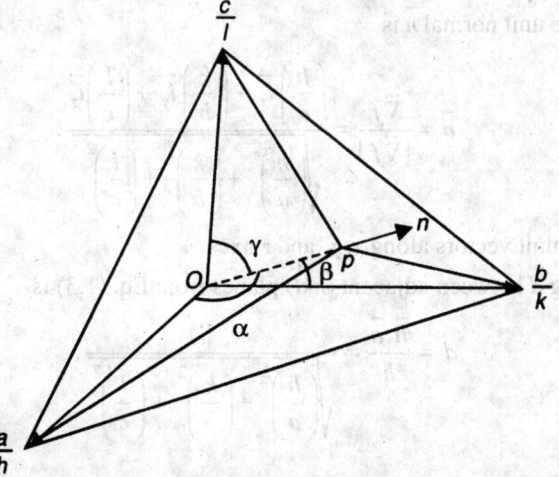

Fig. 1.8. Direction angles of the normal to the (hkl) plane with respect to $\vec{a}, \vec{b}$ and $\vec{c}$ axes.

We consider any lattice point O as the origin and form the $\vec{a}$ and the $\vec{b}$ axes. We take a length $OA = hb$ on the $\vec{b}$-axis and a length $OB = ka$ on the $\vec{a}$-axis (h, k are integers since they denote Miller indices) and consider the parallelogram $OACB$. The area $OACB$ contains hk lattice points since it is made up of hk unit cells each of which has one lattice point. The region $ABCD$ being equal in area to $OACB$ also contains hk lattice points and just hk lattice planes, one for each of the hk lattice points within this area. The hk planes in $ABCD$ intersect the $\vec{b}$-axis in a distance hb. Therefore the difference between the b-intercepts of adjacent planes is $hb/hk = b/k$. Similarly, the area $ABEC$ which is equal in area to $OACB$, also contains hk lattice points and hk planes. The hk planes intersect the $\vec{a}$-axis in a distance ka. Thus the difference between the a-intercepts of adjacent planes is a/h.

Extending the above argument to a set of (hkl) planes in a three-dimensional crystal we find that the *distances between the intercepts of adjacent planes of a system having Miller indices (hkl) on the $\vec{a}$-, $\vec{b}$-, and $\vec{c}$-axes are a/h, b/k, and c/l*, respectively.

Analytical Expression for the Spacing

With any lattice point as the origin, consider the axes in the $\vec{a}$-, $\vec{b}$-, and $\vec{c}$-directions. The actual spacing d between the adjacent planes is then determined by finding the perpendicular distance between the origin and the plane whose intercepts on the three axes are a/h, b/k, and c/l. With reference to Fig. 1.8 we have

$$d = OP = \frac{a \cos \alpha}{h} = \frac{b \cos \beta}{k} = \frac{c \cos \gamma}{l} \qquad \ldots(1.2)$$

where α, β, and γ represent the angles between the normal to the plane and the $\vec{a}$, $\vec{b}$ and $\vec{c}$ axes, respectively. If the unit vector normal to the plane is denoted by $\vec{n}$, then $\vec{n} \cdot \vec{a} = a \cos \alpha$, $\vec{n} \cdot \vec{b} = b \cos \beta$ and $\vec{n} \cdot \vec{c} = c \cos \gamma$. Therefore Eq. (1.2) gives

$$d = \frac{\vec{n} \cdot \vec{a}}{h} = \frac{\vec{n} \cdot \vec{b}}{k} = \frac{\vec{n} \cdot \vec{c}}{l} \qquad \ldots(1.3)$$

In an orthogonal lattice, if the x-axis coincides with $\vec{a}$, the y-axis with $\vec{b}$, and the z-axis with $\vec{c}$, then the equation of the plane having intercepts a/h, b/k and c/l is

$$f(x, y, z) = \frac{hx}{a} + \frac{ky}{b} + \frac{lz}{c} = 1. \qquad \ldots(1.4)$$

If $f(x, y, z) = $ constant is the equation of a surface, then $\vec{\nabla} f$ represents the vector normal to the surface. Therefore, the unit normal $\vec{n}$ is

$$\vec{n} = \frac{\vec{\nabla} f}{|\vec{\nabla} f|} = \frac{\left(\frac{h}{a}\right)\vec{i}_x + \left(\frac{k}{b}\right)\vec{i}_y + \left(\frac{l}{c}\right)\vec{i}_z}{\sqrt{\left(\frac{h}{a}\right)^2 + \left(\frac{k}{b}\right)^2 + \left(\frac{l}{c}\right)^2}}, \qquad \ldots(1.5)$$

$\vec{i}_x$, $\vec{i}_y$, and $\vec{i}_z$ being unit vectors along x, y, and z axes.

Hence the spacing d between adjacent (hkl) planes from Eq. (1.3) is

$$d = \frac{\vec{n} \cdot \vec{a}}{h} = \frac{1}{\sqrt{\left(\frac{h}{a}\right)^2 + \left(\frac{k}{b}\right)^2 + \left(\frac{l}{c}\right)^2}}, \qquad \ldots(1.6)$$

since $\vec{a} = a\vec{i}_x$.

Crystals and Their Properties

For a cubic crystal, $a = b = c$ and Eq. (1.6) becomes

$$d = \frac{a}{\sqrt{h^2 + k^2 + l^2}} \qquad \ldots(1.7)$$

1.8. Anisotropy of The Physical Properties of Single Crystals

Generally, the different physical properties such as electrical and thermal conductivities, elastic deformation, refractive index, electric and magnetic susceptibilities, etc. of single crystals show anisotropy, i.e. they are different along different crystallographic directions. For example, Young's modulus of elasticity of zinc single crystal has a maximum value of 12.6×10^{10} Pa and a minimum value of 6.5×10^{10} Pa along two different directions inside the crystal, reflecting considerable anisotropy.

A crystal may show anisotropy in some physical properties, but it may be isotropic for some other properties. For example, elastic properties of single crystals of the cubic type generally exhibit anisotropy, but the refractive index is isotropic in them. This difference in the properties of different crystals arises from the fact that the arrangement of atoms is not the same in all types of crystals. Let us consider the anisotropy of the electrical conductivity of a single crystal. In general, the electric current density $\vec{J}$ is not in the direction of an externally applied electric field $\vec{E}$. In a rectangular cartesian coordinate system, the x-, y- and z-components of $\vec{J}$, i.e., J_x, J_y and J_z can be written as a linear function of the x-, y- and z-components of $\vec{E}$, i.e. E_x, E_y, and E_z in the following manner:

$$J_x = \sigma_{xx} E_x + \sigma_{xy} E_y + \sigma_{xz} E_z$$
$$J_y = \sigma_{yx} E_x + \sigma_{yy} E_y + \sigma_{yz} E_z$$
and
$$J_z = \sigma_{zx} E_x + \sigma_{zy} E_y + \sigma_{zz} E_z$$

The nine coefficients σ_{lm} ($l, m \equiv x, y, z$) represent the components of the conductivity tensor which is a tensor of the second rank. In actual practice, $\sigma_{lm} = \sigma_{ml}$ ($l \neq m$), so that the conductivity tensor is symmetric with six components in place of nine.

Each crystal possesses three particular directions, called the *principal axes*, for which we have

$$J_x = \sigma_1 E_x, \ J_y = \sigma_2 E_y \text{ and } J_z = \sigma_3 E_z$$

Here E_x, E_y and E_z act along the three principal axes, respectively. It follows that if the electric field $\vec{E}$ is along any one of the principal axes, the current density $\vec{J}$ flows in the direction of $\vec{E}$. Thus, if $\vec{E}$ acts along the principal axis labelled the x-axis, we have $J_x = \sigma_1 E_x$ and $J_y = J_z = 0$. The three coefficients σ_1, σ_2 and σ_3 are referred to as the *principal electrical conductivities* of the crystal.

In a cubic crystal, the electrical conductivity is isotropic, i.e, $\sigma_1 = \sigma_2 = \sigma_3$. For a hexagonal crystal, two of the principal conductivities are equal. The conductivity of a crystal can be expressed in terms of the principal conductivities. The conductivity in a direction having direction cosines (l, m, n) with respect to the principal axes within the crystal is given by $\sigma = l^2 \sigma_1 + m^2 \sigma_2 + n^2 \sigma_3$.

Unless special care is taken, large sizes of single crystals are difficult to grow. Usually, the crystal is a polycrystal containing a large number of minute single crystals oriented at random. Because of this randomness, the physical properties of a polycrystal appear to be isotropic.

1.9. Worked-out Problems

1. Prove that the [hkl] direction is normal to the (hkl) plane in a cubic crystal.

Ans. The cubic crystal is orthogonal. Therefore, $\alpha = \beta = \gamma = 90°$ and $a = b = c$.

We consider x, y, and z axes along the crystallographic axes, and a plane having intercepts a/h, a/k and a/l on the axes. The equation of this (hkl) plane is

$$f(x, y, z) = \frac{hx}{a} + \frac{ky}{a} + \frac{lz}{a} = 1.$$

The normal to the surface is along the vector $\vec{\nabla} f$.

Now, $$\vec{\nabla} f = \frac{h}{a} \vec{i}_x + \frac{k}{a} \vec{i}_y + \frac{l}{a} \vec{i}_z$$

The indices of the direction of $\vec{\nabla} f$ are then h, k, l. This means that the normal to the (hkl) plane is along the $[hkl]$ direction.

2. The lattice constant of an fcc lattice is 6.38 Å. Find (i) the distance between a corner atom at the base and the atom at the centre of the top face, and (ii) the largest distance between two atoms in the cubic cell.

Ans. (i) With reference to Fig. 1.9 the required distance is

$$s = \sqrt{\left(a/\sqrt{2}\right)^2 + a^2}$$

$$= \sqrt{3/2}\, a = 1.225 \times 6.38 \text{ Å} = 7.81 \text{ Å}.$$

(ii) The largest distance between two atoms in the cell is that between the atoms at the ends of a body diagonal. Hence the required distance is $\sqrt{3}\, a = 1.732 \times 6.38$ Å $= 11.05$ Å.

3. The diamond crystal structure has the cube edge of 3.56 Å. Calculate the distance between the nearest neighbours and the number of atoms per cm³.

Ans. The diamond structure consists of two interpenetrating fcc lattices displaced along the body diagonal of the cubic cell by 1/4th the length of that diagonal. The nearest neighbour distance (d) is therefore given by

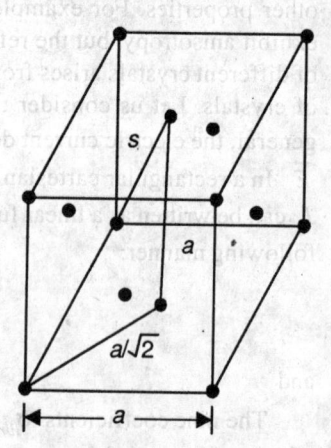

Fig. 1.9. Figure for Prob. 2.

$$d = \frac{1}{4} \times \text{length of the body diagonal}$$

If a is the edge length of the cubic cell, the length of the body diagonal is

$$\sqrt{3}\, a = 1.732 \times 3.56 \text{ Å}$$

Hence $$d = \frac{1}{4} \times 1.732 \times 3.56 \text{Å} = 1.54 \text{ Å}$$

In an fcc structure there are 4 atoms in a cubic cell of volume a^3. Therefore the number of atoms per unit volume is

$$\frac{4}{a^3} = \frac{4}{(3.56)^3} \text{ Å}^{-3} = 0.886 \times 10^{23} \text{cm}^{-3}.$$

Since the diamond structure consists of two interpenetrating fcc lattices, the number of atoms per unit volume of diamond will be

$$2 \times 0.886 \times 10^{23} = 1.772 \times 10^{23} \text{ cm}^{-3}$$

4. (a). Show that the attractive forces between the atoms of a crystal must vary more slowly with distance than the repulsive forces.

(b) The potential energy of a pair of atoms in a crystal is of the form $\frac{A}{r^9} - \frac{B}{r}$, when the separation is r. Interpret these two terms.

Crystals and Their Properties

Assuming the equilibrium separation to be 2.8 Å and the dissociation energy 8×10⁻¹⁹ joule, calculate A and B. **[C.U. 1986]**

Ans. (a) In a solid, two types of forces between the atoms come into play : (i) attractive forces keeping the atoms together, and (ii) repulsive forces which prevent compression of the solid. Let two atoms a and b apply attractive and repulsive forces on each other so that the potential energy of b in the field of a is expressed by:

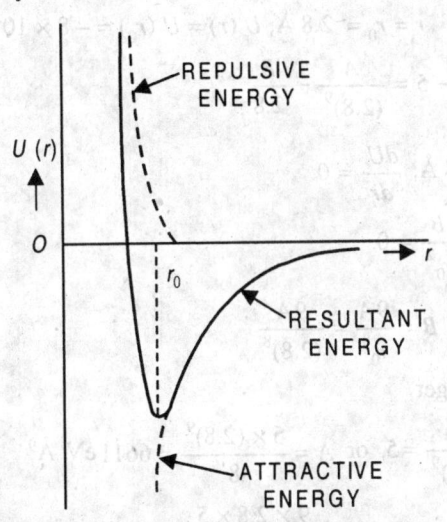

Fig. 1.10. Variation of potential energy with distance.

$$U(r) = -\frac{A}{r^n} + \frac{B}{r^m} \qquad \ldots(i)$$

where r is the distance between the atoms; m, n, A and B are positive quantities. The term $-A/r^n$ is the attractive energy resulting from the attractive forces whereas the term B/r^m is the repulsive energy due to the repulsive forces. As $r \to \infty$, $U(r) \to 0$. A stable crystal will form only if $U(r)$ shows a minimum for a finite value of r, say r_0 (Fig. 10). At this distance, the interatomic force $F(=-dU/dr)$ is zero. The distance between the atoms in the crystal will be r_0 and the energy required to dissociate the atom-pair, i.e. the *dissociation energy* (also called the *binding energy, bond energy or energy of cohesion*) will be equal to the positive quantity $-U(r_0)$.

If $U(r)$ is a minimum at $r = r_0$, then

$$\left.\frac{dU}{dr}\right|_{r=r_0} = 0, \text{ or } r_0^{m-n} = \frac{m}{n} \cdot \frac{B}{A} \qquad \ldots(ii)$$

Also,

$$\left.\frac{d^2U}{dr^2}\right|_{r=r_0} = -\frac{n(n+1)A}{r_0^{n+2}} + \frac{m(m+1)B}{r_0^{m+2}} > 0. \qquad \ldots(iii)$$

Substituting the value of r_0 from (ii) into (iii), we obtain $m > n$. This shows that the attractive forces must vary more slowly with r than the repulsive forces. The repulsive forces are thus *short-range forces*, being appreciable for very small interatomic distances.

(b) The first term $\frac{A}{r^9}$ is the repulsive energy component arising from the repulsive forces between the atoms. The second term $-\frac{B}{r}$ represents the Coulombic attractive energy resulting from the mutual attraction between the atoms.

The attractive forces originate from the interactions between the positive nuclear charge of one atom and the negative electron cloud of the other. The repulsive forces appear when the interatomic distance is so small that the electron clouds of the atoms tend to overlap. The mutual repulsion of the positively charged nuclei is also responsible for such forces.

The potential energy is given by

$$U(r) = \frac{A}{r^9} - \frac{B}{r}.$$

At $r = r_0 = 2.8$ Å, $U(r) = U(r_0) = -8 \times 10^{-19}$ J $= -5$ eV.

Hence
$$-5 = \frac{A}{(2.8)^9} - \frac{B}{2.8} \qquad \ldots(i)$$

Also, at $r = r_0 = 2.8$ Å, $\frac{dU}{dr} = 0$.

i.e.,
$$-\frac{9A}{r_0^{10}} + \frac{B}{r_0^2} = 0$$

or,
$$B = \frac{9A}{r_0^8} = \frac{9A}{(2.8)^8} \qquad \ldots(ii)$$

From (i) and (ii) we get

$$\frac{8A}{(2.8)^9} = 5, \text{ or } A = \frac{5 \times (2.8)^9}{8} = 6611 \text{ eV. Å}^9$$

and $B = \dfrac{9 \times 2.8 \times 5}{8} = 15.75$ eV.Å

5. The potential energy of a system of two atoms in a crystal is $U = -A/r^n + B/r^m$ $(m > n)$, where r is the interatomic distance, and A and B are constants. Show that the interatomic distance r_c at which the interatomic force F is a minimum, is greater than the spacing r_0 at which a stable crystal is formed. What maximum force is required to dissociate the atom-pair? Show graphically the variation of U and F with r.

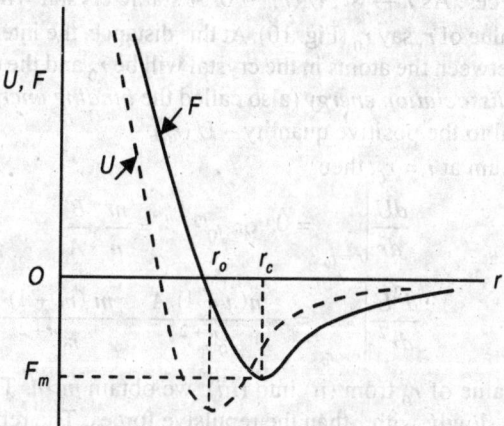

Fig. 1.11. Variation of U (broken curve) and F (continuous curve) with r.

Ans. At $r = r_0$, the potential energy U is a minimum. Here, $F = 0$, the attractive and the repulsive forces balancing each other. We have

$$F = -\frac{dU}{dr} = -\frac{nA}{r^{n+1}} + \frac{mB}{r^{m+1}} \qquad \ldots(i)$$

At $r = r_0$, $F = 0$. giving $r_0 = \left(\dfrac{mB}{nA}\right)^{\frac{1}{m-n}}$

Crystals and Their Properties 189

At $r = r_c$, F is a minimum, i.e. $\dfrac{dF}{dr} = 0$. This gives

$$r_c = \left[\dfrac{m(m+1)B}{n(n+1)A}\right]^{\frac{1}{m-n}} = r_0 \left(\dfrac{m+1}{n+1}\right)^{\frac{1}{m-n}}$$

Since $m > n$, clearly $r_c > r_0$.

The minimum force F_m is obtained by putting $r = r_c$ in Eq. (i). F_m is negative, implying that it is an attractive force. A minimum positive force $F_d = -F_m$ is required to overcome the attraction and dissociate the atom-pair. We have

$$F_d = \dfrac{nA}{r_c^{n+1}} - \dfrac{mB}{r_c^{m+1}}$$

The variation of the potential energy U and the force F with the distance r is displayed in Fig. 1.11.

6. Sodium chloride (NaCl) crystal has a cubic structure. If the molecular weight of NaCl is 58.46 and its density is 2.17 gm per cm^3, find the distance between two adjacent atoms and the lattice constant in the NaCl crystal. **(cf. Burd. U. 1995, C.U. 2004)**

Ans. Mass of a NaCl molecule,

$$M = \dfrac{\text{Molecular weight in gm.}}{\text{Avogadro number}}$$

$$= \dfrac{58.46}{6.02 \times 10^{23}} \text{ gm} = 9.7 \times 10^{-23} \text{ gm.}$$

Number of NaCl molecules per unit volume $= \dfrac{\text{density}}{M}$

$$= \dfrac{2.17}{9.7 \times 10^{-23}} = 2.237 \times 10^{22} \text{ molecules per cm}^3.$$

Since NaCl is diatomic, the number of atoms per unit volume will be $2 \times 2.237 \times 10^{22} = 4.47 \times 10^{22}$ atoms/cm^3.

If a be the distance between two adjacent atoms, the volume a^3 will contain one atom due to the cubic structure. Therefore the number of atoms per unit volume will be $1/a^3$.

Hence, $\dfrac{1}{a^3} = 4.47 \times 10^{22}$

or, $a^3 = 2.237 \times 10^{-23}$ cm^3

or, $a = 2.82$ Å

The lattice constant is $2a = 5.64$ Å

7. Cs metal (atomic weight 130) has a cubic unit cell of side 6Å. If the density of Cs is 2g/cm^3, determine whether the unit cell is simple, face-centered, or body-centered.

(N. Beng. U. 2001)

Ans. Let x be the number of Cs atoms in cubic unit cell of side a. Then the number of atoms per unit volume is x/a^3. If N is Avogadro's number and A is the atomic weight, then the mass of one atom is A/N gm. So, the density is $\rho = \dfrac{x}{a^3} \cdot \dfrac{A}{N}$ g/cm^3, or $x = \dfrac{\rho a^3 N}{A} = \dfrac{2 \times 6^3 \times 10^{-24} \times 6.02 \times 10^{23}}{130} = 2$

So, the unit cell is body-centered.

8. Find the Miller indices of the plane containing the three lattice points $\vec{r_1} = \vec{a} - \vec{b}$, $\vec{r_2} = 2\vec{a} + \vec{c}$, and $\vec{r_3} = 3\vec{b} + \vec{c}$, where $\vec{a}, \vec{b}$ and $\vec{c}$ are the primitive basis vectors.

Ans. The vectors $\vec{r_1} - \vec{r_3}$ and $\vec{r_2} - \vec{r_3}$ lie in the plane in question. So, any point in the plane with position vector $\vec{r}$ satisfies the relationship $\vec{r} = \vec{r_3} + m(\vec{r_1} - \vec{r_3}) + n(\vec{r_2} - \vec{r_3}) = 3\vec{b} + \vec{c} + m(\vec{a} - 4\vec{b} - \vec{c}) + n(2\vec{a} - 3\vec{b}) = (m + 2n)\vec{a} + (3 - 4m - 3n)\vec{b} + (1 - m)\vec{c}$ where m and n are numbers. At the $\vec{a}$ intercept, the coefficients of $\vec{b}$ and $\vec{c}$ are zero, i.e., $3 - 4m - 3n = 0$ and $1 - m = 0$ giving $m = 1$ and $n = -1/3$. So, $m + 2n = 1 - 2/3 = 1/3$, which is the $\vec{a}$ intercept. Similarly, at the $\vec{b}$ intercept, $m + 2n = 0$ and $1 - m = 0$, so that $m = 1$ and $n = -1/2$. Thus the $\vec{b}$ intercept is $3 - 4m - 3n = 3 - 4 + 3/2 = 1/2$. At the $\vec{c}$ intercept, $m + 2n = 0$ and $3 - 4m - 3n = 0$, giving $m = 6/5$ and $n = -3/5$. Hence the $\vec{c}$ intercept is $1 - m = 1 - 6/5 = -1/5$. The reciprocals of the intercepts are 3, 2, and -5, respectively. So the plane is a $(32\bar{5})$ plane.

9. Prove that the density of lattice points (per unit area) in a lattice plane is d/τ; where τ is the volume of the primitive cell and d_1 is the separation between adjacent planes of the family to which the plane in question belongs.

Ans. We construct a primitive cell taking the origin at a lattice point in the plane in question, and drawing two primitive basis vectors in the plane and the third primitive basis vector from the origin to a lattice point in a neighbouring plane. The cell is a parallelepiped, the base of which having the area α lies in the plane in question. At each corner of the cell-base, there is a lattice point which is shared by four parallelograms of the plane meeting at that point. So, 1/4 of each of the four corner points belongs to the cell-base, so that the number of the lattice points in the area α is $4 \times (1/4) = 1$. Thus the number of lattice points per unit area of the plane is $1/\alpha$. The primitive cell volume is $\tau = \alpha d$, where d is the spacing between the plane and the adjacent parallel plane. So, the density of lattice points per unit area of the plane is $\dfrac{1}{\alpha} = \dfrac{d}{\tau}$ (proved).

10. A plane has intercepts of 2, 1 and 3Å on the crystallographic axes with the basis vectors in the ratio 1:2:3. What are the Miller indices of the plane?

Ans. Let the basis vectors be 1, 2, and 3Å, respectively. The intercepts of the plane relative to the basis vectors are 2/1, 1/2 and 3/3 or 2, 1/2, and 1. The reciprocals are 1/2, 2 and 1 or 1, 4, and 2. So, the Miller indices are (142).

11. In the NaCl crystal, the radius of the sodium ion is 0.98 Å and that of the chlorine ion is 1.82Å. The atomic masses of sodium and chlorine are 22.99 amu and 35.45 amu, respectively. Assuming that the sodium and the chlorine ions touch along the cube edges of the unit cell of NaCl, find the packing factor and the density of NaCl.

Ans. If a is the lattice parameter, we have $a = 2$ (radius of Na$^+$ + radius of Cl$^-$) $= 2(0.98 + 1.82)$ $= 5.6$ Å

The volume of the unit cell is $V_0 = a^3 = (5.6)^3$ Å$^3 = 175.6$ Å^3

The volume of the ions in the unit cell is $V = 4\left(\dfrac{4}{3}\pi r_1^3 + \dfrac{4}{3}\pi r_2^3\right)$ where r_1 and r_2 are the radii of Na$^+$ and Cl$^-$, respectively, Thus

$$V = 4 \times \dfrac{4}{3}\pi(0.98^3 \times 1.82^3) = 116.78 \text{ Å}^3.$$

The packing factor is $\dfrac{V}{V_0} = \dfrac{116.78}{175.6} = 0.665$

The mass of the unit cell is

$$M_0 = 4(22.99 + 35.45) \times 1.66 \times 10^{-27} = 388 \times 10^{-27}$$
$$= 388 \times 10^{-27} \text{ kg}$$

The volume of the cell is $V_0 = 175.6 \times 10^{-30}$ m^3.

So, density $= \dfrac{M_0}{V_0} = \dfrac{388 \times 10^{-27}}{175.6 \times 10^{-30}} = 2.21 \times 10^3$ kg / m^3.

12. Show that only one –, two –, three –, four –, and six–fold symmetry axes of rotation are possible in a crystal.

Ans. If a crystal lattice remains unchanged after rotation through a certain angle about an axis, each lattice point must return to its original position after an integral number of like rotations. So the angle of rotation is $2\pi/n$ where n is an integer. If a rotation through $2\pi/n$ leaves the lattice unchanged, the lattice has an n-fold symmetry axis.

Consider a crystal lattice where A and B are two lattice points (Fig. 1.12). If $\vec{T}$ is the lattice translation vector from A to B, we have $AB = T$. Suppose that the lattice is rotated through an angle θ in the clockwise direction about an axis passing through A and perpendicular to the plane of the paper. Let, by such rotation, a lattice point initially at C comes to A'. If the rotation leaves the lattice invariant, there was a lattice print at A' before the rotation. Another lattice point B' is obtained on the line through A' and parallel to AB if the original lattice is rotated through θ about B in the counterclockwise direction. There can be other lattice points between A' and B'. Therefore, generally consistent with the translational symmetry, $A' B' = pT$ where p is an integer, positive or negative (A negative p implies that B' is on the other side of A'). But $A' B' = T + 2T \cos\theta$. Thus

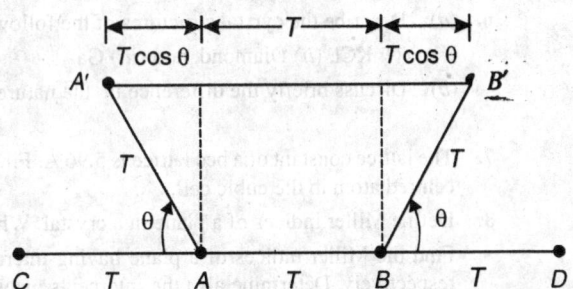

Fig. 1.12. A line of lattice points $A' B'$ parallel to a line of lattice points AB generated by rotation about A and B.

$$\dfrac{A'B'}{AB} = 1 + 2\cos\theta = p$$

or, $\cos\theta = \dfrac{p-1}{2}.$

Since $|\cos\theta| \le 1$, the only possible values of p are $p = -1$ giving $\theta = \pi = \dfrac{2\pi}{2}$, $p = 0$ giving $\theta = \dfrac{2\pi}{3}$, $p = 1$ giving $\theta = \dfrac{\pi}{2} = \dfrac{2\pi}{4}$, $p = 2$ giving $\theta = \dfrac{\pi}{3} = \dfrac{2\pi}{6}$, and $p = 3$ giving $\theta = 2\pi$.

Hence only one–, two–, three–, four–, and six–fold symmetry axes are possible in a crystal.

QUESTIONS

1. (a) What is the difference between a crystal and an amorphous solid ?   (C.U. 1996, 2005)
   (b) What is a single crystal? Name the different classes of crystals on the basis of the binding forces, and compare their basic properties.
2. (a) Describe the differences between ionic, covalent, and metallic binding in solid crystals.
   (C.U. 1983)
   (b) Explain the origin of the van der Waals interaction between two atoms of an inert gas (in the solid state).   (Burd. U. 1996, C.U. 2008)

(c) What are the important features of covalent crystals? **(Burd. U. 1999)**

3. (a) Define the following:

   (i) Bravais lattice, (ii) simple cubic lattice, (iii) body-centred cubic lattice, and (iv) face-centred cubic lattice.

   (b) Mention the number of lattice points per unit cell and the coordination number of (i) an sc lattice, (ii) an fcc lattice, and (iii) a bcc lattice. Name some materials having sc, fcc, and bcc lattice structures.

   (c) Define the various symmetry operations in a crystal lattice.

   (d) Name the seven crystal systems. How many types of cubic crystals are known? Which basic symmetry operation exists for cubic crystals? **(C.U. 2006)**

4. What is a hexagonal close-packed structure? Name some materials having this structure.

5. Define the term 'packing fraction'. What is the packing fraction for (i) a simple cubic lattice, (ii) a bcc lattice, (iii) an fcc lattice, and (iv) a diamond lattice?

6. (a) Describe the crystal structures of the following materials:

   (i) KCl, (ii) Diamond, and (iii) GaAs.

   (b) Discuss briefly the difference in the nature of bonding found in diamond and KBr crystals. **(C.U. 2000)**

7. The lattice constant of a bcc lattice is 5.96 Å. Find the distance between a corner atom and the body-centred atom in the cubic cell. **[Ans. 5.16 Å]**

8. Define Miller indices of a plane in a crystal. What is the utility of Miller indices? **(C.U. 1996)** Find the Miller indices of a plane having intercepts of $6a$, $4b$, and $2c$ on the x–, y–, and z-axes, respectively. Determine also the intercepts of two other planes one on each side of this plane and having these indices. Determine the interplanar spacing when the lattice is a cube of edge length a.

   [Ans. (2 3 6); $3a, 2b, c$; $12a, 8b, 4c$; $a/7$]

9. Chromium is a bcc Bravais lattice having the edge length of 2.88 Å for the cubic cell. Calculate the number of chromium atoms per unit volume of the crystal. **[Ans. 8.37×10^{22} / cm^3]**

10. Show that the differences between the intercepts of adjacent (hkl) planes on the $\vec{a}, \vec{b},$ and $\vec{c}$ crystallographic axes are a/h, b/k and c/l, respectively.

11. Derive the expression for the interplanar spacing d of the set of (hkl) planes of a cubic lattice.

    (C.U. 1994, 99, 2007, 08)

12. Determine the number of atoms per unit cell of silver which crystallises on an fcc lattice with one atom at each lattice point. **[Ans. 4]**

13. The density of bcc iron is 7.9 gm cm^{-3} and its atomic weight is 56. Calculate the length of the side of the cubic unit cell and the nearest neighbour distance **(C.U. 1986)**

    [**Hint.**: The bcc lattice contains two atoms in the cubic unit cell of volume a^3, where a is the edge length. Note that one mole, i.e., 56 gm contains 6.02×10^{23} atoms (Avogadro's number). The nearest neighbour distance is that between the body-centre atom and a corner atom, i.e., $\sqrt{3}\dfrac{a}{2}$]

    [Ans. 2.87 Å, 2.48 Å]

14. Polonium has a cubic unit cell of side 3.42 Å. If the atomic weight and density of Polonium are 210 and 8.72 g/cm^3, respectively, show whether the unit cell is simple, body-centered or face-centered.

    [Ans. simple]

15. Aluminium (atomic weight 27) crystallizes in the cubic form with $a = 4.05$ Å. The density of aluminium is 2.7 g/cm^3 Find the type of the unit cell. **[Ans. face-centered]**

16. The cube edge of a simple cubic structure is 5Å. Calculate the number of atoms per m^2 in the (110) plane. **(cf. Burd. U. 1996) [Ans. 2.828×10^{18} / m^2]**

Crystals and Their Properties 193

17. Show that, if a is the unit cell cube edge of an fcc lattice, the concentration of lattice points on a (111) plane is $(4/\sqrt{3}a^2)$, that on a (110) plane is $\sqrt{2}/a^2$, and that on a (100) plane is $2/a^2$.

18. Prove that the density of atoms (per unit area) on a (110) plane of a bcc lattice of cube edge a is $\sqrt{2}/a^2$, one atom occupying each lattice point.

19. In a cubic crystal, θ is the angle between the normals to the planes $(h_1 k_1 l_1)$ and $(h_2 k_2 l_2)$. Show that

$$\cos\theta = \frac{h_1 h_2 + k_1 k_2 + l_1 l_2}{(h_1^2 + k_1^2 + l_1^2)^{1/2}(h_2^2 + k_2^2 + l_2^2)^{1/2}}$$

[**Hint.** In a cubic crystal, the normal to the (hkl) plane is the direction $[hkl]$. So, θ is the angle between the vectors $h_1\vec{i_x} + k_1\vec{i_y} + l_1\vec{i_z}$ and $h_2\vec{i_x} + k_2\vec{i_y} + l_2\vec{i_z}$]

20. Find the angle between the [111] and the [100] directions in a cubic lattice. [**Ans.** 54°44']

21. What are the distances between the adjacent (100), (110), and (111) parallel planes in an fcc lattice of cube edge a? [**Ans.** $a/2, a/(2\sqrt{2}), a/\sqrt{3}$]

22. The smallest distance between any two atoms in an fcc crystal is 0.2 nm. Find the size of the unit cube. Also, calculate the lattice spacing for (100), (110), and (111) planes. (C.U. 1998) [**Ans.** 0.0226 nm³, 0.1414 nm, 0.1 nm, 0.1633 nm]

23. Write the indices of the different planes belonging to the form (110) in a cubic system where x, y and z directions are four-fold axes. (C.U. 1998)

[**Ans.** $(1 1 0), (\bar{1} 1 0), (\bar{1} \bar{1} 0), (1 \bar{1} 0), (0 1 1), (0 \bar{1} 1),$
$(0 \bar{1} \bar{1}), (0 1 \bar{1}), (1 0 1), (\bar{1} 0 1), (\bar{1} 0 \bar{1}), (1 0 \bar{1})$]

24. The potential energy of a pair of atoms is $U = -a/r^4 + b/r^{12}$ where r is the interatomic distance. Find the value of r where a stable bond is formed. Also calculate the energy released when the atoms form a stable bond. [**Ans.** $(3b/a)^{1/8}, (4a^3/27b)^{1/2}$]

25. Show that for a simple cubic lattice $d_{100}: d_{110}: d_{111} = \sqrt{6}:\sqrt{3}:\sqrt{2}$ where d_{hkl} is the separation between adjacent (hkl) parallel planes. (Burd. U. 2000, C.U. 2005)

26. Copper having a density of 8960 kg/m³ and an atomic mass of 63.54 amu possesses an fcc structure. Calculate the lattice constant. [**Ans.** 3.61 Å]

27. KBr crystal has a cubic structure. Its density is 2.75×10^3 kg/m³ and its molecular weight is 119.01. Calculate its lattice constant. (C.U. 2006) [**Ans.** 6.6 Å]

28. At about 1180 K, iron transforms into the fcc structure from the bcc structure which is the structural form at room temperature. Assuming no change in density, find the ratio of the nearest neighbour distance in the fcc structure to that in the bcc structure. (C.U. 2007) [**Ans.** 1.029]

Chapter 2

X-Ray Crystal Analysis

2.1. INTRODUCTION

The discovery of the X-ray diffraction effect for single crystal samples by von Laue in 1912 paved the way for determining the crystal structure by using X-ray diffraction as a technique. According to von Laue, the atoms of a single crystal specimen diffract an incident parallel monochromatic X-ray beam and produce a series of diffracted beams, the directions and intensities of which depend upon the lattice structure and the chemical composition of the crystal.

In this chapter we shall describe two equivalent formulations, one due to Bragg and the other due to von Laue, of the scattering of the X-rays by a perfect periodic structure. Since the von Laue approach is based on the concept of reciprocal lattice we shall also introduce the reciprocal lattice. Finally, the methods of X-ray diffraction will be discussed.

2.2. REASONS FOR USING X-RAY

The choice of X-ray for producing diffraction effects in crystals arises from the following reasons:

(*i*) X-rays can be produced either by the deceleration of electrons in metal targets or by exciting the core electrons in the atoms of the target inelastically. The first method produces a broad continuous spectrum whereas sharp lines are obtained in the second method.

(*ii*) Longer-wavelength radiation (*e.g.* light) gives rise to the familiar effects of optical refraction and reflection and so cannot be used to explore the structure of crystals on an atomic scale.

(*iii*) Radiation of wavelength shorter than X-ray, on the other hand, is diffracted through inconveniently small angles.

(*iv*) The wavelength corresponding to each line of X-radiation can be determined with very high precision.

(*v*) The wavelength of X-rays is comparable to the interatomic distances in actual crystals.

(*vi*) X-rays are scattered elastically without change of wavelength by the charged particles of the atoms.

The energy acquired by the electrons of an X-ray tube which is subjected to a potential of V_0 volt, is eV_0, where e is the magnitude of the electron charge. The most energetic X-ray photon produced by such electrons is that for which the following relation holds.

$$eV_0 = h\nu = \frac{hc}{\lambda}, \qquad \ldots(2.1)$$

where h is Planck's constant ($h = 6.62 \times 10^{-34}$ J.s), c is the velocity of electromagnetic radiation ($c = 3 \times 10^8$ m/s), and λ is the wavelength of the X-ray radiation. The shortest X-ray wavelength is accordingly given by

$$\lambda = \frac{hc}{eV_0}. \qquad \ldots(2.2)$$

Evidently, for a potential $V_0 = 10$ kV, the minimum X-ray wavelength is 1.24×10^{-10} m, or 1.24Å. This is comparable to the atomic spacing of an actual crystal ($\approx 10^{-10}$ m).

2.3. THE BRAGG DIFFRACTION LAW

W.L. Bragg formulated a relationship for the diffraction condition of X-rays incident on a crystalline material on the basis of a simple model. He assumed that the monochromatic X-rays are reflected from successive parallel planes of atoms in the crystal specularly, where the angle of incidence equals the angle of reflection. The strong diffracted beams are obtained when the reflections from the parallel planes of atoms interfere constructively, as shown in Fig. 2.1. In his model, Bragg further assumed that the scattering is elastic so that the wavelength is not changed on reflection.

Figure 2.1 shows a particular family of crystal planes of ions, spaced at a distance d apart. Consider that a plane wavefront is incident at a glancing angle θ. The incident radiation is reflected specularly by the planes of the crystal. The path difference between the incident and the reflected rays from adjacent planes is $2d \sin \theta$. Constructive interference of the X-radiation from the successive planes will occur when this path difference is equal to an integral multiple of the wavelength λ. That is,

$$2d \sin \theta = n \lambda, \qquad ...(2.3)$$

where n is an integer and gives the *order* of reflection. Equation (2.3) is refened to as *Bragg's equation*. The diffraction lines for $n = 1, 2$, etc, are called respectively the first, the second, etc. lines of diffraction. This is *Bragg's law of diffraction*. Equation (2.3) states that X-ray diffraction in terms of inphase reflections from successive planes of atoms of a crystal will occur for a beam of monochromatic X-rays incident at an angle θ on a crystal having interplaner spacing d. For an incident angle other than θ the beam will not be diffracted. This means that the lattice planes do not act like ordinary mirrors because mirror reflection does not occur for such selective angles only. Also, Bragg reflection occurs only for wavelength $\lambda \leq 2d$. As d is typically 10^{-10} m, visible light instead of X-rays cannot be used in crystal structure analysis, as has been pointed out earlier.

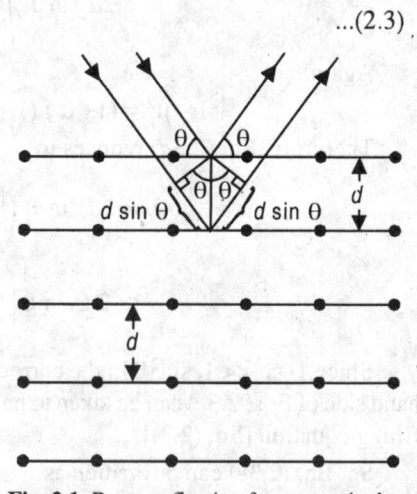

Fig. 2.1. Bragg reflection from a particular family of crystal planes.

Bragg's law is a consequence of the periodic nature of the space lattice. It does not refer to the orientation or basis of lattice atoms. The composition of the basis decides the relative intensity of the different orders n of diffraction from a given set of parallel planes.

Modification of Bragg's equation due to refraction

In the above, we have neglected the refraction of X-rays at the surface of the crystal. For all substances, the refractive index μ for X-rays is very slightly less than unity. This is different from the case of visible light for which μ is greater than unity for a solid, say, glass. In what follows, we seek to find the modified Bragg's equation including the effect of refraction.

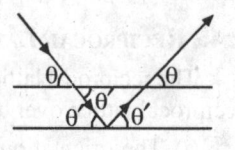

Let a parallel beam of monochromatic X-rays of wavelength λ be incident on a crystal surface at a glancing angle θ (Fig. 2.1A). The glancing angle of refraction θ' is slightly less than θ since μ is slightly less than 1 for X-rays. If λ' is the wavelength of the refracted ray, we have

Fig. 2.1 A. Refraction effect on Bragg's law

$$\mu = \frac{\lambda}{\lambda'} = \frac{\cos \theta}{\cos \theta'} \qquad ...(2.3a)$$

The Bragg equation inside the crystal will be

$$2d \sin \theta' = n \lambda' \qquad ...(2.3b)$$

Now,
$$\sin \theta' = \sqrt{1 - \cos^2 \theta'}$$
$$= \sqrt{1 - \cos^2 \theta / \mu^2}$$
$$= \frac{\sqrt{\mu^2 - \cos^2 \theta}}{\mu} \qquad ...(2.3c)$$

From Eqs. (2.3a), (2.3b), and (2.3c) we have

$$n\lambda' = \frac{n\lambda}{\mu} = \frac{2d}{\mu} \sqrt{\mu^2 - \cos^2 \theta}$$

or,
$$n\lambda = 2d \sqrt{\mu^2 - \cos^2 \theta}$$
$$= 2d \sin \theta \sqrt{1 - \frac{1 - \mu^2}{\sin^2 \theta}} \qquad ...(2.3d)$$

Again,
$$1 - \mu^2 = (1 + \mu)(1 - \mu) \simeq 2(1 - \mu), \text{ since } \mu \simeq 1.$$

Therefore, Eq. (2.3d) reduces to

$$n\lambda = 2d \sin \theta \sqrt{1 - \frac{2(1 - \mu)}{\sin^2 \theta}}$$
$$\simeq 2d \sin \theta \left(1 - \frac{1 - \mu}{\sin^2 \theta}\right). \qquad ...(2.3e)$$

Since $1 - \mu \ll 1$, $\sin^2 \theta$ in the correction term, *i.e.* the last term within parentheses on the right-hand side of Eq. (2.3e) can be taken to have the value $n^2 \lambda^2 / 4d^2$, which is predicted by the unmodified Bragg equation [Eq. (2.3)].

So, Eq. (2.3e) can be written as

$$n\lambda = 2d \sin \theta \left[1 - \frac{4d^2 (1 - \mu)}{n^2 \lambda^2}\right] \qquad ...(2.3f)$$

which is the *modified Bragg equation*. The correction term is usually quite small and becomes progressively smaller at higher orders, *i.e.* for higher values of *n*. Thus the use of unmodified Bragg equation [Eq. (2.3)] does not introduce serious errors.

2.4. Reciprocal Lattice

The reciprocal lattice plays an important role in the analysis of crystal structures. For, the use of reciprocal lattice over direct space lattice has the following advantages:

(*i*) The orientation of the different sets of crystal planes and their spacings are very conveniently represented in terms of reciprocal lattice.

(*ii*) It simplifies considerably the understanding of the diffraction photographs of crystals.

(*iii*) It provides a simple method for obtaining the diffraction condition.

(*iv*) The wave-mechanical behaviour of electrons in periodic crystal lattices is easily understood in terms of reciprocal lattice.

The reciprocal lattice corresponding to a Bravais lattice of points $\vec{R}$ is defined as *the set of all wave vectors $\vec{K}$ that gives plane waves of the form $e^{i\vec{K}\cdot\vec{r}}$ with the periodicity $(\vec{R})$ of the Bravais lattice*. In other words,

$$e^{i\vec{K}\cdot(\vec{r}+\vec{R})} = e^{i\vec{K}\cdot\vec{r}}, \qquad ...(2.4)$$

for any $\vec{r}$ and all $\vec{R}$ in the Bravais lattice. From Eq. (2.4) we obtain

$$e^{i\vec{K}\cdot\vec{R}} = 1. \qquad ...(2.5)$$

Thus the reciprocal lattice is characterized by the set of wave vectors $\vec{K}$ that satisfy Eq. (2.5) for all $\vec{R}$ in the Bravais lattice. Since $e^{i2\pi n} = 1$, where n is an integer, it follows that $\vec{K}\cdot\vec{R} = 2\pi n$. This equation *relates* the reciprocal lattice vector $\vec{K}$ with the corresponding Bravais lattice vector $\vec{R}$.

Primitive Translation Vectors. The primitive translation vectors $\vec{a}*, \vec{b}*$ and $\vec{c}*$ of the reciprocal lattice are defined in terms of the primitive translation vectors $\vec{a}, \vec{b}$ and $\vec{c}$ of a direct space lattice as follows :

$$\vec{a}*\cdot\vec{a} = \vec{b}*\cdot\vec{b} = \vec{c}*\cdot\vec{c} = 2\pi \qquad ...(2.6)$$

and

$$\vec{a}*\cdot\vec{b} = \vec{a}*\cdot\vec{c} = \vec{b}*\cdot\vec{c} = \vec{b}*\cdot\vec{a} = \vec{c}*\cdot\vec{a} = \vec{c}*\cdot\vec{b} = 0 \qquad ...(2.7)$$

The relation $\vec{a}*\cdot\vec{b} = \vec{a}*\cdot\vec{c} = 0$ implies that the vector $\vec{a}*$ is perpendicular to the plane containing $\vec{b}$ and $\vec{c}$. In other words, the vector $\vec{a}*$ is parallel to the vector $\vec{b}\times\vec{c}$. Therefore we can write

$$\vec{a}* = A\,(\vec{b}\times\vec{c}), \qquad ...(2.7a)$$

A being a scalar multiplier. From Eq. (2.6) we get

$$\vec{a}*\cdot\vec{a} = A\,(\vec{b}\times\vec{c})\cdot\vec{a} = 2\pi,$$

or,

$$A = \frac{2\pi}{\vec{a}\cdot\vec{b}\times\vec{c}}.$$

Substituting the value of A in Eq. (2.7a) we obtain

$$\vec{a}* = 2\pi\,\frac{\vec{b}\times\vec{c}}{\vec{a}\cdot\vec{b}\times\vec{c}} \qquad ...(2.8)$$

Proceeding similarly, we find that

$$\vec{b}* = 2\pi\,\frac{\vec{c}\times\vec{a}}{\vec{a}\cdot\vec{b}\times\vec{c}} \qquad ...(2.9)$$

and

$$\vec{c}* = 2\pi\,\frac{\vec{a}\times\vec{b}}{\vec{a}\cdot\vec{b}\times\vec{c}} \qquad ...(2.10)$$

Note that the denominator in each of Eqs. (2.8) through (2.10) viz., $\vec{a}\cdot\vec{b}\times\vec{c}$ is the volume of the primitive cell in the direct lattice.

The vectors in the crystal lattice, *i.e.* $\vec{a}, \vec{b}$ and $\vec{c}$ have the dimension of [length] whereas the vectors $\vec{a}*, \vec{b}*$ and $\vec{c}*$ in the reciprocal lattice have the dimension of [length]$^{-1}$. The crystal lattice represents a lattice in real space; the reciprocal lattice is, on the other hand, a lattice in Fourier space. The term *Fourier space* is motivated by Eq. (2.5).

The points $\vec{R}$ of the crystal lattice are obtained from

$$\vec{R} = m\vec{a} + n\vec{b} + p\vec{c} \qquad ..(2.11a)$$

where $m, n,$ and p are integers.

The reciprocal lattice points are given by the vector $\vec{K}$ where

$$\vec{K} = h\vec{a}* + k\vec{b}* + l\vec{c}* \qquad ...(2.11b)$$

Here $h, k,$ and l are integers.

An important property of the reciprocal lattice is that *every reciprocal lattice vector is normal to a lattice plane of the crystal lattice*. This is proved below.

Consider the reciprocal lattice vector $\vec{K} = h'\vec{a}* + k'\vec{b}* + l'\vec{c}*$ and the (*hkl*) plane of the direct lattice. Here the triad (*hkl*) is obtained by simply dividing the triad (*h'k'l'*) by the longest common factor *n*. In other words

$$\frac{h'}{h} = \frac{k'}{k} = \frac{l'}{l} = n.$$

We have from Fig. 2.2, $\vec{AC} = -\frac{\vec{a}}{h} + \frac{\vec{c}}{l}$,

and $\vec{AB} = -\frac{\vec{a}}{h} + \frac{\vec{b}}{k}$,

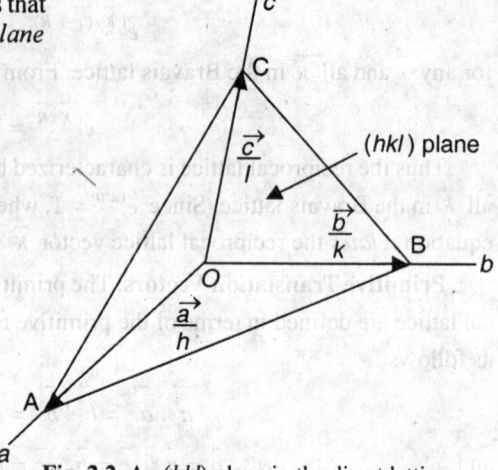

Fig. 2.2. An (*hkl*) plane in the direct lattice.

where both $\vec{AB}$ and $\vec{AC}$ lie in the (*hkl*) plane.

Using Eqs. (2.6) and (2.7) we obtain

$$\vec{K} \cdot \left[-\frac{\vec{a}}{h} + \frac{\vec{c}}{l} \right] = (h'\vec{a}* + k'\vec{b}* + l'\vec{c}*) \cdot \left[-\frac{\vec{a}}{h} + \frac{\vec{c}}{l} \right]$$

$$= 2\pi \left(-\frac{h'}{h} + \frac{l'}{l} \right) = 0 \qquad \ldots(2.12)$$

Similarly, $\qquad \vec{K} \cdot \left[-\frac{\vec{a}}{h} + \frac{\vec{b}}{k} \right] = 0 \qquad \ldots(2.13)$

From Eqs. (2.12) and (2.13) we find that $\vec{K}$ is perpendicular to two linearly independent vectors $\vec{AC}$ and $\vec{AB}$ which are in the (*hkl*) plane. Hence the vector $\vec{K}$ is perpendicular to the (*hkl*) plane.

By using this property it can be shown that the spacing *d* between adjacent (*hkl*) planes of the crystal lattice is given by $d = \frac{2\pi n}{K}$, where *n* is an integer and

$$K = | h'\vec{a}* + k'\vec{b}* + l'\vec{c}* |.$$

Proof : Let $\vec{u}$ be the unit vector normal to the (*hkl*) plane of the direct lattice. Then

$$\vec{u} = \frac{\vec{K}}{K},$$

since $\vec{K}$ is a vector normal to the (*hkl*) plane. Therefore, from Eq.(1.3) we obtain

$$d = \frac{\vec{a} \cdot \vec{n}}{h} = \frac{\vec{a} \cdot \vec{K}}{hK} = \frac{\vec{a} \cdot (h'\vec{a}* + k'\vec{b}* + l'\vec{c}*)}{hK} = \frac{2\pi n}{K}. \qquad \ldots(2.14)$$

2.5. Laue Condition of X-ray Diffraction

Laue examined in detail the phenomenon of X-ray diffraction by assuming that the X-rays are scattered from three-dimensional array of atoms or ions in a manner similar to the diffraction of light by a grating. However, while an ordinary grating consists of parallel lines in a plane, the crystal grating contains the scatterers in a three-dimensional regular arrangement. Sharp and intense peaks are produced only in directions and at wavelengths for which the scattered rays from lattice atoms or ions interfere constructively.

X-Ray Crystal Analysis

To derive the condition of constructive interference, let us consider two identical scattering centres spaced $\vec{x}$ apart (Fig. 2.3). Consider that an X-ray of wavelength λ and wave vector $\vec{k} = \dfrac{2\pi \vec{n_0}}{\lambda}$ is incident from a long distance on the crystal along the direction of the unit vector $\vec{n_0}$. Suppose that the scattered ray of the same wavelength λ and wave vector $\vec{k'} = \dfrac{2\pi \vec{n_1}}{\lambda}$ is observed in the direction of the unit vector $\vec{n_1}$.

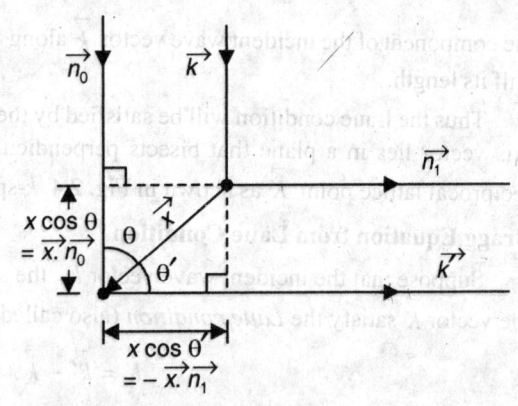

Fig. 2.3. Scattering of X-ray by two identical atoms.

From Fig. 2.3 the path difference between the scattered rays is found to be

$$x \cos\theta + x \cos\theta' = \vec{x}\cdot\vec{n_0} - \vec{x}\cdot\vec{n_1} = \vec{x}\cdot(\vec{n_0} - \vec{n_1}). \qquad ...(2.15)$$

For constructive interference, this path difference must be equal to an integral number of wavelengths. Hence

$$\vec{x}\cdot(\vec{n_0} - \vec{n_1}) = m\lambda, \qquad ...(2.16)$$

where m is an integer. Multiplying the two sides of Eq. (2.16) by $\dfrac{2\pi}{\lambda}$ imposes the following condition on the incident and the scattered wave vectors:

$$\vec{x}\cdot(\vec{k} - \vec{k'}) = 2\pi m. \qquad ...(2.17)$$

Since the lattice points are displaced from one another by the Bravais lattice vector $\vec{R}$, we can extend the above analysis and obtain the general condition for an array of scatterers of a Bravais lattice:

$$\vec{R}\cdot(\vec{k} - \vec{k'}) = 2\pi m \qquad ...(2.18)$$

Equation (2.18) can also be written in the following equivalent form:

$$e^{i(\vec{k'} - \vec{k})\cdot\vec{R}} = 1, \qquad ...(2.19)$$

for all $\vec{R}$ of the Bravais lattice.

Comparing this condition with the definition of the reciprocal lattice given by Eq. (2.5) we obtain the *Laue condition* for constructive interference: *for constructive interference the change in wave vector* $\vec{K} = \vec{k'} - \vec{k}$, *must be a vector of the reciprocal lattice.*

Alternative Form of the Laue Condition

It is appreciated that both $\vec{k'} - \vec{k}$ and $\vec{k} - \vec{k'}$ are reciprocal lattice vectors. Denoting $\vec{k} - \vec{k'}$ by $\vec{K}$ and noting that $\vec{k}$ and $\vec{k'}$ have the same magnitudes, we obtain

$$|\vec{k}| = |\vec{k'}|$$

or,

$$k = |\vec{k} - \vec{K}| \qquad ..(2.20)$$

Squaring both sides of Eq. (2.20) we get

$$\vec{k}\cdot\widehat{K} = \frac{1}{2}K \qquad ...(2.21)$$

where $\widehat{K}$ represents a unit vector along $\vec{K}$. Equation (2.21) indicates that for constructive interference the component of the incident wave vector $\vec{k}$ along $\vec{K}$, *i.e.* along the reciprocal lattice vector must be half its length.

Thus the Laue condition will be satisfied by the incident wave vector $\vec{k}$ provided that the tip of this vector lies in a plane that bisects perpendicularly the line joining the origin of *k*-space to a reciprocal lattice point $\vec{K}$ as shown in Fig. 2.4. *k*-space planes are termed the *Bragg planes*.

Bragg Equation from Laue Condition

Suppose that the incident wave vector $\vec{k}$, the scattered wave vector $\vec{k}'$ and their difference, *i.e.* the vector $\vec{K}$ satisfy the *Laue condition* (also called the *Laue equation*) :

$$\vec{K} = \vec{k}' - \vec{k}.$$

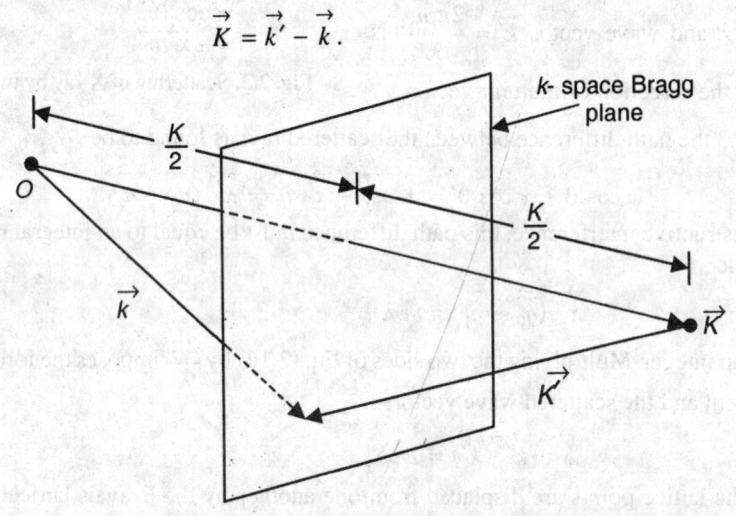

Fig. 2.4. An illustration of the alternative Laue condition.

In Fig. 2.5, drop a perpendicular *CE* to $\vec{K}$ (= $\overrightarrow{BD}$) and consider a lattice plane *PQ* perpendicular to $\vec{K}$. Since the scattering is elastic, $|\vec{k}| = |\vec{k}'|$, *i.e.*, *BC* = *CD*. In the isosceles triangle *BCD*, $\angle BCE = \angle DCE = \theta$, say, Since *EC* ∥ *PQ* and *DC* ∥ *AB*, it follows that $\angle ABP = \angle CBQ = \theta$. Thus the wave vectors $\vec{k}'$ and $\vec{k}$ make the same angle θ with the lattice plane perpendicular to the reciprocal lattice vector $\vec{K}$. So, the scattering can be looked upon as a Bragg reflection, θ being the Bragg angle. The reflection takes place from the family of direct lattice planes perpendicular to $\vec{K}$.

We have from Eq. (2.14)

$$|\vec{K}| = \frac{2\pi n}{d}.$$

From Fig. 2.5 we obtain $|\vec{K}| = 2|\vec{k}|\sin\theta$. Therefore,

$$2|\vec{k}|\sin\theta = \frac{2\pi n}{d}$$

or, $$2d\sin\theta = n\lambda, \qquad \left(\text{since } |\vec{k}| = \frac{2\pi}{\lambda}\right)$$

which is the *Bragg equation*.

From the foregoing discussion we find that a diffraction peak or Bragg reflection for an incident wave vector $\vec{k}$ will occur when and only when the tip of the vector $\vec{k}$ lies on a *k* space Bragg plane. Since the Bragg planes are discrete we cannot expect in general that a fixed X-ray (*i.e.* a fixed incident wave vector) having a fixed incident direction will give rise to a diffraction peak.

X-Ray Crystal Analysis

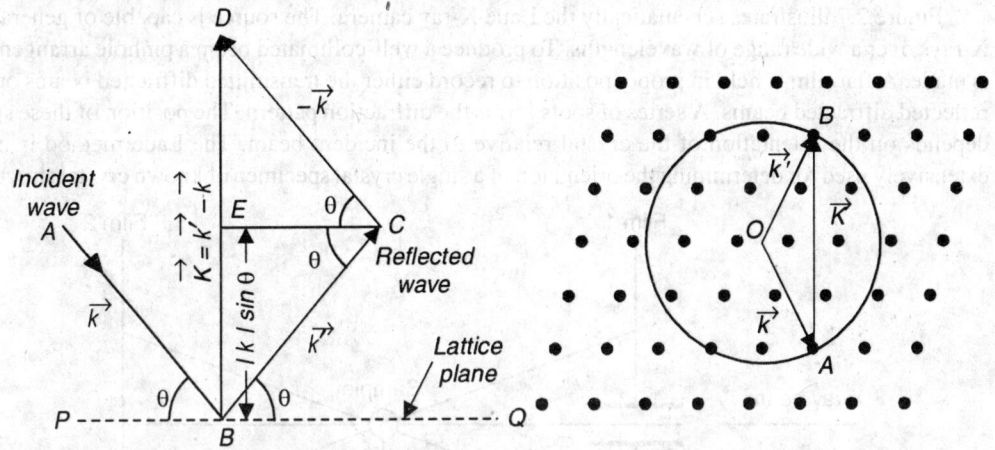

Fig. 2.5. Illustrates the equivalence of Bragg and Laue formulations.

Fig. 2.6. Ewald construction.

Therefore, to form diffraction peaks experimentally one must vary either the magnitude of $\vec{k}$ or the angle of incidence of the X-ray beam. In practice the latter method is used by varying the orientation of the crystal with respect to the incident direction.

Ewald Construction

A simple geometrical construction in k-space due to Ewald may be used to determine whether the diffraction of an X-ray beam of known wavelength will occur with a particular crystal.

A vector $\vec{OA}$, representing the incident beam vector $\vec{k}$ is drawn parallel to $\vec{k}$, A being a reciprocal lattice point. Draw a sphere with O as origin and with a radius equal to k. If this sphere passes through another reciprocal lattice point, e.g. B, then diffraction of the beam $\vec{OA}$ must occur. The diffracted X-ray beam is then given by the wave vector $\vec{k'}$ satisfying the Laue condition:

$$\vec{k'} = \vec{k} + \vec{K}$$

as shown in Fig. 2.6. The reflection planes in the crystal lattice are normal to $\vec{K}$ as shown earlier.

Comparison of Bragg and Laue approaches

The Bragg approach of X-ray diffraction regards the crystal as a set of equispaced parallel planes of ions or atoms. The incident X-rays are specularly reflected by the ions or the atoms in one plane. (In a specular reflection, the angle of incidence and the angle of reflection are equal.) The reflected rays from the successive planes interfere constructively to give the characteristic maxima.

The Laue approach differs from the Bragg approach in that a particular set of parallel lattice planes is not chosen, and specular reflection is not preassumed. Rather, the incident X-radiation is considered to be scattered elastically by the crystal ions or atoms placed at the lattice points in all directions. Maxima will occur only in directions and at wavelengths for which the scattered rays interfere constructively. However, the Laue visualization of constructive interference is equivalent to the Bragg assumption of specular reflection, as we have seen above. Thus the Bragg and the Laue approaches are essentially founded on the same physical principles.

2.6. EXPERIMENTAL METHODS OF X-RAY DIFFRACTION

We shall describe below three standard methods of X-ray diffraction.

(i) **The Laue method** : In this method a single crystal sample is held stationary and an X-ray beam of continuous wavelength, i.e. white radiation is allowed to fall on the crystal. The crystal selects and diffracts the appropriate wavelength λ for which a family of planes of spacing d and incidence angle θ satisfying Bragg's law exists.

Figure 2.7 illustrates schematically the Laue X-ray camera. The source is capable of generating X-rays over a wide range of wavelengths. To produce a well-collimated beam a pinhole arrangement is made. A flat film is held in proper position to record either the transmitted diffracted beams or the reflected diffracted beams. A series of spots forms the diffraction pattern. The position of these spots depends on the orientation of the crystal relative to the incident beam. The Laue method is most extensively used for determining the orientaion of a single crystal specimen of known crystal structure.

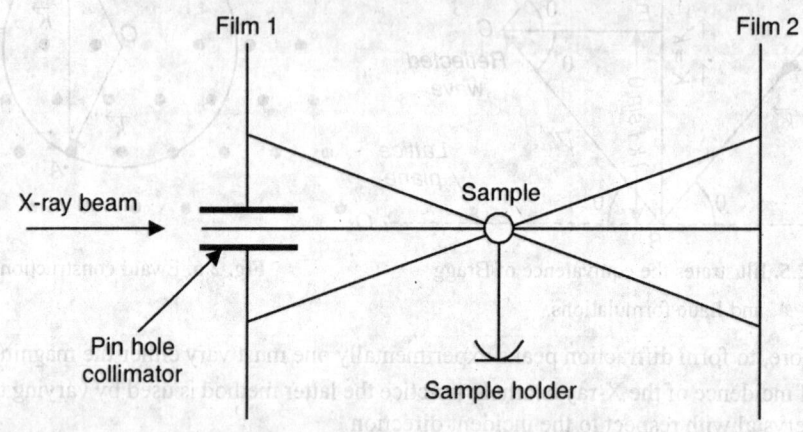

Fig. 2.7. The Laue X-ray camera system.

(*ii*) **The rotating-crystal method :** In this method a single crystal specimen is made to rotate about a fixed axis in a monochromatic beam of X-radiation. The rotation brings different atomic planes into position for reflection. The method is suitable for finding the interatomic distances in a single crystal.

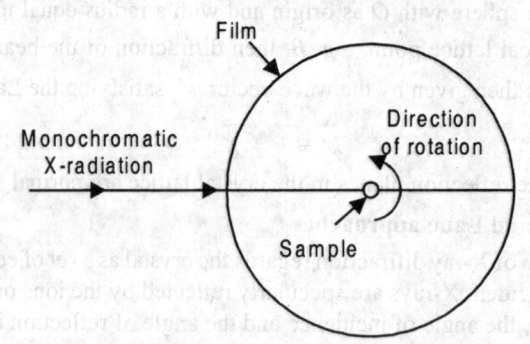

Fig. 2.8. A rotating crystal camera.

Figure 2.8 shows a simple rotating-crystal X-ray camera. The axis of rotation of the crystal coincides with the centre of the cylindrical photographic film. During the course of rotation of the crystal a diffracted beam results from a given set of crystal planes for which the Bragg law is satisfied. Crystal planes parallel to the vertical rotation axis will give rise to diffracted beams that will lie in the horizontal plane. Planes having other orientation will produce reflection in layers above and below the horizontal plane. The spacing between crystal planes is determined by the separation of these layers on the film.

(*iii*) **The powder method :** The powder method of X-ray diffraction introduced in 1916 by Debye and Scherrer, is the most widely used method in crystal structure analysis. The method is convenient since it does not require single crystals.

X-Ray Crystal Analysis   203

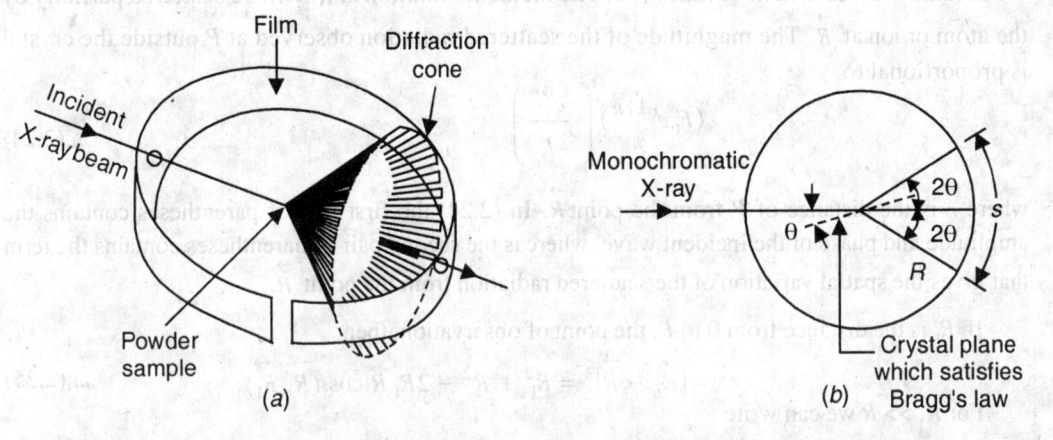

Fig. 2.9. The powder method due to Debye and Scherrer.

The method is illustrated in Fig. 2.9 (a). As the name indicates, the sample consists of a fine crystalline powder which may be coated on a supporting fibre or contained in a thin-walled capillary tube. A parallel beam of monochromatic X-rays is allowed to fall normally on the sample mounted centrally, and a film in the form of a short cylinder coaxial with the axis of the sample is used to record the diffracted beams. Since the sample is polycrystalline, there may be found some crystallites which will produce Bragg diffraction for a particular set of planes. As a result, the diffracted beams in the form of cones coaxial with the incident beam direction, will intersect the film in a series of concentric rings. For any particular reflection, two arcs symmetrically displaced on either side of the central beam direction will be formed [Fig. 2.9(b)]. In the experiment, the separation between these arcs (say, s) and the radius of the film cylinder (say, R) are measured, and the Bragg angle θ is calculated from $\theta = \dfrac{s}{4R}$ radian. Different (hkl) planes will have different values of s, and hence of θ. By carefully analysing the photograph, it is possible, to determine the lattice constants of the crystal.

2.7. Amplitude of the Scattered Wave

Consider a small crystal containing a number of primitive cells. Suppose $\vec{a}$, $\vec{b}$ and $\vec{c}$ represent the primitive vectors of the space lattice and that a plane wave is incident on the crystal.

Assume that the incident beam is not significantly disturbed by the crystal either due to the refractive index of the crystal or due to the energy loss through scattering. Then, with respect to the chosen origin O in the crystal, the form of the incident wave at $\vec{R}$ within the crystal can be written as (Fig. 2.10)

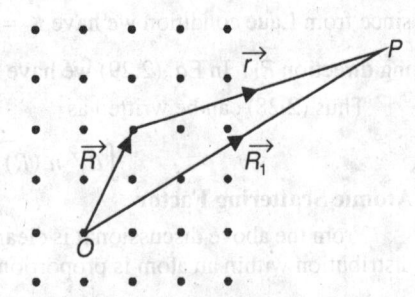

Fig. 2.10. Geometry for the calculation of the scattered wave amplitude.

$$F(\vec{R},t) = F_0 \exp i\, (\vec{k}\cdot\vec{R} - \omega t), \qquad \ldots(2.22)$$

where $\vec{k}$ is the incident wave vector ($|\vec{k}| = 2\pi/\lambda$) and ω the angular frequency.

From Eq. (2.22) we find that the spatial variation $F(\vec{R})$ is given by

$$F(\vec{R}) = F_0 e^{i\vec{k}\cdot\vec{R}} \qquad \ldots(2.23)$$

We assume that the instant of time $t = 0$. The incident radiation at $\vec{R}$ will be scattered partially by the atom or ion at $\vec{R}$. The magnitude of the scattered radiation observed at P outside the crystal is proportional to

$$(F_0 e^{i\vec{k}.\vec{R}}) \left(\frac{e^{ikr}}{r} \right) \qquad ...(2.24)$$

where r is the distance of P from the point $\vec{R}$. In (2.24) the first pair of parentheses contains the amplitude and phase of the incident wave, whereas the second pair of parentheses contains the term that gives the spatial variation of the scattered radiation from the point $\vec{R}$.

If $\vec{R}_1$ is the distance from 0 to P, the point of observation, then

$$r^2 = |\vec{R}_1 - \vec{R}|^2 = R_1^2 + R^2 - 2R_1 R \cos(\vec{R}, \vec{R}_1). \qquad ...(2.25)$$

For $R_1 \gg R$ we can write

$$r \cong R_1 \left[1 - \frac{2R}{R_1} \cos(\vec{R}, \vec{R}_1) \right]^{1/2} \cong R_1 - R \cos(\vec{R}, \vec{R}_1) \qquad ...(2.26)$$

Therefore (2.24) reduces to

$$\frac{F_0}{r} e^{[i\vec{k}.\vec{R} + ikR_1 - ikR\cos(\vec{R}, \vec{R}_1)]} \qquad ...(2.27)$$

The electrons in the atoms mainly scatter the beam; therefore we may assume that the amplitude of the wave scattered from an elemental volume dV around the point $\vec{R}$ in the crystal is proportional to the concentration of the electrons $n(\vec{R})$ at $\vec{R}$. Hence the amplitude of the scattered radiation at the observation point P will be proportional to the integral

$$\int dV \, n(\vec{R}) \, e^{[i\vec{k}.\vec{R} - ikR\cos(\vec{R}, \vec{R}_1)]} \qquad ...(2.28)$$

In (2.28) we have dropped the factor e^{ikR_1} since it is a constant over the volume.

Again, $i\vec{k}.\vec{R} - ikR\cos(\vec{R}, \vec{R}_1) \equiv i\vec{R}.(\vec{k} - \vec{k}') = -i\vec{R}.\vec{K}$ ...(2.29)

since from Laue condition we have $\vec{K} = \vec{k}' - \vec{k}$, where $\vec{k}'$ represents the wave vector for the scattering direction $\vec{R}_1$. In Eq. (2.29) we have assumed the scattering to be elastic, i.e., $|\vec{k}'| = |k|$.

Thus (2.28) can be written as

$$\int dV \, n(\vec{R}) \, e^{-i\vec{R}.\vec{K}} \qquad ...(2.30)$$

Atomic Scattering Factor

From the above discussion it is clear that the amplitude of the radiation scattered by the electron distribution within an atom is proportional to the quantity f, where

$$f = \int dV \, n(\vec{R}) \, e^{-i\vec{R}.\vec{K}} \qquad ...(2.31)$$

Here $n(\vec{R})$ represents the electron concentration at $\vec{R}$. If $\vec{R}$ makes an angle α with $\vec{K}$ then $\vec{R}.\vec{K} = RK\cos\alpha$. Assuming the electron distribution to be spherically symmetric about the origin we obtain $dV = 2\pi R^2 \, dR \sin\alpha \, d\alpha$. Therefore,

$$f = 2\pi \int\limits_{\alpha=\pi} \int_R R^2 \, dR \sin\alpha \, d\alpha \, n(R) \, e^{-iRK\cos\alpha}$$

$$= 2\pi \int\limits_R \int\limits_{\alpha=0}^{\pi} R^2 \, dR \, d(\cos\alpha) \, n(R) \, e^{-iRK\cos\alpha}$$

X-Ray Crystal Analysis

$$= 2\pi \int R^2\, n(R)\, \frac{e^{iRK} - e^{-iRK}}{iRK}\, dR$$

$$= 4\pi \int R^2\, n(R)\, \frac{\sin KR}{RK}\, dR \qquad \ldots(2.32)$$

If the total electron concentration were localized at the origin, then $R = 0$, and

$$f_0 = 4\pi \int n(R)\, R^2\, dR \quad \left[\text{since, as } R \to 0,\ \frac{\sin KR}{KR} \to 1\right]$$

$$= Z, \qquad \ldots(2.33)$$

where Z is the atomic number. *We define the atomic scattering factor f as the ratio of the radiation amplitude scattered by the actual electron distribution in an atom to that of a point electron at the lattice point.*

Geometrical Structure Factor

We shall now consider the amplitude of an X-ray beam diffracted by a unit cell containing a number of atoms (or ions).

Suppose that the position vector of the nucleus of the jth atom is

$$\vec{r}_j = u_j\, \vec{a} + v_j\, \vec{b} + \omega_j\, \vec{c}, \qquad \ldots(2.34)$$

with respect to a lattice point as the origin of $\vec{r}_j$. If f_j represents the atomic scattering factor of the jth atom then the total scattered amplitude is proportional to

$$S = \sum_j f_j\, e^{-i\vec{r}_j \cdot \vec{K}} \qquad \ldots(2.35)$$

where the summation extends over all the atoms in the unit cell. The quantity S is termed the *geometrical structure factor*. In analogy with the atomic scattering factor, S may be defined as *the ratio of the amplitude of the wave scattered by all the atoms in the unit cell to that scattered by a free electron for the same incident beam.*

For the reflection from (hkl) planes, called the (hkl) reflection, we have

$$\vec{r}_j \cdot \vec{K} = (u_j\, \vec{a} + v_j\, \vec{b} + \omega_j\, \vec{c}) \cdot (h\vec{a}^* + k\vec{b}^* + l\vec{c}^*) \qquad \ldots(2.36)$$

$$= 2\pi (u_j h + v_j k + \omega_j l). \qquad \ldots(2.37)$$

Hence

$$S = \sum_j f_j\, e^{-i 2\pi (u_j h + v_j k + \omega_j l)} \qquad \ldots(2.38)$$

The intensity is proportional to S^*S, where S^* is the complex conjugate of S.

We now consider a bcc crystal which contains two atoms per unit cell; one corner atom and one body-centre atom. Also, all the atoms in the crystal are identical. Let us assign this corner atom of the unit cell as the origin with coordinates $(0, 0, 0)$ and exclude from consideration other corner atoms. The coordinates of the body centre atom are $\left(\frac{1}{2}, \frac{1}{2}, \frac{1}{2}\right)$. The diffraction amplitude or S is then given by

$$S = f \sum_j e^{-i 2\pi (u_j h + v_j k + \omega_j l)} \qquad \ldots(2.39)$$

where f_j has been replaced by f because the atoms are identical. Hence, summing the contribution from the corner atom at the origin and the body-centre atom we get

$$S = f\{1 + \exp[-i\pi(h + k + l)]\}. \qquad \ldots(2.40)$$

It is obvious from Eq. (2.40) that the geometrical structure factor S disappears for any (hkl) reflection for which $(h+k+l)$ is an odd integer. Thus, for the bcc structure certain (hkl) reflections will be absent although these reflections would be present for a simple cubic structure having the same edge dimension. For example, in the bcc lattice (100) reflection will be absent but (200) reflection will be present. Likewise, there will be no (111) reflection although (222) reflection will be present. This can be explained physically as follows :

Consider a simple cubic structure and the (100) reflection. For this, the top and the bottom faces of the unit cell will give reflected beams that will differ in phase by 2π for the first order diffraction line. In the case of a bcc structure, in addition to the top and bottom cube faces there is another plane of atoms formed by the body-centre atoms of the unit cells. This additional plane of atoms is parallel to, and equidistant from the top and bottom cube face planes of the unit cell, as shown in Fig. 2.11. The concentration of atoms in this intermediate plane is equal to that of the top and bottom cube face planes, and so this plane will give diffracted beams of equal intensity.

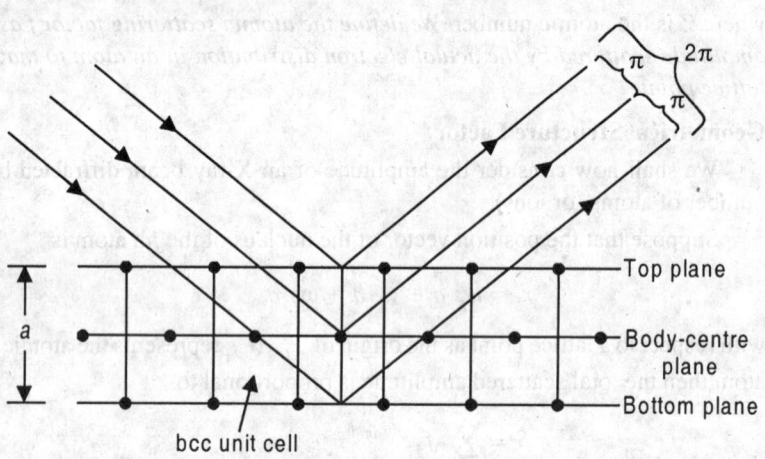

Fig. 2.11. Phase relations for the (100) reflections from a bcc lattice.

However, the phase of the diffracted beam from this plane will lag that from the top plane by π. The diffracted beams from the top and the middle planes of atoms therefore interfere destructively, resulting in zero intensity. However, the reflection corresponding to (200) will be present since for this, the top and bottom plane reflections will differ in phase by 4π whereas the top and the middle plane reflections will differ in phase by 2π, giving constructive interference. Clearly, the first order (200) reflection appears at the position of the second order (100) reflection.

Now we consider an fcc crystal which has four identical atoms in the cubic unit cell. With one corner atom of the unit cell as the origin of co-ordinates (0, 0, 0), the other three atoms are at the centre of the faces and have the co-ordinates $\left(\frac{1}{2}, 0, \frac{1}{2}\right)$, $\left(\frac{1}{2}, \frac{1}{2}, 0\right)$, and $\left(0, \frac{1}{2}, \frac{1}{2}\right)$. Then the diffraction amplitude, obtained from Eq. (2.39), is

$$S = f\{1 + \exp[-i\pi(h+l)] + \exp[-i\pi(h+k)] + \exp[-i\pi(k+l)]\} \qquad ...(2.40a)$$

From Eq. (2.40a) we find that the geometrical structure factor S is nonzero only when h, k, l are *all even or all odd*. Hence reflections such as (111), (200), (220) are present, whereas (100), (110), (211) reflections are absent for an fcc crystal.

2.8. Steps for Analysing A Crystal Structure

Generally a structure is pre-assumed and its possible diffraction pattern is predicted. This prediction is then compared with the experimentally observed diffraction pattern. Usually the observed pattern, given in the form of a map in K-space, provides the values of $\vec{K}$, i.e., $(\vec{k'} - \vec{k})$ for which diffracted beams are obtained.

(i) Select for the assumed structure a set of basis vectors $\vec{a}$, $\vec{b}$ and $\vec{c}$. Determine $\vec{a}^*$, $\vec{b}^*$ and $\vec{c}^*$. Plot $\vec{K} = h\vec{a}^* + k\vec{b}^* + l\vec{c}^*$ for different integral values of h, k, l. Note the points which coincide with the points of the experimental map in K-space. If no point is found to coincide, then adjust the magnitudes of $\vec{a}, \vec{b}$ and $\vec{c}$ so that some points of $\vec{K}$ coincide with the observed map of $(\vec{k'} - \vec{k})$. The new values of $\vec{a}, \vec{b}$ and $\vec{c}$ are then taken to define the crystal lattice.

(ii) Every $(\vec{k'} - \vec{k})$ now coincides with some $\vec{K}$ points, but some $\vec{K}$'s will not coincide with a $(\vec{k'} - \vec{k})$ if the structure factor S vanishes for those values of $\vec{K}$. Using the assumed basis vectors $\vec{a}, \vec{b}$ and $\vec{c}$ we determine S, and see if the zeros of S coincide with those values of $\vec{K}$ for which no diffraction beam was found. In the assumed basis we alter the positions u, v, and w so that the zeros of S are located in the positions of the absent reflections. The corresponding u, v, and w values are then taken. Obtain the values of the atomic scattering factor f from published tables and calculate $|S|^2$. This gives a measure of the intensity of the diffracted beam and may be compared with the observed intensity.

2.9. DETERMINATION OF CRYSTAL STRUCTURES OF KCl AND NaCl FROM BRAGG'S LAW.

To study crystal structures using his law, Bragg designed an X-ray spectrometer, known as the *Bragg spectrometer*. In this instrument a collimated beam of X-rays, produced by the target of an X-ray tube, falls on the cleavage plane of the crystal under test. The crystal is placed on a table capable of rotation so that the glancing angle θ of the crystal face with respect to the incident beam of X-rays can be changed. The angle θ can also be measured with the help of a circular scale and vernier. The reflected beam of X-rays enters an ionization chamber : ionization current produced by the beam is measured by an electrometer E (Fig. 2.12). The angle of setting of the arm carrying the ionization chamber can be changed and measured with the help of a vernier and the circular scale. For each setting of the ionization chamber, the angle of reflection is made equal to the angle of incidence. The ionization current is recorded for different values of θ.

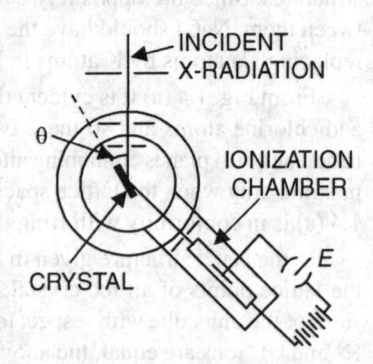

Fig. 2.12. Bragg's spectrometer.

The Bragg spectrometer can be used to study the crystal structures of sylvine (KCl) and rock-salt (NaCl), which are cubic. The prominent K_α line of X-ray from a palladium target gives peak ionization currents at $\theta = 5°23'$, $10°49'$ and $16°20'$ when the (100) face of KCl is exposed to the X-ray. These three maxima correspond to the first three orders, *i.e.* $n = 1, 2$, and 3 in the Bragg equation $2d\sin\theta = n\lambda$. The value of θ with $n = 1$, corresponding to the (110) and the (111) planes are $7°37'$ and $9°25'$, respectively. If the lattice spacing for the (hkl) planes is d_{hkl}, then from Bragg's law we have for KCl

$$d_{100} : d_{110} : d_{111} = (1/\sin 5°23') : (1/\sin 7°37') : (1/\sin 9°25').$$

or, $\quad d_{100} : d_{110} : d_{111} = 1 : 0.7076 : 0.5731.$ ...(2.41)

For a cubic crystal, using Eq. (1.7) we get

$$d_{100} : d_{110} : d_{111} = 1 : 1/\sqrt{2} : 1/\sqrt{3}$$

$$= 1 : 0.7071 : 0.5773. \quad ...(2.42)$$

Comparing Eqs. (2.41) and (2.42) we conclude that the KCl crystal is cubic.

For NaCl the peak intensity of the reflected beam in the first order ($n = 1$) occurs at $\theta = 6°$ for (100) planes. Therefore

$$\frac{(d_{100})_{NaCl}}{(d_{100})_{KCl}} = \frac{\sin 5°23'}{\sin 6°} = \frac{1}{1.115}.$$

The glancing angle θ for $n = 1$ with the (110) planes of NaCl, obtained experimentally, gives

$$\frac{(d_{110})_{NaCl}}{(d_{110})_{KCl}} = \frac{1}{1.115},$$

which is the expected result. However, the experimentally obtained value of $\frac{(d_{111})_{NaCl}}{(d_{111})_{KCl}}$ is found to be twice the value obtained above. The reason for this discrepancy is that the efficiencies of the different planes of atoms in the crystal to scatter X-rays are not the same.

The efficiency of an atom to scatter X-rays depends on the number of electrons in the atom. The atomic numbers, *i.e.* the numbers of orbital electrons for K, Cl and Na atoms are respectively 19, 17 and 11. Thus K and Cl atoms will be approximately equally efficient to scatter X-rays, but Na atom will be much less efficient than Cl atom. If, for NaCl crystal, parallel planes of atoms contain alternately all sodium and all chlorine atoms, then prominent diffraction maxima would not be expected due to reflections from these planes. This is because the condition for destructive interference to give minimum intensities between the maxima will not be obeyed by waves of markedly different amplitudes. On the basis of these arguments it is concluded that both NaCl and KCl will have similar cubic structures. Since the oppositely charged ions occur in pairs with the smallest possible distance between them. NaCl should have the structure shown in Fig. 1.4 (*a*). The KCl structure is obtained by replacing Na atoms by K atoms in Fig. 1.4 (*a*).

From Fig. 1.4 (*a*) it is evident that (100) planes and (110) planes contain equal numbers of metal and chlorine atoms and so these two sets of planes scatter X-rays by equal amounts. On the other hand, the (111) planes containing alternately all metal atoms and all chlorine atoms, can give prominent maxima only when the lattice spacing is $2d_{111}$ for NaCl and d_{111} for KCl. Thus the structure of Fig. 1.4 (*a*) is in conformity with Bragg's experimental results.

In the NaCl structure given in Fig. 1.4 (*a*), the atoms of any one type (in the ionized state) are at the lattice points of an fcc crystal. The fcc lattice for the Cl ions is displaced by half the length of sides of the unit cube with respect to the fcc lattice for the Na ions. Since the scattering efficiencies of K^+ and Cl^- ions are equal, the above two sets of fcc lattices, one for the K ions and the other for the Cl ions, are indistinguishable and are hence considered to form a simple cubic lattice for KCl.

It is possible to calculate the absolute value of the lattice spacing for the cubic crystal from a knowledge of the molecular weight and density of the crystal and the Avogadro number (see, for example, worked-out problem no. 6 of the previous chapter, where we have shown that for NaCl, $d_{100} = 2.82$Å).

Putting $n = 1$ and $\theta = 6°$ for the (100) planes of NaCl we obtain the wavelength of the K_α line of X-rays from the palladium target by using Bragg's equation:

$$\lambda = 2d_{100} \sin \theta = 2 \times 2.82 \times \sin 6° \text{ Å}$$
$$= 0.59 \text{ Å}$$

Thus by using the Bragg equation one can determine

(*i*) the lattice spacing *d* of the crystal with a known wavelength, and

(*ii*) the wavelength of an unknown incident wavelength with a known lattice spacing.

Wigner-Seitz cell. The primitive cell in a reciprocal lattice space can be constructed in a two dimensional situation by drawing a sufficient number of reciprocal lattice vectors from the origin O (Fig. 2.13). Lines perpendicular to these reciprocal lattice vectors at their mid-points are drawn. The smallest area enclosed by these lines (shown shaded in Fig. 2.13) is the Wigner-Seitz cell in a two dimensional reciprocal lattice.

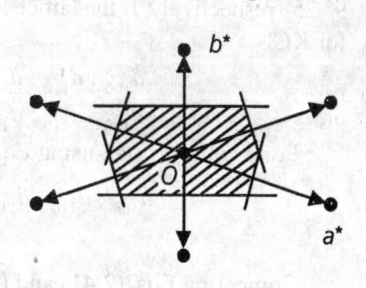

Fig. 2.13. First Brillouin zone for a two dimensional reciprocal lattice.

X-Ray Crystal Analysis

In three dimensions, we draw similar reciprocal lattice vectors from the origin. The primitive cell of the reciprocal lattice is taken as the smallest volume enclosed by planes normal to each of the shorter reciprocal lattice vectors at its mid-point. Each such cell contains one lattice point, the one at the centre of the cell. This primitive cell is the *Wigner-Seitz cell* of the reciprocal lattice, also called the *first Brillouin zone*.

2.10. WORKED-OUT PROBLEMS

1. The first order (100) reflection angle is 18° for a cubic crystal using X-rays of wavelength 1.54Å. Determine the distance between the (100) planes and the (111) planes of the crystal.

(C.U. 2005)

Ans. In this case, Bragg's law gives $2d \sin \theta = \lambda$, where d is the separation between the (100) planes

$$d = \frac{\lambda}{2 \sin \theta} = \frac{1.54}{2 \sin 18°} = \frac{1.54}{0.618} = 2.49 \text{Å}.$$

The separation (d) between the (100) planes gives the cube edge length of the unit cell, *i.e.*, a. The distance between the (111) planes is

$$d' = \frac{a}{\sqrt{h^2 + k^2 + l^2}} = \frac{2.49}{\sqrt{3}} = 1.44 \text{Å}$$

2. The wavelength of the K_α line of Ag is 0.563Å. The radiation from an Ag target is analysed with a Bragg spectrometer using a calcite crystal (a simple cube of lattice constant 3.02945Å). Determine the angle of reflection for the first order. What is the highest order for which this line may be observed?

(C.U. 1990)

Ans. For the first order ($n = 1$), Bragg's law gives $2d \sin \theta = \lambda$. Here $d = 3.02945$Å and $\lambda = 0.563$Å.

Hence $\sin \theta = \lambda / (2d) = 0.563/ 6.0589 = 0.0929$. Therefore the angle of reflection is $\theta = \sin^{-1} 0.0929 = 5.33°$.

For the nth order reflection, we have $2d \sin \theta = n \lambda$. As n increases, θ increases. For the highest value of θ, *i.e.* 90°, we have $n = 2d/\lambda \approx 6.0589/0.563 = 10.8$. The highest value of n is the integer below 10.8, *i.e.* 10.

3. A simple cubic crystal illuminated with X-rays of wavelength 0.09 nm, is rotated and the first order Bragg reflection occurs at a minimum glancing angle of 8.8°. Which set of crystal planes are responsible for this reflection? Find the spacing between these planes. Also, find the angle for the first order reflection from the (110) crystal planes.

Ans. The minimum glancing angle corresponds to the largest spacing of planes. So, the set of planes in question will be the (100) planes. We have $2d_{100} \sin \theta = \lambda$

or, $$d_{100} = \frac{\lambda}{2 \sin \theta} = \frac{0.09}{2 \sin 8.8°} = 0.294 \text{ nm}.$$

The spacing between the (110) planes is

$$d_{110} = \frac{d_{100}}{\sqrt{1^2 + 1^2 + 0}} = \frac{d_{100}}{\sqrt{2}} = \frac{0.294}{\sqrt{2}} = 0.208 \text{ nm}$$

So, $2d_{110} \sin \theta_{110} = \lambda$, or $\sin \theta_{110} = \dfrac{0.09}{2 \times 0.208} = 0.2163$ whence $\theta_{110} = 12.5°$.

4. X-rays suffer first-order Bragg reflection from the (100) planes of a simple cubic crystal, the density of which is known with an error of two parts in 10^4. If the glancing angle of the incident

X-rays with the set of crystal planes is 8° and is measured with an error of 3 minutes of arc, determine the maximum proportional error in the evaluation of the X-ray wavelength.

Ans. The Bragg equation for the first-order is $2d \sin \theta = \lambda$.

The maximum proportional error in λ is obtained by logarithmic differentiation:

$$\left|\frac{\Delta \lambda}{\lambda}\right| = \left|\frac{\Delta d}{d}\right| + \left|\frac{\Delta(\sin \theta)}{\sin \theta}\right| = \left|\frac{\Delta d}{d}\right| + |\cot \theta \, (\Delta \theta)|$$

The gram molecular weight is given by $M = N_0 \rho d^3$ where ρ is the mass density of the crystal and N_0 is Avogadro's number, the volume of the unit cell being d^3. Again, by logarithmic differentiation we obtain

$$3\left|\frac{\Delta d}{d}\right| = \left|\frac{\Delta \rho}{\rho}\right| \quad \text{or,} \quad \left|\frac{\Delta d}{d}\right| = \frac{1}{3}\left|\frac{\Delta \rho}{\rho}\right|.$$

Hence
$$\left|\frac{\Delta \lambda}{\lambda}\right| = \frac{1}{3}\left|\frac{\Delta \rho}{\rho}\right| + |\Delta \theta \cot \theta|.$$

Since θ is a small angle, we can write $\cot \theta = \dfrac{\cos \theta}{\sin \theta} \simeq \dfrac{1}{\theta}$.

So,
$$\left|\frac{\Delta \lambda}{\lambda}\right| = \frac{1}{3}\left|\frac{\Delta \rho}{\rho}\right| + \left|\frac{\Delta \theta}{\theta}\right| = \frac{1}{3} \times \frac{2}{10^4} + \frac{3/60}{8} = 6.9 \times 10^{-4}$$

5. The spacing of the planes in a crystal is 1.2Å and the angle for the first order reflection is 30°. Determine the energy in eV of the X-rays used. If the spacing of the crystal planes changes by ± 0.01 Å, what is the spread in energy in the diffracted beam?

Ans. The wavelength of the X-rays is given by
$$\lambda = 2d \sin \theta = 2 \times 1.2 \times \sin 30° = 1.2 \text{Å}.$$

The energy of the X-rays is $E = \dfrac{hc}{\lambda} = \dfrac{6.62 \times 10^{-34} \times 3 \times 10^8}{1.2 \times 10^{-10}}$ J

$$= \frac{6.62 \times 10^{-34} \times 3 \times 10^8}{1.2 \times 10^{-10} \times 1.6 \times 10^{-19}} \text{ eV} = 10.3 \text{ keV}$$

From $\lambda = 2d \sin \theta$, we have

$$\Delta \lambda = 2(\Delta d) \sin \theta = 2 \times (\pm 0.01) \sin 30° = \pm 0.01 \text{ Å}$$

The corresponding spread in energy is $\Delta E = -\dfrac{hc}{\lambda^2} \Delta \lambda \quad \left[\text{since } E = \dfrac{hc}{\lambda}\right]$

$$= -E \frac{\Delta \lambda}{\lambda} = -\frac{10.3 \times 10^3 \times (\pm 0.01)}{1.2} \text{ eV} = \mp 85.8 \text{ eV}$$

6. For the element polonium, Bragg first-order reflections appear at $\sin \theta = 0.225$, 0.317, and 0.389 for reflections from (100), (110), and (111) planes respectively. The wavelength of X-ray used is 1.54 Å. Show that the unit cell is cubic and calculate the length of the side of the unit cell.

Ans. The values of $\sin \theta$ are in the ratio $0.225 : 0.317 : 0.389 = 1 : 1.41 : 1.73 = 1 : \sqrt{2} : \sqrt{3}$.

The corresponding interplanar distances are in the ratio $d_{100} : d_{110} : d_{111} = 1 : 1/\sqrt{2} : 1/\sqrt{3}$. This shows that the unit cell is cubic.

If a is the length of the side of the cubic unit cell, we have using the data for reflection from (100) planes in Bragg's law

$$a = d_{100} = \frac{\lambda}{2 \sin \theta} = \frac{1.54}{2 \times 0.225} = 3.42 \text{ Å}$$

7. If V is the volume of a primitive unit cell in the direct lattice, show that the volume of a primitive unit cell in the reciprocal lattice is $8\pi^3/V$. **(Burd. U. 1996)**

Ans. The primitive unit cell is defined by the primitive basis vectors. In the direct lattice, these vectors are $\vec{a}, \vec{b}$ and $\vec{c}$, whereas in the reciprocal lattice they are $\vec{a}^*, \vec{b}^*$ and $\vec{c}^*$ given by Eqs. (2.8), (2.9) and (2.10). The volume of a primitive unit cell in the direct lattice is $V = \vec{a} \cdot \vec{b} \times \vec{c}$, and that in the reciprocal lattice is $V^* = \vec{a}^* \cdot \vec{b}^* \times \vec{c}^*$. We have, using Eqs. (2.8) through (2.10),

$$V^* = \frac{8\pi^3}{V^3} (\vec{b} \times \vec{c}) \cdot (\vec{c} \times \vec{a}) \times (\vec{a} \times \vec{b})$$

$$= \frac{8\pi^3}{V^3} (\vec{b} \times \vec{c}) \cdot \vec{A} \times (\vec{a} \times \vec{b})$$

where $\vec{A} = \vec{c} \times \vec{a}$. We have the vector identity

$$\vec{A} \times (\vec{a} \times \vec{b}) = \vec{a} (\vec{A} \cdot \vec{b}) - \vec{b} (\vec{A} \cdot \vec{a})$$

$$= \vec{a} (\vec{c} \times \vec{a} \cdot \vec{b}) - \vec{b} (\vec{c} \times \vec{a} \cdot \vec{a})$$

$$= \vec{a} (\vec{c} \times \vec{a} \cdot \vec{b}), \text{ since } \vec{c} \times \vec{a} \cdot \vec{a} = 0$$

Thus

$$V^* = \frac{8\pi^3}{V^3} (\vec{b} \times \vec{c}) \cdot \vec{a} \, (\vec{c} \times \vec{a} \cdot \vec{b}) = \frac{8\pi^3}{V}, \text{ (proved)}$$

since $V = \vec{a} \cdot \vec{b} \times \vec{c} = \vec{b} \cdot \vec{c} \times \vec{a}$

8. Prove that the reciprocal lattice of a bcc lattice is an fcc lattice. **(Burd. U. 2000)**

Ans. The primitive basis vectors of a bcc lattice are (see Sec. 1.5)

$$\vec{a}' = \frac{a}{2} (\vec{i} + \vec{j} - \vec{k}), \vec{b}' = \frac{a}{2} (-\vec{i} + \vec{j} + \vec{k}), \text{ and } \vec{c}' = \frac{a}{2} (\vec{i} - \vec{j} + \vec{k}),$$

where a is the cube edge and $\vec{i}, \vec{j}, \vec{k}$ are unit vectors along the cube edges. The corresponding reciprocal lattice vectors are

$$\vec{a}^* = 2\pi \frac{\vec{b}' \times \vec{c}'}{\vec{a}' \cdot \vec{b}' \times \vec{c}'} = 2\pi \frac{a^2/2}{a^3/2} (\vec{i} + \vec{j}) = \frac{2\pi}{a} (\vec{i} + \vec{j}),$$

$$\vec{b}^* = 2\pi \frac{\vec{c}' \times \vec{a}'}{\vec{a}' \cdot \vec{b}' \times \vec{c}'} = \frac{2\pi}{a} (\vec{j} + \vec{k}), \text{ and}$$

$$\vec{c}^* = 2\pi \frac{\vec{a}' \times \vec{b}'}{\vec{a}' \cdot \vec{b}' \times \vec{c}'} = \frac{2\pi}{a} (\vec{k} + \vec{i}).$$

The primitive basis vectors of an fcc lattice are $\vec{a}' = \frac{a}{2}(\vec{i} + \vec{j})$, $\vec{b}' = \frac{a}{2}(\vec{j} + \vec{k})$, and $\vec{c}' = \frac{a}{2}(\vec{k} + \vec{i})$.

Comparing $\vec{a}^*, \vec{b}^*, \vec{c}^*$ with $\vec{a}', \vec{b}', \vec{c}'$, we find that $\vec{a}^*, \vec{b}^*, \vec{c}^*$ are the primitive basis vectors of an fcc lattice of which the cube edge is $4\pi/a$. So, the reciprocal lattice of a bcc lattice is an fcc lattice.

9. In a powder diffraction experiment with monochromatic X-rays of wavelength 1.54Å, first order Bragg reflections are observed from a monatomic cubic crystal for the following values of the angle 2θ (in degree): 34.8, 40.6, 58.6, 70.2 and 73.8. Find the Miller indices of the diffracting planes, the crystal structure, and the lattice parameter.

Ans. Bragg's law for the first order (hkl) reflection for the lattice parameter a is $2d \sin \theta = \lambda$,

or, $2 \dfrac{a}{\sqrt{h^2 + k^2 + l^2}} \sin \theta = \lambda$

Thus, $\sin^2 \theta = \dfrac{\lambda^2}{4a^2} (h^2 + k^2 + l^2)$ or, $\sin^2 \theta \propto (h^2 + k^2 + l^2)$

For the given 2θ values, the values of θ are 17.4, 20.3, 29.3, 35.1, and 36.9 degree. The corresponding $\sin^2 \theta$ values are 0.09, 0.12, 0.24, 0.33 and 0.36 (approx). The values of $\sin^2 \theta$ are in the ratio 3:4:8:11:12. The Miller indices (hkl) giving these ratios are (111), (200), (220), (311), and (222). The indices are either all odd or all even. Hence the structure is fcc.

For the (111) reflection, we have $\theta = 17.4°$. With $\lambda = 1.54$ Å, we get

$$a = \dfrac{\lambda}{2 \sin \theta} \sqrt{h^2 + k^2 + l^2} = \dfrac{1.54 \sqrt{3}}{2 \sin 17.4°} = 4.46 \text{ Å}$$

QUESTIONS

1. Why is X-ray used in crystal structure analysis?
2. (a) State Bragg's law of diffraction and discuss its importance in crystal structure analysis.
   (cf. C.U. 1988)
   (b) How is Bragg's law modified due to refraction at the crystal surface? **(C.U. 2001)**
3. (a) What is a reciprocal lattice? State the advantages of using reciprocal lattice over direct space lattice in crystal structure analysis.
   (b) Prove that any reciprocal lattice vector is normal to a lattice plane of the crystal lattice. Also show that the spacing d of (hkl) planes of a crystal lattice is equal to $2\pi n / K$, where $K = |h\vec{a}^* + k\vec{b}^* + l\vec{c}^*|$ and n is an integer. **(c.f. Burd. U. 1995)**
4. (a) Deduce Bragg's law and use it to determine the crystal structure of NaCl. **[C.U. 1982]**
   (b) Show from Bragg's ideas and experiments that in the NaCl structure, Na and Cl atoms are arranged in fcc lattices.
5. (a) Derive Laue's equation and hence deduce Bragg's law of reflection. Consider X-rays of wavelength 1.54Å incident on a single cubic crystal of lattice constant 4 Å. Calculate the angle for the first order ($n = 1$) reflection. **(C.U. 1984) [Ans. 11°]**
   (b) How does the Laue approach differ from the Bragg approach? **(C.U. 2004)**
6. (a) Describe 'Ewald construction' and explain its importance.
   (b) On the basis of Ewald's construction, find the geometrical significance of the Bragg equation in the reciprocal lattice space. **(Burd. U. 1996)**
7. Discuss (i) the Laue method and (ii) the rotating crystal method of X-ray diffraction. Mention specific applications of each of them.
8. (a) Why, even though crystals are all randomly aligned, are X-ray diffraction patterns obtainable from powdered salt?

(b) Describe the basic principles underlying the powder method in X-ray crystallography.

(C.U. 1983)

9. Define the terms: *(i)* atomic scattering factor and *(ii)* geometrical structure factor. Consider a bcc crystal in which all the atoms are identical and derive an expression for the geometrical structure factor. Explain the usefulness of this expression.

10. Describe the steps for the analysis of a crystal structure using X-rays.

11. The cubic unit cell dimension of fcc copper is 3.6 Å. Calculate the longest wavelength of X-rays which will produce diffraction from the most closely packed planes. (C.U. 1986) [Ans. 4.17 Å]

[**Hint.** : For the fcc lattice, the (111) plane is most closely packed].

12. What is a Wigner-Seitz cell ? How would you construct such a cell ?

13. The molecular weight of KBr (a simple cube) is 119 and its density is 2.75 gm/cm^3. Taking Avogadro's number as 6.022×10^{23} per mole, calculate the grating space of the Laue grating for this crystal. If a narrow beam of K_α radiation ($\lambda = 1.541$ Å) from copper is reflected from (100) planes of such a crystal, calculate the glancing angle corresponding to the first order spectrum. (C.U. 1992)

[Ans. 4.157Å, 10.69°]

[**Hint.**: If N is the Avogadro number, d the required grating space, ρ the mass density, then $Nd^3 \rho =$ the gram molecular weight.]

14. Show that the reciprocal lattice of an fcc lattice is a bcc lattice. (C.U. 2008)

15. Prove that the reciprocal lattice for a simple cubic structure of lattice constant a is a simple cube of side $2\pi/a$.

16. The primitive basis vectors of a lattice are $\vec{a} = \vec{i} + 2\vec{j}$, $\vec{b} = 4\vec{j}$ and $\vec{c} = \vec{k}$. What are the primitive translation vectors of its reciprocal lattice ?

[Ans. $\vec{a}^* = 2\pi \vec{i}$, $\vec{b}^* = \frac{\pi}{2}(-2\vec{i} + \vec{j})$, $\vec{c}^* = 2\pi \vec{k}$]

17. Obtain the geometrical structure factor for the fcc crystal having identical atoms. Determine the possible and the missing X-ray reflections for the crystal.

18. The first order Bragg reflection from the most closely packed planes of a bcc crystal of lattice parameter 2.87 Å appears at the glancing angle of 22.3°. Find the wavelength of X-rays used.

[**Hint.** For the bcc lattice, (110) planes are most closely packed] [Ans. 1.54Å]

19. Given a square piece of X-ray film 10cm × 10cm, copper radiation of $\lambda = 1.52$Å, and powdered NaCl (cubic) with lattice parameter 4.57Å, devise a diffraction experiment in such a fashion that the rays from the (111) planes will produce a circle of diameter 10 cm. (sin 16.75° = 0.288, tan 33.5° = 0.662).

(C.U. 1999)

[**Hint.** See 'the powder method' of Sec. 2.6. For the first order, Bragg's law gives $2d \sin\theta = \lambda$. Here $d = 4.57 / \sqrt{3}$ Å and $\lambda = 1.52$ Å, so that

$$\sin\theta = \frac{\lambda}{2d} = \frac{1.52 \times \sqrt{3}}{2 \times 4.57} = 0.288 = \sin 16.75°$$

So, $\theta = 16.75°$. The X-ray film is placed normally to the incident X-radiation at a distance D from the powder sample such that the diffraction cone intersects the film in a circle of radius 5 cm. We have

$\tan 2\theta = 5/D$; or, $D = \dfrac{5}{\tan 33.5°} = \dfrac{5}{0.662} = 7.55$ cm.]

20. The second order Bragg reflection from a set of parallel planes in a crystal with monochromatic X-ray appears at a glancing angle of 31.2°. What is the glancing angle for the first order reflection?

[Ans. 15°]

21. The line *A* of an X-ray beam gives a first order reflection maximum when the glancing angle is 30° to the smooth face of a crystal. Line *B* with $\lambda = 0.97$Å gives a third order reflection maximum at an angle of 60° from the same face of the same crystal. Determine the wavelength of the line *A*. (K.U. 2004)

[Ans. 1.68 Å]

Chapter 3

Band Theory of Solids

3.1. INTRODUCTION

The problem of determining the electron wave functions and the energy levels in a solid is extremely complicated owing to the presence of a large number of interacting particles. A simplified quantum mechanical picture of an electron in a crystal can however be obtained by assuming that the atomic nuclei are at rest in the crystalline state and that the electron is in a periodic potential which has the periodicity of the lattice. The periodic potential may be thought to be caused by the fixed nuclei plus some average potential due to all other electrons. The solution of Schrödinger's equation for this potential gives a set of states which may be occupied by the electrons in accordance with the Pauli principle. We shall see below that all the energies of an electron are not permitted; rather, bands of possible energies, separated by regions of forbidden energies, exist. We shall also see that this picture leads to a natural distinction between metals, insulators, and semiconductors.

3.2. A SIMPLE DISCUSSION ON THE FORMATION OF ENERGY BANDS

Consider an isolated hydrogen atom. We know that the wave function of an electron in the atom in the ground state is [see Eq. (6.49), Quantum Mechanics]

$$u(r) = (\pi a_0^3)^{-1/2} \exp(-r/a_0), \qquad ...(3.1)$$

where a_0 is the first Bohr radius. Assume that two such atoms are brought together. Let u_1 and u_2 denote the electronic wave functions for the two atoms when they are far apart and are not influenced by one another. As the separation between the atoms is decreased, the wave functions u_1 and u_2 overlap, and the resultant electronic wave function due to the two nuclei may be either $u_1 + u_2$ or

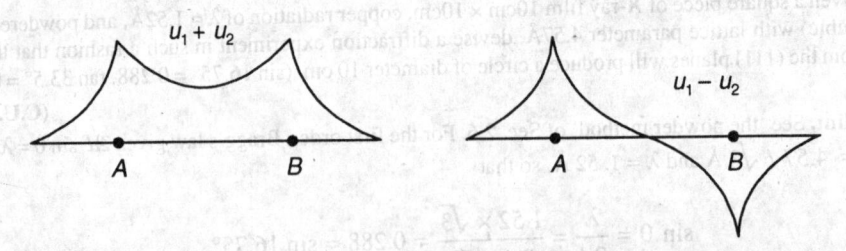

Fig. 3.1. Resultant electronic wave functions for the nuclei at positions marked A and B.

$u_1 - u_2$. The resultant wave functions are shown in Fig. 3.1. For the wave function $u_1 + u_2$, the electron has a finite probability of existing midway between the two nuclei. In this region, the binding energy increases since the electron experiences the binding force of both the nuclei. For the wave function $u_1 - u_2$, the probability density is zero midway between the two nuclei, so that the extra contribution of energy disappears. Thus there is a difference in energy between the states $u_1 + u_2$ and $u_1 - u_2$. This means that as the two atoms are brought close together, each energy state splits into two distinct energy states. If N number of atoms are brought close together, each energy state splits into N number of energy states. When N is large, the separation between these energy states is very small and they may be thought to produce a quasi-continuous band. That is, each energy level may be considered to

Band Theory of Solids

split into a band of energy levels. The band formation for the 1s and 2s states, as the separation between the atoms is reduced, is shown in Fig. 3.2. The width of a band depends on the strength of interaction and the overlap between the neighbouring atoms.

We shall now discuss the formation of allowed and forbidden bands of energies in a periodic potential.

3.3. Periodic Potential in a Crystalline Solid

Consider an electron in the vicinity of an atomic nucleus of charge Ze. The potential energy of the electron in SI units is given by

$$V = -\frac{Ze^2}{4\pi\varepsilon_0 r} \qquad \ldots(3.2)$$

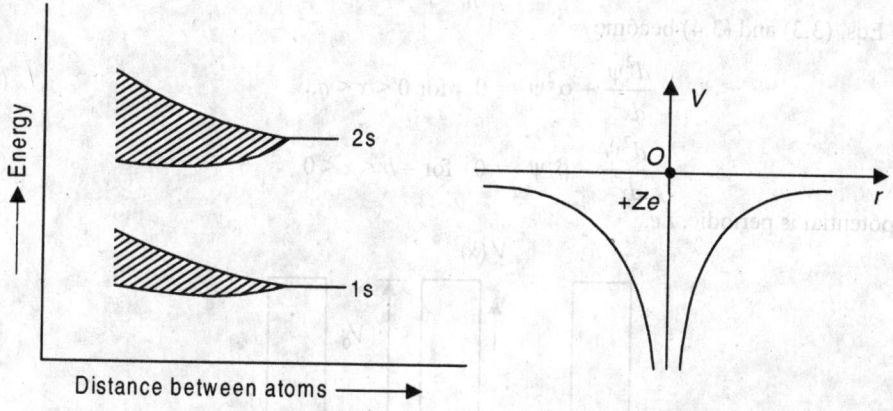

Fig. 3.2. Formation of bands for 1s and 2s states.

Fig. 3.3. Variation of potential energy with distance for an isolated atom.

where ε_0 is the free-space permittivity and r is the distance of the electron from the nucleus. The variation of the potential energy with the distance r is shown in Fig. 3.3. When a number of such atomic nuclei are brought close together to form a crystal, the potential energy of an electron is obtained by summing up the potential energies due to the individual nuclei. Thus the potential energy as a function of distance for an infinite one-dimensional crystal would be as shown in Fig. 3.4. Note that the potential energy is a periodic function of distance, the atomic nuclei being equispaced.

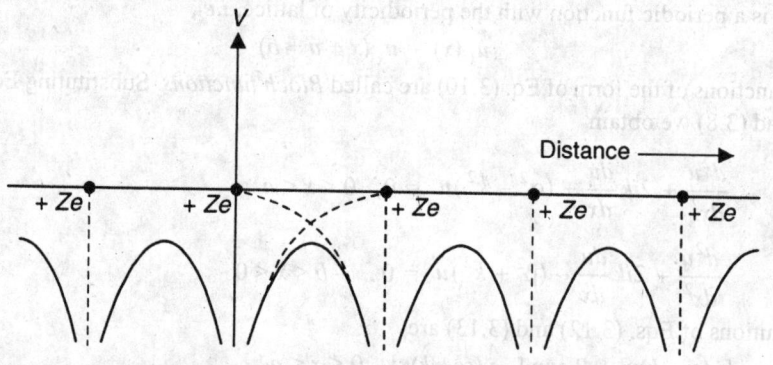

Fig. 3.4. Variation of potential energy with distance for a one-dimensional crystal.

3.4. The Kronig-Penney Model

In this model, for the sake of convenience, the periodic potential-energy curve of Fig. 3.4 is approximated by square wells, as shown in Fig. 3.5. Let the period of the periodic potential be $(a+b)$.

For, $-b < x < 0$, the potential energy is V_0 while for $0 < x < a$, the potential energy is zero. The Schrödinger equations are

$$\frac{d^2\psi}{dx^2} + \frac{2m}{\hbar^2} E\psi = 0 \quad \text{for } 0 < x < a \qquad \text{...(3.3)}$$

and

$$\frac{d^2\psi}{dx^2} + \frac{2m}{\hbar^2}(E - V_0)\psi = 0 \quad \text{for } -b < x < 0 \qquad \text{...(3.4)}$$

We assume that the electron energy E is less than V_0. Let

$$\alpha^2 = \frac{2mE}{\hbar^2}, \qquad \text{...(3.5)}$$

and

$$\beta^2 = \frac{2m}{\hbar^2}(V_0 - E) \qquad \text{...(3.6)}$$

Then Eqs. (3.3) and (3.4) become

$$\frac{d^2\psi}{dx^2} + \alpha^2\psi = 0 \quad \text{for } 0 < x < a, \qquad \text{...(3.7)}$$

and

$$\frac{d^2\psi}{dx^2} - \beta^2\psi = 0 \quad \text{for } -b < x < 0 \qquad \text{...(3.8)}$$

The potential is periodic, i.e.,

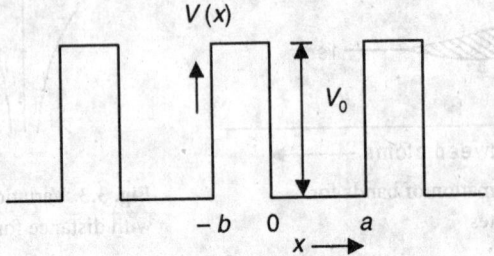

Fig. 3.5. The Kronig-Penney model potential.

$$V(x) = V(x + a + b) \qquad \text{...(3.9)}$$

Bloch's theorem states that for a periodic potential of the form of Eq. (3.9), the solutions of the Schrödinger equations are of the form

$$\psi = u_k(x) \exp(ikx) \qquad \text{...(3.10)}$$

where $u_k(x)$ is a periodic function with the periodicity of lattice, i.e.,

$$u_k(x) = u_k(x + a + b) \qquad \text{...(3.11)}$$

Wave functions of the form of Eq. (3.10) are called *Bloch functions*. Substituting Eq. (3.10) into Eqs. (3.7) and (3.8) we obtain

$$\frac{d^2u}{dx^2} + 2ik\frac{du}{dx} + (\alpha^2 - k^2)u = 0, \quad 0 < x < a \qquad \text{...(3.12)}$$

and

$$\frac{d^2u}{dx^2} + 2ik\frac{du}{dx} - (\beta^2 + k^2)u = 0, \quad -b < x < 0 \qquad \text{...(3.13)}$$

The solutions of Eqs. (3.12) and (3.13) are

$$u_1 = A \exp[i(\alpha - k)x] + B \exp[-i(\alpha + k)x], \quad 0 < x < a \qquad \text{...(3.14)}$$

and $u_2 = C \exp[(\beta - ik)x] + D \exp[-(\beta + ik)x], \quad -b < x < 0 \qquad \text{...(3.15)}$

where A, B, C and D are constants. These constants are determined in such a way that the wave function ψ and its normal derivative $d\psi/dx$ are single-valued and continuous. That is,

$$u_1(0) = u_2(0), \quad u_1(a) = u_2(-b),$$

Band Theory of Solids

$$\left(\frac{du_1}{dx}\right)_{x=0} = \left(\frac{du_2}{dx}\right)_{x=0},$$

$$\left(\frac{du_1}{dx}\right)_{x=a} = \left(\frac{du_2}{dx}\right)_{x=-b} \qquad ...(3.16)$$

Using these conditions in Eqs. (3.14) and (3.15) we get

$$A + B = C + D, \qquad ...(3.17)$$

$$i(\alpha - k) A - i(\alpha + k) B = (\beta - ik) C - (\beta + ik) D, \qquad ...(3.18)$$

$$A \exp[i(\alpha - k) a] + B \exp[-i(\alpha + k) a] = C \exp[-(\beta - ik) b] + D \exp[(\beta + ik) b], \qquad ...(3.19)$$

and $i(\alpha - k) A \exp[i(\alpha - k) a] - i(\alpha + k) B \exp[-i(\alpha + k) a]$
$$= (\beta - jk) C \exp[-(\beta - ik) b] - (\beta + ik) D \exp[(\beta + ik) b)] \qquad ...(3.20)$$

Equations (3.17) through (3.20) are homogeneous and will have a solution other than $A = B = C = D = 0$, provided the determinant of the coefficients A, B, C and D vanishes. Expanding this determinant and simplifying, we obtain

$$\frac{\beta^2 - \alpha^2}{2\alpha\beta} \sinh \beta b \sin \alpha a + \cosh \beta b \cos \alpha a = \cos k(a+b) \qquad ...(3.21)$$

For $E > V_0$, β, as defined by Eq. (3.6), is imaginary. For such energies we put $\beta = i\gamma$ and note that $\sinh i\gamma b = i \sin \gamma b$ and $\cosh i\gamma b = \cos \gamma b$. Then we obtain from Eq. (3.21)

$$-\frac{\alpha^2 + \gamma^2}{2\alpha\gamma} \sin \gamma b \sin \alpha a + \cos \gamma b \cos \alpha a = \cos k(a+b) \qquad ...(3.22)$$

It is convenient to use Eq. (3.21) when $0 < E < V_0$ and Eq. (3.22) when $E > V_0$. Equations (3.21) and (3.22) may also be written respectively in the forms

$$\left[1 + \frac{(\beta^2 + \alpha^2)^2}{4\alpha^2 \beta^2} \sinh^2 \beta b\right]^{1/2} \cos(\alpha a - \theta) = \cos k(a+b), \qquad ...(3.23)$$

where
$$\tan \theta = \frac{\beta^2 - \alpha^2}{2\alpha\beta} \tanh \beta b;$$

and
$$\left[1 + \frac{(\alpha^2 - \gamma^2)^2}{4\alpha^2 \gamma^2} \sin^2 \gamma b\right]^{1/2} \cos(\alpha a - \theta) = \cos k(a+b), \qquad ...(3.24)$$

where
$$\tan \theta = -\frac{\alpha^2 + \gamma^2}{2\alpha\gamma} \tan \gamma b$$

In Eqs. (3.23) and (3.24) the left-hand sides are cosine functions multiplied by factors which are greater than unity. The multiplying factor is a maximum for $\alpha = 0$, i.e. for $E = 0$, and approaches unity for large energies when $\alpha^2 = -\beta^2$.

Equations (3.23) and (3.24) are transcendental in nature and may be solved graphically. The left-hand side of either of these equations is plotted as a function of E in Fig. 3.6. The right-hand side of each equation is $\cos k(a+b)$ which lies between +1 and –1. Thus physically acceptable solutions are possible when the left-hand side of Eq. (3.23) or (3.24) lies between +1 and –1. The corresponding energy values are permitted. Outside these limits, physical solutions are not possible and the corresponding energies are forbidden. Thus, allowed and forbidden energy bands are alternately produced. The forbidden energy bands are shown as shaded regions in Fig. 3.6.

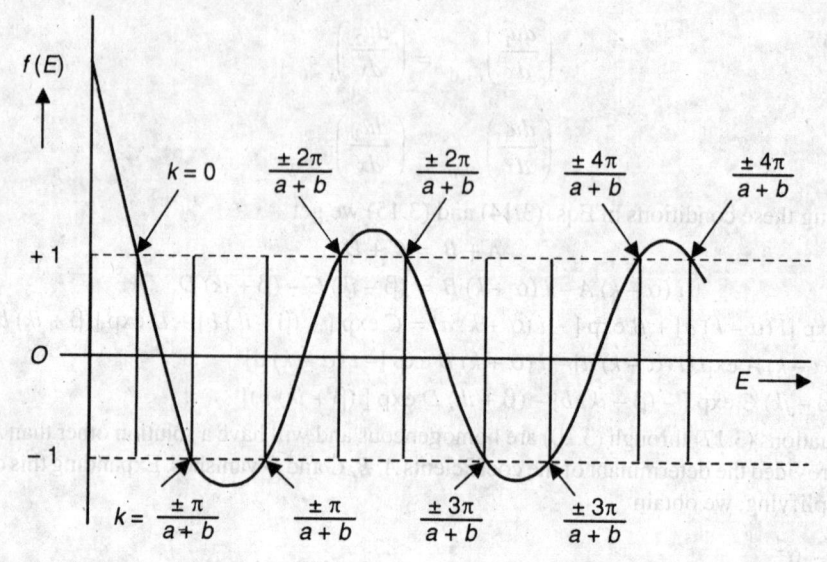

Fig. 3.6. Variation of the left - hand side of Eq. (3.23) or 3.24) as a function of energy.

With the help of Fig. 3.6, *i.e.* Eqs. (3.23) and (3.24), one can plot the energy E as a function of the wave vector k. Such a plot is shown in Fig. 3.7. The edges of the allowed bands occur when $\cos k (a + b) = \pm 1$, *i.e.*, when $k = \pm \dfrac{n\pi}{a+b}$, where n is zero or an integer. The $E - k$ diagram has the following important features :

(*i*) As $E \to \infty$, the E-k relationship approaches that for a free electron; *i.e.* we have $E = h^2k^2/2m_0$ which is shown as a dotted curve in Fig. 3.7.

(*ii*) For large energies, the allowed bands are broad and the forbidden bands are narrow.

(*iii*) If the wells are deep and widely spaced, *i.e.* if b and V_0 are large, a low energy electron will be practically bound to one of the wells. Thus the low energy eigenvalues will be nearly equal to those of a single well. If the wells are close together, the levels spread into energy bands. As the separation b is reduced, the allowed bands become wider.

(*iv*) As the barrier height V_0 is increased with a and b remaining fixed, the widths of the energy bands decrease.

(*v*) At the edges of the allowed bands, *i.e.* at $k = \pm \dfrac{n\pi}{a+b}$, the slope of the E-k curve is zero.

Reduced zone representation

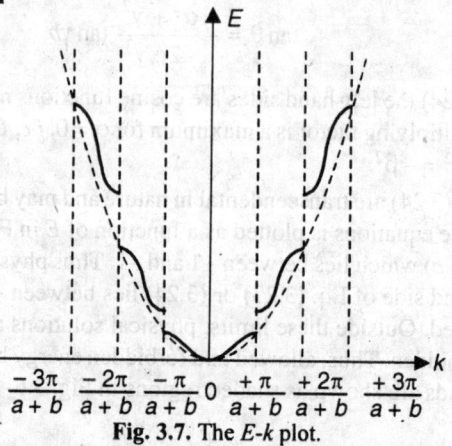

Fig. 3.7. The E-k plot.

Band Theory of Solids

In Fig. 3.7 the region extending from $k = -\dfrac{\pi}{a+b}$ to $k = \dfrac{\pi}{a+b}$ is called the *first Brillouin zone*, that from $-2\pi/(a+b)$ to $-\pi/(a+b)$ and from $\pi/(a+b)$ to $2\pi/(a+b)$ is known as the *second Brillouin zone*, and so on. In Eqs. (3.23) and (3.24) if k is replaced by $k + [2n\pi/(a+b)]$ where n is an integer, positive or negative, the equations are still satisfied since such replacements do not alter the value of $\cos k(a+b)$. Thus k is not uniquely determined, and it is often very useful to translate the various segments of the E-k curve in Fig. 3.7 parallel to the k-axis through distances that are integral multiples of $2\pi/(a+b)$ so that k lies within the region $-\pi/(a+b) \le k \le \pi/(a+b)$. The resulting representation of the E-k diagram, shown in Fig. 3.8, is referred to as the *reduced zone representation*.

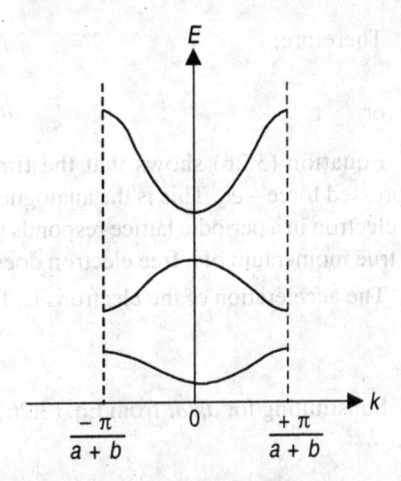

Fig. 3.8. E versus k plot in the reduced zone.

The $E - k$ curve in the extended zone scheme of Fig. 3.7 has discontinuities at $k = n\pi/(a+b)$ where $n = \pm 1, \pm 2$, etc. These k-values give the boundaries of the Brillouin zone. As $k = 2\pi/\lambda$, the Brillouin zone boundaries are defined in terms of λ by $2(a+b) = n\lambda$, which is Bragg's law of reflection for normal incidence. Thus the discontinuities in the $E - k$ curve are associated with the Bragg reflection of the de Broglie waves representing the electrons for these k-values.

We must mention that though the Kronig-Penney model deals with an idealised periodic potential, it brings out the important features of the behaviour of electrons in crystals, even in the three-dimensional case.

3.5. Effective Mass

The quantity $\hbar k$ has the dimensions of momentum. However, we have seen that the choice of the wave vector k in the case of a periodic potential is not unique. The quantity $\hbar k$ is therefore not the true momentum of an electron and is referred to as the *crystal momentum* with k usually lying in the reduced zone.

Let us now investigate the motion of an electron in a crystal in the presence of an externally applied electric field. From the quantum mechanical point of view the velocity of the electron, v, is equal to the group velocity of the "wave packet" representing the electron, *i.e.*,

$$v = \frac{d\omega}{dk},$$

where ω is the angular frequency of the wave and is related to the energy E through the Planck relationship $E = \hbar\omega$. Hence

$$v = \frac{1}{\hbar}\frac{dE}{dk} \qquad \ldots(3.25)$$

If ξ represents the externally applied electric field, the gain in energy of the electron in time dt is

$$dE = -e\,\xi\, v\, dt,$$

where e is the magnitude of the electron charge. Using Eq. (3.25) we write

$$dE = -\frac{e\xi}{\hbar}\frac{dE}{dk}dt$$

Also,

$$dE = \frac{dE}{dk}dk$$

Therefore,
$$dk = -\frac{e\xi}{\hbar}dt$$

or
$$\hbar \frac{dk}{dt} = -e\xi \qquad \ldots(3.26)$$

Equation (3.26) shows that the time rate of change of crystal momentum is equal to the impressed force $-e\xi$. This is the analogue of Newton's law and shows that the crystal momentum of the electron in a periodic lattice responds to an externally applied electric field in the same fashion as the true momentum of a free electron does in vacuum.

The acceleration of the electrons is, from Eq. (3.25),
$$f = \frac{dv}{dt} = \frac{1}{\hbar}\frac{d^2E}{dk^2}\frac{dk}{dt}$$

Substituting for dk/dt from Eq. (3.26) we get
$$f = -\frac{e\xi}{\hbar^2}\frac{d^2E}{dk^2} \qquad \ldots(3.27)$$

In Eq. (3.27), f is the acceleration due to the force $-e\xi$. Hence $-e\xi/f$ represents the *effective mass* m^* of the electron:
$$m^* = -\frac{e\xi}{f} = \frac{\hbar^2}{(d^2E/dk^2)} \qquad \ldots(3.28)$$

Note that the response of an electron in the crystal to an applied field is determined not by its actual gravitational mass but by its effective mass, which depends on the shape of the E-k curve.

For a free electron we have
$$E = \frac{\hbar^2 k^2}{2m_0},$$

where m_0 is the free electron mass. Then $\dfrac{d^2E}{dk^2} = \dfrac{\hbar^2}{m_0}$ and so
$$m^* = \frac{\hbar^2}{d^2E/dk^2} = m_0$$

The relation between E and k is practically parabolic near the bottom and top of the allowed energy bands, as shown in Fig. 3.8. Hence we may write
$$E = Ak^2,$$

where A is a constant. The effective mass is then obtained as $m^* = \dfrac{\hbar^2}{d^2E/dk^2} = \dfrac{\hbar^2}{2A}$. Therefore $A = \dfrac{\hbar^2}{2m^*}$, and the energy wave-vector relationship near the bottom or the top of a band may be written as
$$E = \frac{\hbar^2 k^2}{2m^*} \qquad \ldots(3.29)$$

Thus we find that the dynamical behaviour of the electron is the same as that of a free particle of effective mass m^*. The free-electron picture of a metal is thus justified provided the proper effective mass is used for the electron mass. The value of the effective mass, as already mentioned, depends on d^2E/dk^2 and may be greater or less than m_0. For metals, m^* does not differ greatly from m_0. Hence for metals no great errors occur if m_0 is used in place of m^*. For semiconductors, however, m^* can differ significantly from m_0. Hence in such materials the proper effective mass must be used for accurate results.

Depending on the actual E-k relationship in the solid crystal, the effective mass of the electrons

Band Theory of Solids

can be *negative* also. Then the electrons drift in the direction of the applied electric field. The conventional current is thus in a direction opposite to that of the field, giving a *negative resistance*. This happens in some semiconductors and is exploited in so-called negative-mass amplifiers and oscillators.

3.6. Number of Electrons in a Band

To obtain the number of electrons that can be accommodated in a band we consider periodic or cyclic boundary conditions. We assume that the lattice has the form of a closed ring of N atoms. If a is the separation between two consecutive atoms, then $\psi(x) = \psi(x + Na)$. Using Eq. (3.10) we obtain

$$u_k(x) \exp(ikx) = u_k(x + Na) \exp[ik(x + Na)],$$

or,
$$\exp(ikNa) = 1, \qquad \ldots(3.30)$$

since $u_k(x + Na) = u_k(x)$, u_k being a periodic function.

From Eq. (3.30) we get

$$k = \frac{2\pi n}{Na}, \qquad \ldots(3.31)$$

where $n = 0, 1, 2, \ldots, (N-1)$. Thus the number of possible values of k, or states, is N. When N is large the allowed values of k, given by Eq. (3.31), are so close that they may be regarded as forming a quasi-continuous band. Since each state may be occupied by two electrons with opposite spin, an allowed energy band can accommodate $2N$ electrons where N is the number of atoms in the crystal. This result, though worked out above for a one-dimensional lattice, holds good also for a three-dimensional lattice.

Observation

The velocity of an electron in a band is given by Eq. (3.25) whereas its effective mass is determined by Eq. (3.28). For free electrons, $E = \hbar^2 k^2/(2m_0)$ and hence Eq. (3.25) yields $v = \hbar k/m_0 = p/m_0$, where p is the electron momentum. In the band theory, however, E is not proportional to k^2 over the entire zone, as clearly seen in Fig. 3.8. For the E-k curve of Fig. 3.8A (a), the velocity v as a function of k is represented by the curve of Fig. 3.8A (b) in accordance with Eq. (3.25). In contrast, for a free electron, v is proportional to k. At the top or the bottom of the band $dE/dk = 0$, and so $v = 0$ at these points. The absolute value of v, i.e., $|v|$ is a maximum at the points of inflection I_1 and I_2 of the E-k curve. Thereafter, $|v|$ decreases with increasing energy, which is not expected for a free electron.

The effective mass m^* of an electron, given by Eq. (3.28), is determined by d^2E/dk^2. For the E-k curve of Fig. 3.8A (a), the variation of m^* with k is depicted in Fig. 3.8A (c). Below the inflection points of the E-k curve, m^* is positive, whereas above the inflection points m^* is negative. At the inflection points of the E-k curve, m^* is infinite. The negative value of m^* in the upper half of the energy band implies that there the electron behaves as a positively charged particle. Consider an electron at $k = 0$. If an electric field is applied, $|k|$ increases linearly, following Eq. (3.26), with time. Till $|v|$ attains its maximum the electron is accelerated by the field. Beyond that, the electron is decelerated, $|v|$ decreasing

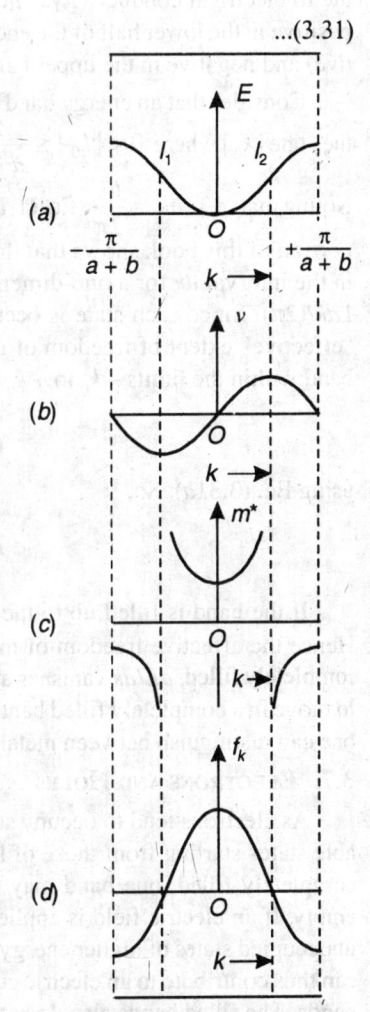

Fig. 3.8A. Plots of energy (E), velocity (v), effective mass m^*, and f_k as a function of k. I_1 and I_2 are the inflection points of the E-k curve.

with increasing $|k|$. Hence m^* must be negative in the upper postion of the energy band. This point is discussed further in Sec. 3.7.

The Bragg reflection at $k = \pm \dfrac{\pi}{a+b}$ accounts for the negative m^*. The Bragg reflection means that a force applied in the forward direction can produce a gain in momentum in the backward direction, leading to the negative m^*.

The *extent of freedom* of an electron at the state k is determined by a factor f_k where

$$f_k = \frac{m_0}{m^*} = \frac{m_0}{\hbar^2}\frac{d^2E}{dk^2} \qquad (3.31a)$$

A small value of f_k occurs when m^* is large. The electron then behaves as a *heavy* particle and its freedom of movement is low. On the other hand, a large value of f_k corresponding to a small value of m^*, shows that the electron acts as a *light* particle and has a greater freedom of movement to contribute to electrical conductivity. The variation of f_k with k, depicted in Fig. 3.8A (d), shows that f_k is positive in the lower half of the energy band (where m^* is positive) and negative in the upper half (where m^* is negative).

Consider that an energy band is filled with electrons up to the state $|k_0|$ where $0 < |k_0| \le \dfrac{\pi}{a+b}$, as shown in Fig. 3.8 B. Noting that $p = \hbar k = \dfrac{hk}{2\pi}$, Eq. (1.18) of 'Statistical Mechanics' portion of this book shows that the number of possible states in the interval dk for a one-dimensional lattice of length L is $Ldk/(2\pi)$. Since each state is occupied by two electrons, the "effective" extent of freedom of the electrons populating the band within the limits $-k_0$ to $+k_0$ is

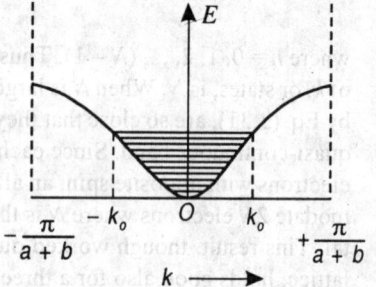

Fig. 3.8B. Energy band in the reduced zone filled up to state $|k_0|$.

$$F_{eff} = 2\frac{L}{2\pi}\int_{-k_0}^{k_0} f_k\, dk = \frac{2Lm_0}{\pi\hbar^2}\int_0^{k_0}\left(\frac{d^2E}{dk^2}\right)dk$$

using Eq. (3.31a). So,

$$F_{eff} = \frac{2Lm_0}{\pi\hbar^2}\left(\frac{dE}{dk}\right)_{k=k_0} \qquad (3.31b)$$

If the band is filled up to the inflection points of the E-k curve, $(dE/dk)_{k=k_0}$ is a maximum. Hence the effective freedom of movement of the electrons attains a maximum then. If the band is completely filled, dE/dk vanishes at the top of the band, and so $F_{eff} = 0$. Thus the electrons are not free to move in a completely filled band, and so cannot contribute to electrical conductivity. On this basis, one can distinguish between metals, insulators, and semiconductors (see Sec. 3.8).

3.7. ELECTRONS AND HOLES

As electrons tend to occupy states of lowest energies, the electrons in a crystal fill up the available states starting from those of lowest energy. Thus in a crystal, the lowermost energy bands are completely filled, one band may be partially filled, and the upper allowed bands are completely empty. If an electric field is applied, the electrons in a partially filled band can move into adjacent unoccupied states of higher energy and momentum by absorbing energy from the applied field. They can thus contribute to an electric current. It is obvious that no current is contributed from the unfilled bands. The filled bands also do not contribute anything to the current flow since the electrons do not find any adjacent vacant state to move into.

If n is the number of electrons per unit volume in a partially filled band, then the current density arising from the band is

$$J = -nev, \qquad \ldots(3.32)$$

where v, the average velocity of the electrons, can be written as

$$v = \frac{1}{nV}\sum_0 v_0 \qquad ...(3.33)$$

Here the sum over 0 indicates the sum over all the occupied states, and V is the crystal volume. From Eqs. (3.32) and (3.33) we write

$$J = -\frac{e}{V}\sum_0 v_0 = -\frac{e}{V}\left[\sum_c v_c - \sum_u v_u\right] = +\frac{e}{V}\sum_u v_u, \qquad ...(3.34)$$

where the sum over c gives the sum over all the states in the band and the sum over u gives that over all the unoccupied states. The sum over c, taken over the complete band, will be zero as discussed above. From Eq. (3.34) we find that the remaining sum over the unoccupied states gives a current that can be produced by a corresponding number of positively charged carriers. Thus the current in a nearly filled band can be obtained from the motion of a relatively small number of empty electronic states or *holes* which behave as positively charged particles. The velocity of a hole is the same as that of an electron in the empty state which normally occurs near the top of the band. In this region, the E-k curve is concave downward so that d^2E/dk^2 is negative. Hence the electron effective mass, given by Eq. (3.28), is negative. A negatively charged particle having a negative effective mass would experience an acceleration in the same direction as the applied electric field. It would thus behave as a positively charged particle with a positive mass, which is referred to as a hole.

3.8. METALS, INSULATORS, AND SEMICONDUCTORS

Insulators are materials that do not conduct electricity. In such materials the number of electrons is just enough to completely fill a number of allowed energy bands. Above these bands there exists a series of completely empty bands. The forbidden energy gap E_g between the topmost filled band (known as the *valence band*) and the lowermost empty band (known as the *conduction band*) is so large that at ordinary temperatures electrons cannot be thermally excited across this gap from the valence to the conduction band. As the bands are either completely full or empty, no electric current can flow. The band diagram of an insulator is depicted in Fig. 3.9 (a).

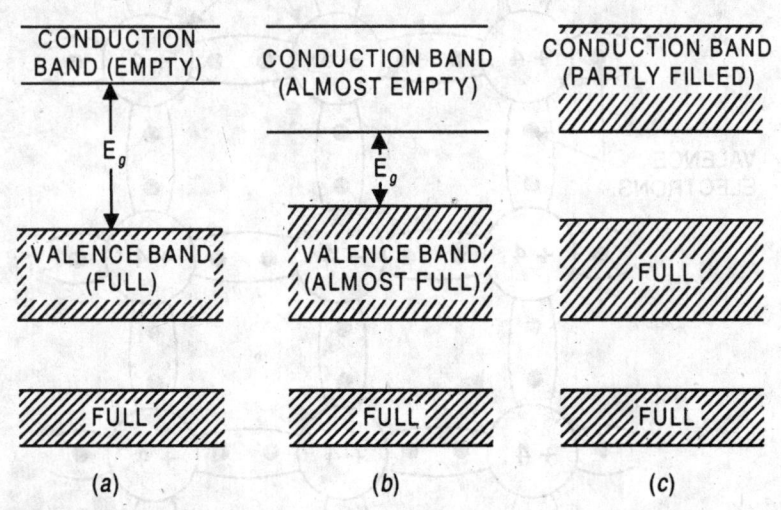

Fig. 3.9. Energy band diagram of (*a*) an insulator, (*b*) a semiconductor, and (*c*) a metal.

A semiconductor is a material for which the energy gap E_g between the valence and the conduction band is relatively small. In this case an appreciable number of electrons can be thermally excited across the gap from the states near the top of the valence band to states near the bottom of the conduction band. Free electrons thus exist in the almost empty conduction band to carry electric currents. Also, the vacancies left near the top of the almost full valence band behave as holes and contribute to electric current. The band structure of a semiconductor is given in Fig. 3.9(b). A semiconductor exhibits an electrical conductivity intermediate between that of a metal and an insulator. The electrical conductivity of a semiconductor decreases with increasing E_g and with decreasing temperature, since the probability of thermal excitation decreases. As the temperature approaches absolute zero, the thermal excitation becomes vanishingly small and therefore all semiconductors behave as insulators at such temperatures.

If the uppermost energy band is partially filled (Fig. 3.9c) or if the uppermost filled band and the next unoccupied band overlap in energy, the crystal is known as a metal. In this case the electrons in the uppermost band find adjacent vacant states to move into, by absorbing energy from an applied electric field. These electrons thus behave as free electrons and conduct electric currents. The electrical conductivity of a metal, at room temperature, is of the order of 10^6 mho cm^{-1}, that of a semiconductor lies in the range 10^3 to 10^6 mho cm^{-1}, and that of a good insulator is of the order of 10^{-12} mho cm^{-1}. The band gap E_g of a typical insulator, such as diamond, is about 6 eV while that of semiconductors lies in the range 0.2 eV to 2.5 eV.

3.9. INTRINSIC SEMICONDUCTORS

When the electrons and the holes are generated solely by the thermal excitation across the band gap, the semiconductor is referred to as a *pure* or *intrinsic semiconductor*. This situation has been considered in Sec. 3.8. The holes and the electrons, so generated, are called *intrinsic charge carriers* and the resulting electrical conductivity is known as the *intrinsic conductivity*. In an intrinsic semiconductor the numbers of electrons and holes must be the same. Thus, if n_i and p_i respectively denote the concentration of electrons and holes in an intrinsic semiconductor, we must have $n_i = p_i$. The quantity n_i (or p_i), known as the *intrinsic carrier concentration*, increases with increase in temperature.

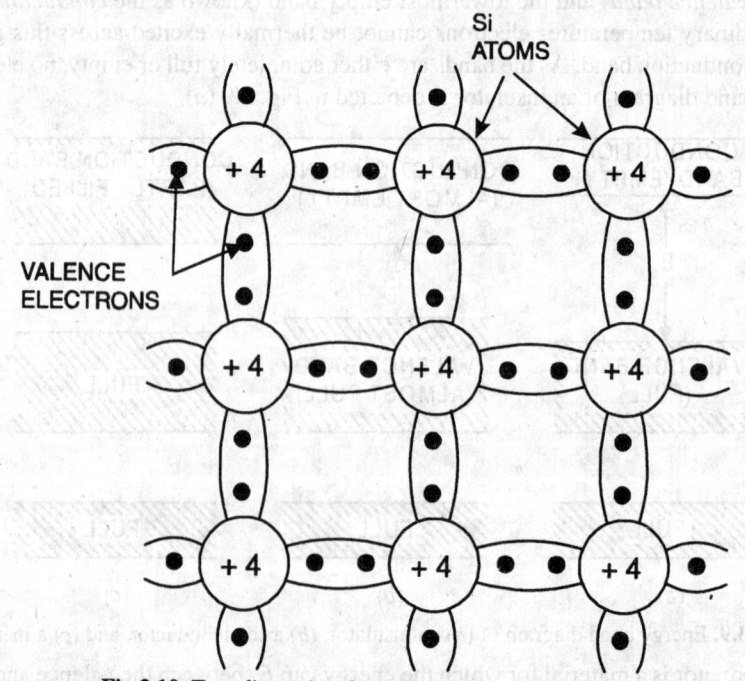

Fig.3.10. Two-dimensional representarion of a Si crystal.

For covalent semiconductors like germanium (Ge) and silicon (Si), the creation of free electrons and holes can be understood with reference to Fig. 3.10, which shows a two-dimensional representation

Band Theory of Solids 225

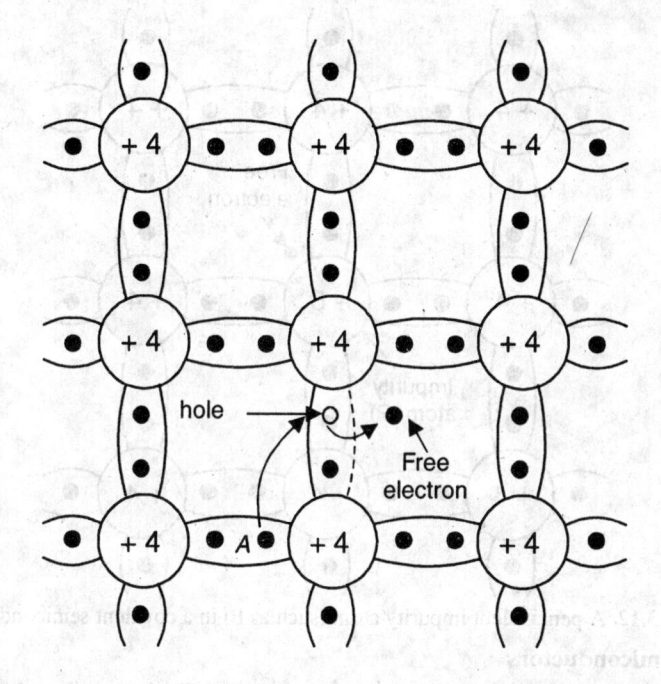

Fig. 3.11. Production of free electrons and holes in a covalent semiconductor.

of the crystal. Si (or Ge) is a group IV element of the periodic table, each atom having four valence electrons. The valence electrons of a particular atom are held by covalent bonds with the valence electrons of four neighbouring atoms. When a valence electron gets sufficient thermal energy it breaks its covalent bond and becomes free. The vacancy left behind by the electron serves as a hole. A free electron and a hole are thus created, as shown in Fig. 3.11. An adjacent valence electron, such as that in position A, may acquire sufficient thermal energy and jump into the position of the hole to reconstruct the broken bond, shown by the dashed line in Fig. 3.11. In doing so, it breaks its own covalent bond; effectively this means a movement of the hole to the position A. Thus, not only the free electron, but also the hole it creates, can move about in the crystal.

3.10. EXTRINSIC SEMICONDUCTORS

We have seen that excitation across the band gap generates free electrons and holes in a semiconductor. Such excitations may be produced thermally or optically. Another and probably the most convenient way of creating free electrons or holes is to introduce small amounts of selected impurities in the crystal. Such introduction of impurity atoms in a semiconductor is called *doping*, and the impurity that is introduced is called *dopant*. A semiconductor containing impurity atoms is referred to as a *doped*, an *impurity* or an *extrinsic semiconductor*. The impurity atoms produce excess free electrons or holes which control the conductivity of the extrinsic semiconductor. Depending on the nature of the dopant, an extrinsic semiconductor may be classified as an *n*-type or a *p*-type semiconductor. An element belonging to group III of the periodic table, introduced in a covalent semiconductor like Ge or Si, generates excess holes and makes it *p*-type. Boron, gallium, indium, and aluminium are common examples of such group III dopants. On the other hand, dopants such as phosphorus, antimony or arsenic, which are the elements of group V, create excess free electrons in Ge or Si and make the semiconductor *n*-type. We consider below in some detail *n*-type and *p*-type semiconductors.

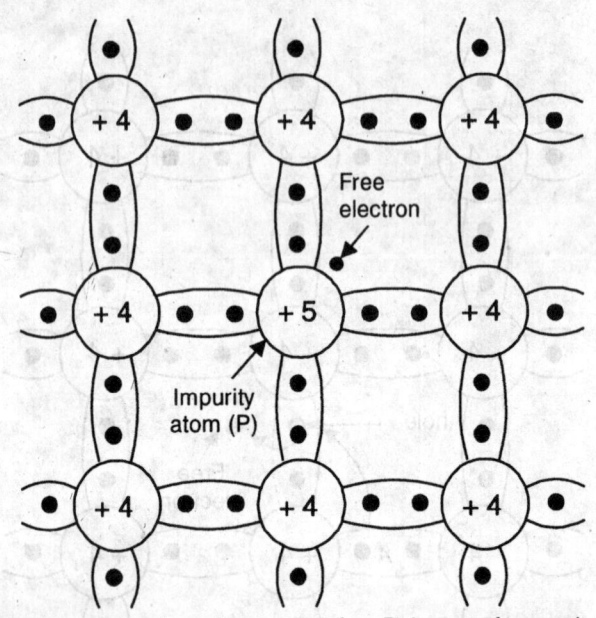

Fig. 3.12. A pentavalent impurity atom (such as P) in a covalent semiconductor.

(a) n-type semiconductors

Suppose that an element of group V, such as phosphorus (P), is introduced in small amounts to a semiconductor of group IV, such as germanium or silicon The size of the impurity atoms is comparable to that of the Ge or Si atoms. An impurity atom may therefore occupy the position of an atom of the host crystal, say, silicon (Fig. 3.12). Four of the five valence electrons of the impurity atom will form covalent bonds with the valence electrons of the four neighbouring Si atoms. The fifth valence electron of the dopant atom is loosely bound to its nucleus. It can, therefore, be separated and made free by expending an amount of energy much less than that required to rupture a covalent bond. This small amount of energy is readily obtained from the thermal agitation of the crystal.

The energy level of the fifth valence electron of the dopant atom is in the forbidden energy gap and lies just below the conduction band. This energy level, shown by the dashed line in Fig. 3.13, is referred to as the *donor level*. The donor level lies about 0.01 eV below the condition band edge for Ge, and about 0.05 eV below the conduction band edge for Si. As this energy is much less than the forbidden energy gap, the thermal energy required to detach the electrons from the pentavalent impurity atoms is much less than that required to excite the electrons from the valence band across the band gap. Thus even at low temperatures most of the dopant atoms will contribute to electrons in the conduction band. When the fifth valence electron is detached from the dopant atom and transferred to the conduction band, the dopant atoms become positively charged. Such impurity elements are called *donors* and the semiconductor having donor-type impurities is referred to as the *n*-type semiconductor. This is because the generated current carriers are negative charges, *i.e.* electrons. Note that while the electrons contributed by the donor atoms to the conduction band are free and mobile, the ionized donor atoms are immobile. The free electrons created in this way are called *excess electrons*.

Fig. 3.13. Donor level in energy diagram.

In an *n*-type semiconductor the electrons are *majority carriers*, and the thermally created holes are the *minority carriers*. If *n* and *p* are respectively the concentrations of free electrons and holes, then it can be shown that

$$np = n_i^2, \qquad \qquad ...(3.35)$$

Band Theory of Solids

where n_i is the intrinsic carrier concentration. Equation (3.35) is known as the *mass-action law*. For each donor atom there is a free electron in the conduction band but there is no corresponding positive hole in the valence band. Addition of donors makes $n > n_i$. Hence in an *n*-type semiconductor, p must be less than n_i according to Eq. (3.35).

(b) p-type semiconductors

Assume that an element of group III, such as aluminium, is added in small amounts to germanium or silicon. For an atom of the group III element, occupying the position of an atom of the host crystal (say, Si), there is an unfilled bond, shown by the dashed line in Fig. 3.14. The trivalent dopant atom can accept an electron, as shown by the small open circle B in Fig. 3.14. A valence electron from an adjacent Si atom (such as at position A) requires a small amount of energy to jump into the position B to complete the unfilled bond. Such a small energy is readily provided by the thermal agitation of the crystal. When the electron jumps from A to B, it breaks its own covalent bond and creates a *hole* at A.

The unfilled energy state created by positions such as B is located just above the valence band of the semiconductor and is indicated by the dashed line in Fig. 3.15. The energy difference between this state, known as the *acceptor level*, and the top of the valence band is about 0.01 eV for Ge and about 0.05 eV for Si. This energy difference is much smaller than the band gap. Electrons from the valence band may therefore be easily agitated thermally to move into the vacant acceptor level. In so doing they leave behind holes in the valence band and make the trivalent dopant atoms negatively charged ions. While the holes are mobile and can carry electric currents, the ionized impurity atoms are immobile and do not contribute to currents.

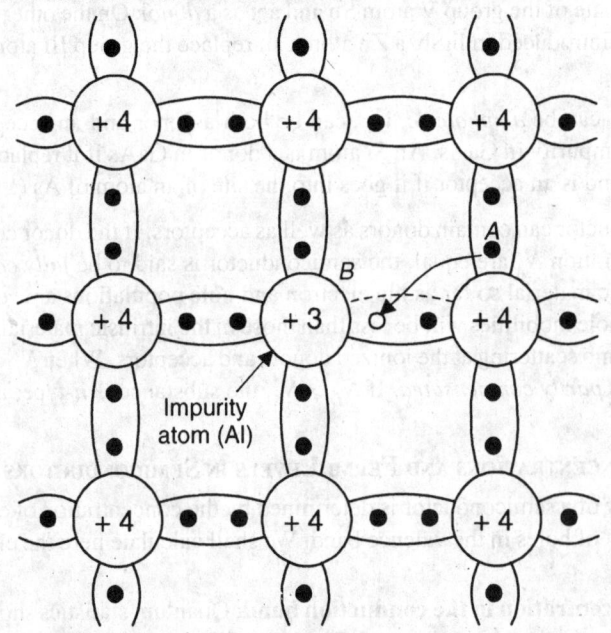

Fig. 3.14. A trivalent impurity atom, such as Al, in a convalent (say, Si) crystal.

The dopant atoms that produce holes in a semiconductor are called *acceptors* since they accept electrons. Semiconductors having acceptor type impurities are referred to as *p-type semiconductors* since the current carriers are positive charges, *i.e.* holes. The holes created in this way are called *excess holes*.

In a *p*-type semiconductor the holes are the

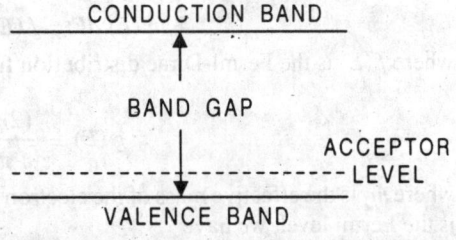

Fig. 3.15. Acceptor level in energy diagram.

majority carriers and thermally produced electrons are the minority carriers. A *p*-type semiconductor also satisfies Eq. (3.35). For each dopant atom in such semiconductors there is a hole in the valence band but no corresponding free electron in the conduction band. Addition of acceptors make $p > p_i$ so that $n < n_i$, according to Eq. (3.35).

Effect of temperature on extrinsic semiconductors

Consider an *n*-type semiconductor and assume that the donors are completely ionized. If the temperature of the material is raised, the number of thermally generated electrons and holes will increase. At a sufficiently high temperature, the number of thermally generated free electrons will be much greater than the number of donor atoms. Then the numbers of free electrons and holes will be nearly equal and the semiconductor will behave intrinsically. The same conclusion can be arrived at for a *p*-type semiconductor. Thus as the temperature is increased, an extrinsic semiconductor becomes an intrinsic one.

Observations

(*i*) The elements Ge and Si are the basic semiconductors. III-V compounds like gallium arsenide (GaAs) indium phosphide (InP), indium antimonide (InSb) etc. and II-VI compounds like cadmium telluride (CdTe), cadmium sulphide (CdS) etc., and alloys like gallium arsenide phosphide (Ga$_x$ P$_{1-x}$), indium gallium arsenide (In$_{1-x}$ Ga$_x$ As), cadmium mercury telluride (Cd$_x$Hg$_{1-x}$Te) etc. also show semiconducting properties.

If a group VI element like Te is introduced in a III-V compound semiconductor, say, InSb, a Te atom can go into the site of the group V atom Sb and act as a *donor*. On the other hand, when a group II element like Zn is introduced in InSb, a Zn atom can replace the group III atom In and serve as an *acceptor*.

(*ii*) An impurity can be *amphoteric*, i.e., can be both a donor and an acceptor. Si (a group IV element) is such an impurity in GaAs. An Si atom is a donor in GaAs if it replaces an atom of Ga (a group III element), and is an acceptor if it goes into the site of an atom of As (a group V element).

(*iii*) A semiconductor can contain donors as well as acceptors. If the donor concentration N_D and the acceptor concentration N_A are equal, the semiconductor is said to be *fully compensated* and behaves like an intrinsic material so far as the electron and hole populations are concerned. However, the electron and the hole mobilities will be less than those in the intrinsic material because the current carriers suffer Coulomb scattering at the ionized donors and acceptors. When N_D and N_A are unequal, the semiconductor is *partly compensated*. If $N_D > N_A$, the substance is *n*-type; and if $N_A > N_D$, it is *p*-type.

3.11. CARRIER CONCENTRATIONS AND FERMI LEVELS IN SEMICONDUCTORS

The conductivity of a semiconductor is determined by the concentration of electrons in the conduction band and that of holes in the valence band. We shall calculate here the electron and the hole concentrations.

(*i*) **Electron concentration in the conduction band**: Quantum statistics shows that the concentration of free electrons in the energy interval E and $E + dE$ in the conduction band is given by (see 'Statistical Mechanics', Chap. 2).

$$n(E)\, dE = f(E)\, g(E)\, dE, \qquad \ldots(3.36)$$

where $f(E)$ is the Fermi-Dirac distribution function and $g(E)$ is the density of states :

$$g(E) = \frac{(2m_e^3)^{1/2}}{\pi^2 \hbar^3} (E - E_c)^{1/2}, \qquad \ldots(3.37)$$

where m_e is the effective mass of the electron and E_c is the lowest energy in the conduction band. If E_F is the Fermi level, we have

Band Theory of Solids

$$f(E) = \frac{1}{1 + e^{(E - E_F)/k_B T}} \qquad ...(3.38)$$

where k_B is the Boltzmann constant and T is the absolute temperature. Substituting for $f(E)$ and $g(E)$ in Eq. (3.36), we find that the concentration of electrons in the conduction band is

$$n = \int_{E_c}^{E_u} n(E)\, dE = \frac{(2m_e^3)^{1/2}}{\pi^2 \hbar^3} \int_{E_c}^{E_u} \frac{(E - E_c)^{1/2}\, dE}{1 + e^{(E - E_F)/k_B T}} \qquad ...(3.39)$$

Here E_u is the highest occupied energy in the conduction band.

Let $x = (E - E_c)/k_B T$. Since E_F usually lies below the conduction band and E is in the conduction band above E_c, we have $(E - E_F)/k_B T \gg 1$. Thus we can neglect unity in the denominator of the integrand in Eq. (3.39) and write

$$n = \frac{(2m_e^3)^{1/2}}{\pi^2 \hbar^3} (k_B T)^{3/2} e^{(E_F - E_C)/k_B T} \int_0^\infty x^{1/2} e^{-x}\, dx \qquad ...(3.40)$$

Here the upper limit of integration has been shifted to infinity because $x^{1/2}$ is a slowly varying function of x whereas e^{-x} decreases rapidly with increasing x. Since $\int_0^\infty x^{1/2} e^{-x}\, dx = \Gamma(3/2) = \sqrt{\pi}/2$, we obtain

$$n = n_c\, e^{-(E_c - E_F)/k_B T} \qquad ...(3.41)$$

where $n_c = \dfrac{1}{\sqrt{2}} \left(\dfrac{m_0 k_B}{\pi \hbar^2}\right)^{3/2} \left(\dfrac{m_e}{m_0}\right)^{3/2} T^{3/2} = 4.82 \times 10^{21} \left(\dfrac{m_e}{m_0}\right)^{3/2} T^{3/2} \qquad ...(3.42)$

Here m_o is the electron rest mass. The quantity n_c is the value of n_e when $E_F = E_c$, i.e., when the Fermi level coincides with the bottom of the conduction band, n_c is expressed in m^{-3} in Eq. (3.42).

(ii) **Concentration of holes in the valence band:** Let E_v be the energy at the top of the valence band, so that the density of states is

$$g(E) = \frac{(2m_p^3)^{1/2}}{\pi^2 \hbar^3} (E_v - E)^{1/2} \qquad ...(3.43)$$

Here m_p is the effective mass of a hole and $E < E_v$. Since a hole signifies an empty energy state, the Fermi function for a hole is $1 - f(E)$, which is the probability for the level E to be unoccupied. So, the number of holes in the energy interval E and $E + dE$ per unit crystal volume is given by

$$p(E)\, dE = \frac{(2m_p^3)^{3/2}}{\pi^2 \hbar^3} (E_v - E)^{1/2} [1 - f(E)]\, dE \qquad ...(3.44)$$

We have

$$1 - f(E) = \frac{e^{(E - E_F)/k_B T}}{1 + e^{(E - E_F)/k_B T}} \qquad ...(3.45)$$

As $(E_F - E) \gg k_B T$ for $E \leq E_v$, the exponential in the denominator can be neglected. Therefore, we have

$$1 - f(E) = e^{(E - E_F)/k_B T} \qquad ...(3.46)$$

Substituting for $[1 - f(E)]$ in Eq. (3.44) we find that the hole concentration in the valence band is given by

$$p = \frac{(2m_p^3)^{3/2}}{\pi^2 \hbar^3} \int_{E_L}^{E_v} (E_v - E)^{1/2} e^{(E - E_F)/k_B T}\, dE$$

where E_L is the lowest unoccupied energy in the valence band. As before, we put $x = (E_v - E)/k_B T$ and obtain

$$p = \frac{(2m_p^3 k_B T)^{3/2}}{\pi^2 \hbar^3} e^{(E_v - E_F)/k_B T} \int_0^\infty x^{1/2} e^{-x} dx \qquad ...(3.47)$$

Since e^{-x} rapidly decreases with increasing x, the upper limit of integration is taken to be infinity in Eq. (3.47). Thus

$$p = p_v e^{-(E_F - E_v)/k_B T} \qquad ...(3.48)$$

where

$$p_v = \frac{1}{\sqrt{2}} \left(\frac{m_0 k_B}{\pi \hbar^2}\right)^{3/2} \left(\frac{m_p}{m_0}\right)^{3/2} T^{3/2} = 4.82 \times 10^{21} \left(\frac{m_p}{m_0}\right)^{3/2} T^{3/2} \qquad ...(3.49)$$

Note that $p = p_v$ when $E_F = E_v$, i.e., when the Fermi level lies at the top of the valence band. p_v is expressed in m^{-3} in Eq. (3.49).

I. Intrinsic semiconductor

In this case, the electron and the hole concentrations are equal, i.e., $n = p$. So, Eqs. (3.42) and (3.48) give

$$m_e^{3/2} e^{-(E_c - E_F)/k_B T} = m_p^{3/2} e^{-(E_F - E_v)/k_B T}$$

Taking the logarithm of both sides, we get

$$\frac{3}{2} \ln\left(\frac{m_e}{m_p}\right) = \frac{E_c + E_v - 2E_F}{k_B T}$$

or,

$$E_F = \frac{E_c + E_v}{2} - \frac{3}{4} k_B T \ln\left(\frac{m_e}{m_p}\right) \qquad ...(3.50)$$

This equation gives the position of the Fermi level in an intrinsic or pure semiconductor. If $m_e = m_p$, we have

$$E_F = \frac{E_c + E_v}{2}, \qquad ...(3.51)$$

i.e., the Fermi level lies at the centre of the forbidden gap (Fig. 3.16).

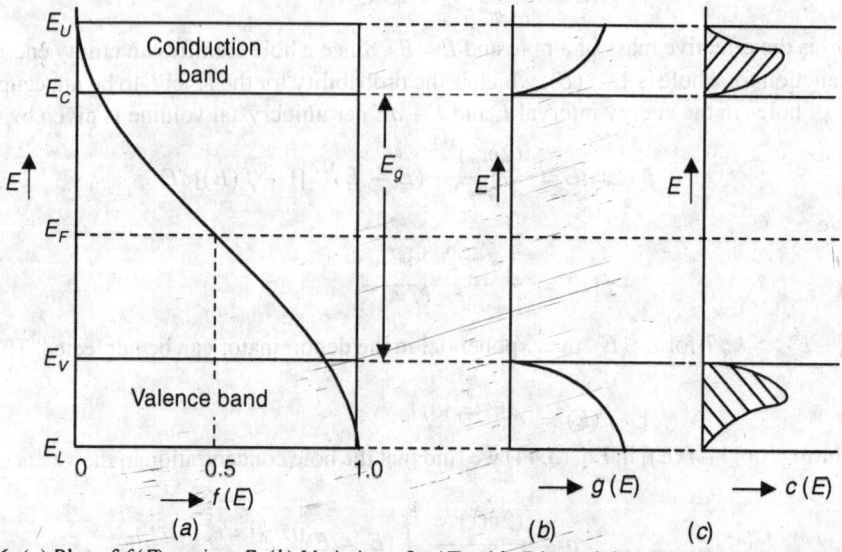

Fig. 3.16. (a) Plot of $f(E)$ against E, (b) Variation of $g(E)$ with E in each band, (c) Plot of $c(E)$ against E, where $c(E) = n(E)$ for electrons and $c(E) = p(E)$ for holes.

Band Theory of Solids

In general, $m_e < m_p$. So E_F lies closer to the conduction band and rises slowly as the temperature T increases.

The intrinsic carrier concentration is:

$$n_i = \sqrt{np} = \frac{1}{\sqrt{2}} (m_e m_p)^{3/4} \left(\frac{k_B T}{\pi \hbar^2}\right)^{3/2} e^{-E_g/(2k_B T)} \quad ...(3.52)$$

using Eqs. (3.41) and (3.48). Here $E_g (= E_c - E_v)$ is the band gap. The smaller the band gap and the higher the temperature, the larger the intrinsic carrier concentration.

II. Extrinsic semiconductor

Let N_D and N_A be respectively the concentrations of the ionized donor and acceptor atoms in an extrinsic semiconductor. Since the semiconductor is electrically neutral, the sum of the concentrations of the electrons (n) and the negatively charged acceptor ions must be equal to the sum of the concentrations of the holes (p) and the positively charged donor ions. That is,

$$n + N_A = p + N_D$$

For an *n-type* material, we put $p = 0$ and $N_A = 0$. Therefore, $n = N_D$, and we obtain from Eq. (3.41)

$$N_D = n_c e^{-(E_c - E_F)/k_B T} \quad ...(3.53)$$

Solving for E_F, we get

$$E_F = E_c - k_B T \ln\left(\frac{n_c}{N_D}\right) \quad ...(3.54)$$

Usually, $n_c > N_D$, so that $E_F < E_C$. Thus, the Fermi level lies in the forbidden gap slightly below the conduction band in an *n*-type semiconductor (Fig. 3.17 *a*). As the donor concentration (N_D) increases, the Fermi level moves up and enters the conduction band. The semiconductor then becomes degenerate and behaves as a metal. As the temperature increases, the Fermi level moves down, and the semiconductor tends to become intrinsic.

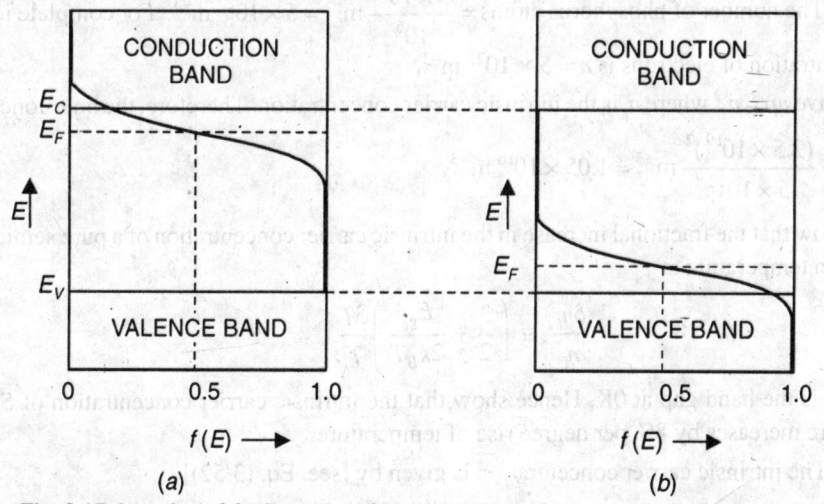

Fig. 3.17. Location of the Fermi Level in (*a*) *n*-type, and (*b*) *p*-type semiconductors.

For a *p*-type semiconductor, we put $n = 0$ and $N_D = 0$, so that Eq. (3.48) gives

$$N_A = p = p_v e^{-(E_F - E_v)/k_B T} \quad ...(3.55)$$

Hence

$$E_F = E_v + k_B T \ln\left(\frac{p_v}{N_A}\right) \quad ...(3.56)$$

Usually, $p_v > N_A$. Consequently $E_F > E_v$, i.e., the Fermi level lies in the forbidden gap slightly above the top of the valence band (Fig. 3.17b). If N_A is increased, the Fermi level moves down and enters the valence band. As the temperature rises, the Fermi level moves towards the centre of the band gap, and the material becomes intrinsic.

If $N_A = N_D$, i.e., the semiconductor is fully compensated, then $n = p$ and the Fermi level is determined by Eq. (3.50).

3.12. Worked-out Problems

1. Taking the origin at the bottom of the conduction band, calculate the crystal momentum for a free electron of energy 0.02 eV. Given, effective mass of the electron = $0.2\, m_0$.

Ans. Taking the origin at the bottom of the conduction band, the energy (E) versus the wave-vector (k) relationship is

$$E = \frac{\hbar^2 k^2}{2m^*}$$

Here, $E = 0.02$ eV $= 0.02 \times 1.6 \times 10^{-19}$ J,
and $m^* = 0.2 m_0 = 0.2 \times 9.11 \times 10^{-31}$ kg

Hence, crystal momentum $= \hbar k = (2m^* E)^{1/2}$
$= (2 \times 0.2 \times 9.11 \times 10^{-31} \times 0.02 \times 1.6 \times 10^{-19})^{1/2}$
$= 3.416 \times 10^{-26}$ kg.m/s

2. The intrinsic carrier concentration of germanium at room temperature is 2.5×10^{19} m^{-3}. It is doped with phosphorus at a rate of 1 phosphorus atom per million atoms of germanium. If the concentration of germanium atoms is 5×10^{28} m^{-3}, calculate the hole concentration. Assume complete ionization of phosphorus atoms.

Ans. The number of phosphorus atoms $= \dfrac{5 \times 10^{28}}{10^6}$ m^{-3} $= 5 \times 10^{22}$ m^{-3}. For complete ionization, the concentration of electrons is $n = 5 \times 10^{22}$ m^{-3},

We have $np = n_i^2$ where n_i is the intrinsic carrier concentration. Therefore, the hole concentration,

$$p = \frac{n_i^2}{n} = \frac{(2.5 \times 10^{19})^2}{5 \times 10^{22}} \text{ m}^{-3} = 1.05 \times 10^{16} \text{ m}^{-3}.$$

3. Show that the fractional increase in the intrinsic carrier concentration of a pure semiconductor with rise in temperature is

$$\frac{\delta n_i}{n_i} = \left(\frac{3}{2} + \frac{E_{g0}}{2k_B T} \right) \frac{\delta T}{T}$$

where E_{g0} is the band gap at 0K. Hence show that the intrinsic carrier concentration of Si at room temperature increases by 8% per degree rise of temperature.

Ans. The intrinsic carrier concentration is given by [see. Eq. (3.52)]

$$n_i = CT^{3/2} e^{-E_g/(2k_B T)}, \qquad \ldots(i)$$

where C is a constant. The band gap Eg decreases linearly with temperature obeying the relationship $E_g = E_{g0} - \alpha T$, where E_{g0} is the energy gap at 0K. Substituting this relationship into (i), we obtain

$$n_i = C' T^{3/2} e^{-E_{g0}/(2k_B T)} \qquad \ldots(ii)$$

Band Theory of Solids   233

where C' is a new constant. Logarithmic differentiation of Eq. (*ii*) shows that the fractional increase in n_i is given by

$$\frac{\delta n_i}{n_i} = \left(\frac{3}{2} + \frac{E_{g0}}{2k_B T}\right)\frac{\delta T}{T}$$

For Si, $E_{g0} = 1.21$ eV. At room temperature ($T = 300$ K), we have $k_B T = 26$ meV $= 26 \times 10^{-3}$ eV. So, for Si at 300K with $\delta T = 1$K we obtain

$$\frac{\delta n_i}{n_i} = \left(\frac{3}{2} + \frac{1.21}{0.052}\right)\frac{100}{300}\% = 8\%$$

4. Silicon contains 5×10^{28} atoms per m^3. In an *n*-type Si sample the donor concentration is 1 atom per 2.5×10^7 Si atoms. Assuming that the electron effective mass is equal to the true mass, calculate the temperature at which the Fermi level lies at the bottom of the conduction band.

Ans. The donor concentration is $N_D = \dfrac{5 \times 10^{28}}{2.5 \times 10^7} = 2 \times 10^{21}$ m^{-3}

Also, $n_c = 4.82 \times \left(\dfrac{m_e}{m_0}\right)^{3/2} T^{3/2} = 4.82 \times 10^{21} T^{3/2}$

(since $m_e = m_0$). The Fermi level is

$$E_F = E_c - k_B T \ln\left(\frac{n_c}{N_D}\right)$$

Since $T \neq 0$, $E_F = E_C$ when $n_C = N_D$, i.e.,

$$4.82 \times 10^{21} T^{3/2} = 2 \times 10^{21}$$

whence $T = 0.56$ K

5. In Problem 4, find the position of the Fermi level at room temperature.

Ans. We have

$$E_F = E_c - k_B T \ln\left(\frac{n_c}{N_D}\right)$$

Here

$n_c = 4.82 \times 10^{21} T^{3/2} = 4.82 \times 10^{21} \times (300)^{3/2} = 2.504 \times 10^{25}$ m^{-3} and $N_D = 2 \times 10^{21}$ m^{-3}. So,

$$k_B T \ln\left(\frac{n_c}{N_D}\right) = 26 \ln\left(\frac{2.504 \times 10^{25}}{2 \times 10^{21}}\right) \text{meV} = 245.3 \text{ meV}$$

Hence, $E_c - E_F = 0.24$ eV, i.e., the Fermi level lies 0.24 eV below the conduction band.

6. The energy wave vector dispersion relation for a one-dimensional crystal of lattice constant a is given by $E(k) = E_0 - \alpha - 2\beta \cos ka$, where E_0, α, β are constants.

  (*i*) Find the value of k at which the velocity of an electron is a maximum.
  (*ii*) Find the difference between the top and the bottom of the energy band.
  (*iii*) Obtain the effective mass m^* of the electron at the bottom and at the top of the band.

(cf. C.U. 2003)

Ans. (*i*) The velocity of an electron is

$$v = \frac{1}{\hbar}\frac{dE}{dk} = \frac{2\beta a}{\hbar}\sin ka$$

v is a maximum when $\sin ka = 1$, or $ka = \dfrac{\pi}{2}$ or $k = \dfrac{\pi}{2a}$.

(ii) At the top of the band $E(k)$ is a maximum, and so $\cos ka$ is a minimum, i.e. $\cos ka = -1$.
So, $$E_{top} = E_0 - \alpha + 2\beta$$
At the bottom of the band $E(k)$ is a minimum, i.e., $\cos ka$ attains a maximum value of $+1$. Hence
$$E_{bot} = E_0 - \alpha - 2\beta.$$
The difference in energy is
$$E_g = E_{top} - E_{bot} = 4\beta.$$

(iii) $m^* = \dfrac{\hbar^2}{d^2 E/dk^2} = \dfrac{\hbar^2}{2\beta a^2 \cos ka}$

At the bottom of the band, $\cos ka = 1$; so $m^* = \dfrac{\hbar^2}{2\beta a^2}$

At the top of the band, $\cos ka = -1$; so $m^* = -\dfrac{\hbar^2}{2\beta a^2}$.

7. The concentrations of ionized donor and acceptors in an n-type compensated semiconductor at temperature T are N_D and N_A, respectively. If n_i is the intrinsic carrier concentration at that temperature, show that the concentrations of free electrons and holes in the semiconductor are

$$n = \frac{1}{2}(N_D - N_A) + \frac{1}{2}\sqrt{(N_D - N_A)^2 + 4n_i^2}$$

and
$$p = -\frac{1}{2}(N_D - N_A) + \frac{1}{2}\sqrt{(N_D - N_A)^2 + 4n_i^2}$$

respectively.

Ans. As the semiconductor is electrically neutral, we have
$$N_D + p = N_A + n \qquad \ldots(i)$$

for an n-type semiconductor, $N_D > N_A$. Also, $p = n_i^2/n$, So Eq. (i) becomes

$$N_D + \frac{n_i^2}{n} = N_A + n$$

or, $$n^2 - (N_D - N_A)n - n_i^2 = 0$$

or, $$n = \frac{1}{2}(N_D - N_A) \pm \frac{1}{2}\sqrt{(N_D - N_A)^2 + 4n_i^2}$$

Since n is positive, the negative sign on the right-hand side must be rejected. Hence
$$n = \frac{1}{2}(N_D - N_A) + \frac{1}{2}\sqrt{(N_D - N_A)^2 + 4n_i^2}, \text{ proved.}$$

Similarly, putting $n = n_i^2/p$ on the right-hand side of Eq. (i) and proceeding in the same way, we obtain
$$p = -\frac{1}{2}(N_D - N_A) + \frac{1}{2}\sqrt{(N_D - N_A)^2 + 4n_i^2}.$$

QUESTIONS

1. Explain in a simple way how the atomic energy levels split into bands when a number of atoms are brought close together to form a crystal.

2. How does the potential energy of an electron vary in an infinite one-dimensional crystal? How is this potential represented in Kronig-Penney model?

3. What are the main conclusions of the Kronig-Penney treatment? Draw schematically the energy-wave vector diagram for a crystal. What do you mean by reduced zone representation?

4. Describe the basic principles for the band structure of electronic states in a crystal. **(C.U. 1982)**

5. (a) Define the term "effective mass of an electron". Is it different from the free electron mass? What is "crystal momentum"? **(cf. C.U. 1997)**

    (b) What is the significance of negative effective mass of electrons? **(Purvanchal Univ. 2003)**

6. Show that a crystal containing N atoms can accommodate $2N$ electrons in an energy band.

7. Explain the concept of "holes".

8. Why does a completely filled band contribute nothing to an electric current in the presence of an electric field?

9. Distinguish between metals, insulators, and semiconductors. **(C.U. 1983. N. Beng. U. 2001)**

10. What is an intrinsic semiconductor? How does its electrical conductivity depend on the temperature and the band gap? When does an intrinsic semiconductor behave as an insulator?

11. Explain the following terms:

    (a) intrinsic carrier concentration, (b) doping, (c) dopant, (d) donor, (e) acceptor, and (f) extrinsic semiconductors.

12. Describe the basic principles of n and p type semiconductors. **(C.U. 1982)**

13. What are impure semiconductors? How are they classified? "With an increase of temperature an impure semiconductor behaves as a pure one". Explain.

14. Give a qualitative interpretation of the band structure of the electronic energy levels in a semiconductor. Discuss n and p-type semiconductors. **(C.U. 1984)**

15. The energy versus wave vector relationship for a conduction electron in a semiconductor is $E = 5\hbar^2 k^2/m_0$. Determine the electron effective mass. **[Ans. 0.1 m_0]**

16. Deduce expressions for the concentration of electrons in the conduction band and the concentration of holes in the valence band of an intrinsic semiconductor at a temperature T. Show that the product is independent of the Fermi level. What is the implication of this result? [You may assume that the density of states of free electrons is $g(E) = \dfrac{1}{2\pi^2}\left(\dfrac{2m}{\hbar^2}\right)^{3/2} E^{1/2}$]

    Also, find the location of the Fermi level in an intrinsic semiconductor. Draw neat diagrams showing the location of the Fermi level in n- and p-type semiconductors. **(Burd. U. 1999)**

17. Show graphically the variation of the energy with the wave number k for the first allowed band in a solid. Hence explain the variation in sign of the effective mass of an electron in the band. What is a hole? **(C.U. 2000)**

18. In the Kronig-Penney model of electrons in a linear lattice, if the height of the periodic potential increases, how will the widths of the allowed energy bands change? **(Purvanchal Univ. 2003)**

19. For the first allowed energy band in a crystalline solid in the reduced zone, plot qualitatively the energy E, the velocity v, the effective mass m^*, and the extent of freedom f_k of an electron against the wave number k. Hence explain the change in sign of m^*. (cf. C.U. 2005)

20. Show that the effective freedom of the electrons in an energy band in a solid is (i) a maximum when the band is filled up to the inflection point of the E-k curve, and (ii) zero when the band is completely filled.

21. (a) Which of the following dopants are donors in (i) Si and (ii) Insb : As, Zn, P, Te, Al?
    (b) What happens if a silicon atom replaces (i) a gallium atom and (ii) an arsenic atom in GaAs ?

22. Show that the effective mass of an electron in a crystal is given by

$$m^* = \hbar^2 \bigg/ \left(\frac{d^2 E}{dk^2}\right)$$

where the symbols have their usual meanings. (C.U. 2007)

23. The energy of a free electron in a crystal is given by $E = A - B\cos ka$, where A and B are constants, k is the wave number, and a is the distance between adjacent atoms. Draw the $E - k$ curve, and find the effective mass. If $a = 2$Å, what is the range of the de Broglie wavelengths in the first Brillouin zone? (C.U. 2007)

[Ans. $\hbar^2/(Ba^2 \cos ka)$, -4Å to $+4$Å]

Chapter 4

Transport Phenomena in Metals and Semiconductors

In this chapter we shall discuss the conduction of heat and electricity in metals which are good conductors. We shall also consider electrical conduction in semiconductors and the Hall effect. These phenomena are known as the *transport phenomena*.

Metallic conduction was first explained by Drude in 1900 on the basis of his free electron theory which was subsequently extended by Lorentz. The Drude model has, however, its limitations which are removed in the quantum theory of solids. We shall first describe below the Drude theory of metals, and thereafter refer to the quantum theory.

4.1. THE DRUDE THEORY

The Drude model assumes that the electrons in a metal are free to move and form an 'electron gas'. Lorentz predicted that the classical kinetic theory of gases could be applied to the free electron gas in the metal. In this theory the free electrons would have the same type of energy distribution as the classical gas particles, namely, the Maxwell-Boltzmann distribution.

The free electrons in a metal are assumed to be the valence electrons of the constituent atoms. When these atoms are brought close together to form the metal, the valence electrons get detached and move freely through the metal; hence they are called *free* or *conduction electrons*. The positively charged metallic ions remain immobile. The concentration of the free electrons is typically of the order of 10^{22} cm^{-3}. Although this concentration is about thousand times larger than those of an ordinary gas at usual temperatures and pressures, the Drude model assumes that the laws of kinetic theory of a dilute gas can be applied to the dense metallic electron gas. Unlike the ordinary gas particles the free electrons in a metal move through the background of immobile ions. The basic assumptions of the Drude model are the following.

(*i*) The electrons are in thermal equilibrium with the environment.

(*ii*) Between collisions, the interactions of an electron with the other electrons and with the ions are negligible. Thus the electron travels uniformly in a straight line between collisions in the absence of applied electromagnetic fields. The motion of electrons in the presence of applied fields is determined by Newton's laws of motion.

(*iii*) The time taken up in a collision is negligible. The velocity of an electron changes abruptly when it undergoes a collision.

(*iv*) The electron-electron collisions are usually unimportant.

(*v*) The metallic ions form the major source of collisions. The exact nature of the scattering is not required; only some broad features associated with the scattering are important. The probability per unit time for an electron to suffer a collision is assumed to be $1/\tau$. The time τ is referred to as the *relaxation time*, the *collision time* or the *mean free time*. For simplicity, τ is taken to be independent of the position and velocity of the electron.

4.2. ELECTRICAL CONDUCTIVITY

In the absence of an electric field, the free electrons in a metal move at random owing to their thermal energy. At each collision, the path of the electron is changed [Fig. 4.1 (*a*)]. The number of electrons moving in one direction is the same as that moving in the opposite direction, so that considering all the electrons the average drift velocity is zero. When an electric field is applied, the electrons are accelerated between collisions. As an electron carries a negative charge the acceleration is opposite to the direction of the field. Immediately after collisions, all the electron velocities are randomised. In equilibrium, all the electrons possess a common drift velocity due to the electric field [Fig. 4.1 (b)]. The drift velocity has a very small magnitude compared to the random thermal velocity on which it is superimposed. The electron drift gives rise to the current flow through the metal.

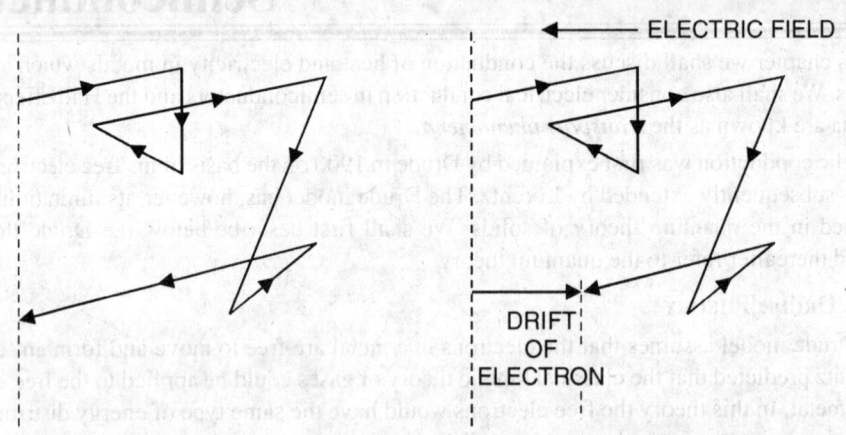

Fig. 4.1. (a) Random electron motion, and (b) electron drift in an applied electric field.

The electrical conductivity σ of a solid is the charge conducted per unit time across unit area of the solid for applied unit electric field. If due to an applied electric field ξ the current density, *i.e.* the current per unit cross sectional area is J, the conductivity σ is given by

$$\sigma = \frac{J}{\xi} \qquad ...(4.1)$$

In SI units J is expressed in A/m², ξ in volt/m, and σ in siemens/m.

Consider a cylindrical metal of length L and having an area of cross section A (Fig. 4.2). If v is the drift velocity of the electrons for an electric field ξ applied along the length of the cylinder, then the time taken by an electron to traverse the distance L is

$$t = \frac{L}{v} \qquad ...(4.2)$$

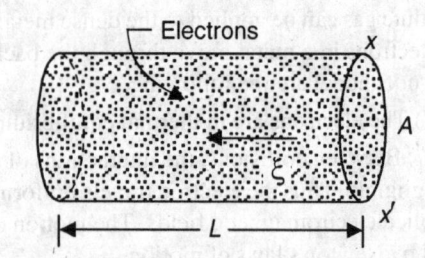

Fig. 4.2. Geometry for calculation of current density.

If n represents the electron concentration then the total number of electrons contained in the cylinder is nAL. The charge inside the cylinder is, therfore, $nALe$, where e is electronic charge. This amount of charge moves through the cross section xx' in time t. Therefore, the current flowing through the metallic cylinder is

$$I = \frac{nALe}{t} \qquad ...(4.3)$$

The current density J will be

$$J = \frac{I}{A} = \frac{nLe}{t} = nev, \qquad ...(4.4)$$

the last step following from Eq. (4.2). To obtain J in A/m², n is expressed in m⁻³, e in coulomb, and v in m/s.

From Eqs. (4.1) and (4.4) we get

$$\sigma = ne\frac{v}{\xi} \qquad \ldots(4.5)$$

The quantity v/ξ, i.e. the drift velocity per unit electric field is termed the *mobility* of the electron. Denoting the mobility by μ we have

$$\sigma = ne\mu \qquad \ldots(4.6)$$

In SI units, μ is expressed in $m^2/(V.s)$. For copper, the value of μ is about $3 \times 10^{-3}\ m^2/(V.s)$.

Mean Free Time

As already mentioned, the probability of an electron suffering a collision in unit time is $1/\tau$. The quantity τ is a constant depending on the system and has the dimension of time.

Consider a group of n_0 electrons at time $t = 0$. Let n denote the number of electrons in the group which do not make a collision in time t. If dn be the change in n due to collision in a further time dt, then

$$dn = -\frac{n}{\tau}dt \qquad \ldots(4.7)$$

Integrating Eq. (4.7) we obtain

$$n = n_0 \exp(-t/\tau) \qquad \ldots(4.8)$$

The mean free time of an electron is given by the total lifetime of all the n_0 electrons divided by n_0. Thus,

$$\text{Mean free time} = -\frac{1}{n_0}\int_0^\infty t\, dn = \frac{1}{\tau}\int_0^\infty t\exp(-t/\tau)\, dt$$

$$= \tau \qquad \ldots(4.9)$$

Thus the quantity τ represents the mean free time of an electron between collisions. The mean free path of an electron is given by

$$l = \tau\bar{v}, \qquad \ldots(4.10)$$

where $\bar{v}$ represents the average velocity of the electron.

During the time between collisions an electron has an acceleration $e\xi/m_e$, where e is the electron charge, ξ is the electric field, and m_e is the mass of the electron. As an electron, observed at a random instant of time, travels freely on the average for a time τ, the average drift velocity v_d of the electron will be

$$v_d = \frac{e\xi\tau}{m_e} \qquad \ldots(4.11)$$

The mobility of the electron is given by

$$\mu = \frac{v_d}{\xi} = \frac{e\tau}{m_e} \qquad \ldots(4.12)$$

An order of magnitude of τ can be obtained by noting that for copper $\mu = 3 \times 10^{-3}\ m^2/(V.s)$. Substituting the values $e = 1.6 \times 10^{-19}$ coulomb and $m_e = 9.11 \times 10^{-31}$ kg, we obtain from Eq. (4.12) $\tau \approx 1.7 \times 10^{-14}$ s. The value is however not very accurate since the effective mass of an electron in metal is not exactly equal to the free electron mass. Also, strictly speaking, τ is not a constant but depends on the electron energy.

From Eqs. (4.6) and (4.12) we express the conductivity σ in the form:

$$\sigma = \frac{ne^2\tau}{m_e} \qquad \ldots(4.12a)$$

4.3. Thermal Conductivity of a Metal

The remarkable success of the Drude theory is the explanation of the empirical law of Wiedemann and Franz. This law states that the ratio of the thermal to the electrical conductivity of a number of metals is proportional to the absolute temperature, the proportionality constant being nearly the same for all metals. In arriving at this law the Drude model assumes that the free electrons in a metal are mainly responsible for the thermal conductivity. The main reason for this assumption is that metals are better conductors of heat than insulators. The contribution of the ions, present in metals and in insulators, to the thermal conductivity is thus much less than that of free electrons, present only in metals.

We shall now develop an expression for the thermal conductivity due to free electrons in a metal. Assume a temperature gradient $\frac{dT}{dx}$ in the x-direction. The thermal current density J_q represents the thermal energy crossing a unit area perpendicular to the direction of flow per unit time. It is given by

$$J_q = -K \frac{dT}{dx} \qquad \ldots(4.13)$$

where K is a positive constant, referred to as the *thermal conductivity*. The negative sign in Eq. (4.13) appears since the heat flow is opposite to the direction of temperature gradient.

After each collision an electron moves with a speed appropriate to the local temperature. Thus an electron emerging from a hotter place of collision will be more energetic. Let $E(T)$ represent the thermal energy of an electron at temperature T. Hence an electron suffering the last collision at the position coordinate x' will carry on the average a thermal energy $E[T(x')]$. The electrons that come to the point x from the high-temperature side had their last collisions at $x' = (x - v_x \tau)$ where v_x is the x-component of the average velocity of the electrons and τ the mean free time. The average thermal energy associated with each such electron is $E[T(x - v_x \tau)]$. If n be the total number of electrons per unit volume, half of this number will come to the point x from the high-temperature side and half from the low-temperature side. The thermal current density at x due to the electrons coming from the high-temperature side will be

$$J_{qh} = \frac{n}{2} v_x E[T(x - v_x \tau)] \qquad \ldots(4.14)$$

Similarly, the thermal current density due to the electrons coming to x from the low temperature side is

$$J_{ql} = \frac{n}{2} (-v_x) E[T(x + v_x \tau)], \qquad \ldots(4.15)$$

since these electrons move in the negative x-direction. The net thermal current at x is

$$J_q = J_{qh} + J_{ql} = \frac{n v_x}{2} \left[E\{T(x - v_x \tau)\} - E\{T(x + v_x \tau)\} \right] \qquad \ldots(4.16)$$

It may be assumed that the variation of the temperature over the mean free path $l = v_x \tau$ is very small. Thus expanding in a Taylor series and neglecting the higher order terms we get

$$J_q = n v_x^2 \tau \frac{dE}{dT} \left(-\frac{dT}{dx} \right) \qquad \ldots(4.17)$$

We shall now replace v_x^2 by its average value $v^2/3$ where v^2 is the mean square velocity. Hence

$$J_q = \frac{1}{3} n v^2 \tau \frac{dE}{dT} \left(-\frac{dT}{dx} \right) \qquad \ldots(4.18)$$

If N be the total number of free electrons and V be the volume of the metal we have

$$n \frac{dE}{dT} = \frac{N}{V} \frac{dE}{dT} = C_v, \qquad \ldots(4.19)$$

Transport Phenomena in Metals and Semiconductors 241

where C_v is the electronic heat capacity per unit volume. Using Eq. (4.19) in Eq. (4.18) we obtain

$$J_q = -\frac{1}{3} v^2 \tau C_v \frac{dT}{dx} \qquad \ldots(4.20)$$

Comparing Eqs. (4.13) and (4.20) we obtain for the thermal conductivity

$$K = \frac{1}{3} v^2 \tau C_v = \frac{1}{3} l v C_v \qquad \ldots(4.21)$$

From Eqs. (4.12a) and (4.21) we obtain

$$\frac{K}{\sigma} = \frac{C_v m_e v^2}{3ne^2} \qquad \ldots(4.22)$$

Applying classical gas laws we get $C_v = \frac{3}{2} nk_B$ and $\frac{1}{2} m_e v^2 = \frac{3}{2} k_B T$ where k_B is Boltzmann's constant. Hence Eq. (4.22) reduces to

$$\frac{K}{\sigma} = \frac{3}{2} \left(\frac{k_B}{e}\right)^2 T \qquad \ldots(4.23)$$

Thus the ratio of the thermal and the electrical conductivity is proportional to T and the proportionality constant depends only on the universal constants k_B and e. This result agrees with the Wiedemann-Franz law. The ratio $\frac{K}{\sigma T}$, known as *Lorenz number*, is found from Eq. (4.23) to be

$$\frac{K}{\sigma T} = \frac{3}{2} \left(\frac{k_B}{e}\right)^2$$

$$= 1.11 \times 10^{-8} \text{ V}^2/\text{K}^2 \qquad \ldots(4.24)$$

This value is about one half the typical value for most metals. Although the classical laws cannot be applied to the eletron gas in a metal, excellent agreement of the Lorenz number with experiments, apart from the factor of 2, is obtained because of cancellation of errors in the electronic contribution to the heat capacity C_v and the mean square electron speed.

4.4. QUANTUM MECHANICAL FREE ELECTRON THEORY

The quantum theory for the assembly of free electrons in a metal was first advanced by Sommerfeld in 1928. In the quantum theory the free electrons are assumed to obey the Pauli exclusion principle and the Fermi-Dirac statistics. The Fermi-Dirac statistics has already been dealt with in detail (see Chapter 2, Statistical Mechanics).

Electrical conductivity

The electrical conductivity has been found to be given by Eq. (4.12a). In the quantum theory a collision of a free electron is allowed only when an unoccupied state corresponding to the energy and momentum of the electron after collision exists. As the Fermi level is the highest energy occupied by an electron at absolute zero of temperature, the electrons at this temperature will occupy states lying within a sphere about the origin of the momentum space. The radius of the sphere, $p_F(0)$, satisfies the relation $\frac{p_F^2(0)}{2m_e} = E_F(0)$, where $E_F(0)$ is the Fermi energy at absolute zero of temperature. The surface of the sphere, referred to as the *Fermi sphere*, is called the *Fermi surface* (Fig. 4.3.). At high temperatures there is partial occupation of electron states over a region of a few times $k_B T$ about E_F, as shown in Fig. 4.3. p_F is referred to as the magnitude of the *Fermi momentum*.

The collisions which the electrons encounter are nearly elastic so that the change in energy upon collision is very small. In the momentum space, an electron suffering a collision is consequently scattered to a state nearly equidistant from the origin. As all the states below a few times $k_B T$ of

E_F are occupied, the collisions for such states are not allowed following the Pauli exclusion principle. Electrons having states close to the Fermi level can take part in collisions since they find empty states to receive them after collision.

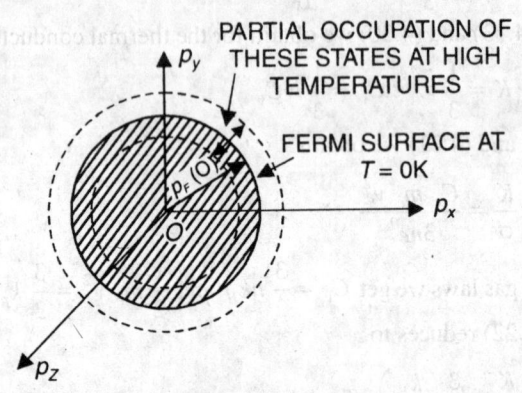

Fig. 4.3. Fermi surface at 0 K.

In the presence of an electric field, the electrons are accelerated during their free flights. As a result, the Fermi sphere of occupied states is slightly shifted in a direction opposite to that of the field (Fig. 4.4). All the free electrons thus take part in electrical conduction and a mean drift velocity can be defined. However, only the electrons with states near E_F encounter collisions, as already pointed out. For example, the electrons with states in the region X can fill the states in region Y after collision.

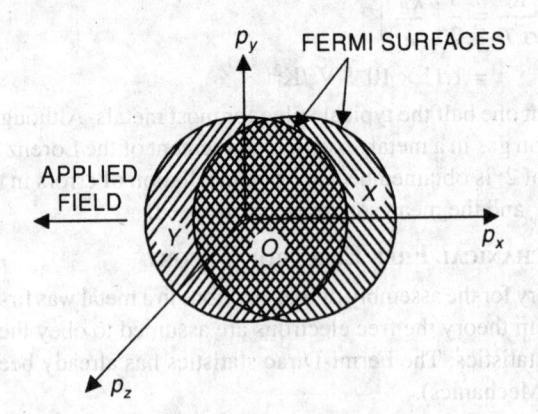

Fig. 4.4. Displacement of the Fermi sphere in the presence of an electric field.

This is possible since the states in the regions X and Y have nearly the same energy. The equilibrium electron distribution tends to be restored by such collisions with a characteristic time, called the *relaxation time*. This relaxation time is the mean free time τ_F of the electrons with energy close to E_F. Thus the mean free time τ_F determines the drift velocity of the whole system of free electrons. The electrical conductivity σ in the quantum theory is consequently given by

$$\sigma = \frac{ne^2 \tau_F}{m_e} \quad \quad ...(4.25)$$

Equation (4.25) is obtained from Eq. (4.12a) by replacing τ by τ_F. The main scattering process contributing to τ_F is due to lattice vibration. The lattice scattering increases with temperature and τ_F is found to be inversely proportional to T at normal and high temperatures. Thus the electron mobility and electrical conductivity σ decrease with increase in temperature. At low temperatures, lattice scattering is small, and the electrons are mainly scattered by impurity atoms and by crystal imperfections.

Transport Phenomena in Metals and Semiconductors

The electron concentration in a metal does not change with temperature. The variation of the electrical conductivity with temperature is therefore essentially due to variation in τ_F with temperature.

Thermal conductivity

As already mentioned, the conduction of heat in a metal is mainly due to the free electrons. Also we have seen that only the electrons with energies near the Fermi level can encounter collisions. The relaxation time τ_F of these electrons controls the drift velocity of the entire free electron system in the metal. On the basis of the Drude model the thermal conductivity K is given by Eq. (4.21). In quantum theory this equation can be used with appropriate values of C_v and the mean free path l. The electronic heat capacity per unit volume C_v is given by*

$$C_v = \frac{\pi^2}{2} n \frac{k_B^2 T}{E_F(0)} \qquad ...(4.26)$$

The mean free path of the electrons between collisions is expressed by

$$l = \tau_F v_F, \qquad ...(4.27)$$

where v_F is the velocity of an electron at the Fermi energy and is termed the *Fermi velocity*.

Using Eq. (4.26) and (4.27) in Eq. (4.21) we obtain

$$K = \frac{\pi^2}{6} n \frac{k_B^2 T}{E_F(0)} \tau_F v_F^2$$

$$= \frac{\pi^2}{3} n \frac{k_B^2 T}{m_e} \tau_F \qquad ...(4.28)$$

where we have used the relation

$$E_F(0) = \frac{1}{2} m_e v_F^2. \qquad ...(4.29)$$

Since τ_F is inversely proportional to T, Eq. (4.28) explains why at ordinary temperatures the thermal conductivity of metals is nearly independent of temperature.

Ratio of K and σ

From Eqs. (4.28) and (4.25) we obtain

$$\frac{K}{\sigma} = \frac{\pi^2}{3} \left(\frac{k_B}{e}\right)^2 T, \qquad ...(4.30)$$

which is the *Wiedemann-Franz Law*. The Lorenz number is given by

$$\frac{K}{\sigma T} = \frac{\pi^2}{3} \left(\frac{k_B}{e}\right)^2 = 2.45 \times 10^{-8} \text{ V}^2/\text{K}^2, \qquad ...(4.31)$$

which is nearly twice the value indicated by Eq. (4.24) and is close to the value for many metals at ordinary temperatures.

4.5. Electrical Conductivity of Semiconductors

In a semiconductor, both the free electrons in the conduction band and the holes in the valence band can carry electric currents.

When an electric field ξ is applied in a semiconductor the electrons are drifted in the direction opposite to ξ since the charge on the electrons is negative (Fig. 4.5). The conventional current due to the electrons is however in the direction of ξ. The holes carry positive charges and therefore they drift in the direction of ξ and contribute a current in the same direction.

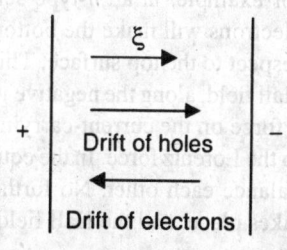

Fig. 4.5. Movement of electrons and holes in an electric field.

---
* Equation (4.26) is obtained from Eq. (2.46) [Statistical Mechanics] by replacing N by n and by noting that $k_B T_F = E_F(0)$.

If σ_n denotes the conductivity of a semiconductor due to the electrons, then we write from Eq. (4.6)

$$\sigma_n = ne\mu_n, \qquad \ldots(4.32)$$

where n is the concentration of the electrons and μ_n is the electron mobility.

The conductivity of a semiconductor due to holes is similarly given by

$$\sigma_p = pe\mu_p, \qquad \ldots(4.33)$$

where p is the concentration of holes and μ_p is the hole mobility. Generally μ_n is greater than μ_p.

Since the electric currents due to the electrons and the holes flow in the same direction, the overall conductivity of the semiconductor is

$$\sigma = \sigma_n + \sigma_p = e\,(n\mu_n + p\mu_p). \qquad \ldots(4.34)$$

In the case of an intrinsic semiconductor, $n = p = n_i$, where n_i is the intrinsic carrier concentration. Therefore, the intrinsic conductivity is

$$\sigma_I = en_i\mu_n\,(1+b), \qquad \ldots(4.35)$$

where $b = \mu_p/\mu_n$. Since $b < 1$, the electrons dominate in the conductivity of an intrinsic semiconductor.

For an n-type semiconductor $n \gg p$, and we get from Eq. (4.34)

$$\sigma_{n-type} \approx ne\,\mu_n. \qquad \ldots(4.36)$$

For a p-type semiconductor, $p \gg n$, and if the product $p\mu_p$ is greater than $n\mu_n$, we obtain from Eq. (4.34)

$$\sigma_{p-type} \approx pe\,\mu_p. \qquad \ldots(4.37)$$

The conductivity due to electrons or holes, σ_n or σ_p, is expressed by Eq. (4.25) with n representing the electron (or hole) concentration, m_e representing the electron (or hole) effective mass and τ_F replaced by an average value of τ. The last step is required since for a semiconductor the Fermi level generally lies inside the band gap, and the distribution function is practically the Maxwell-Boltzmann function. Electrons in all energy states can therefore undergo collisions, and as τ is a function of energy, an average value of τ has to be used.

4.6. Hall Effect

When a piece of conductor (metal or semiconductor) carrying an electric current is placed in a transverse magnetic field, an electric field is developed within the conductor in a direction perpendicular to both the current and the magnetic field. This phenomenon is referred to as the *Hall effect*, and the generated electric field as the *Hall field*.

Assume that a rectangular piece of conductor carrying an electric current I in the positive x-direction is placed in a magnetic flux density B applied in the positive z-direction (Fig. 4.6). The current carriers will experience a Lorentz force in the negative y-direction. Consequently, the carriers will be deflected towards the bottom surface of the sample and will accumulate there. When the current carriers are electrons, for example, in an n-type semiconductor, the accumulation of electrons will make the bottom surface negatively charged with respect to the top surface. This gives rise to an electric field, the Hall field, along the negative y-direction. The Hall field will exert a force on the current-carrying electrons in a direction opposite to the Lorentz force. In the equilibrium condition these two forces balance each other. No further accumulation of electrons then takes place, and the Hall field attains a steady value.

When the current carriers are holes, *i.e.*, when the sample is a p-type semiconductor, the bottom surface becomes positively charged relative to the top surface owing to the accumulation of

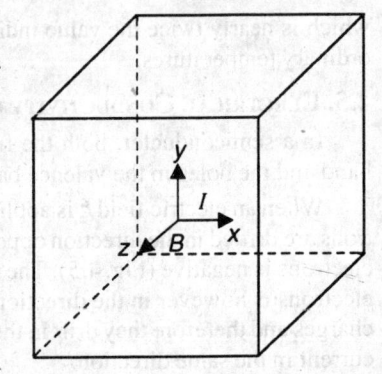

Fig. 4.6. Geometry of Hall effect.

Transport Phenomena in Metals and Semiconductors

holes. The Hall field is now generated along the positive y-direction. The force on the holes due to the Hall field opposes the Lorentz force. In equilibrium, these two forces balance preventing further accumulation of holes. The Hall field then becomes steady.

Let ξ_H be the Hall field in the y-direction. The force exerted on a current carrier of charge e by this field is $e\xi_H$. The average Lorentz force on a carrier is evB, where v represents the drift velocity in the x-direction. In the steady state these two forces balance, i.e.,

$$e\xi_H = evB. \qquad \ldots(4.38)$$

If n_c is the concentration of the current carriers, the current density in the x-direction is given by

$$J = n_c ev. \qquad \ldots(4.39)$$

From Eqs. (4.38) and (4.39) we obtain

$$\xi_H = \frac{JB}{n_c e}. \qquad \ldots(4.40)$$

If ζ is the applied electric field in the x-direction driving the current I, the net electric field in the sample is $\xi_T = \sqrt{\xi^2 + \xi_H^2}$, ξ and ξ_H, being perpendicular to each other. The angle between ξ and ξ_T is called the *Hall angle*, θ_H. We have

$$\tan\theta_H = \frac{\xi_H}{\xi} = \frac{JB}{n_c e\xi} = \frac{Bv}{\xi} = \mu_c B \qquad \ldots(4.40a)$$

where we have used Eqs. (4.40) and (4.39), and the relationship $v = \mu_c \xi$, μ_c being the carrier mobility.

The quantity $\xi_H/(JB)$ giving the Hall field per unit current density per unit magnetic flux density is called the *Hall coefficient*, R_H. That is,

$$R_H = \frac{\xi_H}{JB} \qquad \ldots(4.41)$$

From Eq. (4.40) we get

$$R_H = \frac{1}{n_c e}, \qquad \ldots(4.42)$$

A rigorous analysis shows that R_H is actually given by

$$R_H = \frac{r}{n_c e}, \qquad \ldots(4.43)$$

where r is a numerical constant. In the case of a metal r is precisely equal to unity. In the case of a semiconductor also the value of r does not differ greatly from unity. Therefore, in all the cases r may be put equal to unity without any serious error.

When the carriers are electrons, the charge on the carrier is negative, and so

$$R_H = -\frac{1}{ne}, \qquad \ldots(4.44)$$

where n is the concentration of electrons.

When the carriers are holes, the carrier charge is positive, so that

$$R_H = \frac{1}{pe}, \qquad \ldots(4.45)$$

where p is the concentration of holes. Thus the signs of the Hall coefficient are opposite for n-type and p-type semiconductors.

In the above discussion, we have assumed the presence of either the electrons or the holes. When the contributions of both the electrons and the holes to the current are important, the Hall coefficient may be determined in the following manner.

As the electrons and holes carry opposite charges but they are deflected in the same direction by the magnetic field, the net deflection current density along the y-axis is $(epv_{yh} - env_{ye})$ where v_{yh} and v_{ye} are respectively the y-components of the velocities of the holes and electrons due to the deflection by the magnetic field. In equilibrium, this current density must balance the current density $\sigma\xi_H$ due to the Hall field where σ is the conductivity. That is,

$$\sigma\xi_H = epv_{yh} - env_{ye} \qquad \ldots(4.46)$$

Let v_{xh} and v_{xe} denote respectively the drift velocities of the holes and electrons along the x-axis. The Lorentz force on a hole is eBv_{xh} and this causes the velocity v_{yh}. In equilibrium, the Lorentz force must balance the rate of loss of y-component of momentum due to collisions. Hence

$$eBv_{xh} = \frac{m_h v_{yh}}{\tau_h}, \qquad \ldots(4.47)$$

where m_h is the effective mass and τ_h is the relaxation time for holes. Similarly, if m_e, τ_e are the corresponding quantities for electrons, we write

$$eBv_{xe} = \frac{m_e v_{ye}}{\tau_e} \qquad \ldots(4.48)$$

Since the electron and the hole mobilities, μ_n and μ_p, are given by

$$\mu_n = \frac{e\tau_e}{m_e}$$

and

$$\mu_p = \frac{e\tau_h}{m_h},$$

we get from Eqs. (4.47) and (4.48)

$$v_{yh} = \mu_p B v_{xh}, \qquad \ldots(4.49)$$

and

$$v_{ye} = \mu_n B v_{xe} \qquad \ldots(4.50)$$

Using Eqs. (4.49) and (4.50) in Eq. (4.46) we get

$$\sigma\xi_H = ep\mu_p B v_{xh} - en\mu_n B v_{xe}$$
$$= \frac{e(p\mu_p^2 - n\mu_n^2)}{\sigma} JB \qquad \ldots(4.51)$$

where v_{xh} and v_{xe} have been replaced by $\mu_p \xi$ and $\mu_n \xi$ respectively, ξ being the electric field in the x-direction, and the relation $J = \sigma\xi$ has been used. The Hall coefficient is therefore given by

$$R_H = \frac{\xi_H}{JB} = \frac{e(p\mu_p^2 - n\mu_n^2)}{\sigma^2}$$
$$= \frac{(p\mu_p^2 - n\mu_n^2)}{e(n\mu_n + p\mu_p)^2} \qquad \ldots(4.52)$$

since $\sigma = ne\mu_n + pe\mu_p$. If the electrons predominate, i.e. $n\mu_n \gg p\mu_p$, Eq. (4.52) reduces to Eq. (4.44). If the holes predominate, i.e., $p\mu_p \gg n\mu_n$, Eq. (4.52) goes over to Eq. (4.45).

In metals like Cu, Ag, Au etc. the Hall coefficient is negative indicating that the current carriers are electrons. In metals like Zn, Cd, Fe etc. R_H is however positive. This is because in metals with nearly filled band, the unfilled states near the top of the band effectively carry the current and behave as positive holes. The sign of R_H is thus changed.

Measurement of the Hall Coefficient

A rectangular specimen of the material having thickness d and width ω is taken and an electric current I is passed through it in the x-direction by connecting it to a battery (Fig. 4.7). The sample is

Transport Phenomena in Metals and Semiconductors

positioned between the pole pieces of an electromagnet so that the magnetic flux density B is applied along the z-direction.

The Hall voltage V_H is measured with the aid of two probes placed at the centres of the top and bottom faces of the sample. If ξ_H is the Hall field we have

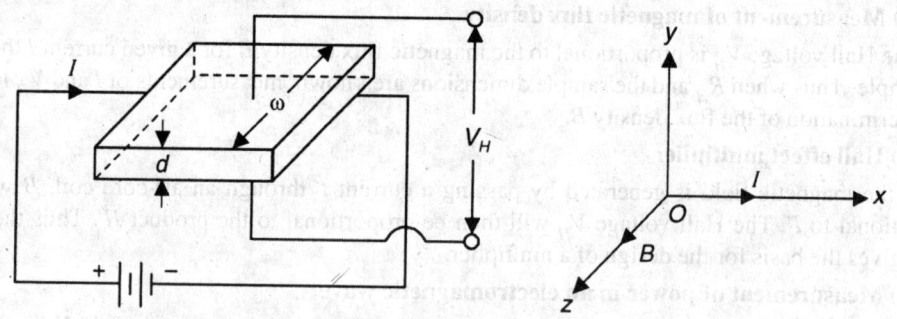

Fig. 4.7. Experimental arrangement for measurement of Hall coefficient.

$$V_H = \xi_H d \qquad \ldots(4.53)$$

Since $\xi_H = R_H JB$, Eq. (4.53) may be expressed as

$$V_H = R_H JBd \qquad \ldots(4.54)$$

The cross-sectional area of the sample is $d\omega$ so that the current density J is given by

$$J = \frac{I}{d\omega} \qquad \ldots(4.55)$$

Using Eq. (4.55) we obtain from Eq. (4.54)

$$V_H = \frac{R_H IB}{\omega}$$

or,

$$R_H = \frac{V_H \omega}{IB} \qquad \ldots(4.56)$$

If V_H is in volt, ω is in metre, I is in ampere, and B is in tesla, the Hall coefficient R_H is determined from Eq. (4.56) in m³/coulomb. Note that the polarity of V_H would be opposite for n- and p-type semiconductors. Therefore R_H will have opposite signs for the two types of semiconductors.

Uses of Hall effect

(1) Determination of semiconductor type

The sign of the Hall coefficient can be used to determine whether a given semiconductor sample is n-type or p-type. R_H is positive for p-type and negative for n-type semiconductors.

(2) Measurement of carrier concentration

The carrier concentration of a metal or a semiconductor can be determined from Eq. (4.44) or (4.45) by measuring R_H. Thus

$$n = \frac{1}{e|R_H|}$$

and

$$p = \frac{1}{R_H e}$$

(3) Determination of mobility

Assuming that there is one type of carriers, say, electrons we have $\sigma = ne\mu_n$ where μ_n is the electron mobility. Using Eq. (4.44) we obtain

$$\mu_n = \sigma|R_H| \qquad \ldots(4.57)$$

Thus, measurements of σ and R_H yield the mobility. Owing to the term r in Eq. (4.43) the mobility determined by R_H is different from the actual drift mobility. The mobility, obtained from Eq. (4.57), is therefore, called the *Hall mobility*. However, since r is nearly equal to unity, the difference between the Hall and the drift mobilities is not large.

(4) Measurement of magnetic flux density

The Hall voltage V_H is proportional to the magnetic flux density B for a given current I through the sample. Thus when R_H and the sample dimensions are known, measurements of I and V_H lead to the determination of the flux density B.

(5) Hall effect multiplier

If the magnetic field is generated by passing a current I' through an air-core coil, B will be proportional to I'. The Hall voltage V_H will then be proportional to the product II'. Thus the Hall effect gives the basis for the design of a multiplier.

(6) Measurement of power in an electromagnetic wave

In an electromagnetic wave in free space the electric field ξ and the magnetic field H are at right angles. If a semiconductor specimen is placed parallel to ξ, a current I will flow in the semiconductor due to ξ. The semiconductor simultaneously feels the influence of a transverse magnetic field H. A Hall voltage, proportional to the product ξH, is therefore developed across the sample. The product ξH is the magnitude of the Poynting vector of the electromagnetic wave. Thus the power flow in an electromagnetic wave can be determined from Hall effect.

4.7. The Boltzmann Transport Equation

The distribution function of the current carriers in a conductor in thermal equilibrium obeys the Fermi-Dirac statistics in the absence of any perturbing field. In the presence of an electric field, the current carriers are accelerated during their free flights. Consequently, the Fermi sphere of occupied states gets shifted and the distribution function deviates from the thermal equilibrium form.

When fields exist, the usual procedure to determine the distribution function is to solve the Boltzmann transport equation (abbreviated BTE). Let the distribution function be denoted by $f(\vec{k}, \vec{r}, t)$. This function gives the probability that the state corresponding to the wave vector $\vec{k}$ at a point in the crystal given by the position vector $\vec{r}$ is occupied by a current carrier at time t. If there is a concentration gradient of the carriers in the crystal, the carriers diffuse. The change in the distribution function in time Δt is

$$\Delta f = f(\vec{k}, \vec{r}, t + \Delta t) - f(\vec{k}, \vec{r}, t). \qquad \ldots(4.58)$$

The carrier occupying $\vec{r}$ at time $t + \Delta t$ was at $\vec{r} - \Delta \vec{r}$ at time t. So, Eq. (4.58) reduces to

$$\Delta f = f(\vec{k}, \vec{r} - \Delta \vec{r}, t) - f(\vec{k}, \vec{r}, t). \qquad \ldots(4.59)$$

Expanding the first term on the right hand side of Eq. (4.59) by Taylor's series and keeping only the first order term, one gets

$$\Delta f = [f(\vec{k}, \vec{r}, t) - \Delta \vec{r} \cdot \vec{\nabla}_r f] - f(\vec{k}, \vec{r}, t) = -\Delta \vec{r} \cdot \vec{\nabla}_r f \qquad \ldots(4.60)$$

In one dimension, the carrier velocity is $(1/\hbar)(dE/dk)$. In three dimensions, the velocity is $\vec{v} = (1/\hbar)\vec{\nabla}_k E$, where E is the carrier energy. Obviously, $\Delta \vec{r} = \vec{v} \Delta t = (1/\hbar)\vec{\nabla}_k E \Delta t$. So, Eq. (4.60) becomes

$$\Delta f = -\frac{1}{\hbar} \vec{\nabla}_k E \cdot \vec{\nabla}_r f \, \Delta t$$

giving

$$\left(\frac{\partial f}{\partial t}\right)_{drift} = -\frac{1}{\hbar} \vec{\nabla}_k E \cdot \vec{\nabla}_r f \qquad \ldots(4.61)$$

The subscript 'drift' indicates the contribution to the time rate of change of f due to a concentration gradient. Note that in the absence of a concentration gradient (*i.e.*, for a homogeneous carrier gas),

$$\vec{\nabla}_r f = 0, \text{ and so } \left(\frac{\partial f}{\partial t}\right)_{drift} = 0$$

Another contribution to the time rate of change of f arises from fields exerting a force $\vec{F}$ on the carrier. The change in the distribution function in time Δt is

$$\Delta f = f(\vec{k}, \vec{r}, t + \Delta t) - f(\vec{k}, \vec{r}, t)$$

$$= f(\vec{k} - \Delta \vec{k}, \vec{r}, t) - f(\vec{k}, \vec{r}, t) \qquad ...(4.62)$$

The second step follows from the fact that the carrier occupying the state $\vec{k}$ at time $t + \Delta t$ was at the state $\vec{k} - \Delta \vec{k}$ at time t. Again, expanding the first term on the right-hand side of Eq. (4.62) in a Taylor's series and retaining only the first order term, one obtains

$$\Delta f = -\Delta \vec{k} \cdot \vec{\nabla}_k f \qquad ...(4.63)$$

Since the crystal momentum is $\hbar \vec{k}$, the force $\vec{F}$, from Newton's second law of motion, is given by $\vec{F} = \hbar \frac{\partial \vec{k}}{\partial t}$. So,

$$\Delta \vec{k} = \frac{1}{\hbar} \vec{F} \Delta t. \qquad ...(4.64)$$

Using Eq. (4.64) in Eq. (4.63) we write

$$\Delta f = -\frac{1}{\hbar} \vec{F} \cdot \vec{\nabla}_k f \Delta t$$

leading to

$$\left(\frac{\partial f}{\partial t}\right)_{field} = -\frac{1}{\hbar} \vec{F} \cdot \vec{\nabla}_k f \qquad ...(4.65)$$

The subscript 'field' gives the contribution to the time rate of change of f due to the force $\vec{F}$.

A third contribution to the time rate of change of f comes from the collision of the carriers. The change in the distribution function from the thermal equilibrium function f_0 is $(f - f_0)$. The collisions tend to restore the thermal equilibrium distribution.

So, if τ is the collision time, one can write

$$\left(\frac{\partial f}{\partial t}\right)_{coll} = -\frac{f - f_0}{\tau} \qquad ...(4.66)$$

The subscript 'coll' accounts for the contribution from collisions.

The net time rate of change of f is obtained by adding the three contributions given by Eqs. (4.61), (4.65) and (4.66). Thus

$$\frac{\partial f}{\partial t} = -\frac{1}{\hbar} \vec{\nabla}_k E \cdot \vec{\nabla}_r f - \frac{1}{\hbar} \vec{F} \cdot \vec{\nabla}_k f - \frac{f - f_0}{\tau} \qquad ...(4.67)$$

Equation (4.67) is referred to as the *Boltzmann transport equation* (BTE). In the steady state when f is independent of time, $(\partial f/\partial t) = 0$. The different transport coefficients, *viz*., the drift mobility, the Hall mobility etc. can be calculated by solving the BTE. Such solutions are beyond the scope of this elementary discussion.

4.8. WORKED-OUT PROBLEMS

1. The density of copper is 8.93 gm/cm³ and its atomic weight is 63. Assuming that each atom contributes one free electron, calculate the concentration of free electrons in copper. If the resistivity of copper is 1.6×10^{-6} ohm. cm, determine the Fermi velocity v_F, the relaxation time τ_F, the mean free path l, and the radius of the Fermi surface, p_F.

Ans. Since one mole contains 6.02×10^{23} atoms (Avogadro number), 63 gm of copper contains 6.02×10^{23} atoms.

Hence 8.93 gm of copper will contain $\dfrac{6.02 \times 10^{23}}{63} \times 8.93 = 8.533 \times 10^{22}$ atoms.

So, the concentration of free electrons is $n = 8.533 \times 10^{22}$ cm^{-3}. Taking $m_e = 9.11 \times 10^{-31}$ kg, the Fermi energy is given by (see Eq. 2.13, 'Statistical Mechanics')

$$E_F = \frac{h^2}{8m_e}\left(\frac{3n}{\pi}\right)^{2/3} = \frac{(6.62)^2 \times 10^{-68}}{8 \times 9.11 \times 10^{-31}}\left(\frac{3 \times 8.533 \times 10^{28}}{3.1416}\right)^{2/3}$$

$$= 11.29 \times 10^{-19} \text{ J}$$

$$= \frac{11.29 \times 10^{-19}}{1.6 \times 10^{-19}} \text{ eV} = 7 \text{ eV}$$

The Fermi velocity v_F is related to Fermi energy by

$$v_F = \sqrt{\frac{2E_F}{m_e}}$$

So,

$$v_F = \sqrt{\frac{2 \times 7 \times 1.6 \times 10^{-19}}{9.11 \times 10^{-31}}}$$

$$= 1.568 \times 10^6 \text{ m/s}$$
$$= 1.568 \times 10^8 \text{ cm/s}$$

The electrical conductivity σ is given by

$$\sigma = \frac{ne^2 \tau_F}{m_e}$$

Hence,

$$\tau_F = \frac{\sigma m_e}{ne^2} = \frac{m_e}{\rho n e^2}, \text{ where } \rho \text{ is the resistivity}$$

or,

$$\tau_F = \frac{9.11 \times 10^{-31}}{1.6 \times 10^{-8} \times 8.533 \times 10^{28} \times (1.6 \times 10^{-19})^2}$$

$$= 0.26 \times 10^{-13} \text{ s}$$

The mean free path l is the mean distance traversed by an electron between two successive collisions. So.

$$l = v_F \tau_F = 1.568 \times 10^8 \times 0.26 \times 10^{-13} = 0.41 \times 10^{-5} \text{ cm}$$

The radius of the Fermi surface is

$$p_F = \sqrt{2m_e E_F} = \sqrt{2 \times 9.11 \times 10^{-31} \times 11.29 \times 10^{-19}}$$
$$= 14.3 \times 10^{-25} \text{ kg. m/s}$$

2. Calculate the conductivity of intrinsic germanium at room temperature (300 K). Given that at 300 K the intrinsic carrier concentration, electron mobility and hole mobility are 2.4×10^{19} m^{-3}, 0.38 m²/(V.s) and 0.19 m² / (V.s), respectively.

Ans. We have

$$\sigma_i = en_i(\mu_n + \mu_p)$$

Putting the values we obtain

$$\sigma_i = 2.4 \times 10^{19} \times 1.6 \times 10^{-19} \times (0.38 + 0.19)$$
$$= 2.19 \text{ mho/m (or, S/m)}$$

3. Calculate the density of donor atoms that has to be added to an intrinsic germanium semiconductor to make *n*-type semiconductor of conductivity 4 mho/cm. Given that the electron mobility of the *n*-type semiconductor is 3900 cm²/(V.s).

Ans. Assuming that the concentration of donor atoms equals the concentration of free electrons we have

$$\sigma_n = e n \mu_n$$

Given $\sigma_n = 4$ mho/cm, $\mu_n = 3900$ cm²/(V.s) and
$$e = 1.6 \times 10^{-19} \text{ coulomb}$$

If N_d represents the donor concentration, then we can write

$$n = N_d = \frac{\sigma_n}{e \mu_n} = \frac{4}{1.6 \times 10^{-19} \times 3900} = 0.64 \times 10^{16} / \text{cm}^3$$

4. An *n*-type Ge strip, 1 mm wide and 1 mm thick, has a Hall coefficient of 10^{-2} m³/coulomb. If for a current of 1 mA the Hall voltage produced inside the strip is 1 mV, calculate the strength of the magnetic field.

Ans. We have from Eq. (4.56)

$$B = \frac{V_H \omega}{R_H I}$$

Substituting the values we get

$$B = \frac{1 \times 10^{-3} \times 1 \times 10^{-3}}{10^{-2} \times 1 \times 10^{-3}} = 0.1 \text{ tesla}$$

5. A semiconductor has an intrinsic concentration of 3.1×10^{13} cm⁻³ at room temperature. It is doped with 10^{14} donor atoms per cm³. Assuming that the donors are all ionised, calculate the electron and hole concentrations and the percentage increase in conductivity of the doped material over the intrinsic value. Given : electron mobility = 3800 cm²/(V.s), hole mobility = 1800 cm²/(V.s).

Ans. From the charge neutrality of the sample we have

$$n - p = N_d, \qquad (i)$$

where n = electron concentration, p = hole concentration and N_d = donor concentration. Also, if n_i is the intrinsic concentration

$$np = n_i^2 \qquad (ii)$$

From Eqs. (*i*) and (*ii*) we have

$$n - \frac{n_i^2}{n} = N_d$$

or $$n^2 - n N_d - n_i^2 = 0$$

or $$n = \frac{N_d \pm \sqrt{N_d^2 + 4 n_i^2}}{2}$$

Since n is positive we have to reject the negative sign before the radical.

Hence, $$n = \frac{N_d}{2} + \sqrt{\left(\frac{N_d}{2}\right)^2 + n_i^2}$$

Substituting $n_i = 3.1 \times 10^{13}$ cm^{-3} and $N_d = 10^{14}$ cm^{-3} we obtain

$$n = 1.09 \times 10^{14} \text{ cm}^{-3}$$

Hence
$$p = \frac{n_i^2}{n} = \frac{(3.1 \times 10^{13})^2}{1.09 \times 10^{14}} = 0.88 \times 10^{13} \text{ cm}^{-3}$$

The conductivity of the doped material is

$$\sigma = e\,(n\mu_n + p\mu_p),$$

where μ_n and μ_p are the electron and hole mobilities, respectively. Substituting the values of n, μ_n, p and μ_p, and putting $e = 1.6 \times 10^{-19}$ C, we have

$$\sigma = 0.068 \text{ mho per cm.}$$

The intrinsic conductivity is

$$\sigma_i = en_i\,(\mu_n + \mu_p) = 0.028 \text{ mho/cm}$$

Hence the percentage increase in conductivity over the intrinsic value is

$$\frac{\sigma - \sigma_i}{\sigma_i} \times 100 = 143\%$$

6. A sample of Si is doped with 10^{11} donor atoms /cm^3 and 3×10^{10} acceptor atoms /cm^3. At 300 K, the resistivity of intrinsic Si is 2.17×10^5 ohm cm. If the applied electric field is 500 V/cm, find the conduction current density in the doped sample. Assume $\mu_p/\mu_n = 5/13$ and $n_i = 1.6 \times 10^{10}$ / cm^3 at 300 K.

Ans. The intrinsic conductivity is

$$\sigma_i = \frac{1}{\rho_i} = n_i e \mu_n \left(1 + \frac{\mu_p}{\mu_n}\right).$$

Given :
$$\rho_i = 2.17 \times 10^3 \text{ ohm m}, \; n_i = 1.6 \times 10^{16} / \text{m}^3, \; \mu_p/\mu_n = 5/13.$$

So,
$$\frac{1}{2.17 \times 10^3} = 1.6 \times 10^{16} \times 1.6 \times 10^{-19} \times \mu_n \times \frac{18}{13}$$

whence $\mu_n = 0.13$ m^2/(V.s). Therefore, $\mu_p = 0.05$ m^2/(V.s).

Let n be the electron concentration and p be the hole concentration in the doped sample. As the sample is electrically neutral, we have

$$N_D + p = N_A + n$$

where N_D is the donor concentration and N_A is the acceptor concentration, assumed to be completely ionized. Also, the mass action law gives $np = n_i^2$. So,

$$N_D + \frac{n_i^2}{n} = N_A + n$$

or, $\qquad n^2 + (N_A - N_D)n - n_i^2 = 0$

or, $\qquad n = \frac{1}{2}\left[N_D - N_A + \sqrt{(N_A - N_A)^2 + 4n_i^2}\right],$

accepting the root giving a positive value of n. Given : $N_D = 10^{17} / m^3$, $N_A = 3 \times 10^{16} / m^3$, and $n_i = 1.6 \times 10^{16} / m^3$.

Hence, $$n = \frac{1}{2}\left[7 \times 10^{16} + 10^{16}\sqrt{7^2 + 4 \times 1.6^2}\right] = 7.348 \times 10^{16} / m^3$$

and $$p = \frac{n_i^2}{n} = \frac{1.6^2}{7.348} \times 10^{16} = 0.3484 \times 10^{16} / m^3$$

The conductivity of the doped sample is

$$\sigma = ne\mu_n + pe\mu_p = 1.6 \times 10^{-3}(7.348 \times 0.13 + 0.3484 \times 0.05)$$

$$= 1.556 \times 10^{-3} \text{ S/m}.$$

It is assumed here that the values of μ_n and μ_p in the doped sample are the same as those in the intrinsic material. This is justified by the low level of doping and by the fact that Coulomb scattering by ionized donors and acceptors is weak at 300 K. The applied electric field is $F = 5 \times 10^4$ V/m. So, the conduction current density is

$$J = \sigma F = 1.556 \times 10^{-3} \times 5 \times 10^4 = 77.8 \text{ A/m}^2 = 0.00778 \text{ A/cm}^2.$$

QUESTIONS

1. Define mean free time, mean free path, electron mobility, and electrical conductivity.
2. State the basic assumptions of the classical Drude theory of metals and deduce the expressions for
    (i) the electrical conductivity, and
    (ii) the thermal conductivity.
3. On the basis of classical ideas obtain the Lorenz number for metals.
4. Describe the basic principles governing the relation between the thermal and the electrical conductivity of metals. **(C.U. 1982)**
5. (a) How is the electrical conductivity of metals described in the quantum free electron theory ?
   (b) Using the free-electron theory, find an expression for the electrical conductivity of a metal in terms of the Fermi velocity of the electron and the mean free path of electrons. **(C.U. 2005)**

   [**Hint :** See Eqs. (4.25) and (4.27).]
6. What is 'Fermi surface' ? Discuss its significance in the calculations of the electrical and the thermal conductivity of metals. Establish the relation

$$\frac{K}{\sigma T} = \frac{\pi^2}{3}\left(\frac{k_B}{e}\right)^2$$

   where the symbols have their usual meanings.
7. Show that the electrical conductivity of a semiconductor is given by $\sigma = e\,(n\mu_n + p\mu_p)$, where the symbols have their usual meanings.
8. Give the theory of Hall effect. How do you use Hall effect for the measurement of magnetic field ?

(C.U. 1984)

9. Describe the basic principles of Hall effect in semiconductors. **(C.U. 1982, 1996, 2001, 04)**

10. For an intrinsic semiconductor show that the Hall coefficient is given by

$$R_H = -\frac{1}{ne}\left(\frac{\mu_n - \mu_p}{\mu_n + \mu_p}\right),$$

where μ_n, μ_p are the mobilities of electrons and holes and n is the intrinsic carrier concentration.

(C.U. 1985)

[**Hint** : Put $n = p$ in $E_q.$ (4.52)]

11. How would you experimentally determine the Hall coefficient ? Mention the uses of Hall effect.

12. What is Hall effect ? Find the Hall coefficient in a metal where the carriers are only electrons. Why is the Hall coefficient positive in some metals ? (C.U. 1992)

13. Sodium has density 0.97 gm/cm^3, relative atomic mass 23, and an electrical conductivity 2.1×10^5 mho/cm. Determine the electron mobility in sodium. [**Ans.** 5.2×10^{-3} m^2/(V.s)]

14. The following data are obtained in an experiment with intrinsic germanium at 300 K:

$$n_i = 2.5 \times 10^{13} \text{ cm}^{-3},$$
$$\mu_n = 3450 \text{ cm}^2/(V.s),$$
and $\quad \mu_p = 1600 \text{ cm}^2/(V.s)$

Calculate the conductivity and resistivity of intrinsic germanium. [**Ans.** 0.02 Ω^{-1}/cm; 50 Ω cm]

15. The intrinsic carrier concentration of germanium at 300 K is 2.5×10^{19}/m^3. It is doped with a pentavalent impurity element at a rate of one pentavalent atom per million atoms of germanium. If the germanium atom concentration is 4×10^{28}/m^3 find the conductivity of the semiconductor. Assume complete ionization of the impurity atoms and $\mu_n = 0.35$ m^2/(V.s). [**Ans.** 2.24×10^3 Ω^{-1}/m]

16. The donor density of an n-type germanium sample is 10^{21}/m^3. If the sample is 3.5 mm thick find the Hall voltage if the magnetic field $B = 0.5$ tesla and the current density $J = 500$ A/m^2.

[**Ans.** 5.47 mV (approx.)]

17. An n-type Si is formed by doping with 10^{17} phosphorus atoms per cm^3. Assuming the electron mobility to be 700 cm^2/ (V.s) and neglecting the hole concentration, calculate the conductivity of the sample. If the sample, 0.1 mm thick, carries a current of 1 mA and is placed in a transverse magnetic field of 1 kilo gauss, what Hall voltage would be produced ? [**Ans.** 11.2 mho/cm, -62.5 μV]

18. Explain why resistivity increases with temperature in metals, while it decreases with temperature in semiconductors. (C.U. 1986)

19. At a given temperature, show that semiconductor has a minimum conductivity σ_m expressed by

$$\sigma_m = 2en_i (\mu_n \mu_p)^{1/2}.$$

[**Hint.** : Note that $\sigma = e (n\mu_n + p\mu_p)$, and that $p = n_i^2/n$. Find $\dfrac{d\sigma}{dn}$ and equate it to zero for a minimum value of σ.]

20. A conductor contains 10^{23} electrons/cm^3. Consider a cube of the conductor of side 0.1 cm carrying a current of 10A flowing perpendicularly from one face to the opposite one. Calculate the current density and the drift velocity. Compare the drift velocity with the Fermi velocity assuming that the Fermi energy is 7eV and electron mass is 9.11×10^{-31} kg.

[**Ans.** 10^3 A/cm^2; 6.25×10^{-2} cm/s, $v_D/v_F \approx 4 \times 10^{-10}$]

21. Write the Boltzmann transport equation, and explain the meaning and the origin of the terms in the equation. (C.U. 1996)

22. (a) Show that the electrical conductivity of a free electron gas is $\sigma = ne^2 \tau/m$ where the symbols have their usual meanings.

(b) Explain why only the electrons near the Fermi surface suffer collisions to contribute to the electrical conductivity of a metal. (**Burd. U. 1995**)

23. The Fermi energy for lithium at 0K is 4.72 eV. Calculate the number of conduction electrons per unit volume in lithium. (C.U. 1997) [Ans. $4.66 \times 10^{22}/cm^3$]

24. When is the Hall coefficient zero in a semiconductor?

    The electron and the hole mobilities in a semiconductor are 0.8 m²/(V.s) and 0.02 m²/(V.s), respectively. The electron concentration in the semiconductor is $2.5 \times 10^{18}/m^3$, and the Hall coefficient is zero. Find the intrinsic carrier concentration.

    [Hint. Refer to Eq. (4.52).]   [Ans. $10^{20}/m^3$]

25. Silver is a monovalent metal with the atoms arranged in an fcc structure as close-packed spheres. If the atomic radius is $r = 1.44$ Å, what is the Fermi energy ?

    [Hint. If a is the lattice constant, $4r = \sqrt{2}\, a$. The volume of the cubic unit cell is $v = a^3 = (2\sqrt{2}\, r)^3$. The number of free electrons in the unit cell is 4. So, the free electron concentration is $n = \dfrac{4}{v} = \dfrac{1}{4\sqrt{2}\, r^3}$]   [Ans. 5.5 eV]

26. The ratio of the hole mobility to the electron mobility in a semiconductor is b. If the Hall coefficient is zero, what fraction of the total current is carried by electrons ?   (cf. N. Beng. U 2001)

    [Ans. $b/(1+b)$]

27. The resistivity of some material is 1.1×10^{-3} Ω m at 20° C. Assuming Maxwell-Boltzmann distribution and a free electron concentration of 10^{22} m⁻³, estimate the mean free path $\bar{l}$ between the collisions of free electrons in the material at 20° C. Given, effective mass of free electrons in the material is 2.733×10^{-31} kg.   [Ans. 1.88×10^{-7} m]

    [Hint : Use $\dfrac{1}{\rho} = \sigma = \dfrac{ne^2 \tau}{m^*}$ and $\bar{l} = \bar{v}\, \tau$ where $\bar{v} = (8 k_B T/\pi m^*)^{1/2}$]

28. A semiconductor sample has a free carrier density of $10^{22}/m^3$ and a mobility of 0.01 m²/(V.s).

    (i) What is the electrical conductivity of the sample ?

    (ii) The charge carriers have an effective mass of $0.1\, m_0$ where m_0 is the mass of a free electron. Calculate the average time between successive scatterings.   (Ans. 16 S/m, 5.7×10^{-15} s)

29. At a temperature of 300 K, the intrinsic carrier concentration is $1.6 \times 10^{10}/cm^3$ for silicon, and the electron and the hole mobilities are 1500 cm²/(V.s) and 500 cm²/(V.s), respectively. Find the resistivity of pure silicon at 300 K. (Given, electronic charge = 1.602×10^{-19} coulomb.)

    (K.U. 2004, cf. Mangalore U. 2005)

    (Ans. 2.17×10^5 ohm. cm)

Chapter 5

Specific Heat of Solids and Lattice Vibrations

When the temperature of a solid is raised, the thermal vibrations of the atoms, called lattice vibrations, become more vigorous. An increase of temperature also increases the kinetic energy \ of free electrons in metals and semiconductors. In this chapter, we shall consider the contribution of the lattice vibrations to the specific heat.

5.1. CLASSICAL CALCULATION OF LATTICE SPECIFIC HEAT

We consider a crystal of N atoms which are free to vibrate about their equilibrium positions. The restoring forces are assumed to obey Hooke's law. The energy of vibration of an atom would be the same as that of three one-dimensional harmonic oscillators, one for each of the three directions of motion. In the classical calculation, each atom is assumed to vibrate independently of the others so that the total vibrational energy of the crystal is obtained by multiplying the average energy of a one-dimensional oscillator by $3N$.

The energy of a one-dimensional harmonic oscillator is given by:

$$E = \frac{p_x^2}{2m} + \frac{1}{2} m\omega^2 x^2, \qquad \ldots(5.1)$$

where p_x is the momentum, m is the mass, ω is the angular frequency, and x is the displacement of the atom from the equilibrium position. The first term on the right-hand side of Eq. (5.1) is the kinetic energy and the second term is the potential energy.

Assuming that the oscillators obey the Maxwell-Boltzmann distribution law, the distribution function giving the probability of an oscillator to have an energy E is,

$$f(E) = A e^{-E/k_B T} = A e^{-\frac{p_x^2}{2mk_B T}} e^{-\frac{m\omega^2 x^2}{2k_B T}} \qquad \ldots(5.2)$$

where A is a constant. According to statistical mechanics, the average energy is

$$\overline{E} = \frac{\displaystyle\int_{p_x=-\infty}^{\infty}\int_{x=-\infty}^{\infty} \left(\frac{p_x^2}{2m} + \frac{1}{2} m\omega^2 x^2\right) e^{-\frac{p_x^2}{2mk_B T}} e^{-\frac{m\omega^2 x^2}{2k_B T}} dp_x dx}{\displaystyle\int_{p_x=-\infty}^{\infty}\int_{x=-\infty}^{\infty} e^{-\frac{p_x^2}{2mk_B T}} e^{-\frac{m\omega^2 x^2}{2k_B T}} dp_x dx} \qquad \ldots(5.3)$$

$$= \frac{1}{2m} \frac{\displaystyle\int_{-\infty}^{\infty} p_x^2 e^{-p_x^2/2mk_B T} dp_x}{\displaystyle\int_{-\infty}^{\infty} e^{-p_x^2/2mk_B T} dp_x} + \frac{m\omega^2}{2} \frac{\displaystyle\int_{-\infty}^{\infty} x^2 e^{-\frac{m\omega^2 x^2}{2k_B T}} dx}{\displaystyle\int_{-\infty}^{\infty} e^{-\frac{m\omega^2 x^2}{2k_B T}} dx} \qquad \ldots(5.3)$$

The first term on the right-hand side of Eq. (5.3) gives the average kinetic energy $\overline{E}_K$ and the second term gives the average potential energy $\overline{E}_P$. Evaluating the integrals by gamma functions we get

Specific Heat of Solids and Lattice Vibrations

$$\overline{E}_K = \overline{E}_P = \frac{1}{2} k_B T, \quad \ldots(5.4)$$

so that
$$\overline{E} = \overline{E}_K + \overline{E}_P = k_B T$$

For a system of N atoms, the total vibrational energy or the total internal thermal energy is given by

$$U = 3N\overline{E} = 3Nk_B T \quad \ldots(5.5)$$

If we consider a mole, the number N is equal to Avogadro's number N_0, and $U = 3N_0 k_B T$. Therefore, the molar specific heat at constant volume is

$$C_v = \left(\frac{\partial U}{\partial T}\right)_v = 3N_0 k_B = 3R, \quad \ldots(5.6)$$

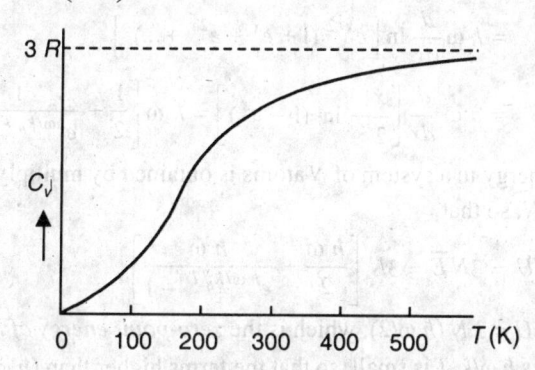

Fig. 5.1. Typical plot of the molar specific heat of a solid as a function of temperature, as observed experimentally.

where R is the gas constant. Equation (5.6) gives the *Dulong-Petit value* of the specific heat at constant volume. The classical calculations thus show that the contribution of the lattice vibration to the molar specific heat is $3R$, i.e., 24.9 JK^{-1} mol^{-1}, which is a constant independent of temperature. This result agrees well with the experimental results of the specific heat of many solids at and above room temperature; but serious disagreements occur at low temperatures. As the temperature approaches zero, the specific heat of all solids is found to decrease towards zero (see Fig. 5.1). For insulators the specific heat varies as T^3 as the temperature T tends to zero, while for metals the specific heat varies as T as the temperature T approaches zero.

5.2. Einstein's Theory of Specific Heat

The drawbacks of the classical theory of specific heat were investigated by Einstein, who could explain qualitatively at least, the variation of the specific heat with temperature, as found experimentally. In Einstein's theory, a solid containing N atoms is assumed to be represented by $3N$ one dimensional *quantum* harmonic oscillators having discrete energy values. The atoms vibrate independently of each other but they have the same angular frequency of vibration owing to their identical surroundings. The energy of an atomic oscillator is given by

$$E_n = \left(n + \frac{1}{2}\right)\hbar\omega \quad \ldots(5.7)$$

where $n = 0, 1, 2, \ldots$ and $\hbar$ is Planck's constant divided by 2π. The term $\hbar\omega/2$ is the *zero-point energy*.

The oscillators are distinguishable from their location at different lattice sites; also there is no restriction on the number of oscillators which may have the same quantum state. Thus although the atoms are quantum oscillators, their energy distribution is given by the Maxwell-Boltzmann statistics. Hence the average energy of an atomic oscillator is

$$\bar{E} = \frac{\sum_{n=0}^{\infty} E_n e^{-E_n/k_B T}}{\sum_{n=0}^{\infty} e^{-E_n/k_B T}} = \hbar\omega \frac{\sum_{n=0}^{\infty} (n+1/2) e^{(n+1/2)x}}{\sum_{n=0}^{\infty} e^{(n+1/2)x}} \qquad \ldots(5.8)$$

where
$$x = -\frac{\hbar\omega}{k_B T} \qquad \ldots(5.9)$$

We can write Eq. (5.8) as

$$\bar{E} = \hbar\omega \frac{(1/2) e^{x/2} + (3/2) e^{3x/2} + (5/2) e^{5x/2} + \ldots}{e^{x/2} + e^{3x/2} + e^{5x/2} + \ldots}$$

$$= \hbar\omega \frac{d}{dx} \ln\left[e^{x/2} (1 + e^x + e^{2x} + \ldots) \right]$$

$$= \hbar\omega \frac{d}{dx}\left[\frac{x}{2} - \ln(1 - e^x) \right] = \hbar\omega \left[\frac{1}{2} + \frac{1}{e^{\hbar\omega/k_B T} - 1} \right] \qquad \ldots(5.10)$$

The total internal energy in a system of N atoms is obtained by multiplying the average energy of an oscillator $\bar{E}$ by $3N$, so that

$$U = 3N\bar{E} = 3N\left[\frac{\hbar\omega}{2} + \frac{\hbar\omega}{e^{\hbar\omega/k_B T} - 1} \right] \qquad \ldots(5.11)$$

At $T = 0$ K, clearly $U = 3N(\hbar\omega/2)$, which is the *zero-point energy of N atoms*.

At high temperatures $\hbar\omega/k_B T$ is small, so that the terms higher than $(\hbar\omega/k_B T)^2$ in the expansion of the exponential term can be ignored. Thus the high-temperature value of U is $3N\left[\frac{\hbar\omega}{2} + \left(k_B T - \frac{\hbar\omega}{2}\right)\right] = 3Nk_B T$, which agrees with the classical result. It may be mentioned here that originally Einstein did not include the zero-point energy term $\hbar\omega/2$ in Eq (5.7). This resulted in an expression for U which differed from Eq. (5.11) only in the absence of the term $\hbar\omega/2$ within parenthesis. The high-temperature value of U was then $3N(k_B T - \hbar\omega/2)$ which did not agree with the classical value.

For one mole of the substance, N is replaced by N_0, the Avogadro number, in Eq. (5.11). Hence the molar specific heat is given by

$$C_v = \left(\frac{\partial U}{\partial T}\right)_v = 3N_0 k_B \left(\frac{\hbar\omega}{k_B T}\right)^2 \frac{e^{\hbar\omega/k_B T}}{(e^{\hbar\omega/k_B T} - 1)^2} \qquad \ldots(5.12)$$

$$= 3R \left(\frac{\Theta_E}{T}\right)^2 \frac{e^{\Theta_E/T}}{(e^{\Theta_E/T} - 1)^2} \qquad \ldots(5.13)$$

where
$$\Theta_E = \frac{\hbar\omega}{k_B} \qquad \ldots(5.14)$$

The quantity Θ_E is known as the *Einstein temperature*. Note that the zero-point energy is independent of temperature, so that it vanishes upon differentiation with respect to temperature and does not contribute to C_v.

At high temperatures when $T \gg \Theta_E$, it is readily found on expanding the exponential term in Eq. (5.13) and retaining only the first order terms that the molar specific heat approaches the classical Dulong-Petit value of $3R$. At low temperatures when $T \ll \Theta_E$, Eq. (5.13) may be approximately written as

$$C_v = 3R \left(\frac{\Theta_E}{T}\right)^2 e^{-\Theta_E/T} \qquad \ldots(5.15)$$

Thus the specific heat at low temperatures varies as $T^{-2}e^{-\Theta_E/T}$. As the exponential term is dominant at very low temperatures, the specific heat falls off more rapidly than the experimental results.

By a proper choice of ω or Θ_E, Einstein's theory may be fitted to the experimental results on the temperature variation of C_v of a given solid fairly well at all but very low temperatures. Numerical values of Θ_E for most solids lie in the range 100–300 K; the corresponding frequencies ($\omega/2\pi$) vary from 2×10^{12} to 6×10^{12} Hz. The discrepancy between Einstein's theory and the low-temperature data on C_v was removed and the general fit to the observed values was improved in Debye's theory. In the latter theory interactions between the atoms are considered, which results in a range of possible values of the frequency of vibration instead of a single frequency ω.

5.3. Debye's Theory of Specific Heat

In the Debye theory, the atomic oscillators in the crystal are assumed to be coupled together and generate elastic waves of frequencies covering a wider range. To determine the vibrational modes of a three-dimensional crystal having fixed boundaries, we shall first consider a one-dimensional crystal or a string of length L fixed at both ends (Fig. 5.2). The deflection of the string at a distance x from one end at time t, denoted by $y(x, t)$, satisfies the one-dimensional wave equation

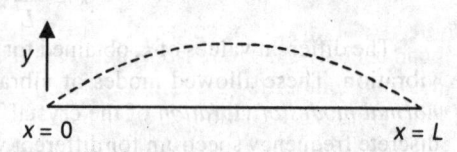

Fig. 5.2. String fixed at both ends.

$$\frac{\partial^2 y}{\partial x^2} = \frac{1}{v^2}\frac{\partial^2 y}{\partial t^2}, \qquad \ldots(5.16)$$

where v represents the velocity of propagation of the waves. If υ is the frequency and λ is the wavelength of the wave, we have

$$v = \upsilon\lambda = \frac{\omega}{k}, \qquad \ldots(5.17)$$

where $\omega = 2\pi\upsilon$ is the angular frequency and $k = 2\pi/\lambda$ is the wave number. Using Eq. (5.17), the wave equation (5.16) may be rewritten as

$$\frac{\partial^2 y}{\partial x^2} = \frac{k^2}{\omega^2}\frac{\partial^2 y}{\partial t^2} \qquad \ldots(5.18)$$

Equation (5.18) is solved by the method of separation of variables assuming that y can be expressed as

$$y(x, t) = X(x)\,T(t) \qquad \ldots(5.19)$$

We then obtain from Eq. (5.18)

$$\frac{1}{T}\frac{d^2 T}{dt^2} = \frac{\omega^2}{k^2}\frac{1}{X}\frac{d^2 X}{dx^2} \qquad \ldots(5.20)$$

The left-hand side of Eq. (5.20) is a function of t only while the right-hand side is a function of x only. The only way in which they can be equal is for each side to be a constant, say, $-\omega^2$. Hence

$$\frac{1}{T}\frac{d^2 T}{dt^2} = \frac{\omega^2}{k^2}\frac{1}{X}\frac{d^2 X}{dx^2} = -\omega^2,$$

so that

$$\frac{d^2 X}{dx^2} + k^2 X = 0, \qquad \ldots(5.21)$$

and

$$\frac{d^2 T}{dt^2} + \omega^2 T = 0 \qquad \ldots(5.22)$$

The solutions of Eqs. (5.21) and (5.22) are
$$X = A_1 \sin kx + A_2 \cos kx, \qquad \ldots(5.23)$$
and
$$T = B_1 \sin \omega t + B_2 \cos \omega t, \qquad \ldots(5.24)$$
where A_1, A_2, B_1 and B_2 are arbitrary constants. As the string is fixed at both ends, the boundary conditions for the problem are $y = 0$ at $x = 0$ and at $x = L$ for all values of t. Since $X = 0$ at $x = 0$, we have from Eq. (5.23), $A_2 = 0$.

Therefore,
$$X = A_1 \sin kx \qquad \ldots(5.25)$$

As $X = 0$ at $x = L$, we must have $\sin kL = 0$, or $kL = n\pi$, where n is an integer excluding zero. For $n = 0$, y is zero at all values of x and t so that the string does not vibrate at all. Hence the value $n = 0$ is discarded. Consequently
$$k = \frac{n\pi}{L} \qquad \ldots(5.26)$$

The different values of k, obtained for different integral values of n, give the different modes of vibration. These allowed modes of vibration due to the boundary conditions are known as the *normal modes of vibration* of the crystal. For a given velocity of propagation of the wave, v, the discrete frequency spectrum for different values of n is given by
$$\omega = kv = \frac{n\pi v}{L} \qquad \ldots(5.27)$$

The deflection y is then expressed as
$$y = XT = (A_1 B_1 \sin \omega t + A_1 B_2 \cos \omega t) \sin\left(\frac{n\pi x}{L}\right) \qquad \ldots(5.28)$$

If the velocity dy/dt at all points along the string is taken to be zero at $t = 0$, we have $A_1 B_1 = 0$. Hence the solution reduces to
$$y = C \cos \omega t \sin\left(\frac{n\pi x}{L}\right), \qquad \ldots(5.29)$$
where $C = A_1 B_2$ = a new constant. Equation (5.29) represents a standing wave.

If n is very large, the relative separation between the allowed values of k or ω would be very small, and the frequency spectrum may be treated as quasi-continuous. The number of possible modes of vibration dn in a frequency interval $d\omega$ is then obtained from Eq. (5.27) as
$$dn = \frac{L}{\pi v} d\omega \qquad \ldots(5.30)$$

We may now generalise the results to find the vibrational modes of a three-dimensional solid. If $\psi(x, y, z, t)$ denotes the deflection in three dimensions, the wave equation becomes
$$\frac{\partial^2 \psi}{\partial x^2} + \frac{\partial^2 \psi}{\partial y^2} + \frac{\partial^2 \psi}{\partial z^2} = \frac{1}{v^2} \frac{\partial^2 \psi}{\partial t^2}, \qquad \ldots(5.31)$$
where v is the velocity of wave propagation, taken to be the same in all directions. We take the solid to be in the form of a cube of edge length L and assume that the faces of the cube are fixed. Then the solutions of Eq. (5.31) are
$$\psi = C \sin\left(\frac{n_x \pi x}{L}\right) \sin\left(\frac{n_y \pi y}{L}\right) \sin\left(\frac{n_z \pi z}{L}\right) \cos \omega t, \qquad \ldots(5.32)$$
which represent standing waves. Here n_x, n_y and n_z are integers other than zero. Substitution of Eq. (5.32) in Eq. (5.31) gives
$$\frac{\pi^2}{L^2} (n_x^2 + n_y^2 + n_z^2) = \frac{\omega^2}{v^2} \qquad \ldots(5.33)$$

Specific Heat of Solids and Lattice Vibrations

The discrete allowed values of ω are determined by the different integral values of n_x, n_y and n_z, called *quantum numbers*. As before, for very large values of n_x, n_y and n_z the frequency spectrum may be considered quasi-continuous. The number of possible modes of vibration dN_s having frequencies between ω and $\omega + d\omega$ is given by the number of sets of values (n_x, n_y, n_z) lying within a spherical shell between radii R and $R + dR$ where $R^2 = n_x^2 + n_y^2 + n_z^2$. As n_x, n_y and n_z are positive, the octant of the shell is to be considered, so that

$$dN_s = \frac{1}{8} 4\pi R^2 \, dR = \frac{\pi R^2}{2} \, dR \qquad ...(5.34)$$

From Eq. (5.33) we have

$$R^2 = \frac{\omega^2 L^2}{\pi^2 v^2}, \qquad ...(5.35)$$

and

$$dR = \frac{L}{\pi v} \, d\omega \qquad ...(5.36)$$

Therefore,

$$dN_s = \frac{\pi}{2} \frac{\omega^2 L^2}{\pi^2 v^2} \frac{L}{\pi v} \, d\omega$$

$$= \frac{V}{2\pi^2 v^3} \omega^2 \, d\omega, \qquad ...(5.37)$$

where $V = L^3$ = volume of the solid. Since for any longitudinal wave there are two transverse waves at right angles to each other, Eq. (5.37) may be put in the general form

$$dN_s = \frac{V}{2\pi^2} \left(\frac{2}{c_t^3} + \frac{1}{c_l^3} \right) \omega^2 \, d\omega, \qquad ...(5.38)$$

where c_t and c_l are the velocities of the transverse and the longitudinal waves. The cut-off frequency ω_D of the waves, called the *Debye frequency*, is determined by equating the total number of possible vibrations to $3N$, where N is the number of atoms in the solid. Thus

$$\int_0^{\omega_D} \frac{V}{2\pi^2} \left(\frac{2}{c_t^3} + \frac{1}{c_l^3} \right) \omega^2 \, d\omega = 3N, \qquad ...(5.39)$$

which gives

$$\omega_D^3 = \frac{18N\pi^2}{V} \left(\frac{2}{c_t^3} + \frac{1}{c_l^3} \right)^{-1} \qquad ...(5.40)$$

If a quantum harmonic oscillator is associated with each vibrational mode of the same frequency, the average energy for the mode of angular frequency ω is given by Eq. (5.10). The total internal energy of the system is then

$$U = \int_0^{\omega_D} \hbar \omega \left[\frac{1}{2} + \frac{1}{e^{\hbar \omega / k_B T} - 1} \right] dN_s$$

$$= \frac{V \hbar}{2\pi^2} \left(\frac{2}{c_t^3} + \frac{1}{c_l^3} \right) \int_0^{\omega_D} \left[\frac{\omega^3}{2} + \frac{\omega^3}{e^{\hbar \omega / k_B T} - 1} \right] d\omega \qquad ...(5.41)$$

Let us put $x = \hbar \omega / k_B T$, $x_m = \hbar \omega_D / k_B T$ and $\Theta_D = \hbar \omega_D / k_B$. The quantity Θ_D is known as the *Debye temperature*. With these substitutions and using Eq. (5.40) we obtain from Eq. (5.41)

$$U = \frac{9}{8} N k_B \Theta_D + 9 N k_B T \left(\frac{T}{\Theta_D} \right)^3 \int_0^{x_m} \frac{x^3 \, dx}{e^x - 1} \qquad ...(5.42)$$

The first term on the right-hand side of Eq. (5.42) is the temperature-independent zero-point energy of N atoms.

At high temperatures, $T \gg \Theta_D$ and $x \ll 1$. Expanding e^x and keeping terms up to x^2 in the expansion, we get from Eq. (5.42)

$$U = \frac{9}{8} Nk_B \Theta_D + 9Nk_B T \left(\frac{T}{\Theta_D}\right)^3 \int_0^{x_m} \left(1 - \frac{x}{2}\right) x^2 \, dx = 3Nk_B T$$

Considering one mole of the solid, we have $U = 3RT$, and the molar specific heat at constant volume is

$$C_v = \frac{\partial U}{\partial T} = 3R$$

which is Dulong–Petit's law. Thus the high-temperature values agree with the experiments and the classical theory.

At low temperatures, $T \ll \Theta_D$ and $x_m \to \infty$. The integral in the second term on the right-hand side of Eq. (5.42) is then standard and has the value of $\pi^4/15$. Equation (5.42) then reduces to

$$U = \frac{9}{8} Nk_B \Theta_D + 9Nk_B T \left(\frac{T}{\Theta_D}\right)^3 \frac{\pi^4}{15}$$

The lattice molar specific heat at constant volume is then given by

$$C_v = \left(\frac{\partial U}{\partial T}\right)_v = \frac{12\pi^4}{5} R \left(\frac{T}{\Theta_D}\right)^3$$

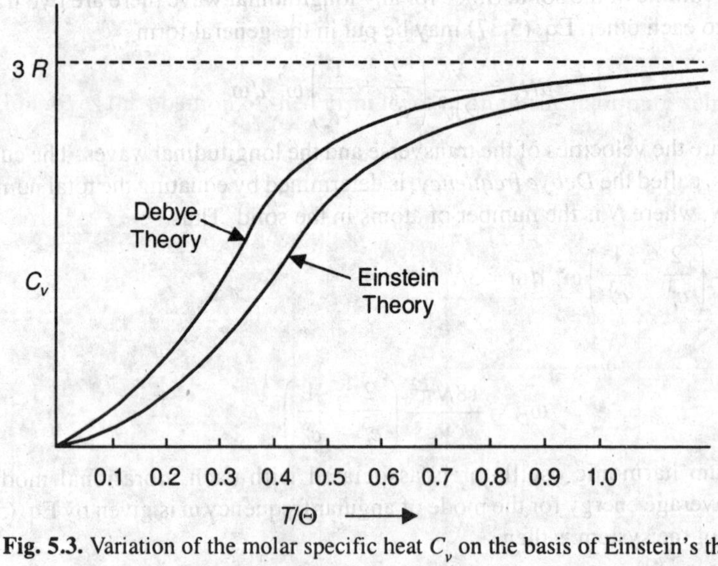

Fig. 5.3. Variation of the molar specific heat C_v on the basis of Einstein's theory (C_v versus T/Θ_E) and Debye's theory (C_v vs. T/Θ_D).

This is the celebrated "Debye T^3 law". The T^3 variation of C_v is observed for many solids at low temperatures ($T < \Theta_D/20$). In metals there is an additional contribution to the specific heat due to the free electrons. This free-electron contribution is negligible at temperatures above about 5 K. Below this temperature, the free electron contribution dominates and gives a linear variation of the specific heat with temperature.

The general expression for C_v at any temperature T can be obtained from Eq. (5.42):

$$C_v = \left(\frac{\partial U}{\partial T}\right)_v = 9Nk_B \left(\frac{T}{\Theta_D}\right)^3 \int_0^{x_m} \frac{x^4 e^x \, dx}{(e^x - 1)^2} = 9R \left(\frac{T}{\Theta_D}\right)^3 \int_0^{x_m} \frac{x^4 e^x \, dx}{(e^x - 1)^2}$$

The integral on the right-hand side of the above equation is a function of $x_m = \hbar\omega_D/k_B T = \Theta_D/T$. Thus we can write

$$C_v = 3RF_D(\Theta_D/T),$$

where the *Debye function* F_D is

$$F_D(\Theta_D/T) = 3\left(\frac{T}{\Theta_D}\right)^3 \int_0^{\Theta_D/T} \frac{x^4 e^x \, dx}{(e^x - 1)^2}$$

To obtain C_v at different temperatures, the integral in the above equation is evaluated numerically for different values of $x_m (= \Theta_D/T)$.

The Debye temperature Θ_D can be calculated by using the values of the velocity of sound waves in the solid in Eq. (5.40). However, the best procedure is to fit the calculated results to the experimentally observed C_v values by treating Θ_D as an adjustable parameter. For most materials, Θ_D generally lies in the range 100 – 500 K. The temperature variations of C_v on the basis of Einstein's theory and Debye's theory are shown in Fig. 5.3.

COMPARISON BETWEEN EINSTEIN'S AND DEBYE'S THEORIES

(a) Similarities

(i) In both theories, the crystal atoms are regarded as quantum oscillators.

(ii) The energy distribution of the atomic oscillators follows Maxwell – Boltzmann statistics.

(iii) Both theories show that the molar specific heat at constant volume is given by the Dulong-Petit value of $3R$ at high temperatures, but the specific heat drops off at low temperatures.

(b) Dissimilarities

(i) Einstein's theory assumes that the atoms of the solid vibrate independentaly with the same angular frequency ω. The theory is characterized by Einstein temperature $\Theta_E = \hbar\omega/k_B$. For most solids Θ_E lies in the range 100 – 300 K.

In Debye's theory, the atomic oscillators are assumed to be coupled together, producing elastic waves with a range of frequencies from 0 up to a cut off frequency ω_D, called the Debye frequency. The Debye temperature, given by $\Theta_D = \hbar\omega_D/k_B$, lies in the range 100 – 500 K for most solids.

(ii) At low temperatures, the specific heat, according to Einstein's theory, varies as $T^{-2} \exp(-\Theta_E/T)$. Thus the specific heat falls off faster than the experimental results which show that the specific heat for insulators varies as T^3 as the temperature T approaches zero.

Debye's theory predicts the T^3 variation of the specific heat at low temperatures in agreement with experiments. Generally, Einstein's theory gives a lower value of the specific heat than Debye's theory. The latter theory includes the low frequency modes which have higher vibrational energies at low temperatures and so account for the larger specific heat predicted by the Debye model.

Limitations of Debye's theory

The limitations of the Debye theory are the following :-

(i) The Debye theory, given above, assumes that the cutoff frequency ω_D is the same for both the transverse and the longitudinal waves. Actually, since c_t and c_l are unequal, two different values of ω_D should be used for the transverse and the longitudinal modes. When this is done, the agreement with experiments improves in some cases.

(ii) The Debye calculation assumes a continuum representation of a crystal and neglects the dispersion of the waves, *i.e.*, assumes that the phase velocity of the waves does not vary with wavelength. Allowing for the dispersion of the waves leads to a complicated frequency distribution of the modes of vibration.

(iii) In the Debye theory, the solid is assumed to contain identical atoms. In crystals where the

5.4. VIBRATIONS OF A ONE-DIMENSIONAL LATTICE

The possible modes of vibration of a crystal can be studied by first considering an infinitely long one-dimensional array of identical atoms. Let m be the mass of each atom and a be the separation between any two consecutive atoms under equilibrium conditions (Fig. 5.4). We consider only the nearest-neighbour interactions and assume the validity of Hooke's law. If the displacements of the $(n-1)$th, nth, and $(n+1)$th atoms from their equilibrium positions be x_{n-1}, x_n and x_{n+1} respectively, then the equation of motion of the nth atom is

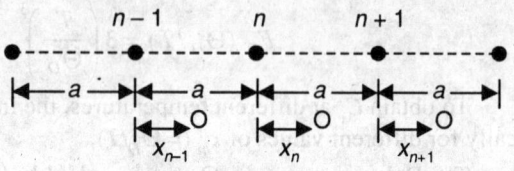

Fig. 5.4. An infinite linear lattice. The solid circles are equilibrium positions and the open circles are the displaced positions of the vibrating atoms.

$$m \frac{d^2 x_n}{dt^2} = -\beta(x_n - x_{n-1}) - \beta(x_n - x_{n+1})$$
$$= \beta(x_{n+1} + x_{n-1} - 2x_n), \qquad ...(5.43)$$

where β represents the nearest-neighbour force constant. The solution to Eq. (5.43) is obtained by putting

$$x_n(t) = A e^{i(qna - \omega t)} \qquad ...(5.44)$$

where A is a constant, na is the equilibrium position of the nth atom with respect to the origin, and q is the wave vector:

$$q = \frac{\omega}{v} = \frac{2\pi}{\lambda}, \qquad ...(5.45)$$

v being the phase velocity of the wave and λ, the wavelength.

Substitution of Eq. (5.44) into Eq. (5.43) yields

$$m\omega^2 = -\beta(e^{-iqa} + e^{iqa} - 2) = 4\beta \sin^2\left(\frac{qa}{2}\right), \qquad ...(5.46)$$

or,

$$\omega = \omega_{max} \left| \sin\left(\frac{qa}{2}\right) \right|, \qquad ...(5.47)$$

where

$$\omega_{max} = 2\left(\frac{\beta}{m}\right)^{1/2} \qquad ...(5.48)$$

is the maximum value of the angular frequency. Since only the positive values of the angular frequency ω are physically significant, the magnitude of $\sin(qa/2)$ is taken in Eq. (5.47).

The variation of the frequency ω with the wave vector q, as given by Eq. (5.47), is represented by the solid curve in Fig. 5.5. Note that for a continuous string, the frequency ω is equal to qv, i.e. the variation of ω with q would be linear, as shown by the broken lines in Fig. 5.5.

Equation (5.47) shows that when $qa \ll 1$, $\omega \approx \omega_{max} qa/2$, i.e. ω becomes proportional to q. Thus the array of atoms would behave as a continuous string only when $qa < 1$, i.e. only when the wavelength is

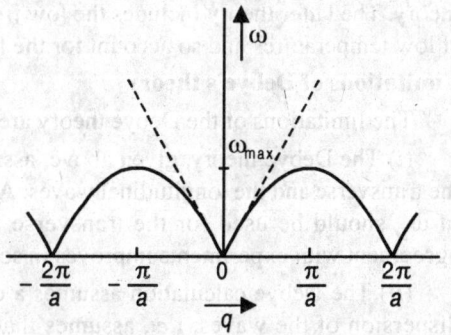

Fig. 5.5. ω as a function of q. The solid curve follows Eq. (5.47). The broken lines are for a continuous string.

Specific Heat of Solids and Lattice Vibrations

much larger than the interatomic spacing. The assumption of a continuoum representation of the crystal, as made in the Debye theory, is thus not correct in the high frequency regime.

Interestingly, there exists a maximum frequency ω_{max} that can propagate through the array of atoms. The array thus behaves as a *lowpass filter* allowing transmission of frequencies in the range from zero to ω_{max}. The value of ω_{max} is about 10^{13} rad/s.

Note that if q in Eq. (5.47) is replaced by q_n given by $q_n = q + 2\pi n/a$ where $n = \pm 1, \pm 2, ...$, then the values of the frequency are unaltered. Equation (5.44) also shows that the solutions are the same for q and q_n. Thus the state of vibration remains unchanged when q is shifted by $2\pi n/a$. It is thus convenient to consider values of q within a range of $2\pi/a$. Ordinarily the values of q within the range $-\dfrac{\pi}{a} \ll q \ll \dfrac{\pi}{a}$ are chosen. This range is known as the *first Brillouin zone*. The positive values of q represent waves moving in one direction while the negative values of q represent waves moving in the opposite direction.

BOUNDARY CONDITIONS

We shall now consider for a finite lattice the boundary conditions which lead to the possible modes of vibrations. These conditions may be put in two ways. One way of introducing the boundary conditions was proposed by Born and von Karman; these are known as *periodic* or *cyclic boundary conditions*. The other form of boundary conditions gives stationary waves. We shall discuss below these two types of boundary conditions.

(i) Periodic or cyclic boundary conditions

Think for the moment of a circularly shaped chain of atoms with interatomic spacing a. Suppose that the number of atoms in the chain is N, where $N \gg 1$. The length of the chain is $L = Na$. Labelling the atoms $1, 2, 3, ..., N$, going around the circle, the boundary condition becomes

$$x_n(t) = x_{n+N}(t) \qquad ...(5.49)$$

The reason for writing Eq. (5.49) is that the subscripts n and $n + N$ denote the same atom. Using Eq. (5.49) in Eq. (5.44) we get

$$e^{iqna} = e^{iq(n+N)a}$$

or
$$e^{iqNa} = 1$$

or
$$q = \dfrac{2\pi s}{Na} = \dfrac{2\pi}{L} s, \qquad ...(5.50)$$

where s is an integer. Considering the first Brillouin zone q is restricted in the region between $-\pi/a$ and $+\pi/a$. Thus the allowed values of s are

$$s = \pm 1, \pm 2, \pm 3, ..., \pm N/2. \qquad ...(5.51)$$

We have discarded the value $s = 0$ since this leads to $q = 0$, which means that all the atoms are at rest. The total number of different s or q values is therefore equal to N. Thus there are as many modes of vibrations as there are *vibrating atoms*. As for each value of q we have a value of the frequency ω, the frequency spectrum has N discrete lines. Since N is large these lines are close enough to form a quasicontinuous spectrum.

In practice, we are interested in a linear chain of atoms rather than a circular chain. However, since $N \gg 1$ the boundary condition (5.49) can also be used for a linear chain. For example, an infinite one-dimensional lattice may be divided into sections of length $L = Na$. As $N \gg 1$, each atom will feel the same configuration, the atomic interactions being confined to very small lengths. Consequently, each linear section will have properties identical to that of a circular chain of length L.

The length of the Brillouin zone is $2\pi/a$ along the q-axis. The number of q values contained in this length is N. Therefore, in a length δq of the q-axis the number of possible q values would be

$\dfrac{N}{(2\pi/a)}\delta q = \dfrac{Na}{2\pi}\delta q = \dfrac{L}{2\pi}\delta q$. In three dimensions, the number of possible q-values in the interval δq, is $\dfrac{V}{(2\pi)^3}\delta q$, where V is the crystal volume.

(ii) Boundary conditions giving stationary waves

In an array of $(N + 1)$ identical atoms, let the two atoms at the ends be fixed so that the remaining $(N - 1)$ atoms are free to move. We now have two waves propagating in two opposite directions, so that

$$x_n(t) = (Ae^{iqna + i\alpha_1} + Be^{-iqna + i\alpha_2})e^{-i\omega t}, \qquad \ldots(5.52)$$

where A and B are the amplitudes and α_1, α_2 are the phase angles. The boundary conditions will be that the displacements of the end atoms are zero for all values of time t, i.e. $x_n(t) = 0$ at all t for $n = 0$ and N. The first condition, when used in Eq. (5.52), gives $A = -B$ and $\alpha_1 = \alpha_2$. Since the phase angles are equal we can put $\alpha_1 = \alpha_2 = 0$. Considering the real part of the solution we get

$$x_n(t) = 2A \sin qna \sin \omega t. \qquad \ldots(5.53)$$

Equation (5.53) represents a standing wave. The second boundary condition, when used in Eq. (5.53), gives

$$\sin qNa = 0$$

so that

$$q = \dfrac{\pi}{Na}r, \qquad \ldots(5.54)$$

where r is an integer. The value $r = 0$ has not been considered since it gives $q = 0$, which means that the atoms do not vibrate at all. The values of q are positive and are confined here in the range from 0 to π/a. The maximum value of q, i.e. π/a is obtained when $r = N$, but this value has also to be discarded for the same reason as for $r = 0$. Thus the possible values of r are

$$r = 1, 2, 3, \ldots, (N-1). \qquad \ldots(5.55)$$

This shows that the number of possible modes of vibrations, i.e. the number of q values is equal to $(N - 1)$, the number of vibrating atoms. Thus the standing wave picture leads to the same conclusion as obtained in the running wave picture by applying cyclic boundary conditions.

5.5. Vibrational Mode as a Linear Harmonic Oscillator

We shall now show that for the one-dimensional lattice, the energy for a particular mode of vibration is the same as that of a harmonic oscillator of the same frequency.

The energy of a one-dimensional harmonic oscillator with mass M and angular frequency ω is given by:

$$E = \dfrac{p_x^2}{2M} + \dfrac{1}{2}M\omega^2 x^2, \qquad \ldots(5.56)$$

where p_x is the momentum and x is the displacement from the mean position. Since $p_x = M\dfrac{dx}{dt}$, Eq. (5.56) can be put in the form

$$E = \dfrac{1}{2}M\left(\dfrac{dx}{dt}\right)^2 + \dfrac{1}{2}M\omega^2 x^2. \qquad \ldots(5.57)$$

The first term on the right-hand side of Eq. (5.57) gives the kinetic energy E_{ko} and the second term gives the potential energy E_{po}. Writing $x = A \sin \omega t$ we obtain

$$E_{ko} = \dfrac{1}{2}M\left(\dfrac{dx}{dt}\right)^2 = \dfrac{1}{2}MA^2\omega^2 \cos^2 \omega t, \qquad \ldots(5.58)$$

and

$$E_{po} = \dfrac{1}{2}M\omega^2 x^2 = \dfrac{1}{2}M\omega^2 A^2 \sin^2 \omega t. \qquad \ldots(5.59)$$

Specific Heat of Solids and Lattice Vibrations

We now consider a finite one-dimensional lattice vibrating in a mode corresponding to a standing wave

$$x_n(t) = C \sin qna \sin \omega t. \quad ...(5.60)$$

The kinetic energy of the atoms in this vibrational mode is

$$E_k = \frac{1}{2} M \sum_n \left(\frac{dx_n}{dt}\right)^2 = \frac{1}{2} m C^2 \omega^2 \cos^2 \omega t \sum_n \sin^2 qna \quad ...(5.61)$$

Here m is the mass of each atom and the summation Σ extends over all the atoms.

From Eq. (5.43) we find that the total force on the nth atom is $\beta(x_{n-1} + x_{n+1} - 2x_n)$. Hence, if $E_p(x_0, x_1, ... x_n, ...)$ is the potential energy of the system in the vibrational mode q, then

$$-\frac{\partial E_p}{\partial x_n} = \beta(x_{n-1} + x_{n+1} - 2x_n)$$

or, $$E_p(x_0, x_1, ... x_n, ...) = \frac{1}{2} \beta \sum_n (2x_n^2 - x_n x_{n+1} - x_n x_{n-1}) \quad ...(5.62)$$

Using Eq. (5.60) in Eq. (5.62) we get

$$E_p = 2C^2 \beta \sin^2\left(\frac{qa}{2}\right) \sin^2 \omega t \sum_n \sin^2 qna$$

$$= \frac{C^2}{2} m\omega^2 \sin^2 \omega t \sum_n \sin^2 qna, \quad ...(5.63)$$

where Eq. (5.46) has been used.

Note that E_k given by (5.61) is identified with the kinetic energy E_{ko} of a simple harmonic oscillator given by Eq. (5.58), and E_p given by Eq. (5.63) is identified with E_{po} given by Eq. (5.59), if

$$C^2 m \sum_n \sin^2 qna = A^2 M. \quad ...(5.64)$$

Equation (5.64) establishes the equivalence of the vibrational mode and a harmonic oscillator.

The modes of vibrations are quantised and each quantum is known as a *phonon*. A phonon can be treated as a particle with energy $\hbar\omega_q$ and momentum $\hbar\omega_q/v_s$, where ω_q is the angular frequency for mode q travelling with velocity v_s. Phonons are indistinguishable particles and obey the Bose-Einstein statistics.

5.6. Vibrational Modes of A Crystal

The theory given above can be extended to actual crystals. For a linear atomic chain, both longitudinal and transverse waves can exist. The transverse modes consist of two independent vibrations in different planes. In an actual crystal, therefore, two transverse branches which may or may not coincide, occur. Also the longitudinal vibrations are there and the velocity c_l for the long-wavelength longitudinal waves in a crystal will be different from the corresponding velocity c_t for the transverse waves. Hence there will be different longitudinal and transverse branches of the ω verses q curves of a crystal (see Fig. 5.6). In the long-wavelength limit, *i.e.* for $q \to 0$, the waves travel with the velocity of sound of corresponding polarisation; these branches are consequently called *acoustic modes*.

If the atoms in the linear chain are not identical, *e.g.*, if the alternate atoms in the array have different masses, then, in addition to acoustic modes, another branch in the $\omega - q$ plot appears. The frequencies for this branch are higher and for $q \to 0$ the

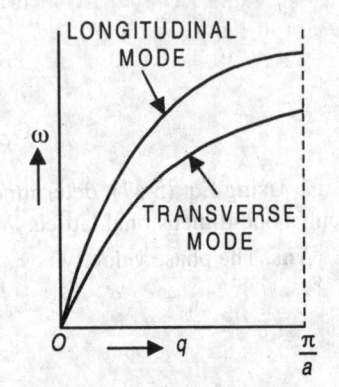

Fig. 5.6. ω versus q curves for the transverse and longitudinal modes.

frequency for this branch approaches an angular frequency $\omega_0 \neq 0$. The frequency ω_0 falls in the infra-red range, so that this new mode of vibration is called *optical mode*. The optical modes may be transverse as well as longitudinal. There is a range of forbidden frequencies between the acoustic and optic modes (Fig. 5.7).

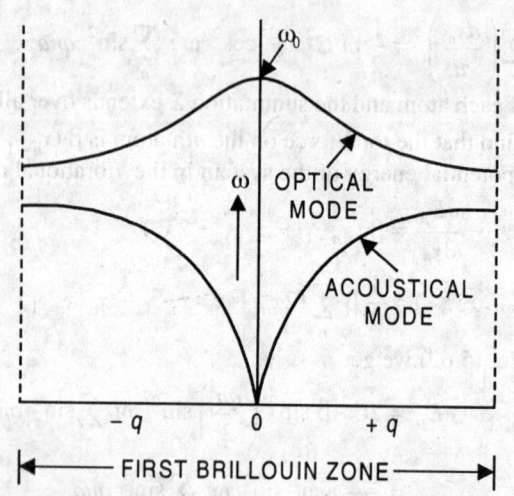

Fig. 5.7. Acoustic and optic vibrational modes for a linear array containing alternate atoms of different masses.

5.7. WORKED-OUT PROBLEMS

1. The velocities of longitudinal and transverse waves in aluminium are 6374 and 3111 ms^{-1}, respectively. Calculate the Debye frequency and the Debye temperature for aluminium. (Assume that there are 6.02×10^{28} atoms per m^3 in aluminium).

Ans. We know that

$$\omega_D^3 = \frac{18N\pi^2}{V}\left(\frac{2}{c_t^3} + \frac{1}{c_l^3}\right)^{-1}$$

Here $N/V = 6.02 \times 10^{28}$, $c_t = 3111$ m/s, and $c_l = 6374$ m/s. Hence

$$\omega_D^3 = 18 \times 6.02 \times 10^{28} \times (3.1416)^2 \left[\frac{2}{(3111)^3} + \frac{1}{(6374)^3}\right]^{-1}$$

$$= 1.5215 \times 10^{41}$$

$$\therefore \quad \omega_D = 5.338 \times 10^{13} \text{ rad/s}$$

or, $\quad v_D = \dfrac{\omega_D}{2\pi} = 8.4956 \times 10^{12}$ Hz.

or, $\quad \Theta_D = \dfrac{\hbar \omega_D}{2\pi} = \dfrac{1.054 \times 10^{-34} \times 5.338 \times 10^{13}}{1.38 \times 10^{-23}} = 407.6$ K.

2. Using Eq. (5.47), determine the phase velocity and the group velocity of the wave motion along a one-dimensional lattice. What happens to the group velocity when $qa = \pm \pi$?

Ans. The phase velocity is

$$v_p = \frac{\omega}{q} = \frac{\omega_{max}}{q}\left|\sin\left(\frac{qa}{2}\right)\right| = v_0 \left|\frac{\sin\left(\dfrac{qa}{2}\right)}{\dfrac{(qa)}{2}}\right|$$

where $\quad v_0 = (a/2)\omega_{max}$

The group velocity is

$$v_g = \frac{d\omega}{dq} = v_0 \left|\cos\left(\frac{qa}{2}\right)\right|.$$

v_0 is the long-wavelength limit of both v_p and v_g. When $qa = \pm \pi$, $v_g = 0$, and the phase of vibration of neighbouring atoms differ by π radians corresponding to a standing wave.

3. If the Einstein temperature for a material is 157 K, find the value of C_v, for that material at 100 K in calorie per mole per K using Einsteins's formula. Also, calculate Einstein's frequency.

(cf. N. Beng. U. 2001)

Ans. Einstein's formula for C_v is

$$C_v = 3R \frac{x^2 e^x}{(e^x - 1)^2}$$

where $x = \Theta_E/T$. Here $\Theta_E = 157$ K and $T = 100$ K. Therefore, $x = 157/100 = 1.57$. Since $R = 1.99$ cal. per mole/K, we have

$$C_v = \frac{3 \times 1.99 \times (1.57)^2 \, e^{1.57}}{(e^{1.57} - 1)^2} = 4.88 \text{ cal. mol}^{-1}.\text{K}^{-1}$$

Einstein's frequency is

$$\nu = \frac{k_B \Theta_E}{h} = \frac{1.38 \times 10^{-23} \times 157}{6.6 \times 10^{-34}} = 3.28 \times 10^{12} \text{ Hz}$$

4. The molar specific heat of a solid at constant volume is 2.77 J. K^{-1} mol^{-1} at 36.8 K. Determine the Debye temperature of the solid.

Ans. Since the given temperature ($T = 36.8$ K) is quite low, we can apply Debye's T^3 law:

$$C_v = \frac{12\pi^4}{5} R \left(\frac{T}{\Theta_D}\right)^3$$

where Θ_D is the Debye temperature. Substituting the numerical values we get

$$2.77 = \frac{12 \times (3.1416)^4}{5} \times 8.31 \times \left(\frac{36.8}{\Theta_D}\right)^3$$

whence
$$\Theta_D = 327 \text{ K.}$$

QUESTIONS

1. Describe the quantum theory of specific heat of solids emphasising the inadequacy of classical ideas. (C.U. 1983)
2. Derive the low temperature behaviour of the lattice contribution to the specific heat of a solid. (C.U. 1982)
3. State the assumptions made by Einstein in his theory of specific heat of solids. Deduce an expression for C_v of a solid according to this theory and discuss how it agrees with experiments. (C.U. 1989)
4. (a) In what way is Einstein's theory of specific heat superior to the classical theory?
   (b) Define the Einstein temperature Θ_E.
   (c) Obtain the values of the molar specific heat when the temperature T is much larger than Θ_E, and when T is much less than Θ_E.
   (d) Compare between Einstein's and Debye's theories of specific heat of solids. (cf. C.U. 2008)
5. (a) Why is the Debye theory of specific heat more acceptable than the Einstein theory?

(b) Assuming that in a solid of volume V, the number of modes of vibration of atomic oscillators having angular frequencies between ω and $\omega + d\omega$ is given by $dN_s = V\omega^2 \, d\omega/(2\pi^2 v^3)$, where v is the velocity of sound in the solid, obtain the low-temperature behaviour of the lattice contribution to the specific heat on the basis of Debye's theory. **(cf. CU 2002,04)**

(c) What is the Debye temperature?

6. (a) What are the assumptions and the weaknesses of the Debye theory?

(b) Explain whether the addition of the zero-point energy term affects the Einstein and the Debye results of specific heat.

(c) Show that the zero-point energy of one mole of a solid crystal according to Debye's theory is $\frac{9}{8} R\Theta_D$.

7. (a) Obtain an expression for the frequency of vibration of an infinite chain of identical atoms as a function of the wave vector.

(b) Show that the chain behaves as a low pass filter.

(c) What is the first Brillouin zone? Explain its significance.

8. What are periodic boundary conditions and the boundary conditions leading to the formation of stationary waves? Considering a linear chain of atoms show that the number of possible modes of vibrations is equal to the number of vibrating atoms.

9. Discuss qualitatively the concept of phonons for elastic vibrations in a solid. **(C.U. 1997)**

10. Establish the equivalence between a vibrational mode of a linear atomic chain and a harmonic oscillator.

11. Show that at very low temperatures the internal energy of a vibrating lattice following Debye's theory is proportional to T^4. **(cf. C.U. 1985)**

12. At very low temperatures, the specific heat of rock salt is given by the following relationship according to Debye's T^3 law:

$$C_v = A\left(\frac{T}{\theta_D}\right)^3,$$

where $A = 464$ cal.mol^{-1} K^{-1} and $\theta_D = 281$ K. Calculate the quantity of heat necessary to raise the temperature of 3 moles of rock salt from 10 to 50 K.

$$\left[\text{Hint. The heat required, } Q = \frac{3A}{\theta_D^3} \int_{10}^{50} T^3 \, dT\right]$$

[Ans. 97.9 Cal.]

13. The Debye temperature of silver is 210 K. Assuming Debye's T^3 law, find the molar specific heat of silver at constant volume at 20 K. Taking the Einstein temperature the same as the Debye temperature, calculate also the molar specific heat at 20 K using Einstein's formula. Comment on the results by comparing them with the experimental value of 1.67 J K^{-1} mol^{-1}.

[Ans. 1.68 J K^{-1} mol^{-1}, 0.076 JK^{-1} mol^{-1}]

14. If c is the velocity of sound in a solid crystal of density ρ and atomic weight M, show that the Debye frequency (in Hz) of the solid is

$$v_D = c\left(\frac{3N_A \rho}{4\pi M}\right)^{1/3}$$

where N_A is Avogadro's number.

Using this relationship, determine the Debye temperature of gold if that of copper having the same structure is 348 K. Given : velocity of sound in gold is 2100 m/s and that in copper is 3800 m/s; the density of gold is 19.3 g/cm^3 and that of copper is 8.96 g/cm^3; atomic weight of gold is 197 and that of copper is 63.54.

[**Hint.** Take $c_l = c_t = c$ in Eq. (5.40)] [**Ans.** 170 K]

15. The Debye temperatures of KCl and NaCl having the same structure are 230 K and 281 K, respectively. The molar specific heat of KCl at 3 K is 6.26 mJ/mol/K. Find the same for NaCl at 3 K and 5 K. [**Ans.** 3.43 and 15.9 mJ/mol/K]

Chapter 6

Dielectric Properties of Solids

6.1. THE STATIC DIELECTRIC CONSTANT

A most useful theorem in electrostatics is Gauss's theorem. It states that in SI units the total electric flux φ emerging from a closed surface is equal to the total charge enclosed by the surface. In mathematical form,

$$\varphi = \oint_s \vec{D} \cdot d\vec{s} = \int_V \rho dV, \qquad \ldots(6.1)$$

where $\vec{D}$ is the *electric displacement vector* or the *flux density*, $d\vec{s}$ is the vector surface element directed outward at the point where $\vec{D}$ is considered, ρ is the volume charge density, and dV is the volume element. The integral $\oint_s \vec{D} \cdot d\vec{s}$ represents a surface integral over the closed surface S enclosing the volume V. In SI units, D is expressed in coulomb/m².

The electric displacement $\vec{D}$ is related to the *electric field strength* $\vec{E}$ at that point through the relationship

$$\vec{D} = \varepsilon_0 \varepsilon_r \vec{E}, \qquad \ldots(6.2)$$

where $\varepsilon_0 = 8.854 \times 10^{-12}$ farad m⁻¹, is called the *permittivity of free space* and ε_r is the *relative permittivity* or the *dielectric constant* of the material. ε_0 is a fundamental conversion factor; it depends on the particular system of units (in this case, the SI). The quantity ε_r depends on the atomic structure of the material and is dimensionless. For free space, *i.e.*, vacuum, $\varepsilon_r = 1$. For all other materials ε_r is larger than unity.

Equation (6.2) holds for isotropic materials for which ε_r has the same value in all directions. If ε_r depends on the direction of measurement, $\vec{D}$ and $\vec{E}$ may not be parallel and the dielectric constant is a tensor quantity. Here we are concerned only with isotropic materials.

The dielectric constant of a material can be determined experimentally by using the material as a dielectric between the plates of a parallel-plate capacitor. If A is the area of the plate and d is the separation of the plates, the capacitance C is given by

$$C = \varepsilon_0 \varepsilon_r \frac{A}{d} \qquad \ldots(6.3)$$

If C_0 is the capacitance when the space between the plates is evacuated, we have

$$C_0 = \varepsilon_0 \frac{A}{d} \qquad \ldots(6.4)$$

From Eqs. (6.3) and (6.4) we obtain

$$\varepsilon_r = \frac{C}{C_0} \qquad \ldots(6.5)$$

Thus ε_r can be obtained experimentally by determining the capacitances C and C_0.

Dielectric Properties of Solids

6.2. Dipole Moment and Polarisation

These two are important concepts leading to a correlation between ε_r and the atomic or microscopic properties of the material.

The electric dipole moment of a neutral system of i number of point charges is a vector quantity $\vec{\mu}$ defined by

$$\vec{\mu} = \sum_i Q_i \vec{r_i} \qquad \ldots(6.6)$$

where $\vec{r_i}$ is the position-vector of the charge Q_i. The unit of μ is coulomb metre.

Consider two point charges $-Q$ and $+Q$, separated by a distance d. If $\vec{r_1}$ and $\vec{r_2}$ are the position-vectors with respect to the origin O of the charges $-Q$ and $+Q$, then according to Eq. (6.6) the dipole moment of the system is

$$\vec{\mu} = -Q\vec{r_1} + Q\vec{r_2} = Q(\vec{r_2} - \vec{r_1}) = Q\vec{d}, \qquad \ldots(6.7)$$

where $\vec{d}$ is the vector distance from $-Q$ to $+Q$ (see Fig. 6.1). The magnitude of the dipole moment is Qd, where d is the separation between the charges. In the simplest case, an electric dipole or simply a dipole is a system of two equal point-charges of opposite sign, separated by a distance which is small compared to the distance of the observer.

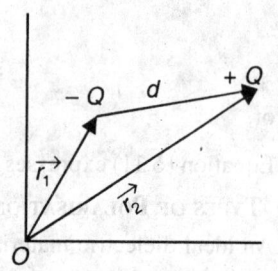

Fig. 6.1. Electric dipole consisting of two point charges $-Q$ and $+Q$.

To introduce the concept of polarisation, we consider a homogeneous isotropic dielectric placed in a homogeneous electric field $\vec{E}$ between two charged parallel plates. If ε_r is the relative permittivity of the material, the flux density is $\vec{D} = \varepsilon_0 \varepsilon_r \vec{E}$. Let us make a small cavity in the dielectric, as shown in Fig. 6.2. The volume of the cavity is $\Delta x \, \Delta y \, \Delta z$, where Δx is perpendicular to the plates. The electric field will be homogeneous in the presence of the cavity if $E_i = E_0 = E$, where E_i and E_0 are respectively the field-strengths inside and outside the cavity. If D_i and D_0 are the flux-densities inside and outside the cavity, then $E_i = D_i/\varepsilon_0$ and $E_0 = D_0/(\varepsilon_0 \varepsilon_r)$. Therefore, for homogeneous fields we must have

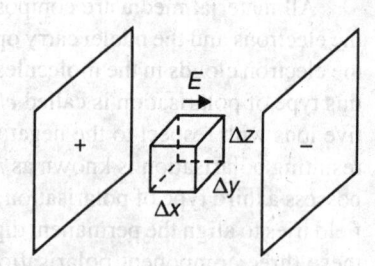

Fig.6.2. Model for calculating the polarisation of dielectric.

$$\frac{D_i}{\varepsilon_0} = \frac{D_0}{\varepsilon_0 \varepsilon_r} = \frac{D}{\varepsilon_0 \varepsilon_r} \qquad \ldots(6.8)$$

Clearly, $D_0 > D_i$, i.e., the flux density is reduced at the surface while moving from outside to inside of the cavity. According to Gauss, this change in flux density must be associated with a surface electric charge. A flux density D_0 is reduced to a flux density D_i in passing across a surface when the surface carries a charge density of $-(D_0 - D_i)$ coulomb m^{-2}. In the present case, therefore, the field would be homogeneous if a negative charge $-(D_0 - D_i) \Delta y \Delta z$ is placed on the left face of the cavity and a positive charge $+(D_0 - D_i) \Delta y \Delta z$ is placed on the right face of the cavity in Fig. 6.2. This system of charges forms a dipole of dipole moment $(D_0 - D_i) \Delta y \Delta z \Delta x$. This dipole moment vector is directed from left to right in Fig. 6.2, i.e., it is parallel to the electric field $\vec{E}$. Since $D_0 = \varepsilon_r D_i$, we can write for the dipole moment

$$D_i (\varepsilon_r - 1) \Delta y \Delta z \Delta x = \varepsilon_0 (\varepsilon_r - 1) E \Delta x \Delta y \Delta z \qquad \ldots(6.9)$$

Since the electric field can also be made homogeneous by inserting the scooped-out material into the cavity, it turns out that the material which initially occupied the cavity had a dipole moment given by Eq. (6.9). The dipole moment per unit volume is called the *polarisation* $\vec{P}$ of the material. From the above discussion we find that a dielectric placed in a homogeneous electric field $\vec{E}$ has a polarisation $\vec{P}$ given by:

$$\vec{P} = \varepsilon_0 (\varepsilon_r - 1) \vec{E} \qquad \qquad ...(6.10)$$

The polarisation is expressed in coulomb m^{-2}. Equation (6.10) shows that P is proportional to E. We may write,

$$\vec{P} = \varepsilon_0 \varepsilon_r \vec{E} - \varepsilon_0 \vec{E}$$
$$= \vec{D} - \varepsilon_0 \vec{E},$$

or,
$$\vec{D} = \varepsilon_0 \vec{E} + \vec{P} \qquad \qquad ...(6.11)$$

Equation (6.11) expresses a fundamental relationship between the three vectors $\vec{D}, \vec{E}$, and $\vec{P}$.

6.3. Types of Polarisation

An ideal dielectric material carries no free charges. When such a material is subjected to an electric field a polarisation is produced in the direction of the field. When the field is removed the polarisation also vanishes. As polarisation is dipole moment per unit volume it is clear that the applied electric field must induce dipoles in the dielectric. There are also materials which contain permanent dipole moments. These dipoles are randomly distributed in the absence of an applied electric field. When an electric field is applied, these dipoles are aligned parallel to the field.

All material media are composed of molecules which contain atomic nuclei and electrons. As the electrons and the nuclei carry opposite charges, the application of an external electric field shifts the electron clouds in the molecules with respect to the nuclei. As a result the dielectric is polarised; this type of polarisation is called *electronic polarisation*. In ionic crystals, the position of the positive ions with respect to the negative ions is influenced by the application of the electric field; the resulting polarisation is known as *ionic polarisation*. Materials having permanent dipole moments possess a third type of polarisation, referred to as the *orientational polarisation*; an applied electric field tries to align the permanent dipole moments in its direction. The total polarisation is the sum of these three component polarisations. In the following sections these three components are discussed in some detail.

6.4. Electronic Polarisation

Electronic polarisation will be considered first on the basis of the classical model of the atom. Quantum mechanical model will then be taken up. Interaction between adjacent dipoles will be neglected. Thus the following treatment will be valid for gases where the interatomic distances are so large that the dipole-dipole interactions can be ignored.

(a) Classical model

Consider an atom the nucleus of which contains a positive charge Ze, where Z is the atomic number and e is the electronic charge. As the atom is electrically neutral, Z electrons would revolve round the nucleus. In the light of quantum mechanics, the electrons do not have fixed positions; instead, there is a finite probability of finding an electron at a certain distance from the nucleus. In the classical model, it is assumed that the total electron charge $-Ze$ outside the nucleus is uniformly distributed inside a sphere of radius r, where r is the radius of the atom (Fig. 6.3). The volume density of electron charge within the sphere is

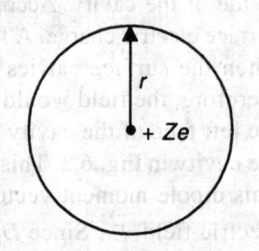

Fig. 6.3. An atom of radius r.

Dielectric Properties of Solids 275

$$\rho = -\frac{Ze}{(4/3)\pi r^3} \qquad \ldots(6.12)$$

Suppose that the atom is placed in an electric field $\vec{E}$. Since the charge of the nucleus is positive, it experiences a force in the direction of the electric field. The force exerted on the electron cloud by the electric field will be in the opposite direction since the electron charge is negative. The electric field then shifts the electron cloud with respect to the nucleus. As they are pulled apart, a Coulomb force appears between them. This Coulomb force tries to restore the electron cloud in its original position when the nucleus is at the centre of the electron cloud. In equilibrium, the Coulomb force is balanced by the force due to the applied electric field, and the nucleus remains shifted with respect to the electron cloud by a small distance d (see Fig. 6.4).

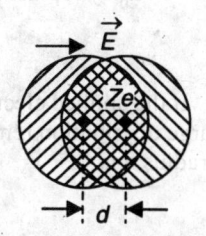

Fig. 6.4. Shifting of the electron cloud relative to the nucleus.

The force exerted by the electric field $\vec{E}$ on the electron cloud is $-Ze\vec{E}$. It follows from Gauss's theorem that only that portion of the electron cloud, which does not surround the nucleus (Ze) experiences the Coulomb force. Thus the Coulomb force is exerted only on the portion of the electron cloud contained in a sphere of radius d. The charge contained within this sphere of radius d is

$$Q = \rho\left(\frac{4}{3}\pi d^3\right) = -Ze\left(\frac{d}{r}\right)^3, \qquad \ldots(6.13)$$

where Eq. (6.12) has been used. The Coulomb force is

$$\frac{Q(Ze)}{4\pi\varepsilon_0 d^2}\vec{a} = -Ze\left(\frac{d}{r}\right)^3 \frac{Ze}{4\pi\varepsilon_0 d^2}\vec{a}, \qquad \ldots(6.14)$$

where $\vec{a}$ is a unit vector in the direction opposite to $\vec{E}$. In equilibrium,

$$ZeE = Ze\left(\frac{d}{r}\right)^3 \frac{Ze}{4\pi\varepsilon_0 d^2}$$

or, $$d = \frac{4\pi\varepsilon_0 r^3}{Ze}E \qquad \ldots(6.15)$$

The shift of the electron cloud is thus proportional to the electric field. This shift of the negative charge from the positively charged nucleus produces a dipole moment

$$\mu = Zed = 4\pi\varepsilon_0 r^3 E = \alpha_e E, \qquad \ldots(6.16)$$

where $\alpha_e = 4\pi\varepsilon_0 r^3$ is called the *electronic polarisability*. If N is the number of atoms per unit volume, the *induced* dipole moment per unit volume or the *electronic polarisation* (P_e) will be given by

$$P_e = N\mu = N\alpha_e E = 4\pi\varepsilon_0 Nr^3 E \qquad \ldots(6.17)$$

Inert gases like He, A and Ne exhibit only electronic polarisation in the presence of an applied electric field. The electronic polarisation depends on the atomic structure and is independent of temperature.

(b) Quantum mechanical model

We consider here, for the sake of simplicity, the quantum mechanical description of the hydrogen atom in the ground state. The volume charge density $\rho(r)$ of an electron at a distance r from the nucleus is obtained by multiplying the probability density uu^* of the electron by the charge $-e$, where $u(r)$ is the electron wave function expressed by Eq. (6.49) of Ch. 6, Quantum Mechanics. Thus

$$\rho(r) = -\frac{e}{\pi a_0^3}\exp(-2r/a_0) \qquad \ldots(6.18)$$

where a_0 is the first Bohr radius. As the charges are confined to very small values of r, so far as the perturbing effect of the applied electric field is concerned, we can approximate Eq. (6.18) for $r \ll a_0$ by replacing the exponential term by unity, and obtain

$$\rho(r) = -\frac{e}{\pi a_0^3} \qquad ...(6.19)$$

If the applied electric field displaces the electron cloud by a distance d, then, assuming that the charge distribution remains unchanged in the presence of the field, the charge confined in a sphere of radius d is

$$Q = \int_0^d \rho(r) \, 4\pi r^2 dr = -\frac{4e}{3a_0^3} d^3 \qquad ...(6.20)$$

The Coulomb force of attraction between the positive nucleus and the displaced electron cloud, has the magnitude

$$\left(\frac{4e}{3a_0^3} d^3\right) \frac{e}{4\pi\varepsilon_0 d^2},$$

since $Z = 1$ for the hydrogen atom. In equilibrium,

$$eE = \frac{4e}{3a_0^3} d^3 \frac{e}{4\pi\varepsilon_0 d^2}$$

or
$$d = \frac{3\pi \varepsilon_0 a_0^3}{e} E. \qquad ...(6.21)$$

A comparison of Eq. (6.21) with Eq. (6.15) shows that the factor 4 in the classical expression for d is replaced by the factor 3 in the quantum mechanical expression. This does not change the order of magnitude of the electronic polarisation.

6.5. Ionic Polarisation

In an ionic compound, the application of an external electric field displaces the positive and the negative ions, causing an ionic polarisation of the medium. Consider a typical ionic compound such as NaCl. For the sake of simplicity, assume that the intermolecular separation is large enough for the dipole-dipole interaction to be neglected, and that the field experienced by the ions is the same as the applied field.

In the compound sodium chloride, the outermost sodium electron spends most of its time in the neighbourhood of the chlorine nucleus, thereby making sodium to be electropositive and chlorine electronegative. The bond between Na^+ and Cl^- is ionic in nature.

Let d_0 be the separation between Na^+ and Cl^- ions in the sodium chloride molecule in the absence of an electric field. The applied electric field $\vec{E}$ exerts a force on the Na^+ ion in the direction of the field, and a force on the Cl^- ion in the opposite direction. The two ions are thus displaced in the opposite directions; the separation between the ions changes and a dipole moment is *induced* (Fig. 6.5). The induced dipole moment is given by an expression similar to Eq. (6.16), i.e.,

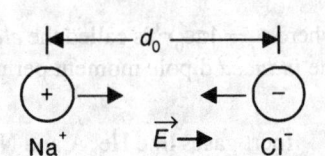

Fig. 6.5. Ionic polarisation in NaCl.

$$\mu = \alpha_i E, \qquad ...(6.22)$$

where α_i is referred to as the *ionic polarisability*. If Δx is the change in the separation between the ions caused by the field E, we have $\beta \Delta x = eE$ where β is the force constant. Thus the ionic polarisability is $\alpha_i = \mu / E = e\Delta x / E = e^2 / \beta$. If N denotes the number of molecules per unit volume, the ionic polarisation will be

$$P_i = N \alpha_i E \qquad ...(6.23)$$

Like the electronic polarisability, the ionic polarisability is independent of temperature.

Dielectric Properties of Solids

The ionic polarisability is important if the electric field is steady or slowly varying with time. If the field is rapidly varying, the ionic movements cannot follow the field variations and so the ionic polarisation is about.

Note that owing to the separation d_0 between Na$^+$ and Cl$^-$ ions in a NaCl molecule, the molecule possesses a *permanent dipole moment*, which exists even in the absence of an applied electric field. The alignment of permanent dipole moments in the presence of an applied electric field is considered below.

6.6. Orientational Polarisation

Consider a medium in which the molecules have a permanent dipole moment. Such molecules are referred to as *polar molecules*. The material consisting of such molecules is called the polar material. In the absence of an external electric field, the dipoles are oriented at random and the vector sum of the dipole moments will be zero. Thus there is no polarisation when there is no applied field.

Let an electric field $\vec{E}$ be applied to this medium containing N dipoles per unit volume at a temperature T. Assume that the permanent dipole moment of a molecule is μ_p. We again assume that the intermolecular separation is large so that the mutual interactions between the dipoles are negligible, and that the field experienced by an atom or a molecule is the applied field $\vec{E}$.

Consider a dipole moment $\vec{\mu}_p$ at an angle θ with the direction of the $\vec{E}$ field (Fig. 6.6). If the dipole contains charges $+Q$ and $-Q$ separated by a distance d, then $\mu_p = Qd$. The forces exerted on these charges by the applied field are of magnitude QE each, but oppositely directed. The components of these forces parallel to the dipole moment are of magnitude $QE \cos θ$; they act in opposite directions and induce a dipole moment on the basic dipole moment (see Sec 6.5). The components of the forces perpendicular to the dipole moment are each of value $QE \sin θ$; they form a couple and tend to rotate the dipole in the direction of the applied field.

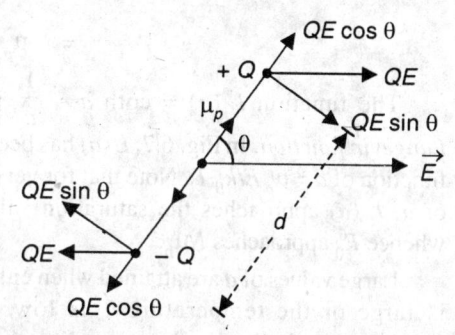

Fig. 6.6. A dipole in an external electric field.

The torque on the dipole is

$$T(θ) = (QE \sin θ) d = \mu_p E \sin θ \qquad ...(6.24)$$

In vector form,

$$\vec{T}(θ) = \vec{\mu}_p \times \vec{E} \qquad ...(6.25)$$

Setting arbitrarily the potential energy of the dipole at θ = 90° to zero, the potential energy of the dipole at an angle θ is given by

$$P.E. = \int_{\pi/2}^{θ} T(θ) \, dθ = \int_{\pi/2}^{θ} \mu_p E \sin θ \, dθ = -\mu_p E \cos θ = -\vec{\mu}_p \cdot \vec{E} \qquad ...(6.26)$$

The potential energy is a minimum when θ = 0, *i.e.*, when the dipole moment is parallel to the field. As every system tends to have its potential energy minimum, the dipole tends to align itself in the direction of the field. This ordering effect is opposed by the thermal energy, which tends to create disorder by distributing the dipoles randomly.

According to Maxwell-Boltzmann statistics, the probability of finding a dipole at an angle θ with respect to the field direction is proportional to $\exp\left(\dfrac{\mu_p E \cos θ}{k_B T}\right)$ where k_B is the Boltzmann constant. Also, the probability for a dipole to lie at angles between θ and θ + dθ is proportional to the solid angle between θ and $\theta + d\theta$, i.e., to $2\pi \sin\theta\, d\theta$. The net probability for a dipole to lie between θ and $\theta + d\theta$ is therefore proportional to $2\pi \sin\theta \exp(\mu_p E \cos\theta/k_B T)\, d\theta$.

As the component of the dipole moment $\vec{\mu}_p$ parallel to the field $\vec{E}$ is $\mu_p \cos\theta$, the dipole moment per unit volume, i.e. the orientational polarisation P_0 is given by

$$P_0 = N \frac{\int_0^\pi \mu_p \cos\theta \sin\theta \exp(\mu_p E \cos\theta/k_B T)\, d\theta}{\int_0^\pi \sin\theta \exp(\mu_p E \cos\theta/k_B T)\, d\theta} \qquad \ldots(6.27)$$

To evaluate the integrals we put $x = a \cos\theta$, where

$$a = \frac{\mu_p E}{k_B T} \qquad \ldots(6.28)$$

and obtain

$$P_0 = \frac{Nk_B T}{E} \frac{\int_{-a}^{-a} x \exp(x)\, dx}{\int_{-a}^{a} \exp(x)\, dx} = N\mu_p \left(\coth a - \frac{1}{a}\right)$$

$$= N\mu_p L(a) \qquad \ldots(6.29)$$

The function $L(a) = \coth a - \dfrac{1}{a}$ is called the Langevin function. In Fig. 6.7, $L(a)$ has been plotted as a function of $a = \mu_p E/k_B T$. Note that for very large values of a, $L(a)$ approaches the saturation value of unity, whence P_0 approaches $N\mu_p$.

Large values of a are attained when either the field E is large or the temperature T is low. Under such conditions all the dipoles are aligned in the field direction, leading to a saturation polarisation of $N\mu_p$.

When the field is not so high and the temperature is not so low, we may make the approximation $a \ll 1$ or $\mu_p E \ll k_B T$. Then

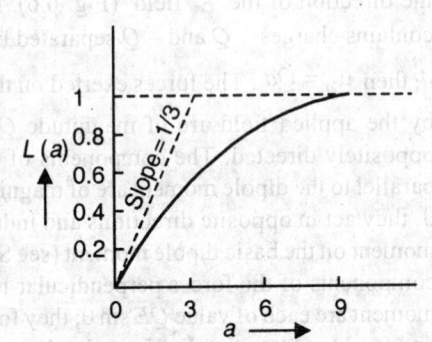

Fig. 6.7. The Langevin function $L(a)$. For $a \ll 1$, $L(a)$ has a slope of 1/3

$$L(a) = \coth a - \frac{1}{a} \approx \frac{1 + a^2/2}{a + a^3/6} - \frac{1}{a}$$

$$= \frac{a + a^3/2 - a - a^3/6}{a(a + a^3/6)} \approx \frac{a}{3}, \qquad \ldots(6.30)$$

when only the terms up to a^3 are retained. Using Eq. (6.30) we have from Eq. (6.29)

$$P_0 = N\mu_p \frac{a}{3} = \frac{N\mu_p^2 E}{3k_B T} \qquad \ldots(6.31)$$

In the case of gases, the saturation polarisation is never attained for all practical values of field and temperature. In such cases, Eq. (6.31) holds for all practical purposes. Equation (6.31) shows that P_0 is proportional to E and μ_p^2, and inversely proportional to T. The quantity μ_p is measured in Debye units, denoted by D. (1 D = 3.33×10^{-30} coulomb metre).

Equation (6.31) gives for the *orientational polarisability* or the *dipolar polarisability* α_0:

$$\alpha_0 = \frac{P_0}{NE} = \frac{\mu_p^2}{3k_B T} \qquad \ldots(6.32)$$

Dielectric Properties of Solids

6.7. Static Dielectric Constant of Gases

In the case of gases, the separation between the molecules is so large that the interactions between the dipoles are absent, and therefore the field seen by a dipole is the applied field. The expressions for the different types of polarisation developed in the preceding sections may therefore be applied for gases. The total polarisation P is the sum of electronic, ionic, and orientational polarisations, *i.e.*,

$$\vec{P} = \vec{P}_e + \vec{P}_i + \vec{P}_o$$
$$= N(\alpha_e + \alpha_i + \alpha_o)\vec{E} \qquad ...(6.33)$$

The total polarisability α is given by

$$\alpha = \alpha_e + \alpha_i + \alpha_o$$
$$= \alpha_e + \alpha_i + \frac{\mu_p^2}{3k_B T} \qquad ...(6.34)$$

Equation (6.34) is referred to as the *Langevin-Debye equation*. The polarisation is related to the dielectric constant ε_r by the equation

$$\vec{P} = \varepsilon_0 (\varepsilon_r - 1) \vec{E} \qquad ...(6.35)$$

Comparing Eqs. (6.33) and (6.35) and using Eq. (6.34) we obtain

$$\varepsilon_0 (\varepsilon_r - 1) = N\left(\alpha_e + \alpha_i + \frac{\mu_p^2}{3k_B T}\right) \qquad ...(6.36)$$

As α_e and α_i are independent of temperature, Eq. (6.36) indicates that the variation of $\varepsilon_0 (\varepsilon_r - 1)$ with $1/T$ is a straight line of slope $N\mu_p^2/3k_B$. Experimentally, the variation of $\varepsilon_0 (\varepsilon_r - 1)$ with $1/T$ for a gas is indeed found to be a straight line. The presence of the permanent dipole moments can be seen from the slope of the straight line. Since N is known from other sources, the dipole moment μ_p can be determined by finding the slope experimentally. The intercept of the straight line on the $\varepsilon_0 (\varepsilon_r - 1)$ axis gives $(\alpha_e + \alpha_i)$ [See Fig. 6.8].

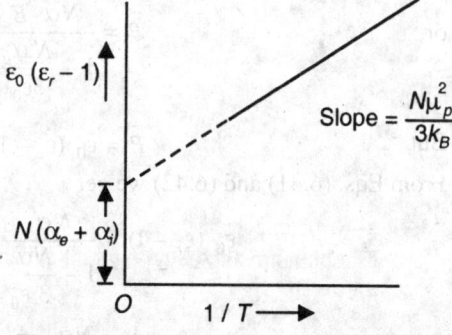

Fig. 6.8. Variation of $\varepsilon_0 (\varepsilon_r - 1)$ with $1/T$ for a gas.

The determination of μ_p in the above way leads to an interpretation of the molecular structure. Measurements for CS_2 in the gaseous state show that $\mu_p = 0$. The molecular structure of CS_2 must therefore be such that the sulphur bonds make an angle of 180° with carbon. For water vapour, μ_p is found to be 1.84 D. In the H_2O molecule, the two OH bonds must therefore make an angle different from 180° with each other.

6.8. Internal Field in Solids

In solids and liquids the distance between the molecules is small, so that a dipole sees not only the externally applied field but also the fields created by the dipoles of other particles. The actual field $\vec{E}_i$ seen by a dipole is therefore different from the externally applied field $\vec{E}$. The field $\vec{E}_i$ is called the *local field* or the *internal field*. We have to replace the field $\vec{E}$ by $\vec{E}_i$ in Eq. (6.33) to obtain

$$\vec{P} = N(\alpha_e + \alpha_i + \alpha_o)\vec{E}_i$$
$$= N\left(\alpha_e + \alpha_i + \frac{\mu_p^2}{3k_B T}\right)\vec{E}_i \qquad ...(6.37)$$

The internal field may be shown to be given by

$$\vec{E_i} = \vec{E} + \frac{\gamma}{\varepsilon_0} \vec{P},$$ ...(6.38)

where γ is called the *internal field constant*. Lorentz in 1909 found that for a cubic crystal structure $\gamma = \frac{1}{3}$, so that for such a structure

$$\vec{E_i} = \vec{E} + \frac{\vec{P}}{3\varepsilon_0}$$ ...(6.39)

which is known as the *Lorentz relation*.

6.9. STATIC DIELECTRIC CONSTANT OF SOLIDS

(*i*) We shall first consider the solids for which the polarisation is solely due to the electronic polarisation. Such materials are elements like diamond, carbon, silicon, and germanium. Here all the atoms are alike and there are no ions or permanent dipoles. Therefore $\alpha_i = \alpha_0 = 0$, and

$$\vec{P} = \vec{P_e} = N\alpha_e \vec{E_i}$$ ...(6.40)

As the crystal structures we are considering are cubic, Eq. (6.39) applies and we obtain from Eq. (6.40)

$$\vec{P} = N\alpha_e \left(\vec{E} + \frac{\vec{P}}{3\varepsilon_0} \right)$$

or,

$$\vec{P} = \frac{N\alpha_e \vec{E}}{1 - \frac{N\alpha_e}{3\varepsilon_0}}$$ ...(6.41)

But

$$\vec{P} = \varepsilon_0 (\varepsilon_r - 1) \vec{E}$$ ...(6.42)

From Eqs. (6.41) and (6.42) we get

$$\varepsilon_0 (\varepsilon_r - 1) = \frac{N\alpha_e}{1 - \frac{N\alpha_e}{3\varepsilon_0}}$$ ...(6.43)

or,

$$\frac{\varepsilon_r - 1}{\varepsilon_r + 2} = \frac{N\alpha_e}{3\varepsilon_0},$$ ...(6.44)

which is known as the *Clausius-Mossotti relationship*. This relationship helps to determine the dielectric constant ε_r from a knowledge of α_e, and *vice versa*, for elemental dielectrics. It may be mentioned that the dielectric constant of the solid materials we are considering remains constant even for frequencies in the visible region.

(*ii*) Next we consider alkali halides, such as LiF, KCl, NaI etc. Such solids have more than one type of atom, but no permanent dipoles. They exhibit electronic and ionic polarisations, but no orientational polarisation. The static dielectric sonstant in this case is given by

$$\varepsilon_0 (\varepsilon_r - 1) \vec{E} = \vec{P_e} + \vec{P_i}$$ ...(6.45)

At high frequencies in the visible region, the ionic displacements in such materials cannot follow the field variations, and so only the electronic polarisation exists. Thus the high-frequency dielectric constant $\varepsilon_{r\infty}$ is different from the static dielectric constant, and is given by

$$\varepsilon_0 (\varepsilon_{r\infty} - 1) \vec{E} = \vec{P_e}$$ ...(6.46)

Dielectric Properties of Solids 281

For LiF, the static dielectric constant is 9.27 while the high-frequency dielectric constant is 1.92.

(*iii*) Finally, we turn to solids possessing permanent dipole moments of molecules. The total polarisation here will be made up of electronic, ionic, and orientational polarisations. The calculation of internal fields in such solids is extremely complicated. Furthermore, the dipoles may not be able to rotate at all or may rotate only to a limited extent. A general quantitative theory for such solids therefore does not exist.

The temperature variation of the static dielectric constant of a dipolar solid like nitro-benzene ($C_6H_5NO_2$) is shown in Fig. 6.9. It is found that as the material freezes from the liquid to the solid state, the dielectric constant abruptly decreases to a low value. In the solid state, ε_r is independent of temperature while in the liquid state ε_r diminishes as the temperature increases. In the liquid state the dielectric constant is determined by the electronic, ionic and orientational polarisations. As the orientational contribution decreases with temperature, the dielectric constant falls with temperature in the liquid state. In the solid state, the permanent dipoles remain frozen and do not respond to the external field. This accounts for the abrupt fall of ε_r at the melting point. The contributions to the dielectric constant in the solid state thus come only from electronic and ionic polarisations. Consequently, the dielectric constant in the solid state does not vary with temperature.

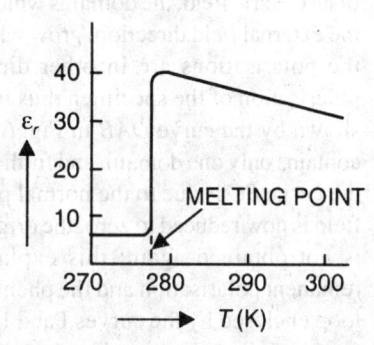

Fig. 6.9. Variation of ε_r of nitrobenzene with temperature.

In the case of solids like HCl, the permanent dipoles may contribute to polarisations. The temperature variation of ε_r in the case of HCl is shown in Fig. 6.10. The density of the material changes as one goes from the liquid to the solid state at the melting point; this causes a small abrupt increase of ε_r upon solidification. In the solid state, a decrease in temperature increases the dielectric constant, owing to orientational polarisation. The dipoles become frozen at a temperature of 100 K whence ε_r decreases abruptly.

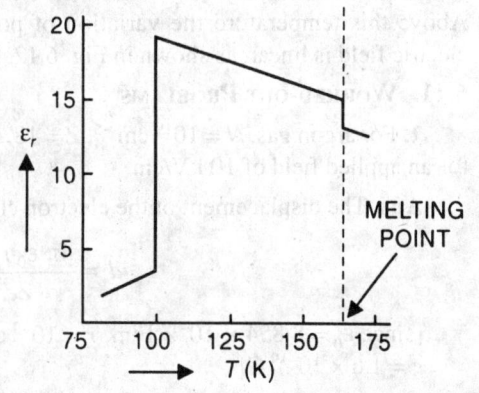

Fig. 6.10. Variation of ε_r with temperature for HCl.

6.10. Ferroelectricity

A class of materials for which the polarisation depends on the history, exhibits hysteresis effects similar to the effect in ferromagnetic materials. These materials are called *ferroelectric materials*; examples are Rochelle salt (sodium potassium tartrate tetrahydrate), dihydrogen phosphate and arsenates of alkali metals such as KH_2PO_4, and oxygen octahedron groups such as barium titanate ($BaTiO_3$).

A hysteresis loop of a ferroelectric material is shown in Fig. 6.11. If a virgin specimen of a ferroelectric material is subjected to an electric field and the field is gradually increased, the polarisation increases along the curve OABC in Fig. 6.11. When the field is decreased, it is found that at $E = 0$ the polarisation does not vanish, but a certain amount of polarisation P_r, called the *residual* or *remanent polarisation*, remains. The existence of P_r means that the material is spontaneously polarised. To make the polarisation zero, an electric field in the opposite direction has to be applied. This electric field E_c is called the *coercive field*.

The zero polarisation of a virgin specimen may be explained with the concept of *domains*. These domains are spontaneously polarised, but the direction of polarisation changes from domain to domain. In a virgin macroscopic specimen, a large number of domains exists, and because of the randomness of the polarisations of the domains, the resultant polarisation of the specimen vanishes. In the presence of an electric field, the domains which have polarisations along the external field direction, grow while the domains for which the polarisations are in other directions, shrink. The net polarisation of the specimen thus increases with the field, as shown by the curve *OAB* in Fig. 6.11. Finally the specimen contains only one domain, and further slow increase of *P* along the curve *BC* is due to the normal polarisation. If the electric field is now reduced to zero, the original domain configuration is not obtained again; this explains the existence of the remanent polarisation and the phenomenon of hysteresis. The loop enclosed by the curves I and II in Fig. 6.11, is called the *hysteresis loop*. Hysteresis means 'to lag behind' – in this case the polarisation lags behind the electric field.

Fig. 6.11. Hysteresis loop for a ferroelectric material.

The spontaneous polarisation vanishes above a temperature θ_f called the *ferroelectric Curie temperature*. Above this temperature the variation of polarisation with electric field is linear, as shown in Fig. 6.12.

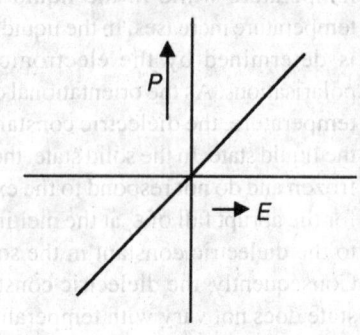

Fig. 6.12. *P-E* relationship above θ_f.

6.11. Worked-out Problems

1. For argon gas, $N = 10^{19}$ cm^{-3}, $Z = 18$, and $r = 10^{-8}$ cm; calculate the electronic polarisation for an applied field of 10 kV/cm.

Ans. The displacement of the electron cloud with respect to nucleus is

$$d = \frac{4\pi \varepsilon_0 r^3 E}{Ze} = \frac{4\pi \times 8.854 \times 10^{-12} \times 10^{-30} \times 10^6}{18 \times 1.6 \times 10^{-19}}$$

[since $\varepsilon_0 = 8.854 \times 10^{-12}$ F/m, $r = 10^{-8}$ cm $= 10^{-10}$ m, $E = 10$ kV/cm $= 10^6$ V/m, $Z = 18$, and $e = 1.6 \times 10^{-19}$ C]

or, $\qquad d = 3.865 \times 10^{-17}$ m

The electronic polarisation is

$$P_e = N\mu = NZed = 10^{25} \times 18 \times 1.6 \times 10^{-19} \times 3.865 \times 10^{-17}$$
$$= 1.1 \times 10^{-9} \text{ C/m}^2$$

2. For a certain gas molecule, the permanent dipole moment is 1.35 Debye unit. Calculate the orientational polarisability at room temperature.

Ans. The given value of μ_p is

$$\mu_p = 1.35 \text{ D} = 1.35 \times 3.33 \times 10^{-30} \text{ C.m}$$

The orientational polarisability is given by

$$\alpha_0 = \frac{\mu_p^2}{3k_B T} = \frac{(1.35 \times 3.33)^2 \times 10^{-60}}{3 \times 1.38 \times 10^{-23} \times 300}$$
$$= 1.63 \times 10^{-39} \text{ F.m}^2$$

3. Silicon has the dielectric constant 12, and the edge-length of the conventional cubic cell of silicon lattice is 5.43Å. Calculate the electronic polarisability of silicon atoms.

Ans. The electronic polarisability α_e is calculated from the Clausius-Mossotti relationship :

$$\frac{\varepsilon_r - 1}{\varepsilon_r + 2} = \frac{N\alpha_e}{3\varepsilon_0}$$

As silicon has the diamond crystal structure, there will be 8 silicon atoms in the conventional cubic cell of volume $(5.43 \times 10^{-10})^3$ m³. The number of atoms per unit volume (N) is therefore given by

$$N = \frac{8}{(5.43 \times 10^{-10})^3} = 4.997 \times 10^{28} \text{ m}^{-3}$$

Since $\varepsilon_r = 12$ and $\varepsilon_0 = 8.854 \times 10^{-12}$ F/m, we get from the Clausius-Mossotti relationship

$$\frac{11}{13} = \frac{4.997 \times 10^{28} \times \alpha_e}{3 \times 8.854 \times 10^{-12}}$$

or,
$$\alpha_e = \frac{11 \times 3 \times 8.854}{13 \times 4.997 \times 10^{40}}$$
$$= 4.5 \times 10^{-40} \text{ F.m}^2$$

QUESTIONS

1. Establish the relationship among the electric displacement, field strength, and polarisation vectors.
2. (a) What are the different contributions to the total polarization of a dielectric material ?
   (Burd. U. 1995)
   (b) Explain electronic polarisation and orientational polarisation in a dielectric. **(C.U. 2008)**
3. Explain the terms : electronic polarisation and electronic polarisability. Find an expression for electronic polarisation of a gas atom of radius R. Does the electronic polarisation vary with temperature ?
4. What are ionic polarisation and ionic polarisability? Is the ionic polarisability dependent on temperature ?
5. What are polar molecules and orientational polarisation ? Obtain an expression for the orientational polarisation neglecting dipole-dipole interactions. Discuss what happens at high and low temperatures.
6. What is Langevin-Debye equation ? Discuss how this equation may be used to obtain information on molecular structures.
7. Why does the field seen by the dipoles in a solid differ from the applied field? Derive the Clausius Mossotti relationship and name the solids for which the equation may be applied.
8. Do the alkali halides exhibit orientational polarisation ? Why does the static dielectric constant in such solids differ from the high frequency dielectric constant ?
9. Illustrate with examples how the static dielectric constant varies with temperature in the liquid and solid states for materials in which the electronic, ionic, and orientational polarisations contribute.
10. Write a note on ferroelectricity.
11. The radius of an argon gas atom is 10^{-10} m. Calculate the electronic polarisability of the atom.
    [Ans. 1.11×10^{-40} F.m²]
12. The permanent dipole moment for a certain gas molecule is 1.35 D. If there are 10^{27} gas molecules per metre³, calculate the orientational polarisation at room temperature for an applied electric field of 10kV/cm. **[Ans. 1.63×10^{-6} C/m²]**
13. The relative permittivity of germanium is 16. The edge length of the conventional cubic cell for germanium lattice is 5.65×10^{-10} m. Calculate the electronic polarisability of germanium atoms.
    [Ans. 4.96×10^{-40} F.m²]
14. Deduce an expression for the electronic polarizability of an atom on the basis of classical theory.
    (Burd. U. 1996)

Chapter 7
Magnetic Properties of Solids

In this chapter we shall consider the observed magnetic properties of solids and the theories to explain these properties. Since the concepts of magnetic dipole moment and magnetisation are essential to understand the magnetic phenomena, they will first be introduced with the help of the magnetic field theory.

7.1. Force on a Current-Carrying Conductor in a Magnetic Field

An electric charge, moving in a magnetic field, experiences a force, termed the *magnetic force* or the *Lorentz force*. The *magnetic flux density B* may be defined in terms of the Lorentz force F experienced by a charge q moving with a velocity v through the relation:

$$\vec{F} = q\vec{v} \times \vec{B}, \qquad \qquad ...(7.1)$$

B is also called the *magnetic induction*. If q is expressed in coulomb, v in m/s and B in tesla, then F is given in newton. The direction of F is normal to both $\vec{v}$ and $\vec{B}$ and is given by the direction in which a right-handed screw advances while rotated from $\vec{v}$ to $\vec{B}$.

We shall apply Eq. (7.1) to calculate the force experienced by a current-carrying conductor placed in a magnetic field. The current density J in the conductor is given by

$$\vec{J} = ne\vec{v}, \qquad \qquad ...(7.2)$$

where n is the number of conduction electrons per unit volume, e is the electronic charge and $\vec{v}$ is the drift velocity of the electrons. Consider a differential volume element of the conductor. If dA represents the area of cross-section of the element and dL its length parallel to $\vec{v}$, then the number of electrons inside the element is $n\,dA\,dL$. The force on this element in the magnetic field is the sum of the forces on all the electrons in it, *i.e.*,

$$\vec{dF} = (ndA\,dL)e\vec{v} \times \vec{B} \qquad \qquad ...(7.3)$$

Using the relation (7.2), Eq. (7.3) can be expressed as

$$\vec{dF} = (dA\,dL)\vec{J} \times \vec{B} \qquad \qquad ...(7.4)$$
$$= dV\,\vec{J} \times \vec{B},$$

since $dA\,dL = dV$, the volume of the element. The force per unit volume of the conductor is therefore given by

$$\frac{\vec{dF}}{dV} = \vec{J} \times \vec{B}, \qquad \qquad ...(7.5)$$

The total force on the conductor is the vector sum

$$\vec{F} = \int \vec{J} \times \vec{B}\,dV, \qquad \qquad ...(7.6)$$

where the integration is taken over the entire volume of the conductor.

7.2. Magnetic Dipole Moment of A Current Loop and Magnetisation

Consider a current loop in the *x-y* plane, the current *I* flowing in the counterclockwise direction (Fig. 7.1). Let $\vec{dL}$ be an elemental length on this loop at a distance $\vec{d}$ from the origin. Then

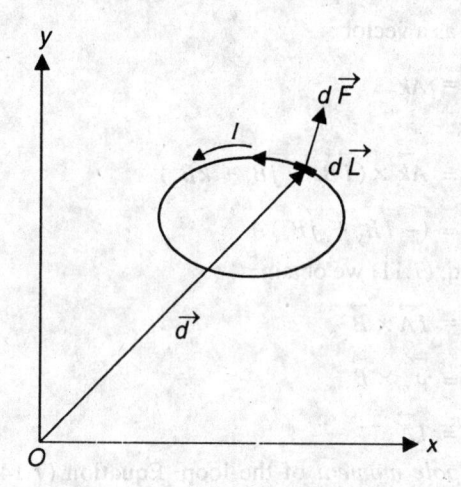

Fig. 7.1. A current loop in a uniform magnetic field.

$$\vec{dL} = \vec{i}\,dx + \vec{j}\,dy, \qquad \text{...(7.7)}$$
$$\vec{d} = \vec{i}x + \vec{j}y,$$
and
$$\vec{B} = \vec{i}B_x + \vec{j}B_y + \vec{k}B_z,$$

where $\vec{i}, \vec{j}$ and $\vec{k}$ are the unit vectors along *x*, *y* and *z* directions, respectively. B_x, B_y and B_z represent the *x*, *y* and *z* components of the magnetic flux density $\vec{B}$ in which the current loop is placed.

The force $\vec{dF}$ on $\vec{dL}$ is

$$\vec{dF} = I\vec{dL} \times \vec{B} \qquad \text{...(7.8)}$$

The torque $\vec{dT}$ about the origin on this element is

$$\vec{dT} = \vec{d} \times \vec{dF}$$

or,
$$\vec{dT} = \vec{d} \times (I\vec{dL} \times \vec{B}) \qquad \text{...(7.9)}$$

Substituting the values of $\vec{d}, \vec{dL}$ and $\vec{B}$ from (7.7) we obtain

$$\vec{dT} = (\vec{i}x + \vec{j}y) \times \{(\vec{i}\,dx + \vec{j}\,dy) \times (\vec{i}B_x + \vec{j}B_y + \vec{k}B_z)\}I$$
$$= (\vec{i}x + \vec{j}y) \times (\vec{k}B_y\,dx - \vec{j}\,dx\,B_z - \vec{k}\,dy\,B_x + \vec{i}\,dy\,B_z)I$$
$$= \{\vec{i}(B_y\,y\,dx - B_x\,y\,dy) + \vec{j}(B_x\,x\,dy - B_y\,x\,dx) - \vec{k}(B_z\,x\,dx + B_z\,y\,dy)\}I,$$
$$\text{...(7.10)}$$

since $\quad \vec{i} \times \vec{i} = \vec{j} \times \vec{j} = \vec{k} \times \vec{k} = 0\,;\; \vec{i} \times \vec{j} = \vec{k},\; \vec{j} \times \vec{k} = \vec{i}, \vec{k} \times \vec{i} = \vec{j}$

and $\quad \vec{i} \times \vec{k} = -\vec{j}\,$ etc.

The total torque T on the loop is given by the vector sum of Eq, (7.10). *i.e.*,

$$\vec{T} = \oint d\vec{T} = (-\vec{i}B_y + \vec{j}B_x)IA, \qquad ...(7.11)$$

where A represents the area of the loop.

Expressing the area A as a vector :

$$\vec{A} = A\vec{k}$$

we obtain

$$\vec{A} \times \vec{B} = A\vec{k} \times (\vec{i}B_x + \vec{j}B_y + \vec{k}B_z)$$

$$= (-\vec{i}B_y + \vec{j}B_x)A \qquad ...(7.12)$$

Using Eq. (7.12) in Eq. (7.11) we obtain

$$\vec{T} = I\vec{A} \times \vec{B} \qquad ...(7.13)$$

$$= \vec{\mu}_m \times \vec{B} \qquad ...(7.14)$$

where

$$\vec{\mu}_m = I\vec{A} \qquad ...(7.15)$$

is called the *magnetic dipole moment* of the loop. Equation (7.14) shows that the torque is independent of the choice of the axis. It is evident from Eq. (7.15) that the dipole moment due to a current loop depends upon the area of the loop and the current I flowing through the loop; the direction of the dipole moment is perpendicular to the plane of the loop and coincides with the direction of advancement of a right-handed screw rotated in the direction of I.

The magnetic dipole moment per unit volume of the material is referred to as the *magnetisation*. For a volume element δV, if the jth dipole moment is denoted by μ_{mj}, then magnetisation $\vec{M}$ is given by

$$\vec{M} = \lim_{\delta V \to 0} \frac{\sum_j \vec{\mu}_{mj}}{\delta V} \qquad ...(7.16)$$

In Eq. (7.16), $\delta V \to 0$ implies dimensions much smaller than those of the sample but much larger than the separation of the individual dipole moments.

7.3. POTENTIAL ENERGY OF A DIPOLE IN A MAGNETIC FIELD

From Eq. (7.14) it is clear that the torque on a dipole is a minimum when $\vec{\mu}_m$ and $\vec{B}$ are parallel and is a maximum when $\vec{\mu}_m$ and $\vec{B}$ are at right angles. That is why a dipole placed in a magnetic field tends to align itself in the direction of the field corresponding to the minimum of potential energy.

From Eq. (7.14) we have for the torque on the dipole

$$T(\theta) = \mu_m B \sin\theta,$$

where θ is the angle between $\vec{\mu}_m$ and $\vec{B}$. If we arbitrarily set the zero of the potential energy of the dipole at $\theta = \pi/2$, then the potential energy at an angle θ is

$$\text{P.E.} = \int_{\theta=\pi/2}^{\theta} T(\theta)d\theta = \int_{\theta=\pi/2}^{\theta} \mu_m B \sin\theta\, d\theta = -\mu_m B \cos\theta$$

$$= -\vec{\mu}_m \cdot \vec{B} \qquad ...(7.17)$$

7.4. Magnetic Field Intensity Due to a Current-Carrying Conductor

The magnetic field intensity at a point due to a current-carrying conductor can be determined by using either the Biot-Savart law or Ampere's law.

The *Biot-Savart law* states that the *magnetic field intensity* $\vec{dH}$ at any point Q due to a current I in a differential vector length $\vec{dL}$ of a conductor (Fig. 7.2) is given by

$$\vec{dH} = \frac{I\vec{dL} \times \vec{r}}{4\pi r^3}, \qquad ...(7.18)$$

where $\vec{r}$ is the distance of Q from the element $\vec{dL}$. The total magnetic field $\vec{H}$ is obtained by integrating Eq. (7.18). SI units have been used in Eq. (7.18) so that $\vec{H}$ is measured in A/m.

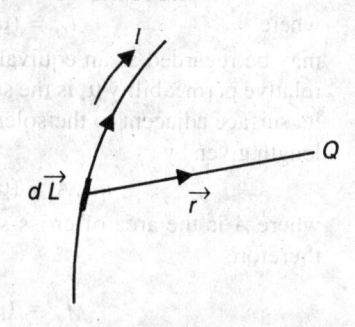

Fig. 7.2. Illustration of Biot-Savart law.

Ampere's law states that the magnetomotive force is equal to the current enclosed by the path of integration, *i.e.*,

$$\oint \vec{H}.\vec{dL} = I, \qquad ...(7.19)$$

where I is the current enclosed.

The equivalence of the Biot-Savart law and Ampere's law can be illustrated by considering an infinitely long straight wire carrying a current I. Either Eq. (7.18) or Eq. (7.19) shows that the magnetic field intensity H at a point located at a distance r from the conductor is given by

$$H = \frac{I}{2\pi r}$$

Relation between B and H

In free space, $\vec{B}$ and $\vec{H}$ are related by

$$\vec{B} = \mu_0 \vec{H}, \qquad ...(7.20)$$

where μ_0 is called the *permeability of free space* and is numerically equal to $4\pi \times 10^{-7}$ H/m.

The relationship between $\vec{B}$ and $\vec{H}$ in a magnetic material of *relative permeability* μ_r is

$$\vec{B} = \mu_r \mu_0 \vec{H} \qquad ...(7.21)$$

Note that μ_r is a dimensionless quantity.

7.5. Magnetic Susceptibility and The Relation between $\vec{B}$, $\vec{H}$ and $\vec{M}$

Let us consider a solenoid whose length is large compared to the diameter. Suppose that a steady current i flows through it and that the solenoid is in vacuum. Using the Biot-Savart law or Ampere's law it can be shown that the magnitude of the magnetic field intensity at an interior point far from the ends is given by

$$H = ni, \qquad ...(7.22)$$

where n represents the number of turns per unit length of the solenoid. The corresponding magnetic induction B is obtained from Eq. (7.20):

$$B = \mu_0 H = \mu_0 ni \qquad ...(7.23)$$

If the solenoid is filled with a medium of relative permeability μ_r, we obtain from Eq. (7.21)

$$B = \mu_0 \mu_r H = \mu_0 \mu_r ni \qquad ...(7.24)$$

Subtracting (7.23) from (7.24) we obtain

$$(\mu_0\mu_r - \mu_0) H = \mu_0 (\mu_r - 1) ni = \mu_0 ni_M,$$

where
$$i_M = (\mu_r - 1) i \qquad \ldots(7.25)$$

may be regarded as an equivalent magnetisation current. Thus the effect of placing a medium of relative permeability μ_r is the same as that obtained when a solenoidal current ni_M per unit length of its surface adjacent to the solenoid is allowed to flow. This produces a magnetic moment per unit length given by

$$ni_M A = (\mu_r - 1) HA, \qquad \ldots(7.26)$$

where A is the area of cross-section of the solenoid. The magnetic moment per unit volume is therefore

$$\vec{M} = (\mu_r - 1) \vec{H} = \chi \vec{H} \qquad \ldots(7.27)$$

where
$$\chi = \mu_r - 1 \qquad \ldots(7.28)$$

The quantity χ is called the *magnetic susceptibility* of the material.

We get from Eq. (7.27), by multiplying both sides by μ_0,

$$\mu_0 \vec{M} = \mu_0 (\mu_r - 1) \vec{H} = \mu_0 \mu_r \vec{H} - \mu_0 \vec{H}$$

or
$$\vec{B} = \mu_0 (\vec{M} + \vec{H}) \qquad \ldots(7.29)$$

Equation (7.29) gives the relationship between $\vec{B}$, $\vec{H}$ and $\vec{M}$.

For truly *linear magnetic materials* (*i.e.*, for which the relationship between B and H is *linear*), an important limitation is that the susceptibility lies in the range $-10^{-5} < \chi < 10^{-3}$ for most of these materials. In dealing with the magnetic or the electromagnetic performance of such materials this small magnetic susceptibility value is usually ignored.

The materials for which $\chi < 0$ are termed *diamagnetic*; those for which $\chi > 0$ and small, are termed *paramagnetic*; and iron-type materials for which $\chi > 0$ and very large, are termed *ferromagnetic*.

7.6. Elementary Magnet

The magnetisation in magnetic materials can be understood by studying the behaviour of electrons in atoms. An *elementary magnet* is formed when an electron of charge $-e$ and mass m moves around a fixed point O in a circular orbit of radius r as shown in Fig. 7.3.

If ω be the angular frequency of the electron, the electron passes any point on its orbit $\dfrac{\omega}{2\pi}$ times per second. This gives rise to a flow of current $-\dfrac{e\omega}{2\pi}$ in the loop.

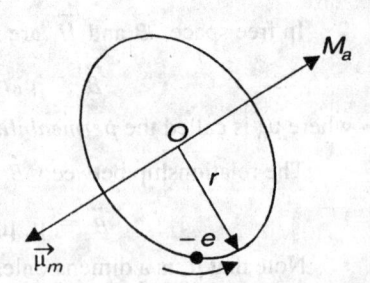

Fig. 7.3. An elementary magnet.

The area of cross-section of the loop is πr^2. Thus, according to Eq. (7.15), the magnetic dipole moment associated with the orbiting electron is

$$\mu_m = -\frac{e\omega}{2\pi}\pi r^2 = -\frac{e\omega r^2}{2} \qquad \ldots(7.30)$$

The electron rotates in the counterclockwise direction, as shown in Fig. 7.3. The conventional current therefore flows in the clockwise direction. The dipole moment is consequently directed downward. The orbiting electron will also possess an angular momentum M_a given by

$$M_a = m\omega r^2 \qquad \ldots(7.31)$$

The direction of the angular momentum is upward since the mass m is orbiting in the counterclockwise direction. The ratio of the magnetic dipole moment and the angular momentum of an orbiting electron, called the *gyromagnetic ratio*, is obtained from Eqs. (7.30) and (7.31) :

Magnetic Properties of Solids

$$\frac{\mu_m}{M_a} = -\frac{e}{2m} \qquad \text{...(7.32)}$$

Equation (7.32) indicates that the ratio is independent of the angular velocity of the electron and the radius of the orbit.

We shall now investigate the influence of an externally applied magnetic field on the behaviour of an orbiting electron.

Suppose that a magnetic field $\vec{B}$ is applied at some angle to the plane of the orbit of an elementary magnet (Fig. 7.4). The torque experienced by the elementary magnet is given by [see Eq. (7.14)]

$$\vec{T} = \vec{\mu}_m \times \vec{B} \qquad \text{...(7.33)}$$

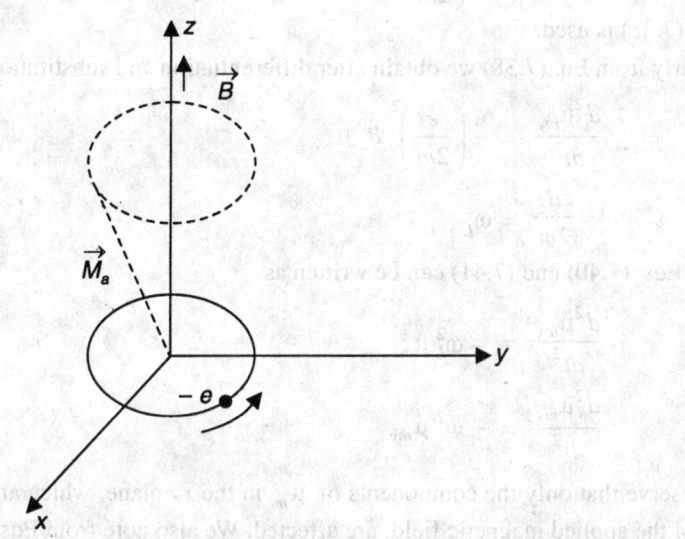

Fig. 7.4. Effect of an external magnetic field on an elementary magnet.

The torque $\vec{T}$ produces a rate of change of the angular momentum of the elementary magnet, and we have

$$\vec{T} = \frac{d\vec{M}_a}{dt} = \vec{\mu}_m \times \vec{B} \qquad \text{...(7.34)}$$

Using Eq. (7.32) in Eq. (7.34) we obtain

$$\frac{d\vec{\mu}_m}{dt} = -\frac{e}{2m}\left(\vec{\mu}_m \times \vec{B}\right) \qquad \text{...(7.35)}$$

Let us now assume that the magnetic field is applied in the z-direction, i.e.

$$\vec{B} = B_z \vec{k}$$
and
$$B_x = B_y = 0 \qquad \text{...(7.36)}$$

Equation (7.35) therefore yields for the components of $\frac{d\vec{\mu}_m}{dt}$:

$$\frac{d\mu_{mx}}{dt} = -\frac{e}{2m}\mu_{my} B_z \qquad \text{...(7.37)}$$

$$\frac{d\mu_{my}}{dt} = \frac{e}{2m}\mu_{mx} B_z \qquad ...(7.38)$$

and $\quad \dfrac{d\mu_{mz}}{dt} = 0$

In order to solve Eqs. (7.37) and (7.38) we proceed as follows :
Differentiating both sides of Eq. (7.37) with respect to t we obtain

$$\frac{d^2\mu_{mx}}{dt^2} = -\frac{e}{2m} B_z \frac{d\mu_{my}}{dt} \qquad ...(7.39)$$

$$= -\left(\frac{e}{2m}\right)^2 B_z^2 \mu_{mx} \qquad ...(7.40)$$

where Eq. (7.38) is used.

Similarly from Eq. (7.38) we obtain after differentiation and substitution :

$$\frac{d^2\mu_{my}}{dt^2} = -\left(\frac{e}{2m}\right)^2 B_z^2 \mu_{my} \qquad ...(7.41)$$

Let $\quad \dfrac{eB_z}{2m} = \omega_L$

Then, Eqs. (7.40) and (7.41) can be written as

$$\frac{d^2\mu_{mx}}{dt^2} = -\omega_L^2 \mu_{mx} \qquad ...(7.42)$$

and $\quad \dfrac{d^2\mu_{my}}{dt^2} = -\omega_L^2 \mu_{my} \qquad ...(7.43)$

We observe that only the components of $\vec{\mu}_m$ in the xy plane, which are at right angles to the direction of the applied magnetic field, are affected. We also note from Eqs. (7.42) and (7.43) that the x and y components of $\vec{\mu}_m$ vary harmonically with time with an angular frequency ω_L, called the *Larmor frequency*.

Suppose $\mu_m \sim e^{i\omega_L t}$. Then from Eqs. (7.39) and (7.42) we get $-\omega_L^2 \mu_{mx} = -\omega_L i\omega_L \mu_{my}$, i.e., $\mu_{mx} = i\mu_{my}$. Thus μ_{mx} and μ_{my} are 90° out of phase with each other. As a result, the resultant vector of μ_{mx} and μ_{my} describes a circle in the xy-plane. Since the z-component of $\vec{\mu}_m$ is unaffected, it follows that the dipole moment of the elementary magnet precesses about the z-axis, the direction of the applied magnetic field, with the Larmor frequency ω_L.

Thus, with the application of an external magnetic field, the plane of the orbit of the electron no longer remains stationary but executes a precessional motion about the magnetic field direction. In dealing with diamagnetism we shall find that this motion induces a dipole moment whose direction will be opposite to the direction of the applied magnetic field.

7.7. Origin of Permanent Magnetic Dipoles in Materials

Some of the properties of magnetic materials are determined by the presence of elementary magnets, called *permanent dipole moments*. We have seen in the foregoing section that if a charged particle has an angular momentum it behaves as an elementary magnet, and so it contributes to the permanent dipole moment. In the case of an atom we find the following three contributions to the angular momentum:

(i) Orbital angular momentum of the electrons in the atom,

(*ii*) Spin angular momentum of the electrons, and

(*iii*) Nuclear spin angular momentum.

The total magnetic dipole moment of an atom is obtained by summing the three contributions. We discuss below each of these three contributions separately.

(*i*) *Orbital angular momentum of the electrons* : In Sec. 7.6 we have seen that an orbital electron with angular momentum will have a permanent dipole moment. We use quantum numbers since the motion of the electrons in the atoms of a material is quantized. The state of motion of an electron in an atom is described by three quantum numbers.

(*a*) The principal quantum number n determines the energy of the orbit; it can take only integer values $n = 1, 2, 3, ...$etc. The corresponding electronic shells are designated by K, L, M, N, ...shells.

(*b*) The angular momentum of the orbit is given by the quantum number l. For a given value of n the quantum number l is restricted to the set of values

$$l = 0, 1, 2, ..., (n-1)$$

The total angular momentum associated with a given value of l is

$$\hbar [l(l+1)]^{1/2},$$

where $\hbar$ is reduced Planck's constant.

Electrons associated with states $l = 0, 1, 2, 3, ...$ are termed respectively $s, p, d, f, g, ...$ electrons. Note that the electrons in an s state have zero angular momentum and hence a zero magnetic moment.

(*c*) The possible components of the angular momentum along any applied magnetic field direction are determined by the magnetic quantum number m_l where m_l is confined to the following set of values

$$m_l = 0, \pm 1, \pm 2, \pm 3, ..., \pm l$$

The value of the component angular momentum associated with the magnetic quantum number m_l is

$$m_l \hbar$$

For a *p*-electron, the possible components of angular momentum along a magnetic field direction are $\hbar, 0, -\hbar$. Again we have

$$\frac{\mu_m}{M_a} = -\frac{e}{2m}$$

Hence the possible components of the magnetic moment along the applied magnetic field direction are

$$-\frac{e\hbar}{2m}, 0, \frac{e\hbar}{2m} \qquad ...(7.44)$$

Figure 7.5 shows the possible orientations of the angular momentum vector for a *p*-electron. The length of the vector is $\hbar [l(l+1)]^{1/2} = \sqrt{2}\hbar$ and its projection on $\vec{B}$ direction is $\hbar, 0$ or $-\hbar$. If there are three *p*-electrons corresponding to $m_l = 0, +1$ and -1 then the vector sum of the three components along $\vec{B}$ vanishes. This results in zero resultant magnetic dipole moment. Thus, a resultant non-vanishing dipole moment can be found in atoms with incompletely filled shells.

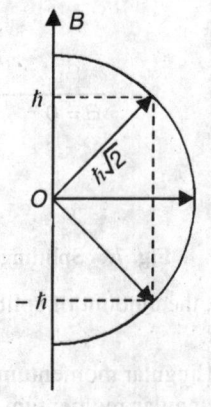

Fig. 7.5. Three possible components of angular momentum in an applied magnetic field for $l = 1$.

Bohr Magneton : The quantity $\dfrac{e\hbar}{2m} = 9.27 \times 10^{-24}$ A.m^2 is referred to as the Bohr magneton. It is used as an atomic unit of magnetic moment and is denoted by μ_B.

(*ii*) *Spin angular momentum of the electrons* : The electron is spinning about its own axis and consequently possesses an angular momentum associated with it. The possible components of the angular momentum of the spin along the direction of an applied field are $\pm \dfrac{\hbar}{2}$. This gives two spin quantum numbers, $m_s = \pm \dfrac{1}{2}$

The component of the magnetic moment μ_{mz} along an external field is generally given by

$$\mu_{mz} = g\left(\dfrac{e}{2m}\right)\dfrac{\hbar}{2}, \qquad \ldots(7.45)$$

where g is termed the *spectroscopic splitting factor*, $g = 2.0023$ for the electron spin. Thus for the electron spin we obtain

$$\mu_{mz} \approx \mu_B$$

That is, the magnetic dipole moment due to electron spin is nearly equal to a Bohr magneton in the direction (or opposite) of an applied field $\vec{B}$. The name 'splitting factor' for g arises for the following reason.

Consider that an electron with spin 1/2 but without angular momentum is subjected to an external magnetic field. For $m_s = +\dfrac{1}{2}$, $|\mu_{mz}| \approx \mu_B$, and the direction of μ_{mz} is antiparallel to the field; while for $m_s = -\dfrac{1}{2}$, $|\mu_{mz}| \approx \mu_B$, and the direction of μ_{mz} is parallel to the direction of the magnetic field. Since the potential energy of a dipole in a magnetic field is $-\vec{\mu}.\vec{B}$, a split up of the energy level of the spin 1/2 electron occurs in the presence of a magnetic field $\vec{B}$ (Fig 7.6). The separation of the split-up energy levels is

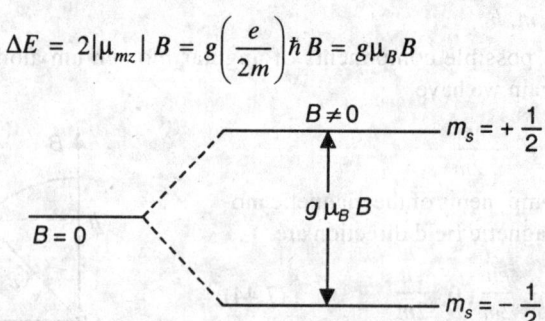

$$\Delta E = 2|\mu_{mz}| B = g\left(\dfrac{e}{2m}\right)\hbar B = g\mu_B B$$

Fig. 7.6. Splitting of energy level of the spin $\dfrac{1}{2}$ electron for $B \neq 0$.

Obviously, the amount of split-up of the levels is determined by g, which is hence termed the 'splitting factor.'

The orbital angular momentum and the spin angular momentum can be vectorially combined to yield the total angular momentum which is determined by the quantum number J. Hence, for an electron with a certain l and spin $\dfrac{1}{2}$, J can assume values $l \pm \dfrac{1}{2}$. For an atom having a number of electrons, the l vectors can be combined to give a resultant L ; similarly the m_s vectors produce a resultant S when combined. This type of combination is referred to as *Russel-Saunders coupling*. The total angular momentum J of the entire electron system of the atom is then obtained by combining the resultants L and S. For such a system the general expression for g is given by *Lande's equation*

$$g = 1 + \frac{J(J+1) + S(S+1) - L(L+1)}{2J(J+1)}$$

Hund's Rules

It has been shown that filled electron shells do not contribute to the magnetic moment of the atom. Therefore the magnetic moment of an atom must arise from partially filled shells. This follows from the Pauli principle which states that no two electrons can have all the quantum numbers n, l, m_l and m_s identical. For the ground state of such atoms Hund's rules for incomplete shells state that :

(1) the electron spins add to yield the maximum possible value of S consistent with the Pauli principle;

(2) the orbital moments combine to give a maximum value of L that is consistent with (1);

(3) for an incompletely filled shell, $J = L - S$, when the shell is less than half occupied; and $J = L + S$, when the shell is more than half occupied.

As an example, we shall apply Hund's rules to find the g value for Cr^{3+}.

The electronic configuration in Cr (atomic number 24) as obtained from the periodic table is

$$1s^2\,2s^2\,2p^6\,3s^2\,3p^6\,3d^5\,4s^1,$$

where $1s^2$ signifies two electrons in the $1s$ state ($n = 1$, $l = 0$)etc. When 3 electrons are removed from Cr it reduces to Cr^{3+} and the corresponding electronic configuration is

$$1s^2\,2s^2\,2p^6\,3s^2\,3p^6\,3d^3,$$

i.e. all the shells exepct the shell $3d$ are completely filled. Therefore only the $3d$ shell contributes to the permanent dipole moment. This shell has 3 electrons.

For a d shell ($l = 2$), the possible m_l values are 2, 1, 0, −1, −2, Each of these can accommodate 2 electrons ($m_s = \pm \frac{1}{2}$), so that the maximum number of electrons in the $3d$ shell is 10. Thus in Cr^{3+}, the $3d$ shell remains less than half occupied.

According to Hund's rule (1) we find $S = \frac{3}{2}$, since there are 3 electrons in the $3d$ shell.

As mentioned, the possible m_l values are 2, 1, 0, −1, −2. If we place the 3 electrons each with a spin $+\frac{1}{2}$ in the first three of these, we find $L = 2 + 1 + 0 = 3$, which is the maximum value for L consistant with the spin distribution.

Applying Hund's rule (3), we obtain

$$J = L - S = 3 - \frac{3}{2} = \frac{3}{2}$$

Therefore the value of g is

$$g = 1 + \frac{\frac{3}{2} \times \frac{5}{2} + \frac{3}{2} \times \frac{5}{2} - 3 \times 4}{3 \times \frac{5}{2}} = 0.4$$

In a similar manner the g value for other atoms or ions can be calculated. The magnetic moment is then obtained from Eq. (7.45).

(iii) Nuclear magnetic moment : The angular momentum associated with the spinning of the nucleus also contributes to the permanent dipole moment of the atom. If the magnetic dipole moment due to the nuclear spin is designated by μ_{mn}, then

$$\mu_{mn} = \frac{e\hbar}{2m_p},\qquad \ldots(7.46)$$

where m_p is the nuclear mass. The mass of the nucleus is approximately 10^3 times larger than the mass of the electron. Consequently, the nuclear magnetic moments are smaller than those associated with the electrons by a factor of about 10^3.

7.8. DIAMAGNETISM

In diamagnetic materials the magnetisation M is proportional to the applied magnetic field intensity H but the susceptibility χ is negative. Materials in which the atoms or molecules have no permanent magnetic moment exhibit the diamagnetic behaviour. In such materials, the number and spatial arrangement of the electron orbits of an atom are such that the vector sum of their magnetic moments is zero.

The basic principle of diamagnetism can be understood as follows :

Consider an electron moving around the nucleus in a circular orbit of radius r with an angular frequency ω_0 in the absence of a magnetic field. When an external magnetic field is applied, the electron experiences a magnetic force, which alters the frequency of motion of the electron. This change of frequency can be looked upon as an induced current that persists without change in magnitude as long as the external field is present. The induced current produces a magnetic moment in a direction opposite to the applied magnetic field, manifesting that the effect opposes the cause, as in Lenz's law of electromagnetism. Obviously then, any atomic orbit gives a negative contribution to the magnetic susceptibility, which is essentially diamagnetism.

In the absence of the external magnetic field, the centrifugal force is balanced by the Coulomb force of attraction between the nucleus and the revolving electron. Thus, we have

$$\frac{mv^2}{r} = \frac{Ze^2}{4\pi\varepsilon_0 r^2}$$

or,

$$mr\omega_0^2 = \frac{Ze^2}{4\pi\varepsilon_0 r^2}, \qquad \ldots(7.47)$$

where Ze represents the charge of the nucleus, e the electronic charge, m the mass of the electron, and ε_0 the permittivity of free space.

When a magnetic field is applied perpendicular to the plane of the orbit an additional Lorentz force acts on the electron. This extra radial force may be inward or outward depending on the direction of the applied magnetic field. Assuming that this force merely accelerates or decelerates the electron without altering the radius of the orbit we can write the new balance equation :

$$mr\omega^2 = \frac{Ze^2}{4\pi\varepsilon_0 r^2} \pm er\omega B, \qquad \ldots(7.48)$$

where ω represents the new angular velocity and B the applied magnetic flux density.

From Eq. (7.48) we obtain with the help of Eq. (7.47)

$$\omega^2 = \omega_0^2 \pm \frac{eB}{m}\omega \qquad \ldots(7.49)$$

Now, $\omega_0 \approx 10^{14}$ to 10^{15} rad/s. The magnetic field (B) generally used in the laboratory is of the order of 1 tesla. Therefore, the quantity $\dfrac{eB}{m}$ is much smaller than ω_0. Hence we can write

$$\omega = \omega_0\left(1 \pm \frac{eB}{m\omega_0}\frac{\omega}{\omega_0}\right)^{1/2}$$

Magnetic Properties of Solids

$$\approx \omega_0 \pm \omega_L \frac{\omega}{\omega_0} \qquad ...(7.50)$$

where $\omega_L \left(= \dfrac{eB}{2m} \right)$ is the *Larmor frequency*.

Solving Eq. (7.50) for ω we get

$$\omega = \frac{\omega_0}{1 \pm \dfrac{\omega_L}{\omega_0}} \cong \omega_0 \left(1 \pm \frac{\omega_L}{\omega_0} \right) = \omega_0 \pm \omega_L$$

The frequency change ω_L produces the induced magnetic moment [see Eq. (7.30)] :

$$\vec{\mu}_{ind} = -\frac{1}{2} er^2 \vec{\omega}_L = -\frac{e^2}{4m} r^2 \vec{B} \qquad ...(7.51)$$

If the orbit is not perpendicular to the applied magnetic field it precesses about $\vec{B}$ with the Larmor frequency. In this case the induced magnetic moment will have a component opposite to the direction of the field $\vec{B}$. This component is given by

$$\vec{\mu}_{ind} = -\frac{e^2}{4m} \vec{B} <r^2>,$$

where $<r^2>$ gives the mean square radius of the projection of the orbit on a plane perpendicular to $\vec{B}$. If n' number of electrons of an atom revolve in orbits which are oriented in various directions, the total induced moment in the atom is given by

$$\vec{\mu}_{ind} = -\frac{e^2}{4m} \vec{B} \sum_{i=1}^{n'} <r^2> \qquad ...(7.52)$$

Considering cartesian coordinate axes OX, OY and OZ where the origin O coincides with the centre of the circular orbits we can write

$$\rho^2 = x^2 + y^2 + z^2,$$

where x, y, z represent the coordinates of any point on an orbit of radius ρ. If OZ coincides with the direction of $\vec{B}$, then

$$r^2 = x^2 + y^2.$$

Since $<x^2> = <y^2> = <z^2>$ we get

$$<r^2> = \frac{2}{3} <\rho^2> \qquad ...(7.53)$$

Here $<\rho^2>$ is the mean square distance of an arbitrary electron from the nucleus. Therefore,

$$\vec{\mu}_{ind} = -\frac{e^2}{6m} \vec{B} \sum_{i=1}^{n'} <\rho^2>. \qquad ...(7.54)$$

If the solid contains N atoms per unit volume, then the magnetic moment per unit volume, *i.e.* the magnetisation $\vec{M}$ is given by

$$\vec{M} = -\frac{e^2 N}{6m} \vec{B} \sum_{i=1}^{n'} <\rho^2>$$

$$= -\frac{e^2 N \mu_0 \mu_r}{6m} \vec{H} \sum_{i=1}^{n'} <\rho^2>, \qquad ...(7.55)$$

where the relationship $\vec{B} = \mu_0 \mu_r \vec{H}$ has been used. The diamagnetic susceptibility χ_{dia} is therefore

$$\chi_{dia} = \frac{M}{H} = -\frac{e^2 N \mu_0 \mu_r}{6m} \sum_{i=1}^{n'} <\rho^2>, \qquad ...(7.56)$$

where μ_r may be considered to be equal to unity. Equation (7.56) indicates that χ_{dia} is negative. In order to determine the order of magnitude of χ_{dia} we take $<\rho^2> \approx 10^{-20}$ m^2, $N \approx 5 \times 10^{28}$/m^3, $e = 1.6 \times 10^{-19}$ C, $m = 9.1 \times 10^{-31}$ kg, $\mu_0 = 4\pi \times 10^{-7}$ H/m and $n' = 1$. With these values Eq. (7.56) gives $\chi_{dia} \cong -0.5 \times 10^{-5}$. The experimental value of susceptibility for copper at room temperature is -0.9×10^{-5}.

Susceptibilities are frequently given as *molar susceptibility*, based on the magnetization per mole instead of per unit volume. Thus the molar susceptibility $\chi^{(molar)}$ is obtained by multiplying χ by the volume of a mole N_A/N, where N_A is Avogadro's number ($N_A = 6.022 \times 10^{23}$). Also, it is customary to define a mean square atomic radius by $<r^2>_a = \frac{1}{n'} \sum_{i=1}^{n'} <\rho^2>$, where n' is the total number of electrons in the atom. Thus

$$\chi_{dia}^{(molar)} = -\frac{n' e^2 N_A \mu_0 \mu_r}{6m} <r^2>_a \qquad ...(7.56\,a)$$

For helium, $\chi_{dia}^{(molar)} = -1.9 \times 10^{-6}$ cm^3/ mole. With $n' = 2$ (for He), we have from Eq. (7.56 a), $<r^2>_a = 0.268 \times 10^{-21}$ m^2 so that the radius of a helium atom is estimated as $\sim \sqrt{<r^2>_a} = 1.64 \times 10^{-9}$ cm.

Equation (7.56) is known as the *Langevin equation*. The equation shows that the outer electrons of atoms make the largest contribution to the diamagnetic susceptibility since χ_{dia} is proportional to ρ^2. The diamagnetic susceptibility increases with the number of atoms but is independent of temperature since the orbital areas of the electrons are independent of temperature. These characteristics are also observed experimentally.

An atom having permanent dipole moment will also have the diamagnetic effect. In this case, however, the permanent magnetic moment will tend to rotate into the direction of the applied magnetic field yielding paramagnetic property. This will mask the diamagnetic action.

The examples of diamagnetic materials are Cu, Au, Ge, Si, NaCl, Al$_2$O$_3$ etc. Inert gases are diamagnetic because their atoms have completely filled electronic shells, and so no permanent magnetic dipole moment.

7.9. Paramagnetism

The permeabilities of many salts of iron and the rare earth families are slightly greater than unity lying in the range 1 and $1 + 10^{-3}$ and are independent of the field strength. These materials do not show hysteresis and are called paramagnetic materials. The paramagnetic property of these materials arises from the permanent magnetic moment of the constituent atoms or molecules. For completely filled electronic shells, the magnetic moments due to orbital and spin motions are self-neutralizing, so that only diamagnetic contribution towards magnetic moment exists. In the case of paramagnetic materials, some of the electronic shells are incomplete. Consequently, there exists a permanent magnetic dipole moment in these materials.

Paramagnetic salts are used for obtaining very low temperatures (less than 1 K) by adiabatic demagnetization. They are also used as essential materials for solid state masers (maser : *m*icrowave *a*mplification by *s*timulated *e*mission of *r*adiation).

Examples of paramagnetic materials are Fe$_2$O$_3$, FeSO$_4$, Cr$_2$O$_3$, NiSO$_4$, MnSO$_4$ etc.

Classical Theory (or Langevin Theory) of Paramagnetism

The classical theory is based on the following two assumptions :

(*i*) The separations of the dipoles are such that their mutual magnetic interaction forces can be neglected.

Magnetic Properties of Solids

(*ii*) In the absence of a field the dipoles can have all possible directions.

When a magnetic field $\vec{B}$ is applied to a paramagnetic material, each dipole experiences a torque $\vec{\mu} \times \vec{B}$, where $\vec{\mu}$ is the dipole moment. This torque tends to establish an order by rotating the dipoles in the direction of the applied field. However, this action of the magnetic field is opposed by the thermal agitational energy which tries to establish a disorder by randomising the positions of the dipoles.

The potential energy of a dipole in a magnetic field of flux density B is $-\mu B \cos \phi$, where ϕ is the angle between $\vec{\mu}$ and $\vec{B}$. According to Maxwell-Boltzmann statistics, the probability of finding the dipole at the angle ϕ, *i.e.*, $p(\phi)$, is proportional to $\exp(\vec{\mu} \cdot \vec{B}/k_B T)$. That is,

$$p(\phi) \propto \exp(\mu B \cos \phi / k_B T),$$

where k_B is Boltzmann's constant and T is temperature.

If $n(\phi) d\phi$ denotes the number of dipole moments in a direction between ϕ and $\phi + d\phi$, then

$$n(\phi) d\phi \propto p(\phi) d\Omega.$$

where $d\Omega$ is the solid angle between two hollow cones having semivertical angles ϕ and $\phi + d\phi$, *i.e.*, $d\Omega = 2\pi \sin \phi \, d\phi$. Therefore,

$$n(\phi) d\phi = 2\pi B_1 \sin \phi \exp\left(\frac{\mu B \cos \phi}{k_B T}\right) d\phi$$

$$= C' e^{a \cos \phi} \sin \phi \, d\phi, \qquad \ldots(7.57)$$

where $C' = 2\pi B_1$ and $a = \dfrac{\mu B}{k_B T}$, B_1 being proportionality constant. $\qquad \ldots(7.58)$

If n represents the number of dipole moments per unit volume, then

$$n = \int_0^\pi n(\phi) \, d\phi = C' \int_0^\pi e^{a \cos \phi} \sin \phi \, d\phi.$$

Putting $\cos \phi = x$, we get

$$n = C' \int_{+1}^{-1} -e^{ax} \, dx = \frac{C'}{a}(e^a - e^{-a}) \qquad \ldots(7.59)$$

or, $\qquad C' = \dfrac{an}{e^a - e^{-a}}.$

Each of the $n(\phi) d\phi$ dipoles will have a component of the magnetic moment, $\mu \cos \phi$, in the direction of the magnetic field. The components perpendicular to B will obviously cancel out.

Therefore the total magnetic moment of the n dipoles, which, by definition, gives the magnetisation M of the material, is

$$M = \int_0^\pi \mu \cos \phi \, n(\phi) \, d\phi$$

$$= \int_0^\pi \mu \cos \phi \, C' e^{a \cos \phi} \sin \phi \, d\phi = -\mu C' \int_1^{-1} x e^{ax} \, dx$$

$$= -\mu C' \left[e^a \left(\frac{1}{a} - \frac{1}{a^2}\right) + e^{-a}\left(\frac{1}{a} + \frac{1}{a^2}\right)\right] \qquad \ldots(7.60)$$

Using Eq. (7.59) we obtain from Eq. (7.60)

$$M = \mu n \left(\coth a - \frac{1}{a} \right). \qquad \ldots(7.61)$$

The function $\left(\coth a - \dfrac{1}{a} \right)$ is termed the *Langevin function* and is denoted by $L(a)$. Hence

$$M = \mu n L(a) \qquad \ldots(7.62)$$

Figure 7.7 shows the variation of $L(a)$ with a. The maximum value of M is given by μ_n and is obtained when all the dipoles are aligned in the direction of $\vec{B}$. This corresponds to the *saturation magnetisation* M_s. Hence the relation between M and M_s is given by

$$M = M_s L(a) \qquad \ldots(7.63)$$

We now specialise to the following two cases :

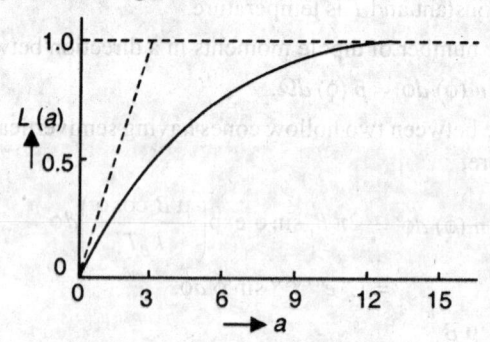

Fig. 7.7. Variation of $L(a)$ with a.

Case 1

Assume that $a = \dfrac{\mu B}{k_B T} \gg 1$. This gives $\coth a \to 1$, and

$$L(a) = \left(\coth a - \frac{1}{a} \right) \to 1$$

Therefore Eq. (7.63) reduces to

$$M \to M_s$$

In other words, the saturation effect occurs when the temperature T is low, the applied magnetic field B is high, and/or the basic magnetic dipole moment μ is large.

Case 2

Assume that $a = \dfrac{\mu B}{k_B T} \ll 1$. In this case,

$$L(a) = \coth a - \frac{1}{a} = \frac{1 + \left(\dfrac{a^2}{2!}\right) + \left(\dfrac{a^4}{4!}\right) + \ldots}{a + \left(\dfrac{a^3}{3!}\right) + \left(\dfrac{a^5}{5!}\right) + \ldots} - \frac{1}{a}$$

$$\approx \left(1 + \frac{a^2}{2!}\right) \frac{1}{a} \left(1 - \frac{a^2}{3!}\right) - \frac{1}{a}$$

$$\simeq \frac{1}{a} \left(1 + \frac{a^2}{2} - \frac{a^2}{6}\right) - \frac{1}{a}$$

$$= \frac{a}{3}$$

Thus the variation of $L(a)$ with a near the origin is linear with a slope of 1/3. The magnetisation in this region is obtained from Eq. (7.63):

$$M = \mu n \frac{a}{3} = \frac{\mu^2 nB}{3k_BT}. \qquad ...(7.64)$$

Therefore the magnetic susceptibility is

$$\chi = \frac{M}{H} = \frac{\mu^2 \mu_0 n}{3k_BT} = \frac{C}{T}, \qquad ...(7.65)$$

where $C = \dfrac{n\mu^2\mu_0}{3k_B}$ is known as the *Curie constant*. Equation (7.65) gives the *Curie-Weiss law* and indicates that the paramagnetic susceptibility is inversely proportional to the absolute temperature. Replacing n by Avogadro's number N_A in Eq. (7.65) we obtain the molar susceptibility.

To estimate the order of magnitude of χ at room temperature we take $\mu = \mu_B = 9.27 \times 10^{-24}$ A. m^2 and $n = 5 \times 10^{28}$/m^3. Then

$$\chi = \frac{(9.27 \times 10^{-24})^2 \times 1.257 \times 10^{-6} \times 5 \times 10^{28}}{3 \times 1.38 \times 10^{-23} \times 300}$$

$$\approx 10^{-3}.$$

In Sec. 7.8 we have seen that the diamagnetic susceptibility is of the order of -10^{-5}. Therefore the paramagnetic susceptibility is numerically two orders of magnitude greater than the diamagnetic susceptibility. The susceptibilities of Fe$_2$O$_3$ and CrCl$_3$ are 1.4×10^{-3} and 1.5×10^{-3}, respectively.

QUANTUM-MECHANICAL THEORY

Langevin theory assumes that in the absence of the magnetic field the large number of magnetic dipoles of the material could be oriented in all possible directions. According to quantum mechanics, however, only certain discrete angles of setting of the atomic magnets are permissible. For simplicity, we shall consider the quantum mechanical theory of a system of paramagnetic atoms which can occupy only two energy levels in a magnetic field. In this case, the dipole moments of the atoms can align themselves either parallel or antiparallel to the direction of the applied field. This is applicable to the situation where the magnetic properties of the material are solely determined by the electron spin system; the contributions of the orbital and nuclear magnetic moments are assumed negligible. The theory based on this assumption is referred to as the *two-level quantum-mechanical theory of paramagnetism*.

Consider a material containing n dipole moments associated with n atoms per unit volume. Suppose that the dipole moment of each atom is represented by μ. When a magnetic field is applied to the material, the magnetic dipole moments align themselves either parallel or antiparallel to the applied field. We further assume that there is no interaction between the dipole moments.

In the presence of a field H, let there be n_p dipole moments parallel to the field and n_a dipole moments antiparallel to the field. Therefore,

$$n = n_p + n_a \qquad ...(7.66)$$

The total magnetisation is given by

$$M = \mu(n_p - n_a) \qquad ..(7.67)$$

The potential energy of a dipole of dipole moment $\vec{\mu}$ in a magnetic flux density $\vec{B} = \mu_0 \vec{H}$ is $-\mu\mu_0 H$ (see Sec. 7.3). For the dipoles aligned parallel to the field, the energy is clearly a minimum and is equal to $-\mu\mu_0 H$. The energy of an antiparallel dipole is $\mu\mu_0 H$ (Fig. 7.8). Thus the energy difference between the antiparallel and the parallel orientations is

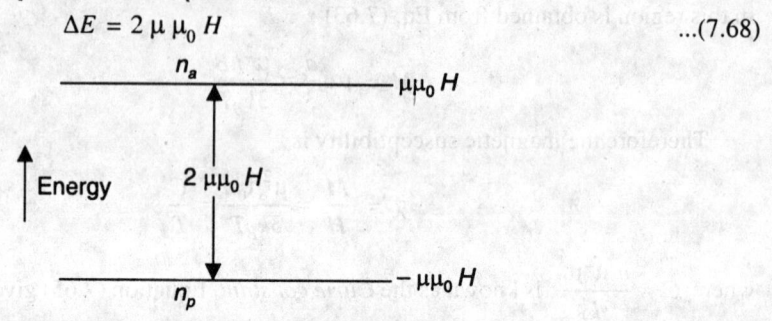

$$\Delta E = 2\mu\mu_0 H \qquad ...(7.68)$$

```
―――――――――――――――  n_a  ―――――――――  μμ_0 H
        ↑
     Energy      ↕ 2μμ_0 H
        
―――――――――――――――  n_p  ――――――――― -μμ_0 H
```

Fig. 7.8. Two-level system with n_p dipoles parallel to the field and n_a dipoles antiparallel to the field.

Both n_a and n_p depend on the temperature of the material and on the field strength H. If we assume that the occupancies of the two energy levels are determined by Boltzmann statistics, then we have

$$\frac{n_a}{n_p} = \exp\left(-\frac{\Delta E}{k_B T}\right)$$

$$= \exp\left(-\frac{2\mu\mu_0 H}{k_B T}\right) \qquad ...(7.69)$$

The quantities n_p and n_a are obtained in terms of n by solving Eqs. (7.66) and (7.69):

$$n_p = \frac{n}{1 + \exp\left(\frac{-2\mu\mu_0 H}{k_B T}\right)} = \frac{n \exp(\mu\mu_0 H/k_B T)}{\exp(\mu\mu_0 H/k_B T) + \exp(-\mu\mu_0 H/k_B T)} \qquad ...(7.70)$$

and

$$n_a = \frac{n}{1 + \exp\left(\frac{2\mu\mu_0 H}{k_B T}\right)} = \frac{n \exp(-\mu\mu_0 H/k_B T)}{\exp(\mu\mu_0 H/k_B T) + \exp(-\mu\mu_0 H/k_B T)} \qquad ...(7.71)$$

Hence the total magnetisation from Eq. (7.67) is

$$M = \mu n \frac{\exp(\mu\mu_0 H/k_B T) - \exp(-\mu\mu_0 H/k_B T)}{\exp(\mu\mu_0 H/k_B T) + \exp(-\mu\mu_0 H/k_B T)}$$

$$= \mu n \tanh(\mu\mu_0 H/k_B T) \qquad ...(7.72)$$

When the temperature $T \to 0$, we have $\tanh\frac{\mu\mu_0 H}{k_B T} \to 1$. Hence from Eq. (7.72) we get

$$M_{T \to 0} = \mu n = M_s \qquad ...(7.73)$$

This indicates that at $T = 0$, all the dipoles are aligned parallel to the magnetic field. The corresponding magnetisation (M_s) in the material is called the *saturation magnetisation*. Hence we may rewrite Eq. (7.72) in the form

$$\frac{M}{M_s} = \tanh\left(\frac{\mu\mu_0 H}{k_B T}\right) \qquad ....(7.74)$$

A plot of the normalized magnetisation, i.e., M/M_s, against $\frac{\mu\mu_0 H}{k_B T}$ is shown in Fig. 7.9. It is evident from the plot that for strong fields and low temperatures, the magnetisation M approaches its saturation value M_s. On the other hand, for usual values of the magnetic field, i.e. when H is weak, and for not too low temperatures, we have $\dfrac{\mu\mu_0 H}{k_B T} \ll 1$ and $\tanh\left(\dfrac{\mu\mu_0 H}{k_B T}\right) \approx \dfrac{\mu\mu_0 H}{k_B T}$. Hence

$$\frac{M}{M_s} \approx \frac{\mu\mu_0 H}{k_B T}$$

or,
$$M \approx \frac{\mu^2 n \mu_0 H}{k_B T} \qquad \ldots(7.75)$$

The paramagnetic susceptibility χ for the above condition is given by

$$\chi = \frac{M}{H} \approx \frac{\mu^2 n \mu_0}{k_B T} \qquad \ldots(7.76)$$

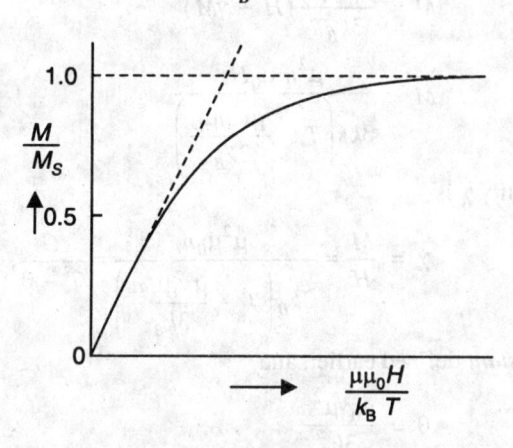

Fig. 7.9. Variation of M/M_s with $\mu\mu_0 H/k_B T$.

Thus, χ is inversely proportional to the absolute temperature T. The magnetic field causes the dipoles of the paramagnetic material to line up parallel to it and, therefore, the magnetisation of the material exists so long the external magnetic field exists. When the field is removed the alignment of the dipoles is destroyed by the random thermal motion of the dipoles due to the temperature of the material.

Comparing Eqs. (7.65) and (7.76) we find that the susceptibility for the two-level system differs by a factor of 3 from that obtained from classical ideas. The quantum-mechanical theory has also been developed for the case where the permanent magnetic moment for an atom or ion occupies a finite set of orientations with respect to the applied field. This theory, discussed in Sec. 7.12, also leads to the result that the paramagnetic susceptibility is inversely proportional to the absolute temperature.

7.10. FERROMAGNETISM

It has been shown in the previous section that the saturation magnetisation in paramagnetic materials occurs at very low temperatures and at high applied magnetic fields. In ferromagnetic materials, the saturation magnetisation is achieved at ordinary temperatures and at ordinary values of the applied magnetic field. Ferromagnetic materials possess spontaneous magnetisation, i.e. magnetisation even in the absence of an external magnetic field below a certain temperature, called the *Curie temperature*. Above this temperature, the behaviour of the material becomes paramagnetic.

Spontaneous Magnetisation and Weiss's Theory

To explain the phenomenon of ferromagnetism, Weiss in 1907 modified the Langevin theory by postulating that the effective magnetising field consists of the externally applied magnetic field H, which orients the molecules, plus an internal or molecular field H_i resulting from the interactions between the neighbouring atomic dipoles. Weiss also assumed that the field H_i is proportional to the magnetisation M, i.e., $H_i = \gamma M$ where γ is a constant, termed the *Weiss factor*. Thus the effective magnetising field H_e is given by

$$H_e = H + H_i = H + \gamma M \qquad ...(7.77)$$

We have seen in Sec. 7.9 that when the temperature T is high, the magnetisation M for a paramagnetic material is

$$M = \frac{\mu^2 n B}{3 k_B T} = \frac{\mu^2 n \mu_0 H}{3 k_B T}$$

According to Weiss, H must be replaced by H_e. Thus,

$$M = \frac{\mu^2 n \mu_0}{3 k_B T}(H + \gamma M)$$

or,

$$M = \frac{\mu^2 n \mu_0 H}{3 k_B \left(T - \dfrac{\gamma \mu^2 n \mu_0}{3 k_B} \right)}$$

Hence the susceptibility χ is

$$\chi = \frac{M}{H} = \frac{\mu^2 \mu_0 n}{3 k_B \left(T - \dfrac{\mu_0 \gamma \mu^2 n}{3 k_B} \right)} = \frac{C}{T - \theta}, \qquad ...(7.78)$$

where C is the *Curie constant* defined earlier, and

$$\theta = \frac{\mu_0 \gamma \mu^2 n}{3 k_B},$$

called the *paramagnetic Curie temperature* or simply the *Curie temperature*, is also a constant for given conditions. Equation (7.78) expresses the *Curie-Weiss law*.

We have from Eq. (7.77)

$$M = \frac{H_e - H}{\gamma} \qquad ...(7.79)$$

Putting

$$a = \frac{\mu_0 \mu H_e}{k_B T}, \qquad ...(7.80)$$

we obtain from Eq. (7.79)

$$\gamma M = \frac{a k_B T}{\mu \mu_0} - H$$

When $H = 0$ or $H_i \gg H$, we get

$$M = \frac{a k_B T}{\mu \gamma \mu_0} \qquad ...(7.81)$$

The saturation magnetisation is $M_s = n \mu$. Therefore,

$$\frac{M}{M_s} = \frac{a k_B T}{\mu_0 \gamma n \mu^2} \qquad ...(7.82)$$

Again, since

Magnetic Properties of Solids

$$\theta = \frac{\mu_0 \gamma \mu^2 n}{3k_B},$$

we get

$$\gamma = \frac{3k_B \theta}{\mu_0 \mu^2 n}$$

Hence Eq. (7.82) reduces to

$$\frac{M}{M_s} = \frac{ak_B T}{3k_B \theta} = \frac{aT}{3\theta} \qquad \qquad ...(7.83)$$

Equation (7.83) shows that the plot of M/M_s against a is a straight line with a slope of $T/3\theta$ for ferromagnetic characteristic, i.e., $M \ne 0$ even when H is zero or nearly so. On the other hand, Langevin's equation is

$$\frac{M}{M_s} = L(a) = \coth a - \frac{1}{a} \qquad \qquad ...(7.84)$$

The condition for spontaneous magnetisation, therefore, requires that Eqs. (7.83) and (7.84) be satisfied simultaneously. This is achieved graphically by plotting M/M_s versus a separately according to both the equations (Fig. 7.10). Both the plots pass through the origin which is a trivial solution. There is a possibility of another intersection of the curves (for example, the point A), if the slope of the straight line $\frac{M}{M_s} = \frac{aT}{3\theta}$ is less than that of the function $L(a)$ at the origin. We know that slope of $L(a)$ is 1/3 at the origin.

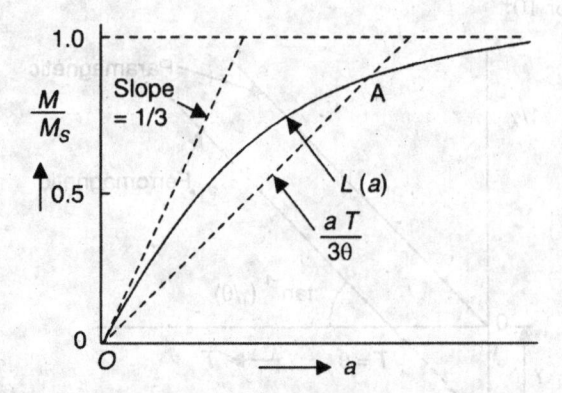

Fig. 7.10. Graphical plots to determine the situation for spontaneous magnetisation.

The intersection at the origin is discarded since it leads to unstable physical conditions. The intersection, such as at A, on the other hand, corresponds to stable physical conditions. The latter intersection point occurs if

$$\frac{T}{3\theta} < \frac{1}{3},$$

or

$$T < \theta$$

Thus, when the temperature is below the Curie temperature, a physically realisable state showing ferromagnetic behaviour (spontaneous magnetisation) occurs. Above this temperature, the thermal agitation becomes so large that the spontaneous magnetisation of the material is destroyed. Above the Curie point, the material shows paramagnetic behaviour. So, the *Curie point θ represents the temperature of the material at which transition from ferromagnetism to paramagnetism* *takes place when the temperature is raised.* As $\theta = \dfrac{\mu_0 \gamma n \mu^2}{3 k_B}$, we find that the Curie point increases with increase in the values of γ and μ. The Curie temperature of Fe is 1043 K, that of Co is 1400 K, and that of Ni is 631 K.

Determination of γ and θ

Noting that

$$C = \frac{\theta}{\gamma},$$

we have for the susceptibility χ from Eq. (7.78)

$$\chi = \frac{\theta}{\gamma (T - \theta)}$$

Hence

$$\frac{1}{\chi} = \frac{\gamma (T - \theta)}{\theta} = \left(\frac{\gamma}{\theta}\right) T - \gamma$$

Thus the plot of $1/\chi$ against T at high temperatures ($T \geq \theta$) is a straight line with a slope of (γ / θ) and an intercept of $-\gamma$ on the $1/\chi$ - axis. This gives a method of determining both γ and θ from the measurements of χ at various temperatures. From the experimental data, $1/\chi$ is plotted against T. The graph would be a straight line. The $1/\chi$ - intercept of the straight line gives γ. θ is then found from the slope (Fig. 7.11). Experimental values of γ for ferromagnetic materials are found to be of the order of 10^3 or 10^4.

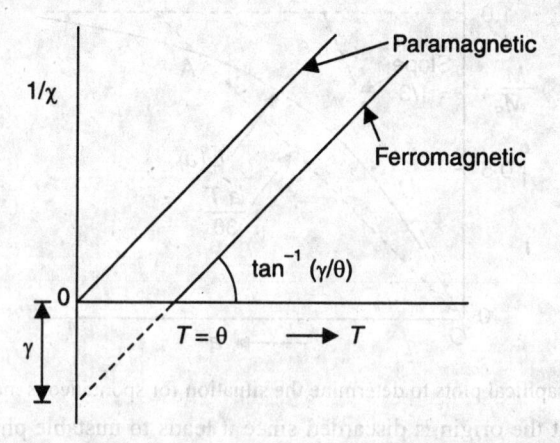

Fig. 7.11. $1/\chi$ versus T for paramagnetic and ferromagnetic materials for $T > \theta$.

For a paramagnetic material the $1/\chi$ versus T plot would be a straight line passing through the origin. The values of both γ and θ for such materials are small so that γ / θ is finite.

Variation of $\dfrac{M}{M_s}$ with T/θ

We know that

$$\frac{M}{M_s} = \coth a - \frac{1}{a}$$

and

$$\frac{M}{M_s} = \left(\frac{T}{\theta}\right) \frac{a}{3}$$

Magnetic Properties of Solids   305

We obtain from these equations by eliminating a

$$\frac{M}{M_s} = \text{Some function of } (T/\theta)$$

Theoretically, this function can be determined by drawing a number of straight lines from the origin with T as a parameter in Fig. 7.10 and finding the corresponding intersection points such as A. The functional relationship between M/M_s and T/θ is shown by the dotted curve t in Fig. 7.12. An experimental curve (solid curve, e) is also shown in Fig. 7.12. The discrepancy between the experimental and the theoretical curves can be reduced by considering the quantum mechanical theory. The discussion of the quantum theory is relegated to Sec. 7.13.

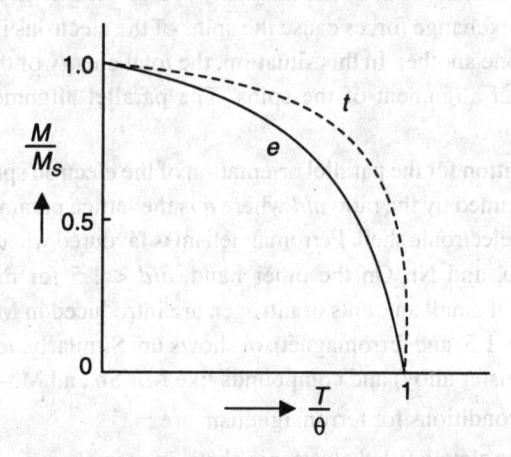

Fig. 7.12. Experimental (e) and theoretical (t) plots of M/M_s versus T/θ.

The internal field in the saturated state is $\gamma M_s = 3k_B\theta/\mu_0\mu$. This follows from the relationship $\theta = \dfrac{\mu_0 \gamma \mu^2 n}{3k_B}$. The typical values of γM_s are of the order of 10^9 A/m.

Origin of the Molecular Field

Atoms with incompletely filled shells can have permanent dipole moments. When the mutual interaction between the atomic dipole moments is negligible, the material is paramagnetic. But, if the mutual interaction between the permanent dipoles is very strong so that the magnetic moments of the neighbouring atoms tend to line up parallel to one another, the material is ferromagnetic.

The first transition group elements Fe, Co, Ni, some rare earth elements Gd and Dy, and also some alloys and compounds show ferromagnetism. All these substances have partially filled atomic shells. In the first three elements, the inner 3d subshells are incompletely filled whereas in the rare earth elements the inner 4f shells are incompletely filled.

Einstein and de Haas, and also others demonstrated experimentally that ferromagnetism originates from the spin magnetic moments of the electrons. In Fe, Co or Ni, the partially filled 3d subshells constitute the outermost subshells. The valence electrons in the outermost 4s subshells form the free electron gas making little contribution to the magnetic properties. The strong electrostatic interaction between the 3d electrons and the neighbouring ions in the crystal lattice causes quenching of the orbital moments of these electrons. Hence the orbital motion of the electrons does not contribute to ferromagnetism of these materials. Consequently, the spin moments of the electrons in the partially filled 3d subshells of these atoms are responsible for ferromagnetism.

In the molecular field H_i, the magnetic energies of the electrons are of the order of $\mu_0 \mu_B H_i \sim k_B T$ $\sim 10^{-21}$ joule, where μ_B is Bohr magneton and k_B is Boltzmann's constant. So, H_i must be of the order of 10^9 A/m. Such large internal fields cannot be accounted for by the mutual magnetic interactions between the atomic dipoles, because such interaction energies being of the order of $\mu_0 \mu_B^2 / a^3$ are about 10^{-23} joule, where $a (\sim 10^{-10}$ m) is the lattice parameter. This value is two orders of magnitude lower than the value given above. In 1928, Heisenberg showed that the large internal molecular fields in ferromagnetic materials originate from quantum mechanical exchange forces.

Due to the electrostatic interaction of the atoms, there is a probability that the electrons in an atom of a ferromagnetic substance can be exchanged with those of a neighbouring atom. Heisenberg showed that the relevant exchange forces cause the spins of the electrons in the neighbouring atoms to be aligned parallel to one another. In this situation, the total energy of the system is less than that in the case of antiparallel alignment of the spins. The parallel alignment of the spins leads to ferromagnetism.

The favourable condition for the parallel orientation of the electron spins under the action of the exchange forces is determined by the ratio a/d where a is the lattice parameter and d is the diameter of the incompletely filled electronic shell. Ferromagnetism is favoured when $a/d > 1.5$. This condition is satisfied for α-Fe, Co, and Ni. On the other hand, $a/d < 1.5$ for manganese which is non-ferromagnetic. However, if small amounts of nitrogen are introduced in Mn, the lattice parameter a is enhanced to make $a/d > 1.5$, and ferromagnetism shows up. Similarly, ferromagnetism appears in the alloy Mn-Cu-Al (Heusler alloy) and compounds like Mn-Sb and Mn-Bi.

Thus the necessary conditions for ferromagnetism are :

(*i*) Existence of incompletely filled electronic shells in atoms,

(*ii*) Values of $a > 1.5\, d$, which cause the electron spins to line up parallel to one another under the action of exchange forces, leading to large internal fields responsible for spontaneous magnetization.

M-H and B-H Curves for Ferromagnetic Materials

Below the Curie temperature a ferromagnetic material shows the well-known *hysteresis* phenomena in the M (or B) versus H curves. Typical natures of the B-H and the M-H curves are shown in Fig. 7.13 (*a*) and (*b*) respectively. For a virgin specimen $B = 0$ and $M = 0$ at $H = 0$, and M and B vary reversibly with H for small fields. With an increase in H from zero, the magnetisation M increases from zero to its saturation value M_s. Although M saturates, B slowly increases in this region with increase in H since $B = \mu_0 (H + M)$. Along the virgin M-H or B-H curve, a relative differential permeability given by $\dfrac{1}{\mu_0}\dfrac{dB}{dH} = 1 + \dfrac{dM}{dH}$ can be defined. The values of the relative differential permeability of iron, nickel, cobalt and some alloys range from 10^3 to 10^6. Figures 7.13 (*a*) and (*b*) show that when the field H is reduced from the saturation region to zero, M and B decrease to finite values M_r and B_r respectively. M_r is termed the *remanent magnetisation*, and B_r, the *remanent flux density*. The presence of M_r and B_r shows that the material is spontaneously magnetised, since $H = 0$. A field $-H_c$ (in the reverse direction), required to reduce the flux density to zero, is known as the **coercive field**.

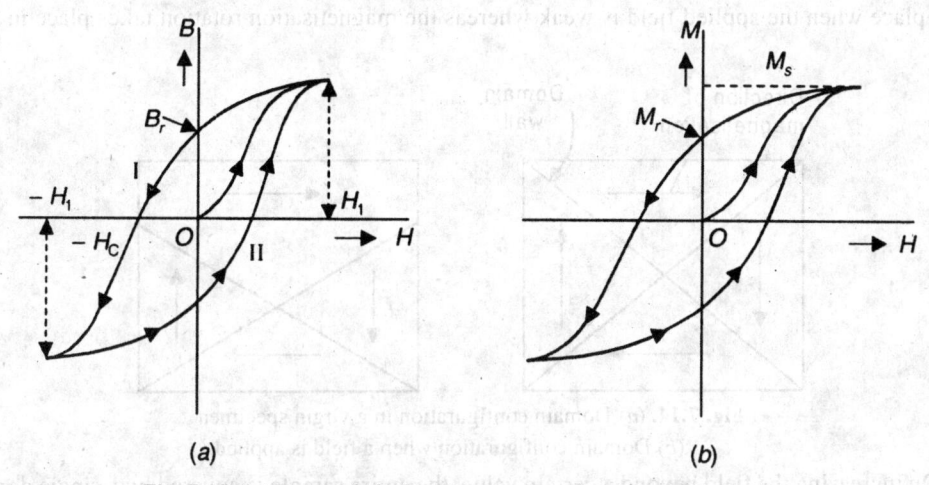

Fig. 7.13. (a) B–H and (b) M–H curves for a ferromagnetic meterial.

When the magnetic field is changed from $+H_1$ to $-H_1$, the flux density B changes along the path marked I in Fig. 7.13 (a). If now the magnetic field is changed from $-H_1$ to $+H_1$, B changes following the path II. Thus, as the magnetic field goes through a cycle, a loop enclosed by the paths I and II is traced in the B-H plane. This shows that in a cyclic variation of the magnetic field the flux density B lags behind the field H: this phenomenon is referred to as *hysteresis* and the loop traced in the B-H plane as the *hysteresis loop* or the *B-H loop*.

When the magnetic field goes through a complete cycle, an amount of energy proportional to the area of the hysteresis loop is dissipated as heat. This loss of energy is known as *hysteresis loss*. Materials for which the B-H loop area is small, are chosen for use in electromagnets and in transformer cores to reduce hysteresis losses. Such materials are called *soft magnetic materials*. The other source of energy dissipation in soft magnetic materials is the eddy current loss, which can be minimised by increasing the electrical resistivity of the material. Iron with 4% silicon, iron-nickel alloys such as, permalloy and supermalloy, ferrites and garnets are used as transformer core materials. Ferrites with a nearly rectangular hysteresis loop are used as memory cores in computers.

The materials used to produce permanent magnets are called *hard magnetic materials*. The hysteresis losses are of no concern here. Such materials must have a large remanent magnetisation and a large coercive field. The best known material for permanent magnets is Al-Ni-Co (alnico) alloy. Pure Fe_2O_3 and with a high coercive field and a high saturation magnetisation is used in magnetic tapes to store information. Thus the study of B-H curves is extremely useful in choosing materials for specific applications.

Domains

In order to explain that for a virgin specimen of a ferromagnetic material $M = 0$ at $H = 0$, and to explain the hysteresis in the B-H curve, Weiss postulated that the material contains a large number of small regions, called *domains*. The size of a domain is about 10^{-6} m or larger, and the material within each domain is magnetised to its full value. The boundary region between two domains is generally called a *domain wall* or *Bloch wall*. The direction of magnetization varies from domain to domain, resulting in a zero net magnetization of the sample [Fig. 7.14 (a)]. The sample is then said to be in a demagnetized state where its *free energy* is a minimum. When an external magnetic field is applied to a ferromagnetic specimen, two distinct processes take place. They are : (*i*) the growth in the size of the domains having favourable direction of magnetisation with respect to the field at the cost of the domains having unfavourable direction of magnetisation. The growth in size occurs with the smooth motion of the domain walls [Fig. 7.14 (b)], and (*ii*) the rotation of the direction of magnetisation within the domain towards the direction of the applied magnetic field. The first process takes place when the applied field is weak whereas the magnetisation rotation takes place in high fields.

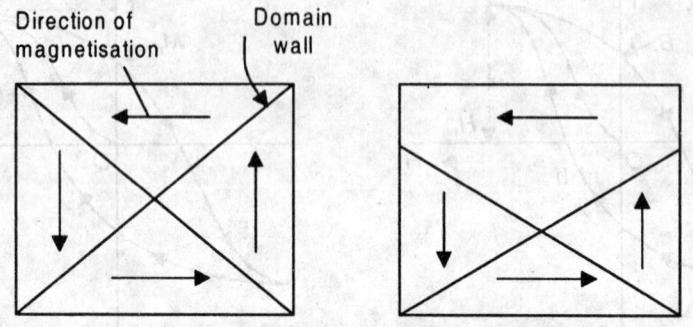

Fig. 7.14. (a) Domain configuration in a virgin specimen
(b) Domain configuration when a field is applied.

On increasing the field beyond a certain value, the entire sample is covered by a single domain which is magnetized in the direction of the applied field. This corresponds to the saturation magnetisation of the sample.

When a very weak field is applied to a virgin specimen, the specimen becomes magnetised and the process of magnetisation is reversible, *i.e.*, when the field is reduced to zero, the domains revert to their original forms. However, when the sample is magnetised with a large field and then the field is reduced to zero, the crystalline imperfections prevent the domain walls from returning to their original unmagnetized configuration. Thus the magnetisation process in a large field is irreversible and causes the phenomenon of hysteresis.

The most direct experimental evidence for the physical existence of domains was provided by F. Bitter in 1931. In this experiment, a drop of colloidal suspension of ferromagnetic particles is taken on a carefully prepared surface of the sample. Owing to the existence of strong local magnetic fields near the vicinity of the domain boundaries, the ferromagnetic particles will assemble there. The domain configuration can be clearly observed with the aid of a microscope.

The domain concept is employed in the fine-particle permanent magnets, termed ESD (*elongated single domain*) magnets. In these magnets, a large number of fine particles separated from one another, are embedded in a resin matrix. The size of these particles is made smaller than the domain size so that each particle behaves as a single domain. A very large field is required to rotate these domains during reversal, since they are not able to reverse by domain growth process. The ESD's thus give high-quality permanent magnets.

7.11. Antiferromagnetism and Ferrimagnetism

Antiferromagnetism. Ferromagnetic behaviour was explained by Heisenberg who showed that quantum mechanical exchange interactions produce large internal fields and the neighbouring dipoles have a tendency to align in the same direction. However, it can be proved quantum mechanically that when the separation between the interacting dipoles is very small, the exchange forces produce a tendency for antiparallel alignment of the spin dipole moments of the neighbouring atoms. Such materials are termed *antiferromagnetic materials*.

Theoretical investigations of antiferromagnetic materials were first made by Neel and Bitter. Antiferromagnetism was discovered experimentally for the first time by Bizette, Squire and Tsai in 1938 as a property of MnO.

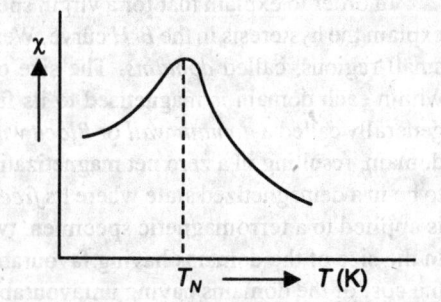

Fig. 7.15. Susceptibility versus temperature of a typical antiferromagnetic material.

Magnetic Properties of Solids

After this discovery, many materials such as MnO_2, MnF_2, FeO etc. were found to show antiferromagnetic properties.

The most characteristic feature of an antiferromagnetic material is that its susceptibility versus temperature curve shows a sharp maximum at a temperature, known as the *Neel temperature* T_N (Fig. 7.15). The significance of the Neel temperature is that above this temperature the material behaviour is paramagnetic. The suspectibility χ is given by

$$\chi = \frac{C}{T + \theta}$$

for $T > T_N$, where C is the Curie constant and θ is the paramagnetic Curie temperature.

Ferrimagnetism : Ferrites

In ferrimagnetic materials, the alignment of the neighbouring dipoles is antiparallel, but the adjacent dipoles are of unequal magnitude. *Ferrites* are the nonmetallic ferrimagnetic materials and are of great importance from the electronic and electrical engineering point of view. They form a class of solid oxides of iron. The general chemical formula of ferrites can be written as $Me^{2+} Fe_2^{3+} O_4$, where Me^{2+} represents a suitable divalent metal ion such as Mn, Ni, Co, Mg, Cu, Cd or Zn. Ni for Me gives Nickel Ferrite. $Fe^{2+} Fe_2^{3+} O_4$ is ferrous ferrite, also commonly known as magnetite (Fe_3O_4). Magnetite is the oldest ferrimagnetic material known to mankind.

A complicated class of ferrites, termed *mixed ferrites*, is formed when Me^{2+} is replaced by a mixture of ions. Mixed ferrites have established their technical importance under the trade name Ferroxcube. MnZn ferrites (Ferroxcube IV) are the most important examples of mixed ferrites. Mixed ferrites give higher saturation magnetisation than simple ferrites.

Most of the ferrites possess magnetic properties similar to those of metallic ferromagnetics. For example, they show hysteresis, spontaneous magnetisation, and related phenomena. The most important property of ferrites is that their d.c. resistivity is 10^4 to 10^{11} times as great as that of iron. As a result, the power losses due to eddy currents in ferrites are much reduced and hence they are well-suited for high-frequency applications, such as, in high-frequency transformers and high-frequency generators.

7.12. Brillouin Theory (or The Quantum Theory) of Paramagnetism

The Langevin theory assumes that the large number of magnetic dipoles involved can be oriented in all possible directions with respect to the magnetic field. The quantum theory permits only certain discrete angles of orientation of the atomic magnets. When an atom is placed in a magnetic field, the definite angles θ at which it can be oriented to the field are given by

$$\cos \theta = M_J/J, \qquad \ldots((7.85)$$

where J is the resultant angular momentum quantum number and M_J is the magnetic quantum number. The possible values of M_J are $J, J-1, J-2, \ldots 0, \ldots, -(J-2), -(J-1), -J$.

The magnetic dipole moment of an atom for a given value of M_J is $M_J g \mu_B$ in the direction of the applied field $\vec{H}$. The potential energy of the atomic dipole in the field H is $-M_J g \mu_B \mu_0 H$, where g is the spectroscopic splitting factor, μ_B is the Bohr magneton, and μ_0 is the free-space permeability.

Classical statistics tells us that the number of atomic dipoles with quantum number M_J is proportional to $\exp[-(-M_J g \mu_B \mu_0)/k_B T] = \exp(M_J g \mu_B \mu_0 H / k_B T)$ where k_B is the Boltzmann constant and T is the absolute temperature. Therefore, the total magnetic moment per unit volume having n atoms, *i.e.* the magnetisation M is expressed by

$$M = n \frac{\sum_{M_J=-J}^{M_J=+J} M_J g \mu_B \exp(M_J g \mu_B \mu_0 H / k_B T)}{\sum_{M_J=-J}^{M_J=+J} \exp(M_J g \mu_B \mu_0 H / k_B T)} \qquad \ldots(7.86)$$

In evaluating M, only the components of atomic dipole moments along the field $\vec{H}$ have to be considered since the perpendicular components cancel out by symmetry.

Let
$$\frac{g\mu_B\mu_0 H}{k_B T} = z$$

Then
$$M = ng\mu_B \frac{\sum_{M_J=-J}^{+J} M_J e^{M_J z}}{\sum_{M_J=-J}^{+J} e^{M_J z}} \qquad ....(7.87)$$

Since $\frac{d}{dz}(e^{M_J z}) = M_J e^{M_J z}$,

Eq. (7.87) can be written as

$$M = ng\mu_B \frac{d}{dz}\left(\ln \sum_{M_J=-J}^{+J} e^{M_J z}\right)$$

$$= ng\mu_B \frac{d}{dz}\left[\ln\left\{e^{Jz} + e^{(J-1)z} + ... + e^{-(J-1)z} + e^{-Jz}\right\}\right]$$

$$= ng\mu_B \frac{d}{dz}\left[\ln \frac{e^{Jz}\{1 - e^{-(2J+1)z}\}}{1 - e^{-z}}\right]$$

$$= ng\mu_B \frac{d}{dz}\left[\ln\left\{\frac{e^{(J+\frac{1}{2})z} - e^{-(J+\frac{1}{2})z}}{e^{z/2} - e^{-z/2}}\right\}\right]$$

$$= ng\mu_B \frac{d}{dz}\ln\left\{\frac{\sinh(J+\frac{1}{2})z}{\sinh(z/2)}\right\}$$

$$= ng\mu_B J\left[\frac{2J+1}{2J}\coth\left(\frac{2J+1}{2J}\right)a - \frac{1}{2J}\coth\left(\frac{a}{2J}\right)\right], \qquad ....(7.88)$$

where $a = Jz = \mu_B Jg\mu_0 H/k_B T$. The quantity within the third brackets on the right-hand side of Eq. (7.88) is referred to as the *Brillouin function*, denoted by $B_J(a)$. Hence Eq. (7.88) can be written as

$$M = ng\mu_B J B_J(a) \qquad ....(7.89)$$

The maximum value of M_J is J, so that the saturation magnetisation is $ng\mu_B J$ corresponding to the alignment of the magnetic dipole moments along the direction of the applied field. Denoting the saturation magnetisation by M_s, Eq. (7.89) gives

$$M = M_s B_J(a) \qquad ....(7.90)$$

At high magnetic fields and low temperatures, $a \gg 1$ and $B_J(a) \to 1$.

When $J \to \infty$, i.e. when all orientations are possible, $(2J+1)/(2J) \to 1$ and $\coth\left(\frac{a}{2J}\right) \to \frac{2J}{a}$.

So, the function $B_J(a)$ is found to approach the Langevin function $L(a) = \coth a - 1/a$. Experimental values of the saturation magnetisation are found to be more accurately described by the Brillouin function than by the Langevin function. This supports the quantum theory.

At high temperatures and low magnetic fields, $a = Jg\mu_B\mu_0H/k_BT \ll 1$. In this case

$$\coth\left(\frac{a}{2J}\right) \cong \frac{2J}{a}\left(1 + \frac{1}{3}\frac{a^2}{4J^2}\right)$$

$$\coth\left(\frac{2J+1}{2J}\right)a \cong \left(\frac{2J}{2J+1}\right)\frac{1}{a}\left[1 + \frac{a^2}{3}\left(\frac{2J+1}{2J}\right)^2\right]$$

and

$$B_J(a) \cong \frac{1}{a} + \frac{a}{3}\left(\frac{2J+1}{2J}\right)^2 - \frac{1}{a}\left[1 + \frac{a^2}{12J^2}\right]$$

$$= \frac{a}{3}\left(\frac{J+1}{J}\right)$$

Therefore, Eq. (7.89) gives

$$M = ng\mu_B(J+1)\frac{a}{3} = \frac{ng^2\mu_B^2\mu_0 J(J+1)H}{3k_BT} \quad \ldots(7.91)$$

Hence the paramagnetic susceptibility is given by

$$\chi = \frac{M}{H} = \frac{ng^2\mu_B^2\mu_0 J(J+1)}{3k_BT} \quad \ldots(7.92)$$

The *molar susceptibility* is obtained by replacing n by Avogadro's number N_A in Eq. (7.92). Equation (7.92) can be compared with the classical result [Eq. (7.65)]

$$\chi_{classical} = \frac{\mu^2\mu_0 n}{3k_BT} \quad \ldots(7.93)$$

Since the total magnetic moment μ_J associated with J is given by

$$\mu_J^2 = g^2 J(J+1)\mu_B^2, \quad \ldots(7.94)$$

Eqs. (7.92) and (7.93) are identical. From the susceptibility measurements in the range where Curie law holds we get μ_J from Eq. (7.92). The ratio μ_J/μ_B will yield the effective number of Bohr magnetons p_{eff}. We have

$$p_{eff} = g[J(J+1)]^{1/2} \quad \ldots(7.95)$$

Determining the values of J and g from Hund's rules and Lande's formula, one can calculate p_{eff} from Eq. (7.95). These calculated values of p_{eff} agree well with the values deduced from the experimental measurements of χ. Thus the quantum theory is favoured.

Observation

The conduction electrons contribute to a temperature independent paramagnetic susceptibility. The associated paramagnetism is called *Pauli paramagnetism*. A magnetic field resolves the degeneracy of the electrons of opposite spin, and thus redistributes the electrons between the two spin orientations in a metal. This effect produces a magnetic moment. The electron with spin parallel to the magnetic field contributes a magnetic moment $-g\mu_B/2$ and the electron with spin antiparallel to the field contributes $g\mu_B/2$ to the magnetic moment of the system of conduction electrons. Pauli susceptibilities are of the order of 10^{-6}, and are therefore about a few hundred times smaller than the Langevin's susceptibility.

Conduction electrons also contribute to a diamagnetic susceptibility. A magnetic field induces an orbital motion producing a magnetization antiparallel to the field. The magnetism so created is called *Landau diamagnetism*. For free electrons; $\chi_{Landau} = -\frac{1}{3}\chi_{Pauli}$. Landau diamagnetism is much smaller than the Larmor diamagnetism.

Nuclear spin also contributes to the magnetic moment of atoms. In paramagnetic substances, the *nuclear paramagnetism* is overshadowed by the electron paramagnetism, because the proton mass is about 1000 times larger than the electron mass. Solid hydrogen exhibits nuclear paramagnetism.

7.13. THE QUANTUM THEORY OF FERROMAGNETISM

The external field H is taken to be negligible compared to the internal field H_i, which is proportional to the magnetisation M. Thus

$$H_i = \gamma M, \qquad \ldots(7.96)$$

where γ is the Weiss factor. From the quantum theory of paramagnetism we have (see Sec. 7.12)

$$\frac{M}{M_s} = B_J(a) = \frac{2J+1}{2J}\coth\left(\frac{2J+1}{2J}\right)a - \frac{1}{2J}\coth\left(\frac{a}{2J}\right) \qquad \ldots(7.97)$$

where

$$a = \frac{\mu_0 \mu H_i}{k_B T}, \qquad \ldots(7.98)$$

and $\mu = J g \mu_B$. Equations (7.96) and (7.98) give

$$M = \frac{H_i}{\gamma} = \frac{a k_B T}{\mu_0 \mu \gamma}, \qquad \ldots(7.99)$$

If n is the number of magnetic particles per unit volume, we have $M_s = n\mu$, where M_s is the saturation magnetisation. Therefore,

$$\frac{M}{M_s} = \frac{a k_B T}{\mu_0 n \mu^2 \gamma} \qquad \ldots(7.100)$$

The point of intersection of the graphs of Eqs. (7.97) and (7.100) where M/M_s is plotted against a in each case, gives M/M_s. The graph of Eq. (7.100) is a straight line whereas the graph of the function $B_J(a)$ is similar to that of the function $L(a)$ shown in Fig. 7.10. For a point of intersection other than the origin to occur, the slope of the graph of Eq. (7.97) at the origin must be larger than the slope of this straight line. Spontaneous magnetisation occurs if this requirement is fulfilled. For $a \ll 1$, the Brillouin function can be expanded, so that Eq. (7.97) gives

$$\frac{M}{M_s} = \frac{J+1}{3J}a - \frac{J+1}{3J}\left(\frac{2J^2+2J+1}{30J^2}\right)a^3 \qquad \ldots(7.101)$$

The slope of Eq. (7.101) at the origin is $(J+1)/(3J)$. For spontaneous magnetisation, we must have

$$\frac{J+1}{3J} > \frac{k_B T}{\mu_0 n \gamma \mu^2}$$

This means that T must be less than a certain temperature θ, the *Curie temperature*, where

$$\frac{J+1}{3J} = \frac{k_B \theta}{\mu_0 n \gamma \mu^2},$$

or,

$$\theta = \frac{\mu_0 n \gamma \mu^2}{k_B}\left(\frac{J+1}{3J}\right) \qquad \ldots(7.102)$$

Using Eq. (7.102), we obtain from Eq. (7.100)

$$\frac{M}{M_s} = \frac{J+1}{3J}\left(\frac{T}{\theta}\right)a \qquad \ldots(7.103)$$

Magnetic Properties of Solids

The quantity a can be eliminated from Eqs. (7.101) and (7.103) to give M/M_s as a function of (T/θ). This theoretical curve agrees much better with experiments than the curve obtained on the basis of the classical theory (see Fig. 7.20).

7.14. Comparison of Diamagnetic, Paramagnetic, and Ferromagnetic Materials

Diamagnetic materials like bismuth, antimony, gold, lead, zinc, water, mercury, hydrogen etc. are repelled by a strong magnet. They have a negative susceptibility of the order of -10^{-5}. The negative sign implies that an inducing magnetic pole induces a similar pole at the near end of the magnetic material. Here $\vec{M}$ opposes $\vec{H}$. The susceptibility of a diamagetic substance is independent of temperature. These materials do not show any remanent magnetization.

If a small light thin bar (B) of a diamagnetic substance is suspended freely in a strong magnetic field, the bar will set itself perpendicularly to the direction of the magnetic field (Fig. 7.16). If the dimagnetic bar is slightly rotated about its point of suspension from this position, the repulsion between the inducing pole and the nearest induced pole of similar nature would tend to set the bar normal to the direction of the magnetic field in equilibrium.

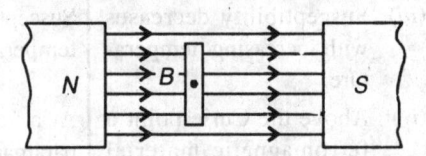

Fig. 7.16. A light thin diamagnetic bar in a strong uniform magnetic field.

Paramagnetic materials like manganese, aluminium, platinum, oxygen etc. are weakly attracted by a strong magnetic field. These materials have a positive susceptibility of the order 10^{-3}. The positive susceptibility means that if a paramagnetic material is placed in a magnetic field its end nearest to the pole piece of the source magnet will attain the opposite polarity. A freely suspended small, light, thin bar of a paramagnetic material (B) in a strong magnetic field will set itself parallel to the field direction (Fig. 7.17). Such substances do not exhibit any remanent magnetization. As the paramagnetic susceptibility is numerically much greater than the diamagnetic susceptibility, the induced diamagnetism does not show up in a paramagnetic material at ordinary temperatures. As the temperature increases, the paramagnetic susceptibility drops off, and so at a sufficiently high temperature a paramagnetic material behaves as a diamagnetic one.

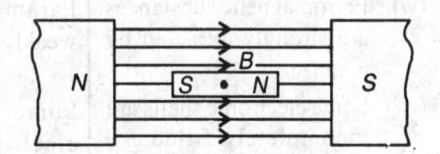

Fig. 7.17. A light, thin paramagnetic bar in a strong uniform magnetic field.

Ferromagnetic materials like soft iron, steel, nickel, cobalt, alnico, permalloy, mumetal etc. are attracted strongly by magnets. The susceptibility of such materials is positive and very large. Hence if a ferromagnetic bar is placed in a magnetic field and parallel to it, the ends of the bar acquire strong polarity, the end nearest to the pole of the source magnetizing field gathering the opposite polarity. Ferromagnetic substances have both remanent magnetization and coervice field. The ferromagnetic susceptibility decreases with increase of temperature. Above a certain temperature called the Curie temperature, a ferromagnetic material behaves as a paramagnetic one.

In diamagnetic substances the atoms or molecules have no permanent magnetic moments. The electronic shell or orbitals for such substances are completely filled. For paramagnetic materials, some of the electronic shells are incompletely filled and so the atoms of such materials have permanent magnetic moments. Atoms of ferromagnets also possess permanent magnetic moments, but here the interaction between the neighbouring magnets are very strong to produce an internal or molecular field. Hence ferromagnets have spontaneous magnetization even in the absence of an external magnetic field.

The basic differences between the ferromagnetic, paramagnetic, and diamagnetic substances are summarised in Table 7.1.

Table 7.1. Differences between ferromagnetic, paramagnetic, and diamagnetic materials.

	Ferromagnetic substance	Paramagnetic substance	Diamagnetic substance
(i)	Susceptibility is positive and very large.	Susceptibility is positive and small.	Susceptibility is negative and small.
(ii)	Relative permeability is very large.	Relative permeability is slightly greater than 1.	Relative permeability is slightly less than 1.
(iii)	Susceptibility decreases with increasing temperature.	Susceptibility varies inversely as temperature.	Susceptibility is independent of temperature.
(iv)	Above the Curie point a ferromagnetic material becomes a paramagnetic substance.	At a very large temperature, a paramagnetic substance behaves as a diamagnetic one.	Temperature changes have no effect on the magnetic property.
(v)	Ferromagnetic materials have remanent magnetiziation and coercive field. They show hysteresis.	Paramagnets do not possess remanent magnetization and coercive field. They do not show hysteresis.	Remanent magnetization, coercive field and hysteresis are absent in diamagnetic materials.
(vi)	Ferromagnetic substances are strongly attracted by magnets.	Paramagnetic materials are weakly attracted by magnets.	Diamagnetic substances are weakly repelled by magnets.
(vii)	Some electronic shells are incompletely filled, so that the atoms have permanent magnetic moments. The strong interaction between neighbouring atoms causes parallel orientation of the electron spins and hence spontaneous magnetization.	Some incompletely filled electronic shells result in permanent dipole moments for the atoms, but there is no spontaneous magnetization.	Atoms or molecules have no permanent magnetic moments since electronic shells are completely filled.

7.15. Worked-out Problems

1. The magnetic field strength in a piece of metal is 10^6 ampere per metre. Find the flux density and the magnetisation in the material. Assume that the magnetic susceptibility of the metal is -0.5×10^{-5}.

Ans. We have
$$M = \chi H = -0.5 \times 10^{-5} \times 10^6 = -0.5 \times 10$$
$$= -5 \text{ A/m}$$
Also,
$$B = \mu_0 (H + M)$$
$$= 4\pi \times 10^{-7} (10^6 - 5)$$
$$\approx 1.257 \text{ tesla}$$

2. The saturation magnetisation of iron is 1.75×10^6 ampere/metre. Assuming that the iron has a body-centered cubic structure with an edge-length of 2.87Å, find the average number of Bohr magnetons contributing to the saturation magnetisation per atom.

Ans. We have for the saturation magnetisation
$$M_s = n\mu,$$
where n is the total number of atoms per unit volume and μ is the dipole moment of each atom.

Now, the volume of each cubic cell is $(2.87 \times 10^{-10})^3$ m^3. For the body-centered cubic cell, there are two atoms per cubic cell. Therefore,
$$n = \frac{2}{(2.87 \times 10^{-10})^3} = \frac{2 \times 10^{30}}{(2.87)^3} \text{ m}^{-3}$$

Hence,
$$\mu = \frac{M_s}{n} = \frac{1.75 \times 10^6 \times (2.87)^3}{2 \times 10^{30} \times 9.27 \times 10^{-24}} \text{ Bohr magnetons,}$$

since 1 Bohr magneton = 9.27×10^{-24} A.m^2.

or, $\mu = 2.23$ Bohr magnetons.

3. The Curie temperature of iron is 1043 K. If each iron atom has a magnetic moment of two Bohr magnetons, calculate the values of the Weiss constant and the Curie constant. Assume that the saturation magnetisation of iron is 1.75×10^6 ampere/meter.

Ans. We have for the Weiss constant
$$\gamma = \frac{3k_B\theta}{\mu_0 n \mu^2} = \frac{3k_B\theta}{\mu_0 M_s \mu}$$

Here the Curie temperatue, $\theta = 1043$ K, $\mu_0 = 4\pi \times 10^{-7}$ H/m, k_B, the Boltzmann constant $= 1.38 \times 10^{-23}$ J/K, $\mu = 2$ Bohr magnetons $= 2 \times 9.27 \times 10^{-24}$ A.m^2, and $M_s = 1.75 \times 10^6$ A/m.

$$\therefore \gamma = \frac{3 \times 1.38 \times 10^{-23} \times 1043}{4\pi \times 10^{-7} \times 1.75 \times 10^6 \times 2 \times 9.27 \times 10^{-24}}$$
$$= 1059$$

The Curie constant C is given by
$$C = \frac{\mu_0 n \mu^2}{3k_B} = \frac{\mu_0 M_s \mu}{3k_B}$$
$$= \frac{4\pi \times 10^{-7} \times 1.75 \times 10^6 \times 2 \times 9.27 \times 10^{-24}}{3 \times 1.38 \times 10^{-23}}$$
$$= 9848 \text{ K}$$

QUESTIONS

1. Define magnetisation and magnetic susceptibility. Derive the relationship between the magnetic flux density B, the magnetic intensity H and the magnetization M.
2. What is an elementary magnet ? Describe in detail the effect of an external field on an elementary magnet.
3. Discuss the origin of permanent dipoles in magnetic materials. What is Bohr magneton ?
   State Hund's rule and apply it to determine the value of the spectroscopic splitting factor (g) for Cr^{3+}.
4. (a) What is a diamagnetic material ? Give some examples of diamagnetic materials. Show that the diamagnetic susceptibility increases with the number of atoms per unit volume but is independent of the temperature of the material.
   (b) Deduce Langevin's formula for the molar diamagnetic susceptibility. **(Burd. U. 1999)**
5. What are paramagnetic materials ? Give some examples. Describe the classical theory of paramagnetism. **(cf. C.U. 2008)**

316   *Solid State Physics*

6. (a) Derive the temperature dependence of the magnetic susceptibility of a paramagnetic substance. **(C.U. 1982)**

   (b) Derive Curie's law of paramagnetism from Langevin' theory. **(C.U. 1984)**

7. A system of paramagnetic atoms (N per unit volume) which can occupy only two energy levels in a uniform external magnetic field H is at a temperature T. If occupancies of these levels are determined by Boltzmann distribution, find the resulting magnetisation and the susceptibility of the system when H is weak. **(C.U. 1985)**

8. (a) Give a comparative study of dia, para, and ferromagnetism.

   (b) Derive an expression showing the temperature dependence of paramagnetic susceptibility. **(C.U. 1983, 1987)**

9. (a) What is a ferromagnetic material? Explain the terms: soft and hard magnetic materials.

   (b) Describe the basic principle of Curie-Weiss law for ferromagnets. **(C.U. 1983)**

   (c) Describe the phenomenon of hysteresis in ferromagnets. **(C.U. 1983, 2008)**

   (d) Explain the usefulness of B-H curves.

   (e) What is spontaneous magnetisation in a ferromagnetic substance? Explain it qualitatively in terms of molecular field hypothesis. Obtain a relation to show the temperature dependence of the spontaneous magnetisation. **(C.U. 1984)**

   (f) Discuss briefly the Weiss theory of ferromagnetism. **(Burd. U. 1999)**

   (g) What are the conditions for the appearance of ferromagnetism in a solid crystal?

10. Derive the Curie-Weiss law of ferromagnetism and obtain an expression for the critical temperature. **(C.U. 1984)**

11. (a) What are domains? Explain physically the existence of domains in a ferromagnetic material. Give a direct experimental evidence in support of your answer. Also, enumerate a practical application of the concept of domains.

    (b) How do domains help in explaining the nature of the B-H curve of a ferromagnetic substance? **(C.U. 2000)**

12. Distinguish between ferromagnetic and antiferromagnetic materials. What are ferrites? Describe the magnetic properties of ferrites. Name some of their applications.

13. What are the characteristics of ferromagnetic substances?
    Derive Curie-Weiss law of ferromagnetic susceptibility and compare the temperature dependence of magnetisation obtained from it with the experimental results. **(C.U. 1988, 99)**

14. Distinguish between para- and ferromagnetic materials. Explain, after Weiss, spontaneous magnetisation and, using quantum theory, obtain an expression of the ferromagnetic Curie temperature. **(C.U. 1990)**

15. (a) Discuss the characteristics of dia-, para-, and ferromagnetic substances. How would you distinguish between them experimentally? **(C.U. 1997)**

    (b) Name an element which exhibits paramagnetism. Explain why the inert gases do not show paramagnetism. **(C.U. 1997)**

16. (a) Find an expression of paramagnetic susceptibility from quantum theory. Under what condition does it reduce to the classical expression of Langevin? **(C.U. 2005)**

    (b) What is Pauli paramagnetism? **(C.U. 2006)**

17. The magnetic field strength in a piece of a material of Fe_2O_3 is 1.2×10^6 A/m. If the magnetic susceptibility of Fe_2O_3 at 300 K is 1.4×10^{-3} determine the flux density and the magnetisation in Fe_2O_3. **[Ans. 1.51 T ; 1.68×10^3 A/m]**

18. The paramagnetic susceptibility of Fe_2O_3 at room temperature is 1.4×10^{-3}. If the dipole moment of each atom is of the order of a Bohr magneton, calculate the number of atoms per unit volume of the material. **[Ans. 1.68×10^{29}]**

19. Given that the diamagnetic susceptibility of He is -1.9×10^{-6} cm^3/ mole, estimate the radius of the He atom. **(Burd. U. 1999)**

    [Ans. 0.0164 nm]

Chapter 8

Superconductivity

8.1. SUPERCONDUCTING STATE

If the temperature approaches 0K, the electrical resistivity of a normal metal in the form of a pure single crystal smoothly approaches zero. It does not go to zero for any temperature above 0K and in the presence of defects it has a nonzero limiting value. Resistivities of some other metals, known as *superconductors*, show a different behaviour. The resistivity of a superconductor, even if it is impure and noncrystalline, vanishes over a range of temperature above 0K. Above the superconducting critical temperature, the metal exhibits normal resistivity.

Superconductivity was first found in mercury in 1911 by Kamerlingh Onnes who observed that, on cooling, the resistance of a mercury sample vanished abruptly at 4.2K (Fig. 8.1). Subsequently, many metallic elements and alloys were found to behave as perfect electrical conductors below a certain low temperature, depending on the material. This transition temperature is referred to as the *critical temperature* (T_C). Usually, T_C is a few kelvin. The nearly disappearing electrical resistance in the superconducting state has been demonstrated in experiments where continuously circulating currents in superconductors maintain their initial values for several years after the removal of the source.

The vanishing electrical resistance in the superconducting state implies that large electric currents can be maintained with negligible heat loss. The phenomenon can be exploited in powerful devices requiring very little electric energy. The disadvantage is that the materials have to be operated at very low temperatures using liquid helium as the coolant. Attaining and maintaining such low temperatures with liquid helium for a long time is difficult and uneconomical. In the later half of 1980s certain ceramics have been developed; such materials have much higher critical temperatures. For example, a critical temperature of 125K has been reported for thallium cuprates. Such ceramics are known as *high – T_C superconductors*. The advantage of these materials is that they require liquid nitrogen as a coolant, and this is more economical than liquid helium. The advent of high - T_C materials has initiated intense researches in superconductivity, and activities are directed to push the critical temperature towards the room temperature.

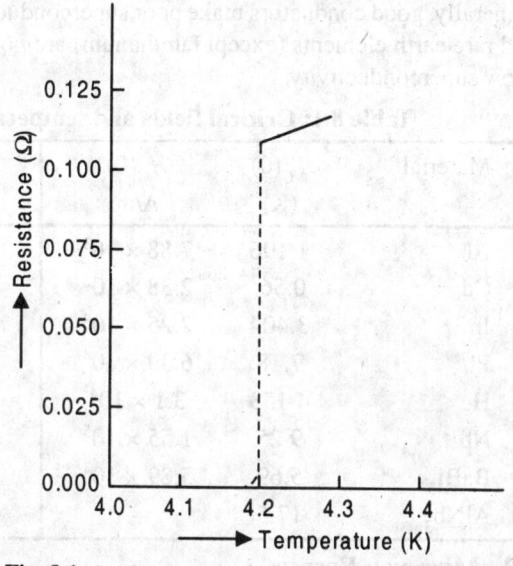

Fig. 8.1. Attainment of superconducting state in mercury

8.2. DESTRUCTION OF SUPERCONDUCTIVITY BY MAGNETIC FIELD

Kamerlingh Onnes also found that superconductivity could be destroyed by applying magnetic fields of sufficient intensity. The *critical magnetic field* for which superconductivity disappears is a function of the absolute temperature T, and is denoted by $H_C(T)$. When $T = T_C$, $H_C = 0$; i.e. at the critical temperature the critical field is zero: $H_C(T_C) = 0$.

The curve giving the variation of the critical field with temperature is nearly parabolic for many materials. A typical curve for mercury is given in Fig. 8.2. The curve is approximately given by

$$\left(\frac{T_C(H)}{T_C(0)}\right)^2 = 1 - \frac{H_C(T)}{H_C(0)}. \quad \ldots(8.1)$$

The region below the curve in Fig. 8.2 is the superconducting phase whereas that outside is the normally conducting phase. Thus, Fig. 8.2 can be looked upon as a *phase diagram*, the transition from the normal to the superconducting phase and vice versa being a thermodynamic phase transition.

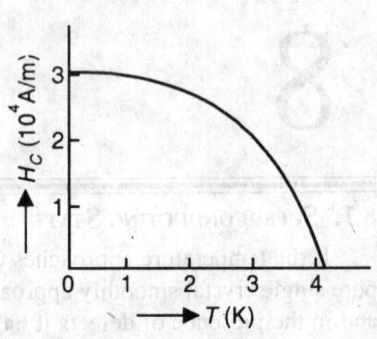

Fig. 8.2. Superconducting–normal phase diagram for Hg

Table 8.1 lists values of $T_C(0)$ and $H_C(0)$ for several superconducting metals and alloys. Generally, good conductors make poor superconductors. Monovalent metals, ferromagnetic metals, and rare earth elements (except lamthanum) are not found to be superconductors. Semiconductors show superconductivity.

Table 8.1: Critical fields and temperatures of several superconductors

Material	$T_C(0)$ (K)	$H_C(0)$ A/m	Material	$T_C(0)$ (K)	$H_C(0)$ A/m
Al	1.196	7.88×10^3	Nb Sn$_2$	2.60	4.93×10^4
Cd	0.56	2.38×10^3	Nb$_3$Ge	23.2	—
In	3.404	2.25×10^4	High - T_C Superconductors :		
Pb	7.18	6.39×10^4	La-Ba-Cu-O	34	—
Hg	4.153	3.1×10^4	Y Ba$_2$ Cu$_3$O$_7$	90	—
Nb	9.25	1.55×10^5	Bi cuprates	108	—
BaBi$_3$	5.69	5.89×10^4	Tl cuprates	125	—
AlNb$_3$	17.5	—			

8.3. MEISSNER EFFECT

Ohm's law gives $\vec{E} = \rho \vec{J}$, where $\vec{J}$ is the electric current density in a material of resistivity ρ due to an electric field $\vec{E}$. In a Superconductor $\rho = 0$, and so $\vec{E} = \vec{0}$. Again, from Maxwell's equation $\vec{\nabla} \times \vec{E} = -\partial \vec{B}/\partial t$, we have $\partial \vec{B}/\partial t = \vec{0}$ for zero resistivity. This shows that the magnetic induction $\vec{B}$ does not vary with time. So, if a sample is kept in a magnetic field and cooled below the transition temperature, the magnetic flux gets trapped in the sample. On the contrary, if the sample is first cooled below the transition temperature, and then the magnetic field is applied, the magnetic flux cannot enter the sample. Thus the magnetic property of a superconductor should depend on its past treatment. This idea was, however, proved false by Meissner and Ochsenfeld who demonstrated that upon cooling in a magnetic field below the transition temperature, the lines of magnetic induction $\vec{B}$ are pushed out from the superconducting sample (Fig. 8.3). This phenomenon is referred to as *flux exclusion* or *Meissner effect*. If a magnetic field is applied prior to cooling, the final result after cooling is the same as that if the sample is first cooled and then placed in the magnetic field. The Meissner effect in a superconductor is thus *reversible*.

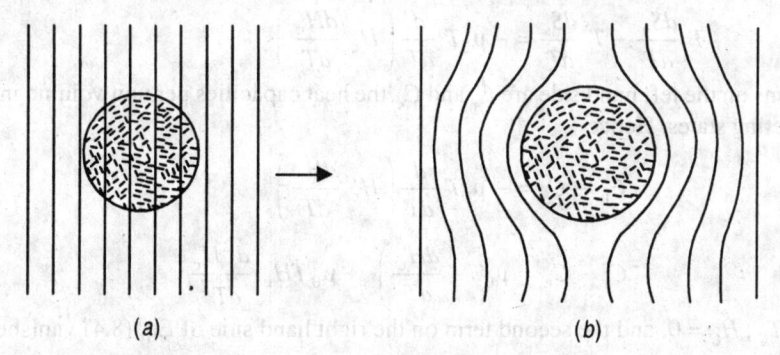

Fig. 8.3. Meissner effect. (*a*) Sphere of a superconducting material cooled in a magnetic field; (*b*) on passing below the transition temperature, the lines of magnetic induction are excluded from the sphere.

Flux exclusion shows that in a superconductor below the transition temperature $\vec{B} = \mu_0 (\vec{H} + \vec{M}) = \vec{0}$, where μ_0 is the permeability of free space, $\vec{H}$ is the magnetic intensity, and $\vec{M}$ is the magnetization. Thus $\vec{M} = -\vec{H}$, so that the susceptibility χ is -1. Therefore a superconductor is a *perfect* diamagnetic material. The perfect conductivity and the perfect diamagnetism are two independent features of superconductors.

8.4. Specific Heat

Figure 8.4 shows the variation of the specific heat with temperature for a normally conducting and a superconducting sample. The remarkable feature is the abrupt change of the specific heat at the transition temperature.

Thermodynamically it is found that the entropy of a superconductor decreases sharply on cooling below the transition temperature. If S_n is the entropy per unit volume of the normal phase and S_s that of the superconducting phase, it can be shown that

$$S_n - S_s = -\mu_0 H_C \frac{dH_c}{dT} \quad \ldots(8.2)$$

Since dH_c/dT is always negative (vide Fig. 8.2), clearly $S_n > S_s$. Entropy being a measure of order, we conclude that the *superconducting electron system is more ordered* than a normal electron system. At $T = 0$, $dH_c/dT = 0$ and at $T = T_c$, $H_c = 0$. Hence at these two points $S_n = S_s$. At all other points, there is a finite entropy change when the transition occurs in a magnetic field. Multiplying the difference in entropy $(S_n - S_s)$ by the temperature we obtain for the latent heat per unit volume in the transition:

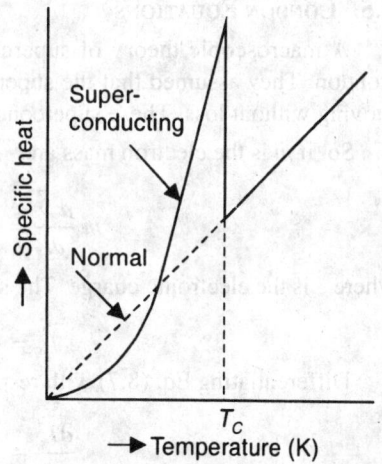

Fig. 8.4. Specific heat versus temperature for normal and superconducting states of a sample. The data for the normal curve is obtained by applying a magnetic field exceeding H_c.

$$Q_{s \to n} = -\mu_0 T H_C \frac{dH_c}{dT} \quad \ldots(8.3)$$

Since $dH_c/dT < 0$, the superconducting-normal transition heat is positive. This heat provides the energy needed to fill the superconductor with the magnetic field when the transition occurs by increasing the temperature in the presence of a magnetic field.

Differentiating both sides of Eq. (8.2) with respect to T and multiplying through by T, we get

$$T\frac{dS_n}{dT} - T\frac{dS_s}{dT} = -\mu_0 T \frac{d}{dT}\left(H_c \frac{dH_c}{dT}\right)$$

The terms on the left hand side are C_n and C_s, the heat capacities per unit volume in normal and superconducting states. Hence

$$C_n - C_s = -\mu_0 T \frac{d}{dT}\left(H_c \frac{dH_c}{dT}\right)$$

or,
$$C_s - C_n = \mu_0 T \left(\frac{dH_c}{dT}\right)^2 + \mu_0 T H_c \frac{d^2 H_c}{dT^2} \qquad ...(8.4)$$

At $T = T_C$, $H_C = 0$, and the second term on the right hand side of Eq. (8.4) vanishes. So,

$$(C_s - C_n)_{T=T_c} = \mu_0 T_c \left(\frac{dH_c}{dT}\right)^2_{T=T_c} \qquad ...(8.5)$$

This shows the increase of the specific heat as the sample is cooled below the critical temperature. Equation (8.5) is referred to as *Rutgers equation*. It is used to find $(dH_c/dT)_{T=T_c}$ from calorimetric measurements of C_s and C_n.

The specific heat data can be fitted to a function like $Ae^{-2\beta\Delta}$, where A and Δ are parameters depending on the material and may depend on temperature. At low temperatures, both A and Δ become independent of temperature, so that the specific heat becomes an exponential function of $1/k_B T$. This indicates an energy gap of about 2Δ in the electron energy spectra.

8.5. LONDON EQUATIONS

A macroscopic theory of superconductivity was developed in 1934 by F. London and H. London. They assumed that the superconductor contains n_s number of electrons per unit volume moving without loss. These superconducting electrons behave as free electrons in an electric field $\vec{E}$. So, if m is the electron mass and $\vec{v}$ is the drift velocity at time t, we have

$$m\frac{d\vec{v}}{dt} = -e\vec{E} \qquad ...(8.6)$$

where e is the electronic charge. The supercurrent density $\vec{J}_s$ is

$$\vec{J}_s = -n_s e \vec{v} \qquad ...(8.7)$$

Differentiating Eq. (8.7) with respect to t and substituting into Eq. (8.6) gives

$$\frac{\partial \vec{J}_s}{\partial t} = \frac{n_s e^2 \vec{E}}{m} \qquad ...(8.8)$$

Equation (8.8) implying the absence of resistance, is known as the *first London equation*. It shows that the presence of an electric field causes the current to change with time. So, steady currents can be set up without an electric field.

We now use the Maxwell equation

$$\vec{\nabla} \times \vec{E} = -\frac{\partial \vec{B}}{\partial t} = -\mu_0 \frac{\partial \vec{H}}{\partial t} \qquad ...(8.9)$$

Here $\vec{B}$ is replaced by $\mu_0 \vec{H}$ since the material has a linear magnetic character. Substituting the curl of Eq. (8.8) into Eq. (8.9), one gets

$$\vec{\nabla} \times \frac{\partial \vec{J}_s}{\partial t} = -\frac{\mu_0 n_s e^2}{m} \frac{\partial \vec{H}}{\partial t} \qquad ...(8.10)$$

Superconductivity

Integrating with respect to t gives

$$\vec{\nabla} \times \vec{J}_s = -\frac{\mu_0 n_s e^2}{m} (\vec{H} - \vec{H}_0) \qquad ...(8.11)$$

The magnetic intensity $\vec{H}_0$ is a constant of integration and represents an arbitrary field at $t = 0$. The Meissner effect precludes frozen-in magnetic fields in the superconductor; so that $H_0 = 0$. Therefore, Eq. (8.11) reduces to

$$\vec{\nabla} \times \vec{J}_s = -\frac{\mu_0 n_s e^2}{m} \vec{H} \qquad ...(8.12)$$

Equation (8.12) is referred to as the *second London equation*.

8.6. Penetration Depth

Taking curl of Eq. (8.12) and using the Maxwell equation

$$\vec{\nabla} \times \vec{H} = \vec{J}_s + \frac{\partial \vec{D}}{\partial t} \qquad ...(8.13)$$

where $\vec{D}$ is the electric displacement, we obtain

$$\vec{\nabla} \times \vec{\nabla} \times \vec{J}_s = -\frac{\mu_0 n_s e^2}{m} \left(\vec{J}_s + \frac{\partial \vec{D}}{\partial t} \right) \qquad ...(8.14)$$

For the dc case, $\partial \vec{D}/\partial t = \vec{0}$. Also we have

$$\vec{\nabla} \times \vec{\nabla} \times \vec{J}_s = \vec{\nabla}(\vec{\nabla} \cdot \vec{J}_s) - \nabla^2 \vec{J}_s,$$

and $\vec{\nabla} \cdot \vec{J}_s = 0$ since no steady accumulation of charge is anticipated. Therefore, Eq. (8.14) reduces to

$$\nabla^2 \vec{J}_s = \frac{\vec{J}_s}{\lambda^2}, \qquad ...(8.15)$$

where

$$\lambda^2 = \frac{m}{\mu_0 n_s e^2} \qquad ...(8.16)$$

The quantity λ having the dimension of distance, is termed the *penetration depth*.

To recognize the significance of λ, we consider the one dimensional form of Eq. (8.15):

$$\frac{d^2 J_s}{dx^2} = \frac{J_s}{\lambda^2},$$

which has the solution

$$J_s = A_1 e^{x/\lambda} + A_2 e^{-x/\lambda} \qquad ...(8.17)$$

The constants A_1 and A_2 are determined from the boundary conditions. We take an infinitely thick superconductor extending in the positive x-direction with the planar surface at $x = 0$. The solution $A_1 e^{x/\lambda}$ involves infinite currents and so it has to be discarded. The constant A_2 is identified with the current density J_0 at the surface. Obviously, J_0 cannot have an x-component (out of the surface). The total current per unit width perpendicular to the direction of the current is

$$I = \int_0^\infty J_0 e^{-x/\lambda} \, dx = J_0 \lambda \qquad ...(8.18)$$

Clearly, the penetration depth λ is the effective thickness of the current carrying layer if the current were uniform with the surface current density J_0. The penetration depth is analogous to the normal skin depth in metals.

With $n_s = 10^{29}$ m^{-3} and $m = m_0$ (the free electron mass), $\lambda \approx 20$ nm. At $T = 0$K, the concentration of superconducting electrons n_s is a maximum. With increasing temperature, n_s decreases and above T_C, n_s becomes zero. Therefore the penetration depth λ is a minimum at 0K and it approaches infinity at $T = T_C$.

To find the penetration of the magnetic field into the superconductor, we assume that the current J_s is in the y-direction (Fig. 8.5). Then $\vec{\nabla} \times \vec{J_s} = \vec{k} \dfrac{dJ_s}{dx}$, where $\vec{k}$ is the unit vector in

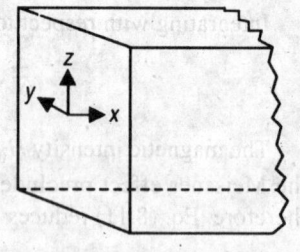

Fig. 8.5. Choice of coordinate axes.

the z-direction. With $J_s = J_0 e^{-x/\lambda}$, we have $\vec{\nabla} \times \vec{J_s} = -\dfrac{J_s}{\lambda}\vec{k} = -\dfrac{1}{\lambda^2}\vec{H}$ using Eq. (8.12). Thus

$$\vec{H} = \lambda J_0 e^{-x/\lambda} \vec{k} \qquad \ldots(8.19)$$

Equation (8.19) shows that H varies with x in the same fashion as J_s does. We at once conclude that the Meissner effect is not complete for thin films of superconductors. If the thickness of the superconductor is of the order of λ, the magnetic field does not drop significantly inside the material. The magnetic field inside the sample is zero at depths of about five times λ.

8.7. Fluxoid

We consider a hole in a superconducting material and a closed loop (indicated by dashed lines) in the material (Fig. 8.6). Using Maxwell's equation $\vec{\nabla} \times \vec{E} = -\dfrac{\partial \vec{B}}{\partial t}$, integrating it over the surface S bounded by the dashed loop, and applying Stoke's theorem of vector analysis, we have

$$\oint \vec{E} \cdot \vec{dl} = -\int_S \dfrac{\partial \vec{B}}{\partial t} \cdot \vec{ds} \qquad \ldots(8.20)$$

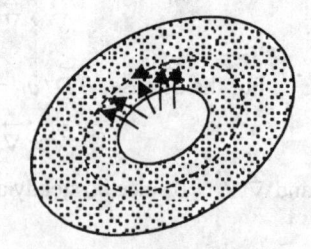

Fig. 8.6. A superconducting ring with magnetic flux through the central hole.

Using the first London equation [Eq. (8.8)], we obtain from Eq. (8.20)

$$\mu_0 \lambda^2 \oint \dfrac{\partial \vec{J_s}}{\partial t} \cdot \vec{dl} = -\int_S \dfrac{\partial \vec{B}}{\partial t} \cdot \vec{ds} \qquad \ldots(8.21)$$

or,

$$\dfrac{\partial}{\partial t}\left(\mu_0 \lambda^2 \oint \vec{J_s} \cdot \vec{dl} + \int_S \vec{B} \cdot \vec{ds}\right) = 0 \qquad \ldots(8.22)$$

Denoting the quantity within brackets by ϕ_s we have,

$$\phi_s = \mu_0 \lambda^2 \oint \vec{J_s} \cdot \vec{dl} + \int_S \vec{B} \cdot \vec{ds} = \text{constant} \qquad \ldots(8.23)$$

If the closed loop (shown by the dashed lines) is entirely within the superconductor, and is sufficiently far removed from the surface, then both $\vec{B}$ and $\vec{J_s}$ are zero at all points on it. For such loops, the first term on the right-hand side of Eq. (8.23) vanishes, so that the total magnetic flux through the loop represented by the second term remains constant. Since very little flux passes through the superconducting material deep inside the surface, this flux is practically confined to the hole. The flux ϕ_s is quantized, being given by

$$\phi_s = S \dfrac{\pi \hbar}{e}, \qquad \ldots(8.24)$$

where S is an integer. The quantum of flux, i.e. $\pi \hbar / e$ (= 2.06×10^{-15} Wb) is termed a *fluxoid*.

Superconductivity

The superconducting ring can trap flux. Consider that the magnetic field is switched on when the superconducting ring (Fig. 8.6) is in the normal state. The ring is now cooled below the transition temperature. Flux is excluded from the ring but passes through the hole. If the applied magnetic field is now switched off, the flux through the hole does not change and remains trapped. The supercurrents produced around the ring can maintain the flux through the hole.

8.8. Type I and Type II Superconductors

The magnetization curves, *i.e.* the plots of the magnetization M against the magnetic intensity H are markedly different for two classes of superconductors, known as type I (or soft) and type II (or hard) superconductors.

Type I superconductors exhibit abrupt superconducting – normal transition at the critical field H_C. Below H_C there is no flux penetration and above H_C, the field penetrates perfectly. In the M–H curve for the type I superconductor (Fig. 8.7), no flux penetration in the superconducting state is shown by the 45° straight line since $M = -H$ here. The critical field $H_C(T)$ decreases as the temperature T increases above 0K, a typical curve being shown in Fig. 8.2. Al, Cd, In, Pb, Hg, Sn, Zn etc. are examples of type I superconductors.

In type II superconductors, there is no flux penetration through the specimen below a lower critical field H_{C_1} (Fig. 8.7). If the applied magnetic intensity exceeds an upper critical field H_{C_2}, the whole sample becomes normally conducting with perfect field penetration. For applied fields between H_{C_1} and H_{C_2}, the flux penetrates the sample partially, so that the sample is in a *mixed state* consisting of both superconducting and normal regions. As the magnetic intensity is reduced from above H_{C_2} the flux does not leave the superconductor reversibly. Thus a *hysteresis* is observed and a remanant magnetization R_m occurs when the applied field drops to zero. Al$_2$CMO$_3$, In$_{0.96}$Pb$_{0.04}$, Nb$_3$Ge, Nb$_3$Sn etc. are examples of type II superconductors. High-T_C ceramic superconductors are also type II superconductors.

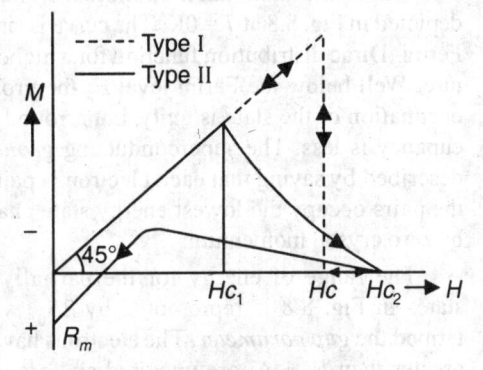

Fig. 8.7. Plot of magnetization M versus the magnetic intensity H for Type I and II superconductors.

For type I superconductors, the critical fields are typically about 10^3 A/m for temperatures well below T_C. For hard type II superconductors, the upper critical field H_{C_2} can exceed 10^6 A/m. Hence type II superconductors are used to construct superconducting motors and magnets.

8.9. Microscopic Theory

The London equations hardly give any idea of the electronic processes leading to superconductivity. A fundamental theory of superconductivity based on quantum mechanics was put forward in 1957 by Bardeen, Cooper, and Schrieffer. This theory, referred to as the *BCS theory* (after the initials of the proposers), and its later modifications could explain the features of superconductivity. The details of the theory will not be given here in view of the complexity of the involved mathematics.

The BCS theory suggests that an attractive force between electrons, sufficiently strong to overcome their mutual electrostatic repulsion, accounts for superconductivity. The origin of such an attractive force is electron-phonon interactions. An electron pulls surrounding ions from their equilibrium positions so that the region around the electron carries more positive charge than in the normal state. After the electron leaves, the region remains positive for a time and hence attracts another electron. The former electron alters the number of phonons which modify the wave function of the latter electron. The attractive force is a maximum when the propagation vectors of the two electrons have the same magnitude and oppositely directed. The spins of the two electrons must be opposite. The pairs of electrons satisfying these requirements are termed *Cooper pairs*. The formation of Cooper pairs creates the ordering suggested by the specific heat data.

The Cooper pair can be regarded as a single particle of zero momentum, zero spin, and charge – $2e$. Cooper pairs obey Bose-Einstein statistics and so any number of them can be in the lowest energy state. The attractive interaction reduces the energy for an electron pair and leads to superconductivity. The electrons in a Cooper pair are thus quasi-bound since their energy is less than that of two free electrons. They are not exactly bound because their net energy is positive with respect to the bottom of the conduction band.

The distribution that minimizes the total energy is depicted in Fig. 8.8 at $T = 0K$. The curve is similar to the Fermi–Dirac distribution function for a higher temperature. Well below the Fermi level E_F the probability of occupation of the state is unity, but around E_F the occupancy is less. The superconducting *ground state* is described by saying that each electron is paired and all the pairs occupy the lowest energy state characterized by zero crystal momentum.

The range of energy for the partially occupied states in Fig. 8.8 is represented by $2\Delta_0$, where Δ_0 is termed the *gap parameter*. The electrons having energy greater than $E_F + \Delta_0$ are unpaired and are the normal electrons. The electrons with energy below $E_F - \Delta_0$ are termed superconducting electrons. For $0 < T < T_C$, the electron system is a mixture of Cooper pairs and normal electrons. At $T = 0K$, all the electrons are superconducting. The *superconducting energy gap* $2\Delta_0$ separating the normally conducting electrons and the superconducting Cooper pairs is about $3.52 k_B T_C$.

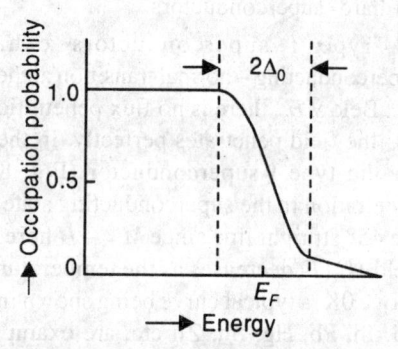

Fig. 8.8. Plot of the occupation probability of electron states of a superconductor at 0K.

The Cooper pairs maintain their coupled motion in the material when separated by a distance called the *correlation length* or the *coherence length* ξ. Typically ξ is about 100 μm.

The materials for which the electron–phonon interaction is strong, are superconducting at higher temperatures than those for which the interaction is weak. The electron–phonon interaction also determines the resistivity of a metal in the normal state, and a strong interaction gives a high resistivity. This explains why good electrical conductors in the normal state make relatively poor superconductors.

The superconducting gap parameter $\Delta(T)$ varies with temperature. It is Δ_0 at $T = 0K$, and decreases with increase in T; it becomes zero for $T > T_C$.

The BCS theory can successfully explain the phenomena of superconductivity. An important experimental finding is that the transition temperature of a superconductor is proportional to $M^{-1/2}$, where M is the atomic weight. Thus if M_1 and M_2 are the atomic weights of two isotopes of the same metal and T_{C_1} and T_{C_2} are their transition temperatures, then $\dfrac{T_{C_1}}{T_{C_2}} = \left(\dfrac{M_2}{M_1}\right)^{1/2}$. This is known as the *isotope effect*. The electron–phonon interaction determining the superconductivity, the BCS theory predicts that the transition temperature is proportional to the Debye frequency which, in turn, is proportional to $M^{-1/2}$. This explains the isotope effect.

The existence of supercurrent, *i.e.* the vanishing of resistance in the superconducting state can also be explained. The Cooper pairs cannot lose kinetic energy in phonon scattering since the phonons cannot supply an energy as high as 2Δ. The absence of phonon scattering means that the resistivity is zero.

When a magnetic field is applied, the kinetic energy of the electron pairs increases because the pairs must move faster to give the current necessary for flux expulsion. If the kinetic energy is

Superconductivity 325

sufficiently large, energy exchange scattering events can occur, splitting the pair into two normally conducting electrons. This accounts for superconducting-normal transition when the applied field exceeds the critical value.

The existence of the superconducting gap 2Δ can be predicted from the analysis of the specific heat data (see Sec. 8.4). The experimental observation of the absorption of electromagnetic radiation in the far infrared region also confirms the existence of the superconducting gap. It is observed that near $T = 0K$ where most of the electrons are superconducting, a sharp absorption edge occurs for radiations of angular frequency ω, where $\hbar\omega \approx 3.5\, k_B T_C$.

This absorption edge is accounted for by the superconducting energy gap $2\Delta_0 \approx 3.52\, k_B T_C$. The states below the superconducting energy gap are occupied by paired electrons. The states above the energy gap are occupied by the normally conducting electrons. By absorbing electromagnetic energy of the appropriate frequency, the superconducting electrons move across the gap and become normally conducting.

8.10. JOSEPHSON JUNCTION

A thin layer of an insulator sandwiched between two superconductors constitutes a Josephson junction (Fig. 8.9). For a thick insulator, the electrons cannot get through it from one superconductor to the other. However, for a sufficiently thin insulating layer, the Cooper pairs can tunnel through the insulator even when there is no potential difference between the superconductors.

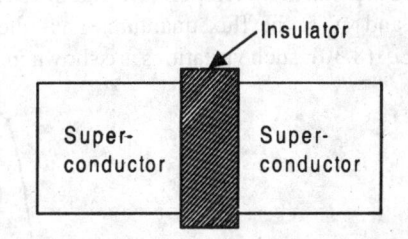

Fig. 8.9. Josephson junction for two superconductors separated by a thin insulator.

Using quantum mechanical concepts, the tunneling current density can be shown to be

$$J = J_0 \sin \delta \qquad ...(8.25)$$

where J_0 is the maximum supercurrent density through the insulator with zero potential difference across it and δ is the change in phase of the order parameter of the Cooper pairs across the junction. For wide insulating layers, J_0 is quite small.

In the presence of a dc voltage V across the insulator, the current density becomes

$$J = J_0 \sin\left(\delta + \frac{q}{\hbar} Vt\right), \qquad ...(8.26)$$

where $q\,(= 2e)$ is the charge of a Cooper pair and t is time. Equations (8.25) and (8.26) give the so called *dc Josephson effect*. For ordinary voltages, $qV/\hbar$ gives a very high frequency, so that the sine function in Eq. (8.26) oscillates rapidly yielding a zero average. Thus we obtain a current for no applied voltage, and no current for an applied voltage V.

The *ac Josephson effect* occurs when a high frequency ac voltage superimposed on a dc voltage is applied across the junction. The applied voltage is thus $V(t) = V_0 + V_a \cos \omega t$, where V_0 is the dc voltage, and $V_a\,(<< V_0)$ is the amplitude of the ac voltage of angular frequency ω. In this case the tunnel current density is

$$J = J_0 \sin\left(\delta + \frac{q}{\hbar} V_0 t\right) + J_0 \left(\frac{q}{\hbar}\right)\left(\frac{V_a}{\omega}\right) \sin \omega t \, \cos\left(\delta + \frac{q}{\hbar} V_0 t\right) \qquad ...(8.27)$$

As in the dc case, for ordinary voltages the first term on the right-hand side of Eq. (8.27) averages to zero. The second term also yields a zero average except for a very special case. To find this case, we ignore the first term in Eq. (8.27) and write it as

$$J = J_0 \left(\frac{q}{\hbar}\right)\left(\frac{V_a}{\omega}\right)\left[\sin \omega t \, \cos \delta \, \cos\left(\frac{q}{\hbar} V_0 t\right) - \sin \omega t \, \sin \delta \, \sin\left(\frac{q}{\hbar} V_0 t\right)\right] \qquad ...(8.28)$$

The second term on the right-hand side of Eq. (8.28) does not average to zero if $\omega = \dfrac{qV_0}{\hbar}$, because then the time functions in it give $\sin^2 \omega t$. So, with

$$\omega = \dfrac{qV_0}{\hbar} \qquad \text{...(8.29)}$$

a current is observed. The phenomenon is analogous to resonance and has been experimentally verified. Since $q = 2e$, the ac Josephson effect can be used to measure the fundamental constant e/h very precisely. It also offers a technique for measuring voltages very accurately.

The tunneling current depends quite sensitively on the magnetic field applied at the junction. It is found that the tunneling current is

$$I = I_0 \dfrac{\sin(\pi \Phi / \Phi_0)}{(\pi \Phi / \Phi_0)}, \qquad \text{...(8.30)}$$

where I_0 is dependent on temperature and the structure of the junction, Φ is the magnetic flux at the junction, and Φ_0 is the flux quantum. Thus the tunneling current varies with the magnetic field obeying Eq. (8.30); such variations are shown in Fig. 8.10.

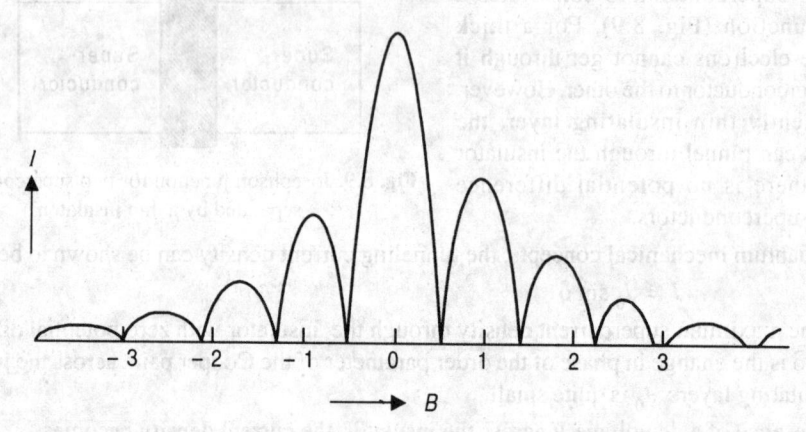

Fig. 8.10. Tunneling current I versus magnetic field B for a Josephson junction.

Interesting magnetic field effects occur in *superconducting quantum interference device* (SQUID) circuits. A SQUID circuit consists of two Josephson junctions in parallel (Fig. 8.11). The circuit is operated below the superconducting transition temperature. A magnetic field is applied through the central region. The currents in the two paths I_1 and I_2 combine to give the total current I entering the point a. If I_0 is the maximum supercurrent through either junction, it can be shown that

$$I = 2I_0 \sin \Phi_0 \cos\left(\dfrac{e\Phi}{\hbar}\right) \qquad \text{...(8.31)}$$

where Φ_0 is a constant and Φ is the magnetic flux through the loop. Equation (8.31) shows that the current is an oscillating function of the flux Φ, attaining a maximum when $e\Phi/\hbar = s\pi$,

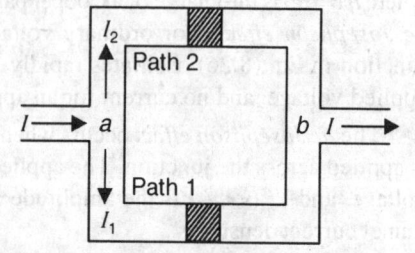

Fig. 8.11. A SQUID circuit.

where s is an integer. Thus the current peaks occur whenever Φ is an integral multiple of $\pi\hbar/e$, the flux quantum. The oscillations originate from the interference of the currents I_1 and I_2.

A simple device based on superconducting quantum interference is a magnetometer, used to measure magnetic fields. The field is measured by increasing the flux from zero and counting the number of peaks in the current. The flux can be measured with a precision less than $\pi\hbar/e$. Hence a SQUID magnetometer is sensitive to very small fields. For instance, for a loop area of 1cm^2 magnetic fields less than 10^{-11} T can be easily measured.

8.11. Applications

(i) Switching devices : A very simple superconducting device for electronic switching applications is the *wound-wire cryotron* (Fig. 8.12). It consists of a superconducting wire of low critical field material (such as Ta) over which several turns of a superconducting wire of a higher critical field material (such as Nb) are wound. If a sufficiently large current passes through the coil of turns, acting as a solenoid, the inner wire becomes normally conducting. The solenoid wire is referred to as the *control wire* and the current I_C through it is called the *control current*. If N is the number of turns in the solenoid of length L, the magnetic intensity produced by the control current is NI_C/L. The inner wire, termed the *controlled wire* or the *gate* carries the current I_g. The circumferential magnetic intensity at the gate surface produced by the gate current I_g is $I_g(2\pi r)$, where r is the radius of the gate. The net magnetic intensity due to I_C and I_g is

$$H = \left[\left(\frac{NI_C}{L}\right)^2 + \left(\frac{I_g}{2\pi r}\right)^2\right]^{1/2} \qquad ...(8.32)$$

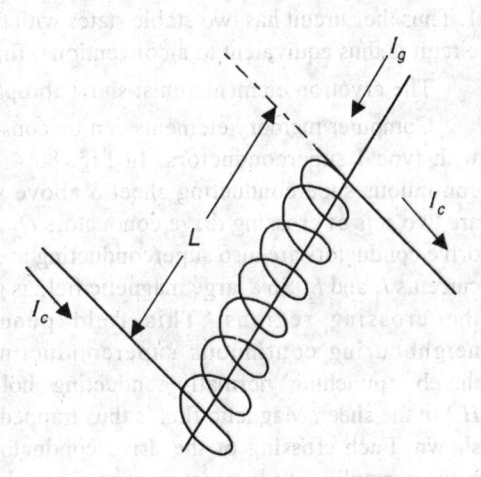

Fig. 8.12. Wound-wire cryotron.

The gate wire turns to a normal conductor when H is greater than H_C. The control wire remains superconducting because the critical field for it is much greater than that for the gate. Primarily the cryotron is a switching device : the gate resistance is either *on* or *off*. Such switches show promise in large-scale computer applications. The simplicity and the small size of the basic elements can offset the cost of maintaining a low-temperature environment.

Figure 8.13 shows a simple cryotron flip-flop circuit. A constant current I passes through the circuit containing two cryotron gate elements. Suppose that both the cryotron gates are superconducting at the time the current is switched on. The current I is divided into the gate currents I_1 and I_2. Equating the voltages across the two arms we have

$$L_1\frac{dI_1}{dt} = L_2\frac{dI_2}{dt}, \qquad ...(8.33)$$

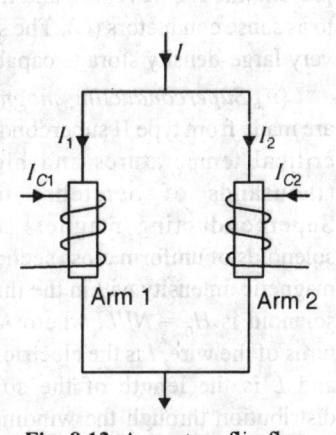

Fig. 8.13. A cryotron flip-flop.

where L_1 and L_2 are the arm inductances. Again,

$$I = I_1 + I_2. \qquad ...(8.34)$$

Equations (8.33) and (8.34) give

$$I_1 = I\frac{L_2}{L_1 + L_2} \text{ and } I_2 = \frac{IL_1}{L_1 + L_2}$$

Usually, $L_1 = L_2$ and hence $I_1 = I_2$. Suppose that the gate in arm 1 is turned resistive by a control current I_{c_1}. The current I then flows totally through arm 2 which remains superconducting. Removal of the control current I_{c_1} does not necessarily shift back the current which remains in arm 2. Only if arm 2 is made resistive momentarily by the control current I_{c_2}, the current does shift to arm 1. Thus the curcuit has two stable states with the current being carried by either arm 1 or arm 2. The circuit is thus equivalent to a conventional flip-flop.

The cryotron elements must show abrupt transitions and so are of type I superconductors.

Computer memory elements can be constructed also with type I superconductors. In Fig. 8.14 is shown a continuous superconducting sheet S above which there are two sets of crossing drive conductors D_1 and D_2. The drive conductors are also superconducting and carry high currents I_1 and I_2. So a large magnetic field is generated at the crossing regions. This field penetrates the neighbouring continuous superconducting sheet S, thereby 'punching' normally conducting 'holes' (H_1 and H_2) in the sheet. Magnetic flux is thus trapped in the loop shown. Each crossing of the drive conductors can thus have normally conducting regions through which the trapped flux always flows in one direction or the other, one representing a binary 0, and the other a 1. The normally conducting holes therefore represent the storage of a binary digit. The memory is monitored by a pair of positive current pulses flowing through the conductors forming a given crossing. If a 1 is present at the crossing, nothing happens. But if a 0 is present, the flux reverses, and this flux change can be sensed by another set of conductors, referred to as sense conductors (C). The sense conductor is normally conducting. The memory has potentially very large-density storage capability.

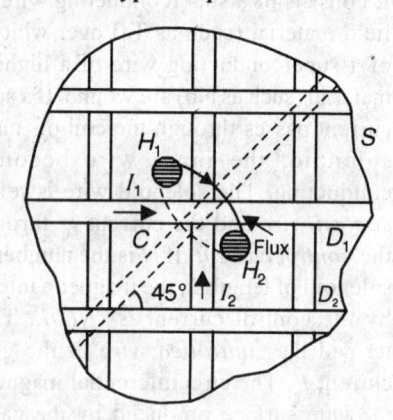

Fig. 8.14. Continuous sheet memory device

(ii) *Superconducting magnets*: Such magnets are made from type II superconductors having high critical temperatures and high critical fields (thousands of oersteds, in some cases). Superconducting magnets are usually long solenoids of uniform cross section (Fig. 8.15). The magnetic intensity within the inner diameter of the solenoid is $H_0 = NI/L$, where N is the number of turns of the wire, I is the electric current in the wire and L is the length of the solenoid. The field distribution through the windings and the central hole of the solenoid is depicted in Fig. 8.16, where H is the magnetic intensity at a radial distance r from the axis of the solenoid. The wire in the innermost layer is subjected to the maximum magnetic intensity. The intensity for the outer layers falls with increasing radius. Finally the intensity becomes zero at the outermost layer. So, to create a high magnetic intensity of some tens of thousand oersted, the inner layer wire must be of a material that can remain superconducting at such high magnetic intensities. For outer layers, the magnetic intensity is less, and so they can use wires of a material of lower critical fields.

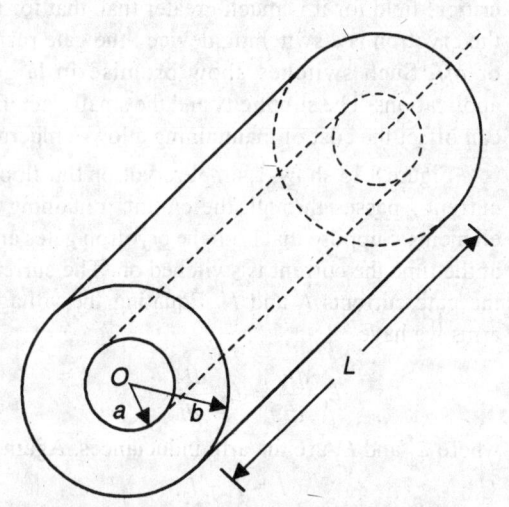

Fig. 8.15. Configuration of a solenoidal form of superconducting magnet.

Superconductivity

Superconducting magnets have the disadvantages that the cost of the raw materials and their processing is very high. There is also the cost of attaining and maintaining the requisite low temperatures. Such magnets are most useful in situations where the high intensities cannot be achieved in another method easily, or where very light weight at the required intensities is desired.

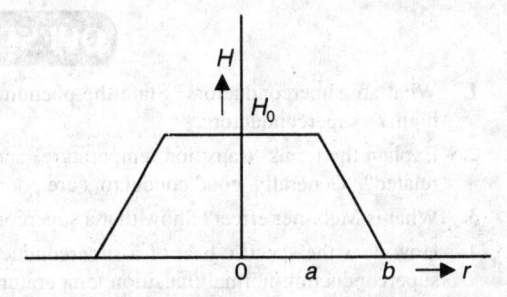

Fig. 8.16. Magnetic intensity distribution in the solenoid.

(*iii*) *Josephson junction devices*: As mentioned in Sec. 8.10, the ac Josephson effect offers a means of measuring the fundamental constant e/h precisely. It also forms the basis for accurate measurement of voltages. The superconducting quantum interference devices (SQUIDs) using parallel Josephson junctions can be operated as magnetometers to measure magnetic fields with a high precision. Medical applications of SQUIDs include the detection of magnetic fields generated by currents circulating in the nervous system and by iron in the digestive system. In geology, SQUIDs are employed to study the magnetic structures underneath the earth's surface.

Sensitive ammeters and voltmeters can also be constructed with SQUIDs. The magnetic field due to an unknown current is measured with a SQUID and compared with the field produced by a known current. The voltmeter application of a SQUID involves the application of an unknown voltage across a known resistance and the measurement of the resulting current.

8.12. Worked-out Problems

1. The critical field H_C of niobium at 0K is 1.55×10^5 A/m. What is the maximum current that can be carried by a niobium wire of diameter 0.1 mm at 0K?

Ans. Applying Ampere's circuital law, it is found that the magnetic intensity at a distance r from the axis of the wire is

$$H = \frac{I}{2\pi r}$$

where I is the current in the wire. Here $r \geq R$, where R is the radius of the wire. The intensity is maximum at $r = R$, and at the maximum current I_m, we have $H(R) = H_C$. Thus the maximum current is

$$I_m = 2\pi R H_C = 3.1416 \times 0.1 \times 10^{-3} \times 1.55 \times 10^5 = 48.7 \text{ A}.$$

2. For lead, the critical field at 0K is 6.39×10^4 A/m and the critical temperature for zero magnetic field is 7.18K. Find the critical field for lead at 4K.

Ans. We have

$$\left(\frac{T_C(H)}{T_C(0)}\right)^2 = 1 - \frac{H_C(T)}{H_C(0)}$$

Here, $H_C(0) = 6.39 \times 10^4$ A/m, $T_C(0) = 7.18$ K, and $T_C(H) = 4$K. So,

$$H_C(T) = H_C(0)\left[1 - \left(\frac{T_C(H)}{T_C(0)}\right)^2\right] = 6.39 \times 10^4 \left[1 - \left(\frac{4}{7.18}\right)^2\right]$$

$$= 4.41 \times 10^4 \text{ A/m}$$

QUESTIONS

1. What are superconductors? State the phenomena associated with superconductivity. Name a few high-T_c superconductors.
2. Explain the terms 'transition temperature' and 'critical field' for a superconductor. How are they related? "Generally, good conductors are poor superconductors" – Why?
3. What is Meissner effect? Show that a superconductor is a perfect diamagnetic material.
4. How does the specific heat of a superconductor vary with temperature? What is observed at the superconducting-normal transition temperature? How is this behaviour related to entropy?
5. Derive London equations. Explain the significance of the term 'penetration depth'.
6. Show that the magnetic flux is confined to a hole in a superconductor. What is fluxoid?
7. Distinguish between type I and type II superconductors. Name some materials belonging to these two types of superconductors.
8. What are Cooper pairs? Which statistics do they obey? Explain: Superconducting ground state, superconducting energy gap, and coherence length.
9. What is isotrope effect? How is it explained by the BCS theory?
10. What is can BCS theory account for vanishing of resistance in the superconducting state, and the superconducting – normal transition in the presence of a magnetic field exceeding the critical value?
11. What is the experimental evidence in support of the superconducting energy gap?
12. What is a Josephson junction? Explain dc Josephson effect and the ac Josephson effect. What is a SQUID?
13. State and explain some properties and applications of superconductors.
14. Write short notes on:
    (*i*) BCS theory
    (*ii*) Superconducting switching devices
    (*iii*) Superconducting magnets
    (*iv*) Josephson junction devices.
15. The critical field of niobium at 0K is 1.55×10^5 A/m and the critical temperature in the absence of a magnetic field is 9.25 K. What is the critical field at 4 K? Calculate also the maximum supercurrent that can be carried by a niobium wire of radius 0.1 mm at 4 K. (**Ans.** 1.26×10^5 A/m, 79.2 A)
16. The critical fields of a superconductor at 6 K and 8 K are 7.616 and 4.284×10^6 A/m, respectively. Find the critical temperature for zero magnetic intensity and the critical field at 0K.

(**Ans.** 10 K, 11.9×10^6 A/m)

17. Mercury having an average atomic mass of 200.59 amu has a critical temperature of 4.153 K. Calculate the critical temperature of the isotope $_{80}Hg^{204}$. (**Ans.** 4.118 K)

MISCELLANEOUS PROBLEMS

1. The position-momentum uncertainty relation is $\Delta x\, \Delta p_x \geq \hbar/2$. A particle oscillates simple harmonically, so that the rms uncertainty in its x-coordinate is a. Show that the rms uncertainty in p_x is $\hbar/(4a)$.

2. The wave function $\psi(x)$ of a one-dimensional physical system is an even function of x. Find the expectation values of the position coordinate x and of the momentum p_x. **(Ans. 0, 0)**

3. A particle of mass m is confined in an infinite square well such that the potential energy V is given by
$$V = 0 \quad \text{for} \quad 0 < x < a$$
$$= \infty \quad \text{otherwise}$$

The wavefunction of the particle at time $t = 0$ is $\psi(x,0) = \dfrac{1}{\sqrt{2}}[u_1(x) + u_2(x)]$, where $u_n(x)$ is the normalized wavefunction of the nth stationary state. The energy eigenvalues are $E_n = n^2\hbar^2\pi^2/(2ma^2)$. Determine the smallest positive time t for which $\psi(x, t)$ would be orthogonal to $\psi(x, 0)$.

[Solution : The wavefunction $\psi_n(x, t)$ is given by
$$\psi_n(x,t) = u_n(x) \exp(-iE_n t/\hbar)$$

By the problem, $\psi(x,0) = \dfrac{1}{\sqrt{2}}[u_1(x) + u_2(x)]$

Therefore, $\psi(x,t) = \dfrac{1}{\sqrt{2}}[u_1(x)e^{-iE_1 t/\hbar} + u_2(x)e^{-iE_2 t/\hbar}]$

where $E_1 = \dfrac{\hbar^2 \pi^2}{2ma^2}$ and $E_2 = \dfrac{4\hbar^2 \pi^2}{2ma^2} = 4E_1$. Putting $e^{-iE_1 t/\hbar} = \beta$, we obtain $e^{-iE_2 t/\hbar} = \beta^4$. So,

$$\psi(x,t) = \dfrac{1}{\sqrt{2}}[\beta u_1(x) + \beta^4 u_2(x)]$$

Here $u_1(x)$ and $u_2(x)$ are orthogonal to each other and are normalized. If $\psi(x, t)$ is orthogonal to $\psi(x, 0)$, we have

$$\int_0^a \psi^*(x,t)\psi(x,0)\,dx = 0$$

giving $\beta + \beta^4 = 0$, or $\beta^3 = -1$, since $\beta \neq 0$. For the smallest positive value of t, we get $\beta^3 = -1 = e^{-i\pi}$ or $\beta = e^{-i\pi/3}$ or $\dfrac{E_1 t}{\hbar} = \dfrac{\pi}{3}$ or, $t = \dfrac{\pi\hbar}{3E_1} = \dfrac{2ma^2}{3\pi\hbar}$ **]**

4. Prove that the commutators $[\hat{x}, \hat{L}_z]$ and $[\hat{x}^2, \hat{L}_z]$, where L_z is the z-component of the angular momentum of a particle of position coordinates (x, y, z), are given by $[\hat{x}, \hat{L}_z] = -i\hbar y$ and $[\hat{x}^2, \hat{L}_z] = -2i\hbar xy$

5. What is the degeneracy of the first excited state of an isolated hydrogen atom? **(Ans. 4)**

6. A system is in a state defined by the wavefunction $\psi(\theta, \phi) = \dfrac{1}{\sqrt{14}}(3Y_4^0 + Y_6^0 - 2Y_6^3)$, where $Y_l^m(\theta, \phi)$ are spherical harmonics. What is the probability of finding the system in a state with $m = 0$? **(Ans. 5/7)**

7. The eigenfunction for the Hamiltonian of a particle of mass m in a one-dimensional potential $V(x)$ is $\psi(x) = Ae^{-bx^2/2}$, where A and b are constants. Show that the x-dependence of $V(x)$ must be x^2.

8. The wavefunction of a particle in one dimension is $\psi(x, t) = a\exp[i(px - Et)/\hbar] + b\exp[-i(px + Et)/\hbar]$. Calculate the probability current density. $\left[\textbf{Ans. } \dfrac{p}{m}(a^2 - b^2)\right]$

9. The expectation value of an observable A in a certain state is $<A> = l$. Also $<A^2> = m$. What is the uncertainty in A ?

   (Ans. $\sqrt{m - l^2}$)

10. ψ_1 and ψ_2 are the wavefunctions of two orthogonal states of a particle having the energy eigenvalues E_0 and $-E_0$, respectively. The measurement of the energy of another state ψ of the particle yields $E_0/2$ for the expectation value of the energy. Express ψ in terms of ψ_1 and ψ_2.

    (Ans. $\psi = \dfrac{\sqrt{3}}{2} \psi_1 + \dfrac{1}{2} \psi_2$)

11. Find the number of electrons per m^3 at the Fermi level at a temperature $T (\neq 0 \text{ K})$ if there are 10^{22} quantum states per m^3 at this level. (Ans. 5×10^{21}/m^3)

12. Two particles are to be distributed in three levels having energies 0, ε and 2ε, so that the total energy is 2ε. Determine the number of ways in which the particles can be disrtributed if they are (i) classical, (ii) identical fermions, (iii) identical bosons. What type of particles has the minimum entropy?

    (Ans. 3, 1, 2; fermions)

13. Find the fraction of electrons in a free electron gas at a temperature of 0 K between the energies $E_F - \varepsilon$ and E_F, where E_F is the Fermi energy and $\varepsilon \ll E_F$. (Ans. $3\varepsilon/2E_F$)

    [Hint : The number of electrons between the energies E and $E + dE$ is $N(E) \, dE = CE^{1/2} f(E)dE$, where C is a constant and $f(E)$ is the FD distribution function. The required fraction is $\int_{E_F - \varepsilon}^{E_F} N(E) \, dE \, / \, \int_0^{E_F} N(E) \, dE$]

14. Three noninteracting bosons occupy a region having energies ε_0, $4\varepsilon_0$, and $9\varepsilon_0$. Find the ground-state energy of the system.   (Ans. $3\varepsilon_0$)

15. In a system of free electrons at 0K, the energy per electron is 3 eV. What is the Fermi energy ?

    (Ans. 5eV)

16. A system consists of a large number of distinguishable particles. Each particle can be in any of the two nondegenerate states differing in energy by 0.1 eV. The system is in thermal equilibrium at room temperature, i.e. 27°C. Show that the ratio of the number of particles occupying the higher energy state to that occupying the lower energy state is approximately $\exp(-4)$.

17. Given, Fermi temperature of copper is 80000 K. Show that the average speed of a conduction electron in copper is of the order of 10^6 m/s.

18. An isolated system of a number of noninteracting distinguishable particles is in thermal equilibrium at an absolute temperature T. Each particle can occupy three possible nondegenerate states of energies 0, ε, and 3ε. If $T \gg \varepsilon/k_B$, where k_B is Boltzmann's constant, calculate the average energy of each particle.   (Ans. $4\varepsilon/3$)

19. N distinguishable particles are distributed among three nondegenerate states having energies 0, $k_B T$ and $2k_B T$, where k_B is Boltzmann's constant and T is the absolute temperature. The total equilibrium energy of the system is $1000 \, k_B T$. Determine N.   (C.U. 2008) (Ans. 2354)

20. First order Bragg reflections taken with copper K_α-line of wavelength 1.542 Å for a cubic solid crystal occur at angles θ such that $\sin^2 \theta$ values are 0.06, 0.16 and 0.22 for three sets of parallel planes. Find the Miller indices of the diffracting planes and the type of the cubic lattice. Also determine the lattice parameter.

    [Solution : The $\sin^2\theta$ values are given by Bragg's law for first order diffraction, i.e. $2d \sin\theta = \lambda$ where $d = a/\sqrt{h^2 + k^2 + l^2}$. Thus $\sin^2\theta \propto (h^2 + k^2 + l^2)$. The given $\sin^2\theta$ values are in the ratio 3 : 8 : 11. The Miller indices $(h \, k \, l)$ of the diffracting planes giving these ratios are (111), (220), and (311). The Miller indices are either all even or all odd. Hence the lattice is fcc. Putting $\lambda = 1.542$ Å, $\sin\theta = \sqrt{0.06}$, and $h = k = l = 1$ in $2a \sin\theta / \sqrt{h^2 + k^2 + l^2} = \lambda$, we get $a = 5.453$ Å].

21. The primitive basis vectors (in Å) of a crystal lattice are $\vec{a} = 2\vec{i} + \vec{j}$, $\vec{b} = 2\vec{i} + \vec{k}$, and $\vec{c} = 2\vec{k} + \vec{i}$. Show that the corresponding reciprocal lattice is a bcc lattice with cube edge πÅ^{-1}.

Miscellaneous Problems

22. When an intrinsic semiconductor is lightly doped with donors, the free electron concentration increases from n to na. If μ_n is the electron mobility and μ_p is the hole mobility, what is the electrical conductivity of the doped semiconductor? **(Ans.** $ne(a\mu_n + \mu_p/a)$.)

23. The Hall coefficient of a bcc metal is -2.36×10^{-10} m³/C. Determine the Fermi energy of the metal in eV. What is the lattice constant of the metallic crystal? Assume that each atom contributes one free electron. **(Ans.** 3.23 eV, 4.22 Å)

24. N number of noninteracting atoms of a solid are subjected to a constant external magnetic field H along the z-direction at an absolute temperature T. If J is the total angular momentum quantum number of each atom, g is Lande's factor, μ_B is the Bohr magneton, and k_B is the Boltzmann constant, show that the partition function of an atom is

$$\frac{\sinh[g\mu_0 H(2J+1)/2k_BT]}{\sinh(g\mu_B\mu_0 H/2k_BT)}$$

[**Solution**: An atom has the magnetic quantum number M_J, the possible values of M_J being $J, J-1, J-2, \ldots 0, \ldots, -(J-2), -(J-1), -J$. The potential energy of an atomic dipole in the field H is $-M_J g\mu_B \mu_0 H$, μ_0 being the permeability of the space. The partition functions is

$$Z = \sum_{M_J=-J}^{+J} \exp[-(-M_J g\mu_B\mu_0 H)/k_BT]$$

Putting $x = g\mu_B\mu_0 H/k_BT$, we have

$$Z = \sum_{M_J=-J}^{+J} e^{M_J x} = e^{Jx} + e^{(J-1)x} + \ldots + e^{-(J-1)x} + e^{-Jx}$$

$$= e^{Jx}\frac{[1-e^{-(2J+1)x}]}{1-e^{-x}} = \frac{e^{(J+\frac{1}{2})x} - e^{-(J+\frac{1}{2})x}}{e^{x/2}-e^{-x/2}} = \frac{\sinh(J+\frac{1}{2})x}{\sinh(x/2)}$$

$$= \frac{\sinh[g\mu_B\mu_0 H(2J+1)/2k_BT]}{\sinh(g\mu_B\mu_0 H/2k_BT)}$$

25. The velocity of the elastic waves in a solid crystal is independent of the polarization of the wave. If the Debye temperature is 450 K, the density of the solid is 8.9 g/cm³ and its atomic weight is 58.71, what is the velocity of the elastic waves? Given, Avogadro number = 6.02×10^{23}.

(Ans. 3370 m/s)

26. A monochromatic beam of X-rays of wavelength 1.54 Å is employed for X-ray diffraction studies of a powder sample of an fcc crystalline solid of lattice constant 4.09 Å. What is the smallest value of the angle θ for which the Bragg reflection will occur?

If, instead of the powder sample, a single crystal of the same solid is used and X-rays are incident along the [100] direction, find the longest wavelength of X-ray that gives diffraction for (311) planes.

[**Solution**: For the fcc lattice, the smallest value of θ occurs for the reflection from (111) planes. The separation between such planes is $d_{111} = \frac{4.09}{\sqrt{1^2+1^2+1^2}} = 2.361$ Å. Since $2d \sin\theta = \lambda$, where $\lambda = 1.54$ Å and $d = 2.361$ Å we have $\sin\theta = (1.54/(2 \times 2.361)) = 0.326$ whence $\theta = 19.03°$.

If the angle between [100] and [311] directions is ϕ, we have $\cos\phi = 3/\sqrt{11}$ whence $\phi = 25.24°$. The angle between the incident X-ray beam and the (311) plane is $\theta = 90° - \phi = 64.76°$. The separation between the (311) planes is $d = 4.09/\sqrt{3^2+1^2+1^2} = 4.09/\sqrt{11}$ Å. The desired wavelength is $\lambda = 2d \sin\theta = 2 \times 4.09 \times 3/11 = 2.23$ Å]

27. A simple cubic crystal of lattice constant 4 Å is used to scatter X-rays of wavelength 4.8 Å. Obtain the Miller indices of all sets of planes which give the first order Bragg reflection.

(Ans. {100}, {110})

28. Find the angles between the primitive basis vectors of (i) a bcc lattice, and (ii) an fcc lattice.

(Ans. (i) 109.47°; (ii) 60°]

29. A number of distinguishable particles are distributed among three energy levels 0, k_BT, and $2k_BT$, where k_B is Boltzmanns's constant and T is the absolute temperature of the system of particles. The degeneracy of the levels are 1, 1, and 2, respectively. What is the partition function?

(Ans. $1 + e^{-1} + 2e^{-2}$)

30. A normalized wavefuction $\psi(x)$ of a linear harmonic oscillator is $\psi(x) = \sum_{n=0}^{\infty} C_n u_n(x)$ where $u_n(x)$ is the stationary state with energy $(n + 1/2)\hbar\omega$. If only C_0 and C_1 are not zero, and the mean value of the energy for this condition is $\frac{5}{4}\hbar\omega$, determine the probability of finding the particle in the ground state.

(Ans. 1/4)

31. A semiconductor has a conduction band whose energy E_{CB} as a function of the wave vector k is given by $E_{CB} = E_1 - E_2 \cos(ka)$. The valence band energy E_{VB} is described by $E_{VB} = E_3 - E_4 \sin^2(ka/2)$, where $E_3 < (E_1 - E_2)$, and $-\frac{\pi}{a} \leq k \leq \frac{\pi}{a}$.

(i) Sketch the variation of energy in the two bands as a function of k in the range $-\frac{\pi}{a} \leq k \leq \frac{\pi}{a}$.

(ii) What is the bandwidth of the conduction band and that of the valence band?

(iii) Find the band gap of the material.

(iv) What is the effective mass of the electron at the bottom of the conduction band?

[Ans. (ii) $2E_2$, E_4; (iii) $E_1 - E_2 - E_3$; (iv) $\hbar^2/E_2 a^2$)]

32. The electron concentration in a metal is 10^{29} m^{-3}. Calculate the number of free electrons per m^3 with energies between 0 and $E_F/3$.

(Ans. 1.92×10^{28} m^{-3})

33. The energy of each particle in a system of N (>>1) distinguishable noninteracting particles can be either zero or E (> 0). There are n_0 particles with zero energy and n_1 particles with energy E. The constant total energy of the system is U. Show that the entropy of the system is

$$S = k_B \left[N \ln N - \frac{U}{E} \ln \frac{U}{E} - \left(N - \frac{U}{E}\right) \ln \left(N - \frac{U}{E}\right) \right].$$

[Hint. The number of ways in which the particles are arranged in the energy levels is $W = N!/(n_0! n_1!)$. Use Stirling's theorem to express the entropy $S = k_B \ln W$. Solve the equations $N = n_0 + n_1$ and $U = n_1 E$ for n_0 and n_1, and substitute in the expression for S.]

34. From the expression for S in problem No. 33, obtain the temperature T of the system. Hence show that $T > 0$ if $n_0 > N/2$.

[Hint. For a fixed number of particles, we have $\frac{1}{T} = \left(\frac{\partial S}{\partial U}\right)_{N = \text{const.}}$

Use the expression for S to obtain $T = E/[k_B \ln (EN/U - 1)] = E/[k_B \ln (n_0/n_1)]$. Clearly $T > 0$ if $n_0 > n_1$, i.e., if $n_0 > N/2$.]

35. A hydrogen atom is in the state represented by the normalized wavefunction $\psi(\vec{r}) = \frac{1}{\sqrt{3}} \psi_{200}(\vec{r}) + \sqrt{\frac{2}{3}} \psi_{210}(\vec{r})$ where $\psi_{nlm}(\vec{r})$ is the energy eigenfunction of the hydrogen atom in the state defined by the quantum numbers n, l, m, and having energy $-m_e e^4/[32\pi^2 \varepsilon_0^2 \hbar^2 n^2]$. Calculate the expectation values of energy, orbital angular momentum, and its Z-component. (C.U. 2000)

(Ans. $-m_e e^4/[128\pi^2 \varepsilon_0^2 \hbar^2]$, $2\hbar/\sqrt{3}$, 0)

36. A particle of mass m moving one dimensionally in a region of potential $V(x)$ is in the energy eigenstate
$$\psi(x) = (A^2/\pi)^{1/4} \exp(-A^2 x^2/2)$$
with energy $E = \hbar^2 A^2/(2m)$. Determine

(i) the mean position and the mean momentum of the particle, and
(ii) $V(x)$.

[Solution : (i) $<x> = \int_{-\infty}^{+\infty} \psi^* \, x \, \psi \, dx = \frac{A}{\sqrt{\pi}} \int_{-\infty}^{+\infty} x \, e^{-A^2 x^2} \, dx = 0$,

the integrand being an odd function of x. Also,

$$<p> = -i\hbar \int_{-\infty}^{+\infty} \psi^* \frac{d\psi}{dx} \, dx = -\frac{iA\hbar}{\sqrt{\pi}} \int_{-\infty}^{+\infty} e^{-A^2 x^2/2} \frac{d}{dx}\left(e^{-A^2 x^2/2}\right) dx$$

$$= \frac{iA^3\hbar}{\sqrt{\pi}} \int_{-\infty}^{\infty} x \, e^{-A^2 x^2} \, dx = 0.$$

(ii) Schrödinger's equation is

$$\left[-\frac{\hbar^2}{2m}\frac{d^2}{dx^2} + V(x)\right]\psi(x) = E\,\psi(x)$$

or, $\quad -\frac{\hbar^2}{2m}\frac{d^2\psi}{dx^2} = [E - V(x)]\psi(x)$

or, $\quad -\frac{\hbar^2}{2m}\frac{d^2}{dx^2}(e^{-A^2 x^2/2}) = [E - V(x)]e^{-A^2 x^2/2}$

or, $\quad \frac{\hbar^2 A^2}{2m}(1 - A^2 x^2) = E - V(x)$

or, $\quad V(x) = \frac{\hbar^2 A^4 x^2}{2m}$, since $E = \frac{\hbar^2 A^2}{2m}$.]

OBJECTIVE-TYPE QUESTIONS

Choose the correct answer :-

1. The principle of complementarity refers to (i) the particle nature of waves, (ii) the wave nature of particles, (iii) the natural exclusiveness of the particle and the wave characters.
2. The de Broglie wavelength of a particle of momentum p is
   (i) h/p (ii) p/h,   (iii) $2h/p$.
3. The Compton effect shows (i) the wave character of photons, (ii) the particle character of electromagnetic waves, (iii) the dual nature of matter.
4. For a nonrelativistic particle moving with a velocity v, the phase velocity of the de Broglie wave is
   (i) v,   (ii) v/2,   (iii) 2v.
5. For a relativistic particle moving with a velocity comparable to c, the velocity of light in free space, the phase velocity of the de Broglie wave is
   (i) greater than c,   (ii) equal to c,   (iii) less than c.
6. For a particle moving with a velocity v, the group velocity of the wave packet representing the particle is equal to
   (i) v/3,   (ii) v,   (iii) 3v.
7. Heisenberg's uncertainty principle relates the uncertainties in
   (i) position and energy,
   (ii) time and energy,
   (iii) linear momentum and angular momentum.
8. The lowest value of the product of momentum and position uncertainties is
   (i) h,   (ii) $2\hbar$,   (iii) $\hbar/2$.
9. The quantum mechanical operator for the linear momentum is
   (i) $-i\hbar\nabla$ (ii) $i\hbar\nabla$,   (iii) $-i\hbar\nabla$.

10. The kinetic energy operator in quantum mechanics is

   (i) $-\dfrac{\hbar^2}{2m}\nabla^2$,   (ii) $-\dfrac{h^2}{2m}\nabla^2$,   (iii) $\dfrac{h^2}{2m}\nabla^2$

11. The quantum mechanical energy operator is

   (i) $-i\hbar\dfrac{\partial}{\partial t}$,   (ii) $-ih\dfrac{\partial}{\partial t}$,   (iii) $i\hbar\dfrac{\partial}{\partial t}$.

12. For a wavefunction ψ, the position probability density is

   (i) $\psi^*\psi$,   (ii) $\int_{-\infty}^{\infty}\psi\,dx$,   (iii) $\int_{-\infty}^{\infty}\psi^*\psi\,dx$.

13. A wavefunction ψ is normalized if

   (i) $\int \psi\,d\tau = 1$,   (ii) $\int \psi^*\psi\,d\tau = 0$,   (iii) $\int \psi^*\psi\,d\tau = 1$

14. An energy eigenfunction represents a stationary state of the particle, if

   (i) $\psi^*\psi$ goes to zero at $\pm\infty$,   (ii) $\psi^*\psi$ is constant in time,   (iii) ψ can be normalized.

15. The eigenvalues of Hermitian operators are

   (i) real,   (ii) complex,   (iii) imaginary conjugates.

16. If $\hat{T}$ and $\hat{V}$ are the kinetic energy and the potential energy operators, respectively, of a particle, the Hamiltonian operator is

   (i) $\hat{T} - \hat{V}$,   (ii) $\hat{T}\hat{V}$,   (iii) $\hat{T} + \hat{V}$.

17. Two observables are compatible, if

   (i)   they are represented by commutating operators,
   (ii) they are represented by noncommutating operators,
   (iii) they obey the uncertainty principle.

18. The probability current density exists, if

   (i) ψ is complex,   (ii) ψ is real,   (iii) none of the above.

19. The energy eigenvalue of a free particle is

   (i) nondegenerate,   (ii) triply degenerate,   (iii) doubly degenerate.

20. A quantum mechanical harmonic oscillator in the ground state has the energy

   (i) 0,   (ii) $\dfrac{1}{2}\hbar\omega_o$,   (iii) $\dfrac{1}{2}h\omega_o$.

21. For a quantum mechanical harmonic oscillator in the ground state, the position probability density is a maximum

   (i)   at the centre,
   (ii) at the ends of the classical limit of motion,
   (iii) none of the above.

22. The position probability density of the electron in the hydrogen atom in the ground state is a maximum at

   (i) $a_o/2$,   (ii) a_o,   (iii) $2a_o$,

   where a_o is the first Bohr radius.

23. Phase space represents

   (i)   vibration states,
   (ii) energy states,
   (iii) wavefronts,
   (iv) combination of position and momentum.

(Purvanchal Univ. 2003)

24. The partition function for a system is given by
    (i) e^{-E_i/k_BT},
    (ii) $\sum_i e^{-E_i/k_BT}$,
    (iii) e^{E_i/k_BT},
    (iv) $\sum_i e^{E_i/k_BT}$.

    (Purvanchal Univ. 2003)

25. The particles obeying the Maxwell-Boltzmann statistics
    (i) are distinguishable particles,
    (ii) are indistinguishable particles,
    (iii) have no special character.

26. Which of the following does not follow Fermi-Dirac distribution law?
    (i) Electron,   (ii) Neutron,
    (iii) Proton,   (iv) Radiation.

    (Purvanchal Univ. 2003)

27. Bose-Einstein statistics does not hold for
    (i) neutrinos,   (ii) pions,   (iii) alpha particles.

28. Miller indices refer to
    (i) two perpendicular planes in a crystal,
    (ii) a particular plane in a crystal,
    (iii) a set of parallel planes in a crystal.

29. Fermi level is defined as
    (i) the lowest filled level at 0K,
    (ii) the highest filled level at 0K,
    (iii) the highest filled level at 300 K,
    (iv) the lowest filled level at 300 K.

    (Purvanchal Univ. 2003)

30. The cubic unit cell of a bcc lattic contains
    (i) one lattice point,   (ii) nine lattice points,   (iii) two lattice points.

31. The number of lattice points in the cubic unit cell of an fcc lattice is
    (i) 4   (ii) 14,   (iii) 6

32. The primitive unit cell of a crystal lattice contains
    (i) 8 lattice points,   (ii) no lattice point,   (iii) 1 lattice point.

33. In a cubic crystal, the [h k l] direction and the (h k l) plane are
    (i) perpendicular,
    (ii) parallel,
    (iii) oriented at an angle of 45°.

34. At the equilibrium separation of a pair of atoms in a crystal,
    (i) the potential energy is a maximum,
    (ii) the potential energy is a minimum,
    (iii) the interatomic force is a minimum.

35. For the Bragg reflection from a set of parallel adjacent planes separated by d, the wavelength of the X-rays must be
    (i) less than d,   (ii) greater than d,
    (iii) greater than $2d$,   (iv) less than or equal to $2d$.

36. Which of the following Bragg reflections are absent for an fcc crystal?
    (i) (100),   (ii) (200),
    (iii) (220),   (iv) (111).

37. In the Kronig-Penney model of a linear lattice, if the strength of the periodic potential increases, the width of the allowed energy bands
    (i) increases,   (ii) decreases,   (iii) remains constant.
    (Purvanchal Univ. 2003)

38. n-type silicon is obtained when intrinsic silicon is doped with
    (i) phosphorus,   (ii) aluminium,
    (iii) gold,   (iv) silver.

39. The Fermi level of an intrinsic semiconductor lies in the
    (i) conduction band,   (ii) valence band,   (iii) forbidden gap.

40. The electrical conductivity of an intrinsic semiconductor
    (i) increases with rise of temperature,
    (ii) decreases with rise of temperature,
    (iii) remains constant with rise of temperature.

41. The unit of Hall coefficient is
    (i) C/m^3,   (ii) $C.m^3$,
    (iii) m^3/C,   (iv) $C.m$.

42. In Debye's theory of specific heat of solids, the atomic oscillators obey
    (i) MB statistics,   (ii) FD statistics,   (iii) BE statistics.

43. The paramagnetic susceptibility of a solid varies with the absolute temperature T as
    (i) T,   (ii) T^2,
    (iii) T^0,   (iv) T^{-1}

44. Which of the following rotational symmetries is not possessed by a crystal ?
    (i) 2 - fold,   (ii) 4 - fold
    (iii) 5 - fold,   (iv) 6 - fold

45. Ferromagnetism is solids originates from
    (i) orbital motions of atomic electrons
    (ii) both orbital and spinning motions of atomic electrons,
    (iii) spinning motion of atomic electrons
    (iv) none of the above.

ANSWERS

1. (iii),	2. (i),	3. (ii),	4. (ii),
5. (i),	6. (ii),	7. (ii),	8. (iii),
9. (i),	10. (i),	11. (iii),	12. (i),
13. (iii),	14. (ii),	15. (i),	16. (iii),
17. (i),	18. (i),	19. (iii),	20. (ii),
21. (i),	22. (ii),	23. (iv),	24. (ii),
25. (i),	26. (iv),	27. (i),	28. (iii),
29. (ii),	30. (iii),	31. (i),	32. (iii),
33. (i),	34. (ii),	35. (iv),	36. (i),
37. (ii),	38. (i),	39. (iii),	40. (i),
41. (iii),	42. (i),	43. (iv),	44. (iii)
45. (iii)			

INDEX

A

Acceptor, 227
  level, 227
Accessible state, 117
Acoustic modes, 267
Allowed energy bands, 217
Amorphous solids, 171
Ampere's law, 287
Amphoteric impurity, 228
Amplitude of scattered wave, 203
Angular momentum, 101
Antiferromagnetism, 308
Associated Laquerre polynomial, 104
  Legendre functions, 100
Atomic scattering factor, 204
Average speed, 127
  value, 30
Azimuthal angular momentum
  quantum number, 103

B

BCS theory, 323
B-H curves, 307
Band theory of solids, 214
Basis, 175
  vectors, 175
Behaviour of thermodynamic systems
  near 0K, 165
Binding energy, 187
Biot-Savart law, 287
Block's theorem, 216
Body-centered cubic lattice, 176
Bohr magneton, 292
  radius, 16, 105

Boltzmann relation, 123
  transport equation, 248
Bond energy, 187
Bose condensation, 153
Bose-Einstein distribution function, 152
  postulates of, 153
  statistics, 151
Boson, 151
Boundary conditions, 40, 265
Bragg's diffraction law, 195
  from Laue condition, 200
Bragg's Spectrometer, 207
Bravais lattice, 176
Brillouin theory, 309

C

Canonical distribution, 125
Centrifugal potential energy, 101
Chemical potential, 129
Classical limit, 154
  Oscillator, 82
  system, 117
Classification of crystals, 171
Clausius-Mossotti relationship, 280
Close packed arrangement, 179
Coercive field, 281, 306
Coherence Length, 324
Collision time, 237
Commutation of operators, 37
Compensated semiconductor, 14
Complementarity principle, 14
Compton effect, 5
  scattering, 5
  wavelength, 6
Conduction band, 223
  electrons, 237

Cooper pairs, 324
Coordination number, 178
Correlation length, 324
Correspondence principle, 3
Covalent crystals, 172
Critical temperature, 317
    magnetic field, 317
Cryotron, 327
Crystal, 171
    momentum, 219
    system, 176
Crystalline solids, 171
Crystallography, 171
Cubic lattice, 176
Curie constant, 299
    temperature, 282, 301
    Weiss law, 299
Cyclic boundary conditions, 265

D

Davisson and Germer experiment, 8
de Broglie wave, 4
Debye frequency, 261
    temperature, 261
    theory of specific heat, 259
Degeneracy, 40, 121
    parameter, 147
Degenerate distribution function, 145
Degrees of freedom, 125
Density of states, 121
Diamagnetic material, 288
Diamond structure, 179
Dielectric constant, 272
Dipole moment, 273
Dirac's constant, 4
    notation, 39
Displacement vector, 272
Dissociation energy, 187
Distribution function, 122
Domains, 307
Donor, 226
    level, 226
Dopant, 225

Doping, 225
Drude theory of metal, 237
Dulong-Petit's law, 262
Dynamical variable, 38

E

E-k diagram, 218
Effective mass, 219
Ehrenfest's theorem, 31
Eigenfunction, 35
    state, 35
    value, 35
Einstein temperature, 258
    theory of specific heat, 257
Electric displacement, 272
Electrical conductivity, 237
Electron diffraction experiment, 8
    gas, 237
Electronic polarization, 274
Elementary magnet, 288
Emission of α-particles, 74
Energy bands, 218
    density, 156
    of cohesion, 187
    operator, 28
Ensemble, 117
    canonical, 128
    microcanonical, 121
Entropy, 123
Equations-θ, ϕ, solution of, 99
Equipartition of energy, 125, 130
Ewald's construction, 201
Expectation value, 30
Extrinsic semiconductor, 225

F

Face-centered cubic lattice, 176
Fermi energy, 145
    level, 145, 228
    momentum, 241
    sphere, 241
    surface, 241

Index

temperature, 151
velocity, 243
Fermi-Dirac distribution function, 145
postulates of, 153
statistics, 144
Fermions, 144
Ferrimagnetism, 309
Ferrites, 309
Ferroelectricity, 281
Ferromagnetic materials, 288, 301
First Brillouin zone, 219, 265
Fluxoid, 322
Forbidden energy bands, 217
Fourier space, 197
Free electron, 237
particle, 52

G

Gap parameter, 324
Gibbs paradox, 133
G.P. Thomson's experiment, 11
Gamma-ray microscope, 14
Geometrical structure factor, 205
Grain boundary, 171
Ground state, 105
Group velocity, 13

H

Hall angle, 245
coefficient, 245
effect, 244
field, 244
mobility, 248
Hamiltonian operator, 35
Hard magnetic material, 307
Harmonic oscillator, 79, 131
Heat of transformation, 166
Heisenberg's uncertainty
principle, 13
application of, 15
Helmholtz free energy, 129
Hermite equation, 81
polynomial, 81

Hermitian operator, 35
properties of, 38
Hexagonal closed packed
structure, 180
Holes, 223
Homopolar crystal, 172
Hund's rules, 293
Hydrogen atom, 103
bonded crystal, 174
Hysteresis, 282, 307
loop, 282, 307
loss, 307

I

Indices of crystallographic
direction, 182
Insulator, 223
Internal energy 128
field, 279
Intrinsic carrier concentration, 224
charge carriers, 224
conductivity, 224
semiconductor, 224
Inversion symmetry, 180
Ionic crystal, 172
polarization, 276
Isolated system, 117
Isometric crystals, 181
Isothermal compressibility, 129
Isotope effect, 324

J

Josephson effect, 325
junction, 325

K

Kinetic energy operator, 36
Kronecker delta function, 38
Kronig-Penne model, 215

L

Landau diamagnetism, 311
Langevin equation, 296

function, 278, 298
theory of paramagnetism, 296
Larmor frequency, 290
Lattice 175
    constant, 176
    parameter, 176
    point, 175
    specific heat, 256
    vibration, 256
Laue condition, 198
    method, 201
Legendre's equation, 99
    polynomial of the first kind, 100
Linear harmonic oscillator, 79
Magnetic material, 288
    operator, 34
Local field, 279
London equations, 320
Lorentz force, 284
Lorenz number, 241, 243
    relation, 280
Lowering operator, 86

M

M-H curves, 306
Macroscopic system, 117
Macrostate, 117
Magnetic dipole moment, 286
    field intensity, 287
    flux density, 284
    induction, 284
    properties of solids, 284
    quantum number, 103
    susceptibility, 288
Magnetisation, 286
Majority carriers, 226, 228
Mass-action law, 227
Mathematical probability, 118
Matrix element, 39
Maxwell-Boltzmann
    distribution function, 122
    limitations of, 134

    postulates of, 153
    statistics, 124
Mean free time, 237, 239
Meissner effect, 318
Metal, 223
Metallic crystals, 173
Microscopic system, 117
Microstate, 117
Miller indices, 181
Minority carriers, 226, 228
Mobility, 239
Molar susceptibility, 296
Molecular crystals, 173
Momentum operator, 28
Most probable speed, 127

N

N-type semiconductor, 226
Natural radioactivity, 74
Neel temperature, 309
Nernst's theorem, 165
Normal modes of vibration, 260
Normalisation of the wave
    function, 29

O

Observables, 30
    compatible, 37
Operator, 30
Optical mode, 268
Orbital angular momentum quantum
    number, 103, 291
Orientational polarisation, 277
Orthogonality, 39
Orthonormal set 40, 82

P

Packing fraction, 179
P-type semiconductor, 227
Paramagnetic material, 288, 296
Parity, 72, 84
Particle confined in an enclosure, 53
    extention to a three-dimensional

box, 55
    infinitely deep potential well, 73
    rectangular potential barrier, 65
    square well, 70
Partition function, 125
Pauli paramagnetism, 311
Penetration depth, 321
    distance, 68
Perfect gas law, 124
Periodic boundary condition, 41, 265
    potential, 215
Permanent dipole moment, 277, 290
Permeability of free space, 287
Permittivity of free space, 272
Phase plane, 119
    point, 119
    space, 119
    trajectory, 119
    velocity, 12
Phonon, 267
Photon, 4
    flux, 21
Planck's constant, 3
    distribution, 155
    law of radiation, 155
    relation of entropy, 123
Polarisation, 274
Polycrystalline substance, 171
Position probability density, 29
Powder method, 202
Primitive basis vectors, 175
    translation vectors, 197
    unit cell, 175
Principal axes, 185
    electrical conductivities, 185
    quantum number, 104
Principle of complementarity, 3
Probability, 117, 145
    current density, 41
Propagation vector, 28

Q

Quantum mechanical operators, 34

    tunneling, 68
Quantum mechanics, 3
    postulates of, 42
Quantum number, 53, 81, 261
    state, 117

R

Radial equation, 101
    quantum number, 104
Rayleigh line, 7
    – Jeans law, 158
Raising operator, 86
Reciprocal lattice, 196
Recursion relation, 80
Reduced Planck's constant, 4
    zone representation, 218
Reflection coefficient, 67
    symmetry, 180
Relative permeability, 287
    permittivity, 272
Relativistic effects on de Broglie wavelength, 11
Relaxation time, 237, 242
Remanent flux density, 306
    magnetisation, 306
    polarisation, 281
Root mean square speed, 127
Rotating crystal method, 202
Rotation symmetry, 180
Rutgers equation, 320

S

Sackur-Tetrode formula, 134
Saturation magnetisation, 298
Schrödinger's wave equation, 27
    its separation, 33, 98
    one-dimensional, 28
    salient features of, 30
    three-dimensional, 28
    time-dependent, 33
    time-independent, 33
Second Brillouin zone, 219

Semiconductors, 223
Short-range forces, 187
Simple cubic lattice, 176
Single crystals, 171
Sodium chloride structure, 178
Soft magnetic material, 307
Solid State Physics, 171
Sommerfeld equation, 146
Space lattice, 175
Spacing between crystal planes, 183
Specific heat of conduction
    electrons in metals, 148
Specific heat of diatomic gases, 131
    monatomic gases, 131
    triatomic gases, 131
    solids, 256
    Dulong-Petit value, 257
Spectroscopic splitting factor, 292
Spherical harmonics, 100
Spherically symmetric potential, 98
SQUID, 326
Static dielectric constant, 272
Stationary state, 35
    waves, 266
Statistical mechanics, 117
    weight, 121
Statistics, classical, 135
    quantum, 135
Stefan Boltzmann law, 157
Stirlings' theorem, 118
Structure of solids, 171
Superconducting energy gap, 324
    ground state, 324
    magnets, 328
Superconductors, 317
    hard, 323
    high-T_c, 317
    soft, 323
    type I, 323
    type II, 323
Symmetry operations, 180

T

Thermal conductivity of a metal, 240, 243
Thermodynamic probability, 122
Third law of thermodynamics, 165
Translation symmetry, 180
Transmission coefficient, 67
Transport phonomena in
    metals and semiconductors, 237

U

Unit cell, 175
Uncertainty principle, 13

V

Valence band, 223
van der Waal's interaction, 174
Velocity distribution curve, 127
Vibrating atoms, 265

W

Wave equation in a force field, 29
Wave function, 27
    mechanics, 27
    packets 12, 56
Weiss factor, 302
Wiedemann-Franz Law, 240, 243
Wien's law, 158
    displacement law, 157
Wigner-Seitz cell, 208

X

X-ray crystal analysis, 194

Z

Zero-point energy, 81
Zincblende structure, 179